Thalamocortical Assemblies

Thalamocortical Assemblies

Sleep Spindles, Slow Waves and Epileptic Discharges

Second Edition

ALAIN DESTEXHE AND
TERRENCE J. SEJNOWSKI

OXFORD
UNIVERSITY PRESS

Great Clarendon Street, Oxford, OX2 6DP,
United Kingdom

Oxford University Press is a department of the University of Oxford.
It furthers the University's objective of excellence in research, scholarship,
and education by publishing worldwide. Oxford is a registered trade mark of
Oxford University Press in the UK and in certain other countries

First Edition published in 2001

Published in the United States of America by Oxford University Press
198 Madison Avenue, New York, NY 10016, United States of America

British Library Cataloguing in Publication Data
Data available

Library of Congress Control Number: 2023948335

ISBN 978-0-19-886499-8

DOI: 10.1093/oso/9780198864998.001.0001

Printed in the UK by
Ashford Colour Press Ltd, Gosport, Hampshire

To Beatrice and Laurence, who helped us in many ways and
for their forbearance with our sleep deprivation

Foreword by M. Steriade

This book, written by two distinguished computational neuroscientists, explores a topic that is close to my heart, thalamocortical oscillations during normal and paroxysmal states. To begin, I should confess that, before the early 1990s, I thought (along with other fellow neuroscientists) that you could obtain anything you wanted with a model. However, it was the privilege of collaborating with the team of Terry Sejnowski at the Salk Institute that made me appreciate the many prerequisites for building realistic models, such as a deep knowledge of intrinsic and synaptic currents of different neuronal types as well as insights into cellular morphology and interneuronal connectivity. This requires intensive readings of many fields in neuroscience, explored both *in vitro* and *in vivo*; however, what computational neurobiology taught us during the past decade is much more than a large-scale inventory of neuronal machinery. Indeed, without this relatively recent development in neuroscience, we wouldn't have been able to reconcile opposing views among researchers and predict the type of experiments that should be done to clarify our ideas.

Let me give only a few examples, taken from the core of this book, illustrating the nature of the models developed by Destexhe and Sejnowski, along with others at the Salk Institute, in collaboration with my group based on intracellular recordings *in vivo*. In a series of studies, combining modelling with experiments, several predictions were advanced to explain the synchronized oscillations observed in an isolated network of inhibitory thalamic reticular neurons. Thus, these neurons should receive a certain proportion of depolarizing inputs arising in the brainstem monoaminergic aggregates and their dendrites must be intact to contain a high density of low-threshold calcium currents. More recently, we developed the idea that an intact collection of thalamic reticular neurons, as *in vivo*, is necessary for initiation of spindle oscillations; the model demonstrated that such synchronized rhythms could arise only with a certain number of interconnected neurons, but not fewer. Some of these predictions are ready to be tested in experiments; others have not yet been fully developed but, overall, the computational work thus far has had a leading role in demonstrating how neuronal populations could become synchronized. This is the topic of one of the best chapters in the book, based on both experimental data and modelling, dealing with the role of the cerebral cortex in the coherence of sleep spindles, a rhythm that was conventionally thought as organized within thalamic circuits.

For those working in the thalamocortical system and related ones, this book is a gem as it contains extremely helpful information about the different ion channels used in biophysical models and the electrophysiological properties of thalamocortical and thalamic reticular cells (local interneurons were much less studied in experiments, but in the next edition of this book they will no doubt find a place). Sleep spindles as well as some paroxysmal oscillations in thalamic and thalamocortical networks are culminating points in this sequence. These models are based firmly on solid intracellular recordings *in vivo* (if I may be allowed to say this) and also very good recordings made in slices. The whole represents an impressive collection of studies that will have, I am sure, a lasting impact on both experimenters and modelers. I highly recommend this book not only to those investigating thalamocortical systems, but also to all who care about the brain because of the valuable lessons learned here about the interplay between the experiments and realistic models.

M. Steriade
Québec
January 2001

Preface

The brain is never at rest. Electrical recordings of weak electrical signals from the scalp, the electroencephalogram (EEG), reveal constantly ongoing activity at low frequencies during relaxed, quiet states that shifts to higher frequencies during attentive states. Single neurons, recorded with a microelectrode from the cerebral cortex, fire a continual background of spikes, called 'spontaneous' or maintained activity. Although much of the research on sensory systems is still dominated by the stimulus-response paradigm, there is growing appreciation that the brain is endogenously active and neural mechanisms exist to support this activity (Llinás, 1988). The responses of neurons evoked by sensory stimuli are mixed with ongoing activity, which makes it difficult to analyse each separately.

Sleep remains one of the greatest mysteries about the brain. Far from decreasing in activity, during sleep the brain enters even more rhythmic states in which millions of neurons begin to fire bursts of action potential in synchrony. The endogenous nature of this activity is more transparent than spontaneous activity during alert states, since the amount of sensory input to the brain during sleep is greatly reduced. Where do these rhythms originate in the brain? How are they related to each other? We have the impression that the experiences during the day influence our dreams, but the nature of this influence remains illusive. Perhaps the biggest mystery is the purpose of sleep itself. In this book we summarize progress that has been made over the last decade toward answering these questions.

We are at an exciting point in our exploration of the brain. The spectacular advances that have taken place at the molecular level are beginning to change the way that we think about problems in systems neuroscience. But before these two levels of investigation can be integrated, new techniques are needed for relating microscopic properties of neurons and their molecular components with the macroscopic behaviour of large networks comprising thousands to millions of nerve cells. This is not an easy problem since spatial scales in the brain span 9 orders of magnitude, from nanometres to metres, and temporal scales span 12 orders of magnitude, from microseconds to years; complex phenomena are studied at many intermediate levels of organization between the molecular and systems levels (Churchland and Sejnowski, 1992).

The approach taken here combines 'bottom-up' data from the molecular and cellular levels of investigation with 'top-down' observations of sleep rhythms that build on the network and systems levels of descriptions for essential constraints. Computational models provide a unique way to bridge the gap that exists between these levels. Although the focus of this monograph is on thalamocortical systems, we start from the basics and introduce the properties of neuronal membranes and synaptic transmission, and outline how biophysical mechanisms can be modelled relatively simply while preserving the essential properties and relevant time scales observed experimentally. Thus, this text can be used as an introduction to cellular biophysics in a setting that quickly takes the student to the frontiers of what is known about one of the most complex neural systems.

The experiments and modelling studies that form the core of this book began more than ten years ago in several laboratories. In Québec, Mircea Steriade had shown experimentally that networks of inhibitory neurons in the reticular nucleus of the thalamus might be the source of spindle rhythms *in vivo* (Steriade et al., 1987). In La Jolla, Terrence Sejnowski was studying the mechanisms underlying 30–70 *Hz* oscillations in the cerebral cortex and showed that the most likely source of synchronizing inputs to cortical pyramidal neurons was from inhibitory neurons (Lytton and Sejnowski, 1991). In Brussels, Alain Destexhe was developing biophysical models of sleep oscillations and showed that synchronized oscillations can arise in networks of thalamic inhibitory neurons (Destexhe, 1992). In New Haven, David McCormick had developed an *in vitro* thalamic slice preparation in which inhibition synchronized thalamic oscillations (von Krosigk et al., 1993). These early studies pointed toward inhibitory neurons as essential to understanding rhythms in thalamocortical assemblies.

The thalamus proved to be a treasure trove for studying the basis of rhythmogenesis, and the deep intuitions of Mircea Steriade and David McCormick were an invaluable guide to every aspect of thalamocortical assemblies. However, there were inconsistencies between some *in vivo* and *in vitro* experimental results that might be reconciled with modelling studies, as we show throughout this book (see also Steriade et al., 1993). Because there are relatively few cell types in the thalamus and their intrinsic and synaptic mechanisms have been well studied, accurate models of thalamic neurons and their interactions could be achieved and used to develop an integrated view of thalamocortical assemblies, as we show in detail here.

During the last decade of collaborative research, strong personal as well as professional relationships were forged. Close friendships crystallized between Alain Destexhe and postdoctoral fellows in these laboratories, especially with Diego Contreras and Thierry Bal. The high quality of their experimental work and the ideas that arose from numerous discussion with them served as the basis for the models that we explored together as a team. John Huguenard, who has done groundbreaking work on thalamic mechanisms, provided invaluable insight into the models. Alex Thomson, who pioneered the exploration of the dynamics of synaptic transmission, was also a generous collaborator. Alexander Borbély, who studies sleep rhythms in humans, brought an important set of constraints on the models and pointed us toward a possible function for slow waves. We have benefited from collaborations and discussions with many others, including Agnès Babloyantz, Maxim Bazhenov, Damien Debay, Arthur Houweling, William Lytton, Zachary Mainen, Michael Neubig, Denis Paré, Igor Timofeev and Daniel Ulrich. This collaborative research was made possible by support from the Howard Hughes Medical Institute, The Human Frontier Science Program, the National Institutes of Health, the Medical Research Council of Canada and the Centre National de la Recherche Scientifique in France.

Much of the inspiration for our research has arisen from spirited conversations with colleagues, and some of those to whom we are deeply indebted for discussions and sharing their data include Ted Abel, Peter Achermann, Thomas Albright, Florin Amzica, Massimo Avoli, Thomas Bartol, Tony Bell, Michael Berridge, Hal Blumenfeld, Fred Gage, Thomas Brown, Gyorgi Buzsáki, Brian Christie, Patricia Churchland, John Clements, Jack Cowan, Francis Crick, Vincenzo Crunelli, Yves De Koninck, Martin Deschênes, Rodney Douglas, Daniel Durstewitz, Jean-Marc Fellous, Yves Frégnac, J. Christopher Gillin, Charles Gray, Nicolas Hô, David Golomb, Neal A. Hessler, Mike Hines, Geoffrey Hinton, Barbara Jones, Ted Jones, Tzyy-Ping Jung, Helmut Kröger,

Nancy Kopell, Nathalie Leresche, Anita Lüthi, Rodolfo Llinás, Scott Makeig, Roberto Malinow, Henry Markram, Kevin Martin, Istvan Mody, Christophe Mulle, Venki Murthy, Tom Otis, Chris Pape, Klaus Pawelzik, Nicholas Priebe, John Rinzel, Raphael Ritz, Christian Rosenmund, Jeremy Seamans, Michelle Rudolph, Charles Stevens, Mavi Sanches-Vives, Murray Sherman, Paul Tiesinga, Roger Traub, Misha Tsodyks, Marcus von Krosigk, Xiao-Jing Wang, Richard Warren, Zixiu Xiang, and Tony Zador.

We thank Paule Bissonnette, Denis Drolet, Diane Lessard, Rosemary Miller, Leslie Shaden and Amie Wakai who provided invaluable assistance during the production of this book.

Gif-sur-Yvette and La Jolla
February 2001

Preface to the Second Edition

For this second edition, we included 'sleep spindles' in the title to acknowledge that a large part of the book is devoted to the physiology and modelling of sleep spindles. Since the publication of the first edition, 'Thalamocortical Assemblies', in 2001, much more is known about sleep spindles and other brain rhythms. Summaries of these advances have been appended to the end of the relevant chapters and include (1) an update on important experimental results about spindle rhythmicity, such as recordings of inhibitory neurons during cortical spindles and the discovery of gap junctions in the thalamic RE nucleus; (2) evidence for a cortical focus for absence seizures in rodents; (3) updated models, such as recent models of spindles with integrate-and-fire neurons, models of RE oscillations with gap junctions, and the genesis of spindles and other rhythmical activity such as Up/Down states in thalamocortical systems; (4) new experimental results on the involvement of sleep spindles in the consolidation of memory traces during sleep by transferring information between hippocampus and cortex.

Mircea Steriade wrote the foreword to the first edition, but did not live to see the second edition. Mircea was a legendary figure in the study of cortical and thalamic oscillations *in vivo*. He directly influenced most of the experimental and theoretical results in this book. He found that the isolated reticular nucleus of the thalamus (RE) oscillates *in vivo*, which led to his 'RE pacemaker' hypothesis. He also discovered the slow oscillation with Up/Down states during slow-wave sleep. Mircea was personally very intense, and held strong opinions. He had a truly impressive memory of facts and numbers and a vast knowledge of the physiological and anatomical literature, making him a 'living encyclopedia'. He was also a good friend and a long-term collaborator, whom we miss enormously.

Gif sur Yvette and La Jolla
January 2023

Contents

Introduction

The neocortex greatly expanded during evolution of the mammalian brain (Allman, 1999). The thalamus evolved in close association with the cerebral cortex, with a general topographic relationship between corresponding thalamic and cortical regions (Jones, 1985; Steriade et al., 1997). In this monograph, we survey some of the key experimental and computational advances over last decade regarding rhythmic activity in the thalamocortical system. The focus is on the biophysical mechanisms underlying the genesis of rhythmicity by single neurons and local circuits, and how these rhythmically active elements interact with each other through synaptic interactions to generate coherent oscillations across millions of thalamic and cortical neurons. The computational models integrate the molecular, cellular and network components and capture the complex dynamical properties of neurons and their synaptic interactions. This approach reveals the complex interplay between intrinsic and synaptic mechanisms in the genesis of thalamocortical oscillations. We begin with a brief historical review of rhythmic activity in the brain.

1.1 Brain rhythmicities

Rhythmic activity was first discovered in electrical recordings from the scalp by Caton (1875), and later by Berger (1929) in humans. They observed that the electrical activity of the brain is dominated by oscillations that have a frequency and amplitude varying widely across different behavioural states. Awake and attentive states are characterized by low-amplitude fast-frequency electroencephalographic (EEG) activity, of which the dominant frequency band (20–80 Hz) has been classified as beta and gamma waves. Large amplitude alpha rhythms (8-12 Hz) appear mostly in occipital cortex in aroused states with eyes closed and diminish with eyes opened (Berger, 1929). The early stages of sleep are characterized by spindle waves (7–14 Hz), which consist of short bursts of oscillations lasting a few seconds and displaying a typical waxing-and-waning appearance. When sleep deepens, slow-wave complexes, such as delta (1–4 Hz) and slower waves (< 1 Hz), progressively dominate the EEG. Slow wave sleep

Thalamocortical Assemblies: Sleep Spindles, Slow Waves and Epileptic Discharges. Second Edition. Alain Destexhe and Terrence J. Sejnowski, Oxford University Press. © Oxford University Press 2023.
DOI: 10.1093/oso/9780198864998.003.0001

is interrupted by periods of rapid eye movement (REM) sleep, during which the EEG activity has a low-amplitude and high frequencies, similar to that during arousal. Finally, there are a number of pathological wave types, such as the 3 *Hz* 'spike-and-wave' complexes typical of several forms of epileptic seizures (Niedermeyer and Lopes da Silva, 1998).

The prominent oscillations in EEG activity led early investigations to search for the underlying mechanisms of rhythmicity, starting in the beginning of the twentieth century (see Section 1.2). Today, with the advent of precise intracellular recording techniques, pharmacological manipulations and molecular biology, the biophysical mechanisms underlying oscillatory activity in neurons have been well characterized (Section 1.3). Despite this progress, the link between these microscopic properties and the global activity of large assemblies of neurons, such as the EEG, is a problem that has not yet been solved. In the present monograph, we attempt to establish such a link using computational models of oscillations in the thalamocortical system.

1.2 Early views on brain rhythmicity

Despite more than one hundred years of research, the origin and function of the EEG oscillations remains a mystery; over the last forty years research has instead emphasized the function of signals carried by single neurons, as recorded by sharp microelectrodes stably positioned in the brain. New techniques are being developed, based on multiple recording electrodes and optical techniques, to observe populations of neurons and theories are being developed for how the brain may represent information by the electrical and chemical signals in these populations (Abbott and Sejnowski, 1999). The explosion of new knowledge about the biophysical properties of neurons and their anatomical organization has raised interest about the origin of brain rhythmicities and their possible significance for the large-scale organization of information processing in the brain.

The earliest theory to explain EEG rhythmicity was the 'circus movement theory', proposed by Rothberger (1931). According to this theory, EEG rhythms are due to action potentials travelling along chains of interconnected neurons. The period of the rhythmicity corresponded to the time needed for a volley of action potentials to traverse a loop in the chain. This theory might be called a 'connectionist' theory of the EEG (Rumelhart et al., 1986) since it emphasizes that the rhythmicity depends more on the cyclic circulation of activity within neuronal networks than on the pacemaker properties in neurons.

Bishop (1936) proposed the concept of 'thalamocortical reverberating circuits'. According to this view, rhythmicity was generated by action potentials travelling back and forth between thalamus and cortex, but local interneurons were also involved. This view was directly inspired from the circus movement theory and was the first time a thalamocortical mechanism was invoked to explain rhythmicity. Although the reverberating circuit theory remained prevalent for several years, subsequent experiments demonstrated that the EEG activity is not generated by action potentials (Renshaw et al., 1940), invalidating a fundamental premise of the circus movement theory.

Bremer (1938b, 1949, 1958) proposed an alternative theory in which rhythmicity is generated by autorhythmic properties of cortical neurons and that the EEG results from the synchronized oscillatory activity of large assemblies of oscillating cortical neurons. His objection to the circus movement theory was that the slow time course of EEG waves does not match the fast time course of action potentials. In addition, a considerable number of synapses would be needed to account for the slow frequency of EEG oscillations, which would result in severe constraints on connectivity. Bremer proposed that the EEG is generated by non-propagated potentials, by analogy with the electrotonic potentials in the spinal cord (Bonnet and Bremer, 1938). Working on motoneurons in the spinal cord, Eccles (1951) provided convincing evidence that the EEG reflects summated postsynaptic potentials. To explain the slow time course of EEG waves, Eccles postulated that distal dendritic potentials, and their slow electrotonic propagation to soma, participate in the genesis of the EEG. This assumption was confirmed by intracellular recordings from cortical neurons, which demonstrated a close correspondence between the EEG and synaptic potentials (Klee et al., 1965; Creutzfeldt et al., 1996a, 1996b). This view of the genesis of the EEG is still widely held (Nunez, 1981).

Bremer (1949) also proposed that oscillations depend on the 'excitability cycle' of cortical neurons; this was the first time that the *intrinsic properties* of cortical neurons were identified as important in generating EEG oscillations. He emphasized that cortical neurons are endowed with intrinsic properties that participate in rhythm generation, and that brain rhythms should not be described as the passive driving of the cerebral cortex by impulses originating from pacemakers (Bremer, 1938a, 1958). Bremer's proposal for the genesis of EEG rhythmicities rested on four core ideas: (i) That EEG rhythmicity is generated by the oscillatory activity of cortical neurons; (ii) that the genesis of these oscillations depends on properties *intrinsic* to cortical neurons; (iii) that EEG oscillations are generated by the *synchronization* of oscillatory activity in large assemblies of cortical neurons; (iv) that the mechanisms responsible for synchronization were due to intracortical excitatory connections. Although not all of Bremer's insights are correct, they were remarkably perspicacious and his concepts continue to influence today's theories of brain rhythmicity (see below).

1.3 Origins of brain rhythmicity

Motoneurons in the spinal cord were among the first to be studied in detail with intracellular recording techniques (Brock et al., 1951). Early views about activity in other parts of the central nervous system, particularly the cerebral cortex, were strongly influenced by studies of motoneurons (Eccles, 1951). Spinal motoneurons integrate synaptic activity and, when a threshold membrane potential is reached, emit an action potential that is followed by a prolonged hyperpolarization. This led to an early model of the neuron based on the concept of 'integrate-and-fire' followed by a reset. Brain activity in different parts of the nervous system was thought to arise by interactions between similar neurons connected in different ways into complex networks. In this connectionist view, the function of a brain area was determined primarily by its pattern of connectivity (Eccles, 1951).

Pioneering research on invertebrate neurons has revealed complex intrinsic firing properties that depart from the traditional integrate-and-fire model (Connor and Stevens, 1971a–c; Kandel, 1976; Adams et al., 1980). Further evidence against the integrate-and-fire view came from studies of small invertebrate ganglia showing that connectivity was insufficient by itself to specify function (Selverston, 1985; Getting, 1989) and that the modulation of intrinsic properties needed to be taken into account (Harris-Warrick and Marder, 1991). However, these idiosyncratic properties were thought to reflect the peculiar evolutionary history of invertebrates, distinct from the large populations of simpler neurons found in vertebrate brains. That this view was fundamentally wrong became apparent when intracellular recording techniques were first applied to slice preparations of vertebrate brains.

Neurons in vertebrate central nervous systems are characterized by complex intrinsic properties similar to those in invertebrate neurons. Many investigators have contributed to our knowledge of the intrinsic neuronal properties in vertebrate brains, but Llinás and his collaborators (Llinás and Sugimori, 1980a, 1980b; Llinás and Yarom, 1981a, 1981b; Llinás and Jahnsen, 1982; Alonso and Llinás, 1989) were largely responsible for discovering the rich repertoire of intrinsic electrophysiological behaviour in central neurons. Interactions among voltage-dependent and calcium-dependent conductances determine a neuron's intrinsic properties along with the geometry of its dendritic trees (Traub and Llinás, 1979; Llinás, 1988; Mainen and Sejnowski, 1996). These properties give rise to rhythmic activity, including the intrinsic ability of some neurons to generate sustained oscillations, as suggested earlier (Bremer, 1938b, 1949).

We now have a repertoire of voltage-dependent and calcium-dependent ionic conductances in central neurons, which are responsible for their intrinsic properties (Llinás, 1988). Understanding how the interactions between these ionic conductances leads to the genesis of cellular rhythms is difficult because these interactions are highly nonlinear. Computational models can make a significant contribution in linking the microscopic properties of ion channels and cellular behaviour by simulating these complex interactions. This approach was used by Hodgkin and Huxley (1952) to understand the genesis of action potentials and the same approach is used here to understand the complex behaviours of thalamic neurons (see Chapters 3 and 4).

In addition to intrinsic properties, neurons interact in various ways, including chemical synaptic transmission, electrical coupling through gap junctions and ephaptic interactions through electric fields. Whole-cell and patch-clamp recording techniques (Sakmann and Neher, 1995) have made it possible to investigate the detailed mechanisms underlying the conductances of the ionic channels involved in synaptic transmission. An extraordinarily rich variety of dynamic properties of synaptic interactions between central neurons have been uncovered on a wide range of time scales. Many neurotransmitters have been identified in the thalamocortical system (McCormick, 1992) as well as a wide variety of receptor types, each of which confer characteristic temporal properties to synaptic interactions. These properties are now well understood for the main receptor types mediating synaptic interactions (see Chapter 5).

In the current view, rhythmicity arises from both intrinsic and synaptic properties (Steriade and Llinás, 1988; Steriade et al., 1993b; Destexhe and Sejnowski, 1997). Some neurons generate oscillations through intrinsic properties and interact with other types of neurons through multiple types of synaptic receptors. These complex interactions generate large-scale coherent oscillations. However, it is difficult to determine experimentally the precise mechanisms underlying synchronization in large populations of neurons. Computational models can help in dissecting these mechanisms (see below). Because of the reciprocal projections between corresponding areas of the thalamus and cortex, it is difficult to separate the thalamic and cortical contributions to the genesis of oscillatory behaviour.

1.4 Identification of the key neuronal structures

Spindle waves are by far the best studied type of rhythmicity in the thalamocortical system, in part because they can be enhanced by anesthetics such as barbiturates (Derbyshire et al., 1936; Andersen and Andersson, 1968). The thalamic origin of spindles was first suggested by Bishop (1936), who observed the suppression of rhythmic activity in the cortex after sectioning connections with the thalamus. The thalamic origin for spindle generation was confirmed in stages by experiments on decorticated animals (Adrian, 1941; Morison and Bassett, 1945), by identifying the cellular events underlying this rhythmic activity *in vivo* (Steriade and Deschênes, 1984; Steriade and Llinás, 1988), and finally by observing spindles in isolated thalamic slices *in vitro* (von Krosigk et al., 1993). The latter preparation allowed the biophysical mechanisms underlying spindle rhythmicity to be explored, particularly the voltage-dependent conductances and receptor types involved, and theories for the genesis and termination of spindle oscillations to be rigorously tested.

Absence seizures, like sleep spindles, are characterized by large-scale synchrony across the entire brain. Jasper and Kershman (1941) were the first to suggest that absence seizures could originate in thalamic nuclei that project widely to cerebral cortex. This hypothesis found support from chronic recordings during absence seizures in humans, showing that signs of a seizure were observed first in the thalamus before it appeared in the cortex (Williams, 1953). The introduction of experimental models of absence seizures, such as the penicillin model in cats (Prince and Farrell, 1969), showed that although the thalamus is critical for generating seizures, it was not sufficient to explain all of their properties. Seizures can be obtained from injection of convulsants limited to cerebral cortex, but not when the same drugs are injected into the thalamus (Ralston and Ajmone-Marsan, 1956; Gloor et al., 1977; Steriade and Contreras, 1998). It is now clear that both the thalamus and cortex are necessary partners in these experimental models of absence seizures, but the exact mechanisms are unknown (Gloor et al., 1988; Danober et al., 1998).

Electrophysiological studies *in vivo* suggested that spindle waves and absence seizures may share common mechanisms (Gloor et al., 1988). *In vitro* techniques have also provided invaluable insights into the biophysical mechanisms involved in seizure generation. Application of convulsants to thalamic and cortical slices can produce discharge patterns similar to that found *in vivo* during epileptic seizures (Chagnac-Amitai and Connors, 1989; von Krosigk et al., 1993); this has allowed the different receptor types involved in these discharges to be identified by physiological and pharmacological techniques. Here, computational models can be used to identify critical parameters involved in the genesis of pathological behaviour, as well as to suggest ways to resolve apparently inconsistent experimental observations (see details of this approach in Chapter 8).

Not only are the cerebral cortex and thalamus capable of endogenous rhythm generation, but they also respond to periodic electrical stimulation, particularly in the frequency range 7–12 Hz. Repetitive stimuli in this range is potent in eliciting cortical and thalamic responses that grow in size during the first few stimuli. This phenomenon, called the augmenting response (Morison and Dempsey, 1943), may be viewed as a kind of resonance due to the intrinsic rhythmogenic capabilities of the thalamocortical system. Early investigators showed that decortication reduced but did not abolish augmenting responses in the thalamus; however, removal of the thalamus abolished the augmenting responses in the cortex evoked by capsular stimulation (Morison and Dempsey, 1943). Later studies reported that stimulation of white matter could elicit responses growing in size in cerebral cortex but their patterns were different from thalamically evoked augmenting waves (Morin and Steriade, 1981). These phenomena also provide important constraints to the models.

The question now is how to organize these many empirical studies and use them to develop an integrated view of normal and pathological states of thalamocortical oscillations? The critical molecular and biophysical building blocks of spindle waves, absence seizures and augmenting responses have been identified, but to convincingly prove that additional components have not been left out, some way to reconstitute these rhythmic states from the identified components is needed. An explanation for these naturally occurring rhythms and seizures should also be consistent with both *in vivo* and *in vitro* experiments, sometimes providing useful hypotheses to resolve apparent inconsistencies between the different preparations or conditions. This is one of the central aims of this book (see Chapters 6–8).

1.5 Thalamocortical assemblies

The computational models introduced here could make definitive contributions to two issues. The first problem is to understand how the complex behaviours observed with intracellular electrophysiological recordings from thalamic and cortical neurons arise from the interaction between intrinsic calcium- and voltage-dependent conductances, and their spatial location within the neurons. The second issue is how rhythms are generated and organized by networks of neurons endowed with complex intrinsic properties and interacting through complex synaptic properties. The ultimate goal of this monograph is to establish a link between the biophysical properties of ion channels and the emergence of oscillatory behaviour in cells and networks. Computational models are relatively new in neuroscience, although the basic methodology was already used by Hodgkin and Huxley (1952) in their seminal model of the action potential in the squid giant axon.

The first step in our program is to build accurate models of biophysical mechanisms such as voltage-dependent or calcium-dependent conductances. Accurate models are possible because *in vitro* experiments have precisely characterized these biophysical mechanisms. We begin by reviewing computational models for the various intrinsic voltage-dependent and calcium-dependent conductances in neurons in Chapter 2.

The second step consists in building computational models of individual neurons. Here also, models benefit from accurate voltage-clamp characterizations of the intrinsic conductances of thalamic and cortical neurons *in vitro*. Models are used to investigate the interactions between these intrinsic conductances in generating the complex dynamical properties of these neurons. This approach is applied in Chapters 3 and 4 to thalamic neurons.

The third step is to develop accurate computational models of synaptic transmission. Here again, *in vitro* experiments have provided precise data on each type of receptor and other biophysical mechanisms associated with synaptic interactions. Models of synaptic interactions mediated by several of the main receptor types are presented in Chapter 5.

To investigate network behaviour, there is a computational tradeoff between the accuracy and speed of a simulation. The most accurate single neuron models are computationally intensive;

but in order to scale up a model to explore the additional properties that interactions between neurons provide, it is necessary to simplify the single neuron and single synapse models to their bare essentials. However, these simplified models of single neurons and single synapses must capture the essential dynamical features of the more complex models, to adequately account for the contributions of single cell properties to dynamical properties at the network level.

In the final chapters, all the pieces of the puzzle will be assembled to explore how oscillations are generated at the network level. These large-scale network models will be constrained mainly by *in vivo* measurements of global oscillatory activity, such as the EEG and local field potentials. It is only at this stage that the value of EEG measurements as an essential constraint becomes clear. In particular, multisite field potentials characterize how oscillatory activity is distributed in different areas or nuclei, how it begins and ends, and under which conditions oscillations are observed. Modelling distributed data is the main theme investigated in Chapters 6–8.

The network model is built up in stages, starting with models of spindle oscillations in thalamic circuits (Chapter 6), the mechanism underlying large-scale coherent oscillations in the thalamocortical system (Chapter 7) and finally, the genesis of pathological behaviour such as absence seizures (Chapter 8). In each case, we attempt to establish the link between the complex intrinsic electrophysiological properties of individual neurons, their interaction through specific types of synaptic receptors, and the emergence of oscillations at the network level. In the final chapter (Chapter 9), we broaden the focus to other types of thalamocortical oscillations and conclude with some implications of these rhythms for the larger computational economy of the brain.

Biophysical models of the membrane potential and ionic currents

The models in this monograph are based on the elementary biophysical mechanisms underlying the electrical behaviour of neuronal membranes. In this chapter, we review the experimental basis for how ion channels establish membrane excitability, how these channels can be activated and how these processes can be modelled accurately. The focus here is on the biophysical models of voltage-dependent and calcium-dependent ion channels; the ion channels mediating synaptic interactions will be considered in Chapter 5. The importance of these biophysical mechanisms and their interplay in generating the complex behaviour of thalamic neurons is covered in Chapters 3 and 4; the network behaviour implying thalamic neurons is considered in Chapters 6–8.

2.1 Ionic bases of neuronal excitability

2.1.1 Membrane potential

The membrane potential of a neuron is established by an uneven distribution of ions on either side of the membrane. The potential across the membrane is maintained by the selective permeability of the membrane to different types of ions (see Appendix A). The selectivity is supported by transmembrane proteins called *ion channels* that contain a pore specific to one or more types of ion. In particular, the resting membrane potential depends on ion channels selective for potassium (K^+), sodium (Na^+) and chloride (Cl^-) ions. The value of the membrane potential at equilibrium is given by the Goldman equation:

Thalamocortical Assemblies: Sleep Spindles, Slow Waves and Epileptic Discharges. Second Edition. Alain Destexhe and Terrence J. Sejnowski, Oxford University Press. © Oxford University Press 2023.
DOI: 10.1093/oso/9780198864998.003.0002

$$V = \frac{RT}{ZF} \ln \left[\frac{P_K[K]_o + P_{Na}[Na]_o + P_{Cl}[Cl]_i}{P_K[K]_i + P_{Na}[Na]_i + P_{Cl}[Cl]_o} \right] \tag{2.1}$$

where P_K, P_{Na} and P_{Cl} are the membrane permeabilities of K^+, Na^+ and Cl^- ions, respectively.

The Goldman equation shows that changing the permeability to a given ion leads to a new equilibrium value of the membrane potential. For example, opening K^+ channels will move the equilibrium membrane potential to more negative values. Opening Na^+ channels will have the opposite effect by drawing the membrane towards more depolarized levels. The regulation or gating of ion channel conductances is a fundamental mechanism for producing dynamic changes in the membrane potential.

2.1.2 Voltage-gated ion channels

Some ion channels are gated by the membrane potential itself. These ion channels are voltage-gated and can open or close when the membrane potential changes. For example, some K^+ channels open with depolarization, a processes that will result in a tendency to hyperpolarize the membrane; these channels tend to counteract the depolarization. By contrast, Na^+ channels that open with depolarization have the opposite effect by amplifying the depolarization. Changes in the conductances of the voltage-gated ion channels have a wide range of time courses and can produce highly complex effects on the membrane potential; these effects are the basis for the diverse intrinsic electrical properties found in neurons (Llinás, 1988). Exploring and understanding the genesis of these properties remains an active area of research and is one of the goals of the present monograph.

Perhaps the best known intrinsic electrical property of neurons is the action potential. The action potential is an all-or-none dynamic electrical event that is caused by voltage-dependent ion channels in the membrane. Cole and Curtis first established in 1939 that a voltage-dependent conductance increase occurs during an action potential in the giant axon of the squid. This axon has a diameter of more than a millimetre, and before its discovery by J. Z. Young was considered to be a blood vessel. By replacing the ionic concentrations of the solutions inside and outside of the axon it was further shown by Hodgkin and Katz in 1949 that the depolarizing current was carried by sodium ions, and that a transient increase of conductance to sodium ions occurred during the action potential.

Hodgkin, Huxley and Katz (1952) clearly established that action potentials are due to the voltage-dependent gating of Na^+ and K^+ ionic currents. In a remarkable series of experiments on the squid giant axon that appeared in 1952, Hodgkin and Huxley determined how voltage-dependent conductance increases generate an action potential. They used the voltage-clamp technique, introduced earlier by Cole (1949), to record the ionic currents generated at different voltages (see Fig. 2.1). They identified the kinetics of two voltage-dependent currents, called the fast sodium current, I_{Na}, and the delayed potassium rectifier, I_K, mediated by Na^+ and K^+ respectively.

Hodgkin and Huxley incorporated their experimental results into a model that successfully accounted for the main properties of the action potential (Fig. 2.2). This model anticipated the later discovery of microscopic currents through voltage-gated ion channels, which are responsible for the macroscopic currents that they measured. Today, the view of how voltage-dependent

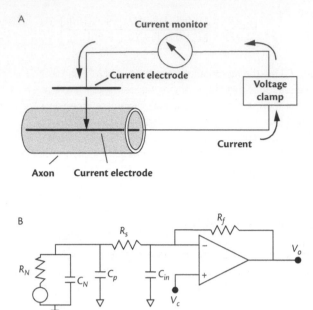

Fig. 2.1: The voltage-clamp. A. In Hodgkin and Huxley (1952), two axial electrodes were placed inside the axon and an extracellular ground electrode. Current injected into one axial electrode was adjusted through a feedback system to maintain or 'clamp' the potential measured by the second axial electrode at a fixed value. The current necessary to clamp the membrane, a measure of the current passing through the membrane, is plotted as a function of time. B. Circuit diagram for single-electrode voltage-clamp in the whole-cell mode (A from Kandel et al., 1995; B from Johnston and Wu, 1995).

ion channels influence the electrical properties of neuronal membranes is still widely influenced by Hodgkin and Huxley's approach to membrane biophysics.

In concert with pharmacological agents that selectively block different ion channels, voltage-clamp recordings were used to dissect the dynamics of the voltage-dependent currents I_{Na} and I_K underlying the fast action potential. As shown in Fig. 2.2, both Na$^+$ and K$^+$ currents are highest at positive voltages. But there is also a major difference between them: the Na$^+$ current is *transient* whereas the K$^+$ current is *sustained*. Thus, when the membrane is stepped from V_r to a positive voltage, I_K activates and stays activated until the voltage is stepped back to V_r. Following the same voltage command, I_{Na} first activates, reaches a peak value and then decays back to zero before the voltage is stepped back to V_r. This decay is an active process called *inactivation*. Inactivation also occurs in other types of channels (see Chapters 3 and 4). A current that does not show inactivation, like I_K in Fig. 2.2, is called *non-inactivating*.

Once the contributions of these two currents were separated, it became possible to examine how their interaction could account for all-or-none action potential: as the membrane is depolarized by current injection, the conductance of I_{Na} starts to increase; as Na$^+$ channels open, the membrane depolarizes still further. This leads to an explosive opening of Na$^+$ channels, accompanied by a rapid increase in the membrane potential. This positive feedback process would lead to saturation of the Na$^+$ channels without some mechanism to terminate the response. There are two such mechanisms in the axon.

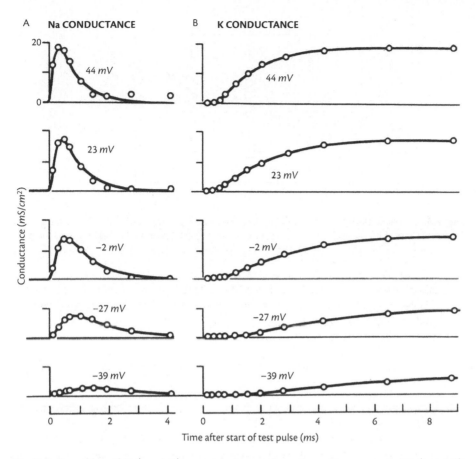

Fig. 2.2: Recordings of Na$^+$ and K$^+$ currents using a voltage clamp. Time courses of Na$^+$ and K$^+$ conductances as revealed in voltage-clamp experiments on the giant axon of the squid. The membrane was clamped to the voltages indicated. Circles are from experimental data and smooth curves are the currents calculated from the Hodgkin–Huxley model (from Hille, 1992; after Hodgkin and Huxley, 1952).

As the membrane potential rises toward its peak, K$^+$ channels activate through similar voltage-dependent mechanisms (Fig. 2.2), and draw the membrane back towards more negative values. The rapid depolarization also begins the process of inactivation of the Na$^+$ channels, which also contributes to the repolarization of the action potential. This inactivation also conserves the total number of ions that flow down their gradients since Na$^+$ and K$^+$ channels do not have to compete against each other. The I_K repolarizes the membrane, and may hyperpolarize beyond the resting membrane potential, called the *after-hyperpolarization*.

2.1.3 The Hodgkin-Huxley model for action potentials

The starting point for the Hodgkin-Huxley equation is a model of the membrane, introduced in Appendix A, that treats the passive electrical properties of the membrane as idealized resistive

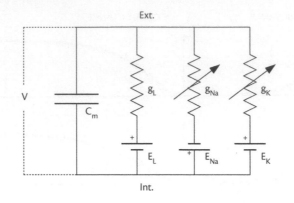

Fig. 2.3: Equivalent electrical circuit model for an active membrane. Two conductances, g_{Na} and g_K, were added to the passive equivalent circuit of Fig. A.3B. g_{Na} and g_K are represented by variable resistances.

and capacitive elements. This equivalent electric circuit model of the membrane is modified to incorporate time- and voltage-dependent conductances, represented by variable resistances in Fig. 2.3. This circuit obeys the equation:

$$C_m \frac{dV}{dt} = -g_L(V - E_L) - g_{Na}(V)(V - E_{Na}) - g_K(V)(V - E_K) \tag{2.2}$$

where the conductances $g_{Na}(V)$ and $g_K(V)$ depend explicitly on the membrane potential and implicitly on time.

The next step is to specify the voltage-dependent activation properties of I_{Na} and I_K. Consider first the simplest case, a channel that can have two configurations, closed and open states, and transitions between them that obey:

$$C \underset{\beta_m(V)}{\overset{\alpha_m(V)}{\rightleftharpoons}} O \tag{2.3}$$

where C and O represents the open and closed states of the channel, and α and β are the forward and backward rate constants of transitions between the open and closed states. Both α and β depend on the membrane potential. If m is defined as the fraction of channels in the open state:

$$m = \frac{[O]}{[C] + [O]}, \tag{2.4}$$

one obtains:

$$\frac{dm}{dt} = \alpha_m(V)(1 - m) - \beta_m(V)\, m. \tag{2.5}$$

The ionic current I_{ion} would then be proportional to m:

$$I_{ion} = \bar{g}_{ion}\, m(V)\, (V - E_{ion}) \qquad (2.6)$$

where $\bar{g}_{ion}$ is the maximal conductance and E_{ion} is the reversal potential.

It is more convenient to write Eq. 2.5 in the form:

$$\frac{dm}{dt} = \frac{1}{\tau_m(V)}\, (m_\infty(V) - m) \qquad (2.7)$$

where

$$m_\infty(V) = \frac{\alpha(V)}{\alpha(V) + \beta(V)} \qquad (2.8)$$

$$\tau_m(V) = \frac{1}{\alpha(V) + \beta(V)} \qquad (2.9)$$

m_∞ is called the *steady-state activation* and τ_m is the *activation time constant* of the current.

When the membrane is clamped to a constant voltage value V_c, Eq. 2.7 reduces to a first-order equation with constant coefficients α and β, which has the solution:

$$m(t) = m(0) + [m_\infty(V_c) - m(t)]\, \exp[-t/\tau_m(V_c)] \qquad (2.10)$$

Therefore, if I_{Na} and I_K were described by Eq. 2.6, the decay of the current should follow a simple exponential time course. Hodgkin and Huxley (1952), however, observed that the Na$^+$ and K$^+$ currents followed multiexponential behaviour. Moreover, the Na$^+$ current, as described by Eq. 2.6 must necessarily be non-inactivating, which is inconsistent with the observed inactivation of I_{Na} in Fig. 2.2. A more complex model must be considered to account for the features of these currents.[1]

Hodgkin and Huxley fit the multiexponential Na$^+$ and K$^+$ currents by assuming that each current was controlled by several independent processes, which could be interpreted as *gates* in a channel (Fig. 2.4). The sodium channel needed three activation gate that open with depolarization, and one inactivation gate that opens with hyperpolarization. The potassium current required four activation gates to open. These gates respectively followed simple open/closed kinetics:

$$m_c \underset{\beta_m(V)}{\overset{\alpha_m(V)}{\rightleftharpoons}} m_o$$

---

[1] Modelling Na$^+$ and K$^+$ currents using Eq. 2.6 can generate action potentials but their features are inconsistent with the experimental data. The currents would not match the observed behaviour in voltage-clamp experiments, nor would the shape and conductance increase of the action potential be correct. However, it is possible to develop simplified models of the action potential that have time courses and conductance changes that are approximately correct (Destexhe et al., 1994d).

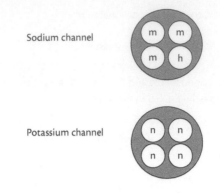

Sodium channel

Potassium channel

Fig. 2.4: Na$^+$ and K$^+$ channels contain several gates. In Hodgkin and Huxley (1952), Na$^+$ and K$^+$ currents depend on the configuration of several charged particles. These charged particles can also be considered as gates within a channel. According to the scheme illustrated here, Na$^+$ channels possess three activation gates (m) and one inactivation gate (h), whereas K$^+$ channels possess four activation gates (n). These gates are independent of each other and all of them must be open simultaneously for the channel to conduct ions.

$$h_c \underset{\beta_h(V)}{\overset{\alpha_h(V)}{\rightleftharpoons}} h_o \tag{2.11}$$

$$n_c \underset{\beta_n(V)}{\overset{\alpha_n(V)}{\rightleftharpoons}} n_o$$

where the subscripts c and o represent the closed and open forms of the channels and the variables m, h and n are defined as the fraction of gates in the open state, in analogy with Eq. 2.3. These variables obey the following set of first-order equations:

$$\frac{dm}{dt} = \alpha_m(V)\,(1-m) - \beta_m(V)\,m$$
$$\frac{dh}{dt} = \alpha_h(V)\,(1-h) - \beta_h(V)\,h \tag{2.12}$$
$$\frac{dn}{dt} = \alpha_n(V)\,(1-n) - \beta_n(V)\,n$$

Note that each equation only involves one of the variables, a reflection of their independence. The currents are then given by:

$$I_{Na} = \bar{g}_{Na}\, m^3 h\, (V - E_{Na}). \tag{2.13}$$
$$I_K = \bar{g}_K\, n^4\, (V - E_K). \tag{2.14}$$

These are the Hodgkin-Huxley (HH) equations for action potential generation (Hodgkin and Huxley, 1952).

As in the simple model, Eqs. 2.12 can be rewritten as:

$$\frac{dm}{dt} = \frac{-1}{\tau_m(V)} \, (m - m_\infty(V))$$
$$\frac{dh}{dt} = \frac{-1}{\tau_h(V)} \, (h - h_\infty(V)) \qquad\qquad (2.15)$$
$$\frac{dn}{dt} = \frac{-1}{\tau_n(V)} \, (n - n_\infty(V)).$$

where τ_m, τ_h, τ_n are time constants, and m_∞, h_∞, n_∞ are and steady state values defined in analogy with Eq. 2.8. This form of the equation is more convenient because the steady state values and time constants can be determined experimentally. The values of these parameters for the giant axon of the squid were determined by Hodgkin and Huxley (1952). They are given in Table 2.1 and are represented graphically in Fig. 2.5.

The steady-state activation of the Na$^+$ current (m_∞) follows a sigmoid-shaped curve that increases with voltage while the steady-state value of the inactivation variable h decreases with voltage (Fig. 2.5). The kinetics of activation are much faster than that for inactivation ($\tau_m < \tau_h$). The behaviour of the current for a step change in voltage can be inferred from these curves: consider a voltage-clamp step to 0 mV from an initial value of $-80\ mV$. Initially, $m = m_\infty(-80) \simeq 0$ and $h = h_\infty(-80) \simeq 1$, therefore $I_{Na} \sim 0$. When the voltage is stepped to 0 mV, m now tends to m_∞ of about 1 (activation) and simultaneously h tends to $h_\infty \sim 0$. Because τ_m is smaller than τ_h, the current quickly reaches $m = 1$ while h has not yet decreased to zero,

Table 2.1 Rate constants of the Hodgkin-Huxley model.

	Forward rate constant $\alpha(V)$	Backward rate constant $\beta(V)$
Hodgkin and Huxley, 1952:		
m	$\alpha_m = \dfrac{-0.1\,(V-V_r-25)}{\exp[-(V-V_r-25)/4]-1}$	$\beta_m = 4\ \exp[-(V-V_r)/18]$
h	$\alpha_h = 0.07\ \exp[-(V-V_r)/20]$	$\beta_h = \dfrac{1}{1+\exp[-(V-V_r+30)/10]}$
n	$\alpha_n = \dfrac{-0.01\,(V-V_r+10)}{\exp[-(V-V_r+10))/10]-1}$	$\beta_n = 0.125\ \exp[-(V-V_r)/80]$
Traub and Miles, 1991:		
m	$\alpha_m = \dfrac{-0.32\,(V-V_T-13)}{\exp[-(V-V_T-13)/4]-1}$	$\beta_m = \dfrac{0.28\,(V-V_T-40)}{\exp[(V-V_T-40)/5]-1}$
h	$\alpha_h = 0.128\ \exp[-(V-V_T-17)/18]$	$\beta_h = \dfrac{4}{1+\exp[-(V-V_T-40)/5]}$
n	$\alpha_n = \dfrac{-0.032\,(V-V_T-15)}{\exp[-(V-V_T-15)/5]-1}$	$\beta_n = 0.5\ \exp[-(V-V_T-10)/40]$

Two sets of rate constants are given for the variables m, n, h as in Eq. 2.12. In the top half panel, the rate constants were estimated by Hodgkin and Huxley (1952) for the squid giant axon at 6°C. In the original study, the voltage axis was reversed in polarity and voltage values were given with respect to the resting membrane potential (V_r here). In the bottom half panel, the rate constants were rescaled to match the kinetics of I_{Na} and I_K in central neurons at 36°C (Traub and Miles, 1991). V_T adjusts the value of the threshold and is of $-55\ mV$, unless stated otherwise.

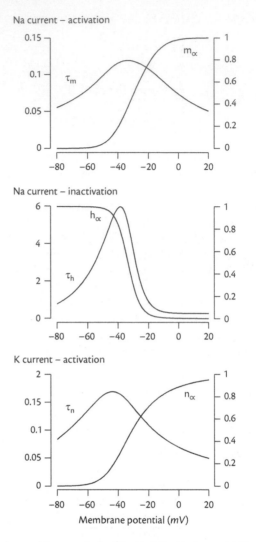

Fig. 2.5: Steady state values and time constants of the Hodgkin-Huxley model. The values of steady-state, activation, inactivation and time constants are represented as functions of the membrane potential. These functions were determined experimentally from the squid giant axon by Hodgkin and Huxley (1952) and are displayed here at 36°C. The rate constants as a function of membrane potential are given in Table 2.1. (Traub and Miles' version shown here.)

resulting in a sudden increase of conductance (proportional to m^3h). After a few milliseconds, however, the inactivation variable h reaches its steady-state value near zero and the current decays back to zero. The Hodgkin-Huxley model therefore produces a transient Na^+ current, as observed experimentally. The quantitative fitting of the model to the experimental data is shown in Fig. 2.2.

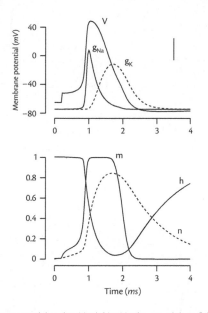

Fig. 2.6: Action potential generated by the Hodgkin-Huxley model at 36 °C. In the top panel, the membrane potential is depicted (continuous curve) together with the conductances for Na$^+$ (g_{Na} – dotted curve) and K$^+$ (g_K – dashed line). The vertical calibration bar indicates 15 mS/cm^2 for g_{Na} and 1.5 mS/cm^2 for g_K. Bottom panel: time course of the activation variables m, n and inactivation variable h during the same action potential. Passive parameters were identical to Fig. A.5; Hodgkin-Huxley model from Traub and Miles in Table 2.1, with $V_T = -63\ mV$, $\bar{g}_{Na} = 100\ mS/cm^2$, $\bar{g}_K = 10\ mS/cm^2$, $E_{Na} = 50\ mV$ and $E_K = -90\ mV$.

Once the parameters of the Hodgkin-Huxley equations have been fit to the time courses of the Na$^+$ and K$^+$ currents in voltage-clamp experiments, they immediately account quantitatively for the time course of the action potential in the squid axon (Hodgkin and Huxley, 1952). A brief electrical stimulation of the excitable membrane elicits an all-or-none action potential, as depicted in Fig. 2.6. The rising part of the spike is coincident with a sharp increase of the Na$^+$ conductance (g_{Na} in Fig. 2.6, top panel). As the Na$^+$ conductance reaches a peak and begins to decrease and inactivate, the K$^+$ conductance is already beginning to increase and repolarize the membrane potential towards resting values. The time courses of activation and inactivation variables (m, h, n) are also shown in Fig. 2.6 (bottom panel).

The Hodgkin-Huxley model also accounts for several other important properties of action potentials. First, a membrane endowed with these Na$^+$ and K$^+$ channels shows a threshold for excitability: electrical stimulation below threshold produces only passive responses (Fig. 2.7, 0.13 *nA*); in contrast, a slight increase in the stimulus amplitude produces an regenerative action potential (Fig. 2.7, 0.14–0.15 *nA*). Further increase of injected current advances the time of action potential firing, and at another threshold level several spikes are elicited (Fig. 2.7, 0.15–0.5 *nA*). The membrane in this regime produces *repetitive firing* whose frequency varies approximately linearly with injected current (Fig. 2.7, 0.25–1 *nA*). These are all familiar properties of neurons; it should be emphasized that they are implicit in the Hodgkin-Huxley equations without any additional fine-tuning of the parameters or any addition mechanisms.

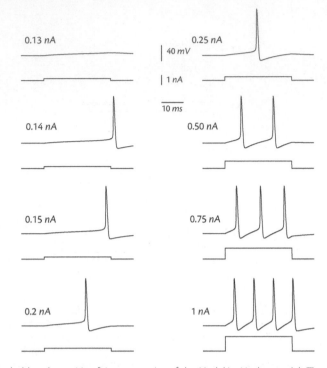

Fig. 2.7: Threshold and repetitive firing properties of the Hodgkin-Huxley model. The response of the model to different amplitudes of injected current is similar to that in Fig. 2.6. The threshold of the membrane was around −57 mV.

Another important property of the action potential captured by the Hodgkin-Huxley model is the refractory period (Fig. 2.8). Immediately following an action potential, it is more difficult to elicit a second action potential; however, the second action potential can be elicited with a higher stimulus amplitude so there is no absolute refractory period. After a few milliseconds, the membrane recovers full excitability (Fig. 2.8, left panel). In Fig. 2.8, the same stimulation protocols with different current amplitudes leads to a different value for the refractory period (left and right panels).

In conclusion, the remarkably successful model introduced by Hodgkin and Huxley almost fifty years ago showed that voltage-clamp methods could be used to study the dynamics of voltage-sensitive membrane currents. The parameters estimated from voltage-clamp experiments are the voltage-dependence of the activation and inactivation processes. The steady-state values and time constants of the gating variables are sufficient to characterize the dynamics of membrane currents that generate action potentials. Since then, similar methods have been applied to the analysis of many other voltage-dependent currents, and a similar approach can be taken to study even more complex cellular phenomena, as illustrated in Chapters 3 and 4.

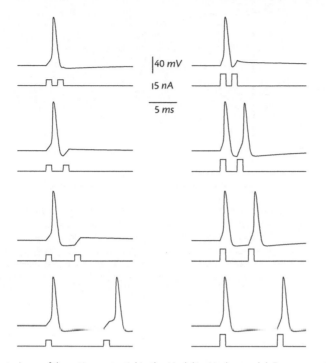

Fig. 2.8: Refractoriness of the action potential in the Hodgkin-Huxley model. Responses of the model to paired current pulses for different values of interpulse interval and pulse amplitude. In the left panel, two identical pulses of 1 *ms* duration and 5 *nA* amplitude were injected at different latencies. In the right panel, the same stimulation was repeated with the amplitude of current pulses twice as large. Model parameters as in Fig. 2.6.

2.1.4 Biophysical bases for voltage-dependence

In the Hodgkin-Huxley model described above, the rate constants $\alpha(V)$ and $\beta(V)$ were fit to the experimental data by using exponential functions of voltage obtained empirically. An alternative approach is to deduce the exact functional form of the voltage-dependence of the rate constants from thermodynamics. These *thermodynamic models* (Eyring et al., 1949; Tsien and Noble, 1969; Hill and Chen, 1972; Stevens, 1978; Borg-Graham, 1991; Hille, 1992; Destexhe and Huguenard, 2000a) provide a plausible physical basis to constrain and parameterize the voltage-dependence of rate constants, which are then used to fit voltage-clamp experiments, as we review below.

A channel is a transmembrane protein that can adopt various configurations, some of which conduct ions through the membrane. Such a large and complex protein is likely to have many charged amino-acid residues throughout its structure. Consequently, an external electrical field applied across the membrane should favour transitions to particular conformational states that are energetically preferred. For a channel that opens with depolarization, the external field may induce a conformational change that opens a pore through the protein.

Within this framework, consider a voltage-sensitive transition between two conformational states, S_i and S_j, described by the following diagram:

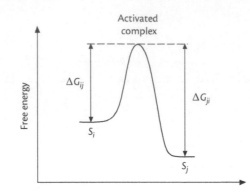

Fig. 2.9: Schematic representation of the free energy profile for the transition between two states of a channel. The diagram represents the free energy of different states involved in a transition: S_i and S_j represent two stable conformational states of the channel, and their transition involves the formation of an activated complex. The rate of the transition will be governed by the *free energy barrier* ΔG, which is the free energy difference between the activated complex and the initial state. Here, ΔG_{ij} is the free energy barrier that determines the rate of the transition $S_i \rightarrow S_j$, while ΔG_{ji} governs the rate of the reverse transition $S_j \rightarrow S_i$.

$$S_i \underset{r_{ji}(V)}{\overset{r_{ij}(V)}{\rightleftharpoons}} S_j \qquad (2.16)$$

where $r_{ij}(V)$ and $r_{ji}(V)$ are voltage-dependent rate constants. According to the theory of reaction rates (Eyring, 1935; Johnson et al., 1974), the rate of the transition depends exponentially on the free energy barrier between the two states (Fig. 2.9). Thus

$$r_{ij}(V) = \exp[-\Delta G_{ij}(V)/RT], \qquad (2.17)$$

where R is the gas constant and T is the absolute temperature in degrees Kelvin. The voltage-dependence of the free energy $\Delta G_{ij}(V)$ is in general quite difficult to ascertain, and may involve both linear and nonlinear components arising from interactions between the channel protein and the electrical field in the membrane. This dependence can be approximated without making any assumptions about underlying molecular mechanisms by a Taylor series expansion of the form (Hill and Chen, 1972; Stevens, 1978):

$$\Delta G(V) = c_o + c_1 V + c_2 V^2 + \ldots \qquad (2.18)$$

which yields a general transition rate function

$$r(V) = \exp[-(c_o + c_1 V + c_2 V^2 + \ldots)/RT], \qquad (2.19)$$

where $c_0, c_1, c_2, \ldots$ are constants that are specific for each transition. The constant c_0 corresponds to energy differences that are independent of the applied field; the linear term $c_1 V$ can be associated with the translation of isolated charges or the rotation of rigid dipoles

(Tsien and Noble, 1969; Hill and Chen, 1972; Stevens, 1978; Andersen and Koeppe, 1992). Linear terms in V will also result if the conformations differ in their net number of charges, or if the conformational change is accompanied with the translation of a freely moving charge inside the structure of the channel (Hill and Chen, 1972; Hille, 1992). Nonlinear terms results from effects such as electronic polarization and pressure induced by V (Hill and Chen, 1972; Stevens, 1978; Andersen and Koeppe, 1992) or mechanical constraints in the movement of charges due to the structure of the ion channel protein (Destexhe and Huguenard, 2000a).

In the 'low field limit' (during relatively small applied voltages), the contribution of the higher-order terms may become negligible (Stevens, 1978; Andersen and Koeppe, 1992). Thus, neglecting these higher-order terms, the voltage dependence takes the form of a simple exponential:

$$r_{ij}(V) = a_{ij} \exp[-V/b_{ij}], \tag{2.20}$$

where a_{ij} and b_{ij} are constants.

Consider again a simple system

$$C \underset{\beta_m(V)}{\overset{\alpha_m(V)}{\rightleftharpoons}} O \tag{2.21}$$

consisting of a transition between the open (O) and closed (C) forms of a channel. If the rate constants α and β obey a simple exponential voltage-dependence, the steady-state activation and time constants can be easily derived (see Eqs. 2.7–2.9) and are given by:

$$m_\infty(V) = \frac{1}{1 + \exp[-(V - V_0)/k_0]} \tag{2.22}$$

$$\tau_m(V) = \frac{1}{\exp[-(V - V_1)/k_1] + \exp[-(V - V_2)/k_2]} \tag{2.23}$$

where V_i and k_i are constants ($i = 0, 1, 2$). These functional forms typically have a sigmoidal shape for $m_\infty(V)$ and a bell-shaped function for $\tau_m(V)$. They are qualitatively similar to the functions used by Hodgkin and Huxley for fitting the kinetics of currents in the squid giant axon (see Fig. 2.5).

Expressing rate constants as exponential functions of the voltage has the drawback that due to their steep rise, they may take unrealistically high values at high voltages. Under these conditions, the linear approximation in Eq. 2.18 is inappropriate and a higher-order approximation is necessary. For example, nonlinear rate functions (Eq. 2.19) were shown to adequately capture the complex voltage-dependence of some channels (Destexhe and Huguenard, 2000a), which was not possible using linear expressions such as Eq. 2.20. In addition to the effect of voltage on isolated charges or rigid dipoles, described in Eq. 2.20, nonlinear expressions (Eq. 2.19) account for more sophisticated effects such as electronic polarization or the deformation of the protein by the electrical field (Hill and Chen, 1972; Stevens, 1978), and perhaps most importantly, they also take into account the presence of mechanical constraints on the gating process (Destexhe and Huguenard, 2000a). If gating depends on the movement of charged residues subject to a mechanical constraint, the force due to this constraint will contribute for a nonlinear term in the free energy. Because charged residues in general are strongly constrained by the structure of the protein and are not moving freely, these nonlinear thermodynamic models are probably more realistic.

Another approximation of Eq. 2.18 is to include nonlinear as well as linear components such that it takes the form:

$$r_i(V) = \frac{a_i}{1 + \exp[-(V - c_i)/b_i]} \tag{2.24}$$

which is sigmoidal in V. The constant a_i sets the maximum transition rate, b sets the steepness of the voltage-dependence, and c_i sets the voltage at which the half-maximal rate is reached. This functional form is seen both in the Hodgkin-Huxley model and in other models (see Keller et al., 1986; Clay, 1989; Chen and Hess, 1990; Borg-Graham, 1991). Through saturation, Eq. 2.24 effectively incorporates voltage-independent interactions that become rate limiting at extreme values of the membrane potential. This type of rate constant is quite useful, as shown below.

2.2 Calcium-dependent ion channels

2.2.1 Intracellular calcium

The concentration of intracellular free Ca^{2+} is highly regulated inside cells and maintained at around 100–200 nM, which is about 10,000 times lower than the 1-2 mM concentration in the extracellular solution. The intracellular Ca^{2+} levels can increase by entry through calcium-selective channels and by release of calcium from intracellular storage.

In the models presented here, the concentration of intracellular Ca^{2+} is determined by:

(i) *Influx of Ca^{2+} due to calcium currents*

Ca^{2+} ions enter through calcium channels and diffuse into the interior of the cell. Only the Ca^{2+} concentration in a thin shell beneath the membrane was modelled. The influx of Ca^{2+} into such a thin shell is governed by:

$$[\dot{Ca}]_i = -\frac{k}{2Fd} I_{Ca} \tag{2.25}$$

where I_{Ca} is the calcium current, $F = 96489\ C\ mol^{-1}$ is the Faraday constant, $d = 1\ \mu m$ is the depth of the shell beneath the membrane and the unit conversion constant is $k = 0.1$ for I_{Ca} in $\mu A/cm^2$ and $[Ca]_i$ in mM.

(ii) *Efflux of Ca^{2+} due to active transport*

The level of intracellular free Ca^{2+} is maintained at a low concentration by calcium transporters in the membrane, calcium binding proteins, and other intracellular stores, such as the endoplasmic reticulum and mitochondria. In the thin shell beneath the membrane, Ca^{2+} can also diffuse to neighbouring shells. Ca^{2+} efflux can be modelled by the following kinetic scheme (Destexhe et al., 1993a):

$$Ca_i^{2+} + P \quad \underset{c_2}{\overset{c_1}{\rightleftharpoons}} \quad CaP \quad \overset{c_3}{\rightarrow} \quad P + Ca_o^{2+} \tag{2.26}$$

where P represents the Ca^{2+} transporter, CaP is an intermediate state, Ca_o^{2+} is the extracellular Ca^{2+} concentration and c_1, c_2 and c_3 are rate constants. Ca^{2+} ions have a high affinity for the transporter P, whereas extrusion of Ca^{2+} follows a slower process (Blaustein, 1988). Therefore, c_3 is low compared to c_1 and c_2, and the Michaelis-Menten approximation can be used for describing the kinetics of the transporter. According to such a scheme, the kinetic equation for the Ca^{2+} transporter is:

$$[\dot{Ca}]_i = -\frac{K_T [Ca]_i}{[Ca_i] + K_d} \tag{2.27}$$

where $K_T = 10^{-4} \, mMms^{-1}$ is the product of c_3 with the total concentration of P, and $K_d = c_2/c_1 = 10^{-4} \, mM$ is the dissociation constant, which can be interpreted here as the value of $[Ca]_i$ at which the transporter is half activated (if $[Ca]_i \ll K_d$ then the efflux is negligible).

The extracellular Ca^{2+} concentration is taken to be $[Ca]_o = 2 \, mM$, as found *in vivo*. The change of $[Ca]_i$ due to the binding of Ca^{2+} to I_h channels is negligible and was neglected, as was the contribution of Ca^{2+} efflux to the net Ca^{2+} current in Eq. 2.25).

The Ca^{2+} reversal potential depends on the intracellular Ca^{2+} concentration and can be calculated with the Nernst relation:

$$E_{Ca} = k' \frac{RT}{2F} \log \frac{[Ca]_o}{[Ca]_i} \tag{2.28}$$

where $R = 8.31 \, Jmol^{-1}{}^{\circ}K^{-1}$, $T = 309^{\circ}K$, and the constant for unit conversion is $k' = 1000$ for E_{Ca} in mV. For $[Ca]_i = 2.4 \, 10^{-4} \, mM$, which is an average value at rest, E_{Ca} was approximately 120 mV.

In some circumstances, the efflux of calcium can be modelled by a first-order decay term (see for example Huguenard and McCormick, 1992):

$$[\dot{Ca}]_i = -\frac{1}{\tau_{Ca}} ([Ca]_i - [Ca]_\infty) \tag{2.29}$$

where τ_{Ca} is the time constant of calcium efflux and $[Ca]_\infty$ is the intracellular concentration at rest.

2.2.2 Constant field equations

The usual Nernst equation describes the near-to-equilibrium behaviour of ion channels in which the current is described by Ohm's law. Due to the nonlinear and far-from-equilibrium behaviour of calcium currents (the internal and external Ca^{2+} concentration differ by about four orders of magnitude), these currents are more adequately represented by constant-field equations, also called Goldman-Hodgkin-Katz equations (Goldman, 1943; Hodgkin and Katz, 1949; see details in Hille, 1992):

$$I_{Ca} = \bar{P}_{Ca} \, m^2 h \, G(V, Ca_o, Ca_i)$$
$$\dot{m} = -\frac{1}{\tau_m(V)} \, (m - m_\infty(V)) \tag{2.30}$$
$$\dot{h} = -\frac{1}{\tau_h(V)} \, (h - h_\infty(V))$$

where the current is described by the maximum permeability of the membrane to Ca^{2+} ions ($\bar{P}_{Ca}$, in cm/s). m and h are, respectively, the activation and inactivation variables similar to the model based on Nernst equations. $G(V, Ca_o, Ca_i)$ is a nonlinear function of voltage and ionic concentrations:

$$G(V, Ca_o, Ca_i) = Z^2 F^2 V/RT \, \frac{Ca_i - Ca_o \exp(-ZFV/RT)}{1 - \exp(-ZFV/RT)} \tag{2.31}$$

where $Z = 2$ is the valence of calcium ions, F is the Faraday constant, R is the gas constant and T is the temperature in Kelvin. Ca_i and Ca_o are the intracellular and extracellular Ca^{2+} concentrations (in M), respectively.

This formalism is more accurate than Nernst relations and will be used in cases where a precise representation is needed for calcium currents, such as the investigation of dendritic calcium currents in Chapters 3 and 4.

2.2.3 Calcium-activated channels

Calcium-dependent currents are gated differently from the voltage-dependent channels analysed above. Calcium-activated channels are opened by the binding of intracellular Ca^{2+} ions. Other examples of ligand-gated channels are considered in Chapter 5.

There are many types of calcium-gated channels, including classes that are voltage-dependent and others that are voltage-independent (reviewed in Latorre et al., 1989; Hille, 1992). The apamin-sensitive slow Ca^{2+}-dependent K^+ current is independent of voltage and is activated by intracellular calcium (Hille, 1992). A simple kinetic scheme to describe this type of channel is:

$$(closed) + n \, Ca_i^{2+} \underset{\beta}{\overset{\alpha}{\rightleftharpoons}} (open) \tag{2.32}$$

Here, the binding of n intracellular calcium ions (Ca_i^{2+}) opens the channel and α and β are rate constants. The ionic current is then given by:

$$I_{K[Ca]} = \bar{g}_{K[Ca]} \, m^2 \, (V - E_K)$$
$$\dot{m} = -\frac{1}{\tau_m([Ca]_i)} \, (m - m_\infty([Ca]_i)) \tag{2.33}$$

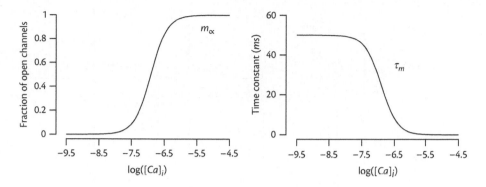

Fig. 2.10: Activation function and time constant for the calcium-dependent potassium current $I_{K[Ca]}$. The activation function (left) and time constant (right) are represented as a function of the logarithm of intracellular calcium concentration $\log([Ca]_i)$. From Eqs. 2.34–2.37 with $n = 4$, $\alpha = 0.2$ $ms^{-1}mM^{-n}$ and $\beta = 0.02$ ms^{-1}.

where $\bar{g}_{K[Ca]}$ is the maximal conductance, E_K is the reversal potential for potassium, m is the activation variable of $I_{K[Ca]}$ (fraction of open channels), and $[Ca]_i$ is the intracellular calcium concentration.

The Ca^{2+}-dependent activation function and time constant are given by:

$$m_\infty([Ca]_i) = \alpha \, [Ca]_i^n \, / \, (\alpha \, [Ca]_i^n + \beta) \tag{2.34}$$

$$\tau_m([Ca]_i) = 1 \, / \, (\alpha \, [Ca]_i^n + \beta) \tag{2.35}$$

These quantities can be re-expressed in the more familiar form:

$$m_\infty(X) = 1 \, / \, (1 + \exp[-(X - X_0)/K]) \tag{2.36}$$

$$\tau_m(X) = 1 \, / \, (\alpha \, \exp(X/K) + \beta) \tag{2.37}$$

where $X = \log([Ca]_i)$, $X_0 = 1/n \log(\beta/\alpha)$ and $K = 1/n$. Here, the activation function is functionally dependent on the logarithm of calcium concentration, similar to the functional dependence on voltage for voltage-dependent channels (compare with Eq. 2.22). In particular, n sets the steepness of the activation function, while α and β set the half-activation value. These functions are graphed in Fig. 2.10.

2.2.4 Oscillations with I_{Ca}

The $I_{K[Ca]}$ current typically generates a long *after-hyperpolarization* (AHP) following a calcium event such as the opening of calcium channels in the membrane. The increase of intracellular calcium activates $I_{K[Ca]}$ and hyperpolarizes the membrane toward the reversal potential for K^+ ions. This behaviour is illustrated in Fig. 2.11: the event labeled 'BURST' is an activation of a calcium current I_{Ca}, which triggers a sharp rise in intracellular calcium ($[Ca]_i$). The rise

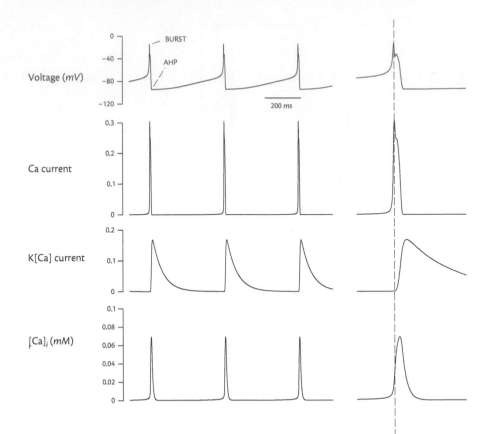

Fig. 2.11: The combination of I_{Ca} and $I_{K[Ca]}$ currents can generate after-hyperpolarizations and oscillations. The voltage, the calcium current (I_{Ca}), the calcium-dependent potassium current ($I_{K[Ca]}$), their normalized conductances, and the intracellular calcium concentration ($[Ca]_i$) represented from top to bottom. The combination of $I_{K[Ca]}$ and I_{Ca} gave rise to low-frequency oscillations consisting of calcium bursts (BURST; fast rise of I_{Ca} and $[Ca]_i$) followed by an after-hyperpolarization (AHP) generated by the activation of $I_{K[Ca]}$. The right panel shows the same events at five-times higher temporal resolution. The low-threshold calcium current was modelled by Hodgkin-Huxley equations ($\bar{g}_{Ca} = 1.75\,mS/cm^2$; see Chapter 4). $I_{K[Ca]}$ was described by Eqs. 2.33–2.37 ($\bar{g}_{K[Ca]} = 10\,mS/cm^2$; $n = 2$, $\alpha = 3.2\,ms^{-1}mM^{-2}$ and $\beta = 0.002\,ms^{-1}$). See Chapter 4 for more details on the underlying mechanisms.

of $[Ca]_i$ subsequently activates $I_{K[Ca]}$, generating an after-hyperpolarization in the membrane potential.

The interaction between I_{Ca} and $I_{K[Ca]}$ may also generate sustained oscillations. As shown in Fig. 2.11, the after-hyperpolarization itself may trigger the next calcium event, resulting in the same burst/AHP sequence as described above and leading to sustained oscillations. This type of behaviour is examined in more detail in Chapter 4 in the case of thalamic reticular neurons.

2.3 Markov models of voltage-dependent ion channels

The Hodgkin-Huxley-type model reviewed above accurately describes the behaviour of many currents, but is not always an adequate description. For example, modelling the behaviour of single ion channels requires a different formalism. *Markov models* which have proven to be more flexible formalism for modelling single-channel currents, are summarized in the next section and compared with the Hodgkin-Huxley model.

2.3.1 Single-channel recordings

A major advance in our understanding of ion channels followed the electrical recording of ionic currents through single ion channels using patch recording techniques (Fig. 2.12; see Sakmann and Neher, 1983, 1995). Recordings from single channels exhibit rapid transitions between conducting and non-conducting states (Fig. 2.13). Voltage-dependent conformational changes of the channel protein, rather than movement of gating particles per se give rise to the voltage-sensitivity of ion currents.

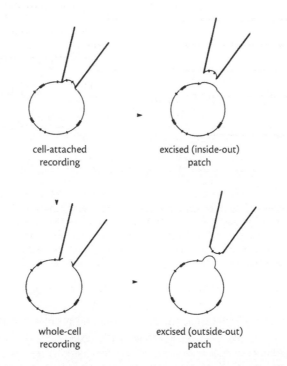

cell-attached
recording

excised (inside-out)
patch

whole-cell
recording

excised (outside-out)
patch

Fig. 2.12: Patch-clamp recording configurations. The electrode can be used in the cell-attached or whole-cell configuration, or a small patch of membrane can be excised from the cell and recorded in isolation. In cell-attached and excised configurations, single-channel recordings can be obtained from an extremely small area of membrane. From Johnston and Wu, 1995.

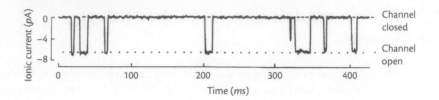

Fig. 2.13: Patch-clamp recording of a single ion channel. Single-channel recorded in an excised patch from a cultured rat myotube. In this recording, the channel was closed most of the time and showed briefs openings. From Hille, 1992; after Sanchez et al., 1986.

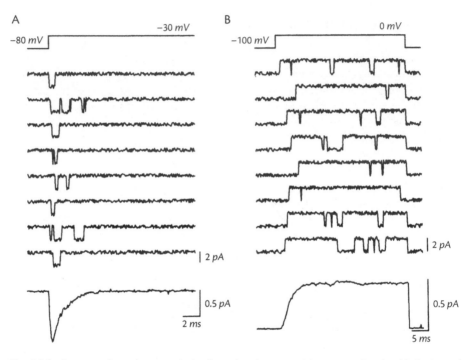

Fig. 2.14: Correspondence between single-channel and macroscopic currents. Simulated behaviour of Na^+ (A) and K^+ channels (B) during a voltage-clamp experiment. The top trace shows the voltage-clamp protocol, the eight next traces shows examples of single-channel currents and the bottom trace shows the ensemble average computed over many single-channel trials. The ensemble average current is similar to the voltage-clamp recordings obtained by Hodgkin and Huxley (1952) (see Fig. 2.2). Figure from Johnston and Wu, 1995.

How are the apparently discontinuous current traces from single channels related to the smooth current traces seen in classical voltage-clamp experiments (Fig. 2.2)? The answer depends on the number of channels recorded. In patch-clamp experiments, a small patch of membrane about 1–3 μm^2 area is excised from the cell and recorded. This size of patch is sufficiently small that only a single functional ion channel is often present in the patch. In contrast, voltage-clamp experiments on intact cells usually register the current through a population of several thousands of channels. As a consequence, they are called *macroscopic* currents compared with the *microscopic* single-channel currents. The correspondence between these two levels of description is illustrated in Fig. 2.14 for Na^+ and K^+ channels.

Single-channel recording techniques, recombinant DNA techniques, and most recently x-ray crystallographic techniques, have revealed of the structure and function of ion channels. The amino acid sequences of most ion channels have been identified and are now used to infer information on the three dimensional structure of ion channels (see Hille, 1992). Significant progress has been made in recent years in relating the function of ion channels to specific elements of their molecular structure (reviewed in Unwin, 1989; Andersen and Koeppe, 1992; Armstrong, 1992; Catterall, 1992; Hille, 1992; Jan and Jan, 1992; Sakmann, 1992; Armstrong and Hille, 1998; Catterall, 2000). Fig. 2.15 schematically illustrates the structure of the Na^+ channel (Marban et al., 1998) and a possible mechanism underlying voltage sensitivity is shown in Fig. 2.16 (Yang et al., 1996). Recently, the crystallographic structure of a voltage-dependent K^+ channel has revealed details of the ionic pore and selectivity mechanism (Doyle, 1998; Gulbis, 2000).

2.3.2 Markov kinetic models

Conformational changes, which are responsible for the gating of ion channels, follow state diagrams analogous to those used to describe chemical reactions. *Markov models* are a class of kinetic schemes based on state diagrams in which the probability of a transition to a new state at a given time depends only on the present state and not on previous states. These models have been used to explore the gating characteristics of many voltage-dependent ion channels and receptors (see Hille, 1992; Sakmann and Neher, 1995; see also Chapter 5). In particular, Markov models can account for many details observed in single-channel recordings. When re-evaluated in light of single channel data, the Hodgkin-Huxley description of the sodium channel has remained accurate in some respects but in many detailed ways has been superseded (Armstrong, 1992). The majority of biophysical studies at the single channel level fit the statistics of channel open and closed times to Markov models. Although the state diagram cannot be uniquely determined from the channel statistics, it is possible to determine equivalence classes (Kienker, 1989).

State diagrams for the gating kinetics of single-channels are of the form:

$$S_1 \rightleftharpoons S_2 \rightleftharpoons \ldots \rightleftharpoons S_n. \tag{2.38}$$

The states $S_1 \ldots S_n$ represent different conformations that protein can adopt and the arrows represent the possible transitions between them. Define $P(S_i, t)$ as the probability of being in a state S_i at time t and $P(S_i \rightarrow S_j)$ as the *transition probability* from state S_i to state S_j. The time evolution of the probability of state S_i is then governed by the *Master equation* (e.g. Colquhoun and Hawkes, 1977, 1981; Stevens, 1978):

$$\frac{dP(S_i, t)}{dt} = \sum_{j=1}^{n} P(S_j, t) \, P(S_j \rightarrow S_i) - \sum_{j=1}^{n} P(S_i, t) \, P(S_i \rightarrow S_j). \tag{2.39}$$

The left term represents the 'source' contribution of all transitions entering state S_i, and the right term represents the 'sink' contribution of all transitions leaving state S_i. In this equation, the time evolution depends only on the present state of the system, and is defined entirely by

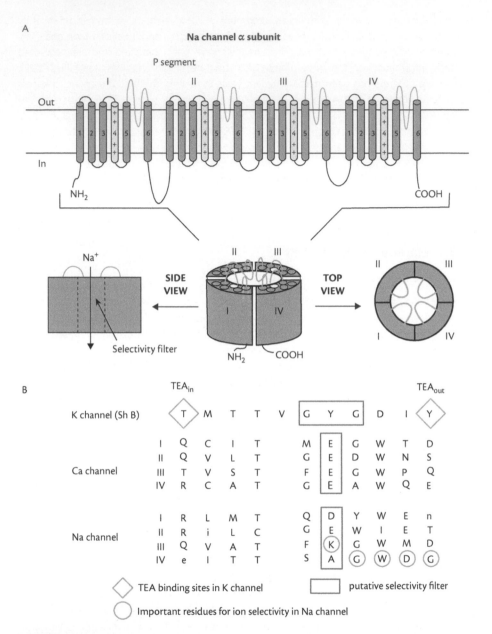

Fig. 2.15: Schematic three dimensional structure of the sodium channel. A. Probable folding of the amino acid sequence forming the channel. The four α subunits are indicated by I–IV (β subunits are not shown), S4 is the charged transmembrane segment possibly involved in voltage sensitivity, and P indicates the segments situated at the mouth of the pore and which may be responsible for ionic selectivity. B. Primary amino acid sequences of the P segment for different channel types: K^+ channels (Shaker B), Ca^{2+} channels (L-type) and Na^+ channels. Diamonds, circles and rectangles indicate the amino acids that have been identified as critical for ligand binding or ionic selectivity. Reproduced from Marban et al., 1998.

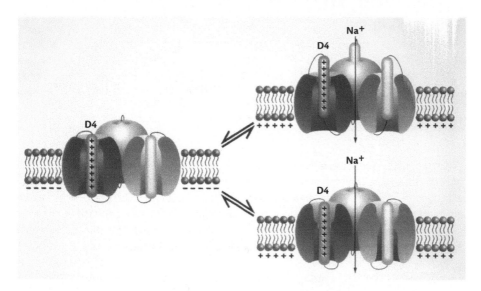

Fig. 2.16: Possible mechanism for the voltage sensitivity of Na$^+$ channels. At hyperpolarized (left) or depolarized (right) membrane potentials, the S4 segment (indicated by positive charges) can move due to the electric field, resulting in opening of the pore. Depolarization could either cause S4 to move outward (top right) or the field to move inward past it (bottom-right). From Yang et al., 1996.

knowledge of the set of transition probabilities. This type of model is called a *Markovian system* and Eq. 2.39 is a *stochastic Markov model*.

For large numbers of identical channels or other proteins, the quantities given in the master equation can be reinterpreted. The probability of being in a state S_i becomes the *fraction of channels* in state S_i, noted s_i, and the transition probabilities from state S_i to state S_j becomes the *rate constants*, r_{ij}, of the reactions

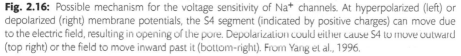

$$S_i \underset{r_{ji}}{\overset{r_{ij}}{\rightleftharpoons}} S_j. \tag{2.40}$$

In this case, the master equation can be rewritten as:

$$\frac{ds_i}{dt} = \sum_{j=1}^{n} s_j \, r_{ji} - \sum_{j=1}^{n} s_i \, r_{ij} \tag{2.41}$$

which is a conventional kinetic equation for the states of the system. We refer to this as a *Markov kinetic model*.

The Markov formalism is more general than the Hodgkin-Huxley formalism and includes it as a special case. In other words, any Hodgkin-Huxley scheme can be written as a Markov kinetic model. However, the translation of a Hodgkin-Huxley system with multiple independent gates into a kinetic equivalent may result in a combinatorial explosion of states. For example, the Markov model corresponding to the Hodgkin-Huxley sodium channel is

$$
\begin{array}{ccccccc}
C_3 & \underset{\beta_m}{\overset{3\alpha_m}{\rightleftharpoons}} & C_2 & \underset{2\beta_m}{\overset{2\alpha_m}{\rightleftharpoons}} & C_1 & \underset{3\beta_m}{\overset{\alpha_m}{\rightleftharpoons}} & O \\[2pt]
\alpha_h \updownarrow \beta_h & & \alpha_h \updownarrow \beta_h & & \alpha_h \updownarrow \beta_h & & \alpha_h \updownarrow \beta_h \\[2pt]
I_3 & \underset{\beta_m}{\overset{3\alpha_m}{\rightleftharpoons}} & I_2 & \underset{2\beta_m}{\overset{2\alpha_m}{\rightleftharpoons}} & I_1 & \underset{3\beta_m}{\overset{\alpha_m}{\rightleftharpoons}} & I
\end{array}
\tag{2.42}
$$

(Fidzhugh, 1965). The states represent the channel with the inactivation gate in the open state (top) or closed state (bottom) and (from left to right) three, two, one or none of the activation gates closed. To reproduce the m^3 formulation, the rates must have a 3:2:1 ratio in the forward direction and a 1:2:3 ratio in the backward direction. Only the O state is conducting. The squid delayed rectifier potassium current modelled by Hodgkin and Huxley (1952) with four activation gates and no inactivation also can be translated into a Markov kinetic system:

$$
C_4 \underset{\beta_m}{\overset{4\alpha_m}{\rightleftharpoons}} C_3 \underset{2\beta_m}{\overset{3\alpha_m}{\rightleftharpoons}} C_2 \underset{3\beta_m}{\overset{2\alpha_m}{\rightleftharpoons}} C_1 \underset{4\beta_m}{\overset{\alpha_m}{\rightleftharpoons}} O .
\tag{2.43}
$$

(Fitzhugh, 1965; Armstrong, 1969).

The Hodgkin-Huxley model assumes that each gate activates or inactivates independently of each other, resulting in a considerable simplification of the equations of the model. Markov models can treat in addition the more general cases of ion channels where the gates are not identical, or when there is coupling between opening/closing of different gates, or where the opening of the channel simply depends on a series of conformational changes of the protein, leading to the open state.

One of the simplest Markov kinetic models is a two-state scheme (Eq. 2.3) augmented by a single inactivated state, giving

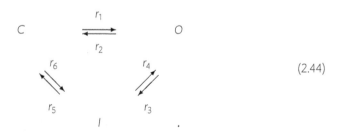

All six possible transitions between the three states are allowed, giving this kinetic scheme a *looped* form. The transition rates may follow voltage-dependent equations in the general form of Eq. 2.18 or some of these rates may be taken as either zero or independent of voltage to yield even simpler models (see below).

Additional closed or inactivated states may be necessary to fit more accurately the time courses of channel openings or gating currents. The squid sodium channel model of Vandenberg and Bezanilla (1991) is a good example of a biophysically derived multistate Markov model. Parameters for a variety of Markov schemes were fit by least squares constrained by a combination of single channel data, macroscopic ionic currents, and gating currents. The nine-state diagram

$$C \underset{r_6}{\overset{r_5}{\rightleftharpoons}} C_1 \underset{r_6}{\overset{r_5}{\rightleftharpoons}} C_2 \underset{r_6}{\overset{r_5}{\rightleftharpoons}} C_3 \underset{r_2}{\overset{r_1}{\rightleftharpoons}} C_4 \underset{r_4}{\overset{r_3}{\rightleftharpoons}} O$$

$$r_{10}\downarrow\uparrow r_8 \qquad\qquad\qquad r_9\downarrow\uparrow r_7$$

$$I_1 \underset{r_3}{\overset{r_4}{\rightleftharpoons}} I_2 \underset{r_1}{\overset{r_2}{\rightleftharpoons}} I_3$$

$$(2.45)$$

was found to be optimal by maximum likelihood criteria. The voltage-dependence of the transition rates was assumed to be a simple exponential function of voltage (Eq. 2.20). To complement the detailed sodium model of Vandenberg and Bezanilla, the six-state scheme for the squid delayed rectifier used by Perozo and Bezanilla (1990) was examined:

$$C \underset{r_2}{\overset{r_1}{\rightleftharpoons}} C_1 \underset{r_4}{\overset{r_3}{\rightleftharpoons}} C_2 \underset{r_4}{\overset{r_3}{\rightleftharpoons}} C_3 \underset{r_4}{\overset{r_3}{\rightleftharpoons}} C_4 \underset{r_6}{\overset{r_5}{\rightleftharpoons}} O \qquad (2.46)$$

where again rates were described by a simple exponential function of voltage (Eq. 2.20).

These models were compared to the original Hodgkin-Huxley model in voltage-clamp (Fig. 2.17). These two models were compared to a simple three-state scheme for the Na$^+$ current:

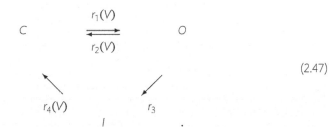

$$(2.47)$$

This scheme was chosen to have the fewest possible number of states (three) and transitions (four) while still reproducing the essential behaviour of the more complex models. The form of the state diagram was based on a looped three-state model with several transitions eliminated to give an irreversible loop (Bush and Sejnowski, 1991). The latter model incorporated voltage-dependent opening, closing, and recovery from inactivation, while inactivation was voltage-independent. For simplicity, neither opening from the inactivated state nor inactivation from the closed state was permitted. Although there is evidence for the latter (Horn et al., 1983), under the conditions of the present simulations including this transition was unnecessary.

The three models gave qualitatively similar conductance time courses in response to a voltage-clamp step (Fig. 2.17; Destexhe et al., 1994d). For all three models, closed states were favoured at hyperpolarized potentials. Upon depolarization, forward (opening) rates sharply increased while closing (backward) rates decreased, causing the channels to migrate in the forward direction toward the open state. The three closed states in the Hodgkin-Huxley model and the five closed states in the detailed (Vandenberg-Bezanilla) model gave rise to the characteristic sigmoidal shape of the rising phase of the sodium current (Fig. 2.17D). In contrast,

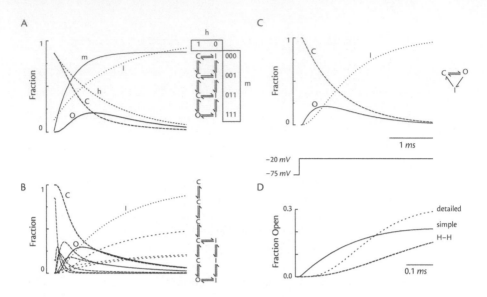

Fig. 2.17: Three kinetic models of squid axon Na$^+$ channel produce qualitatively similar conductance time courses. A voltage-clamp step from rest, $V = -75$ mV, to $V = -20$ mV was simulated. The fraction of channels in the open state (O, solid lines), closed states (C, dashed lines), and inactivated states (I, dotted lines) are shown for the Hodgkin-Huxley model, a detailed Markov model, and a simple Markov model. A. An equivalent Markov scheme for the Hodgkin-Huxley model is shown (right insert, Eq. 2.42). Three identical and independent activation gates (m, thin solid line) yield a scheme having three closed states (corresponding to zero, one and two activated gates) and one open state (three activated gates). The independent inactivation gate (h, thin dotted line) adds four corresponding inactivated states. Voltage-dependent transitions were calculated using the original equations and constants of Hodgkin and Huxley (1952; Table 2.1, top panel). B. The detailed Markov model of Vandenberg and Bezanilla (1991) (Eq. 2.45; $a_1 = 11490$ s^{-1}, $b_1 = 59.19$ mV, $a_2 = 8641$ s^{-1}, $b_2 = -5860$ mV, $a_3 = 31310$ s^{-1}, $b_3 = 17.18$ mV, $a_4 = 2719$ s^{-1}, $b_4 = -51.54$ mV, $a_5 = 33350$ s^{-1}, $b_5 = 74.58$ mV, $a_6 = 1940$ s^{-1}, $b_6 = -21.03$ mV, $a_7 = 863.1$ s^{-1}, $b_7 = 27050$ mV, $a_8 = 1538$ s^{-1}, $b_8 = 27050$ mV, $a_9 = 7.992$ s^{-1}, $b_9 = -27.07$ mV, $r_{10} = r_8 \, r_9 \, / \, r_7$). Individual closed and inactivated states are shown (thin lines) as well as the sum of all five closed and all three inactivated states (thick lines). C. A simple three-state Markov model adjusted fit the detailed model (Eq. 2.47; $a_1 = 1500$ s^{-1}, $a_2 = 200$ s^{-1}, $a_4 = 150$ s^{-1}, $b = 5$ mV, $c_1 = c_2 = -27$ mV, $c_4 = -65$ mV, $r_3 = 3000$ s^{-1}). D. Comparison of the time course of open channels for the three models on a faster time scale shows differences immediately following a voltage step. The Hodgkin-Huxley model (dashed line) and detailed Markov modelled (solid line) give smooth, multiexponential rising phases, while the simple Markov model (dotted line) gives a single exponential rise with a discontinuity in the slope at the beginning of the pulse. Reproduced from Destexhe et al., 1994d.

the simple model, with a single closed state, produced a first-order exponential response to the voltage step. As expected, the addition of one or more closed states to the simple model led to a progressively more sigmoidal rising phase (data not shown).

A range of complexity was also introduced to model the non-inactivating delayed-rectifier potassium channel models. The main difference was in the number of closed states, from six for the detailed Markov model of Perozo and Bezanilla (1990; Eq. 2.46), to four for the original (Hodgkin and Huxley, 1952) description of the potassium current, to just one for a minimal model (Eq. 2.3) with rates having sigmoidal voltage dependence (Eq. 2.24).

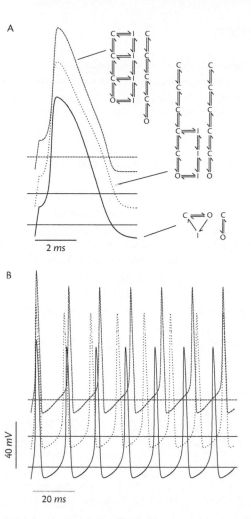

Fig. 2.18: Action potentials produced using three different kinetic models of squid Na^+ and K^+ channels. A. Single action potentials in response to 0.2 ms, 2 nA current pulse are elicited at similar thresholds and produce similar waveforms using three different pairs of kinetic models: Hodgkin-Huxley (dashed line; Hodgkin and Huxley, 1952), detailed Markov (dotted line; Vandenberg and Bezanilla, 1991; Perozo and Bezanilla, 1990), and simple Markov (solid line). B. Repetitive trains of action potentials elicited in response to sustained current injection (0.2 nA) have slightly different frequencies. Sodium channels were modelled as described in Fig. 2.17. The detailed Markov potassium channel model had six states (Perozo and Bezanilla, 1990) (Eq. 2.46; $a_1 = 484.5\ s^{-1}$, $b_1 = 112\ mV$, $a_2 = 19.23\ s^{-1}$, $b_2 = -8.471\ mV$, $a_3 = 1757\ s^{-1}$, $b_3 = 25.83\ mV$, $a_4 = 569\ s^{-1}$, $b_4 = -491.0\ mV$, $a_5 = 672.9\ s^{-1}$, $b_5 = 212\ mV$, $a_6 = 784.4$, $b_6 = 0$). The simple Markov potassium channel had two states (Eq. 2.3 with rates given by the sigmoid function of Eq. 2.24; $a_1 = 100\ s^{-1}$, $a_2 = 240\ s^{-1}$, $b = 5\ mV$, $c_1 = c_2 = -27\ mV$). Reproduced from Destexhe et al., 1994d.

The three corresponding models for the Na^+ and K^+ channels also produced comparable action potentials and repetitive firing despite the significant differences in their complexity and formulation (Fig. 2.18).

2.4 Discussion

In this chapter, we have focused on modelling macroscopic ionic currents resulting from large populations of ion channels. Two types of kinetic equations were used: first, the model introduced by Hodgkin and Huxley (1952) and second, Markov kinetic models. The advantages and limitations of these two types of models are discussed here.

2.4.1 Hodgkin-Huxley models of ion channels

The model of the fast sodium current and the delayed rectifier potassium current introduced by Hodgkin and Huxley (1952) was remarkably forward looking. The general framework, based on four independent gating particles, is echoed in the multiple subunit channel structure revealed by molecular techniques. The Hodgkin-Huxley description not only accounted quite well for the conductances of the squid giant axon, but has been widely applied, with very minor alterations, to describe nearly all of voltage-dependent currents (see e.g. Yamada et al., 1989; Borg-Graham, 1991; Lytton and Sejnowski, 1991; Koch, 1999).

One great advantage of Hodgkin-Huxley models is that they describe the behaviour of ion channels using quantities (activation functions, time constants) that are observable experimentally with relatively simple voltage-clamp protocols. This may explain why this type of models is so widely used. Because it reproduces the most salient features of current-clamp behaviour, such as action potentials, the Hodgkin-Huxley model should continue to be the basis for modelling macroscopic currents for many more years.

Nevertheless, the Hodgkin-Huxley model is inadequate for some purposes. First, there is considerable physiological and molecular evidence that channel gating is organized differently from what Hodgkin-Huxley kinetics requires (see Armstrong, 1992). Activation and inactivation are clearly coupled in a multistate sequence, as shown by Na^+ channel measurements (Armstrong, 1981; Aldrich et al., 1983; Bezanilla, 1985), which is in contrast with the independence of these processes in the Hodgkin-Huxley model. K^+ channels may also show an inactivation which is not voltage-dependent, as in the Hodgkin-Huxley model, but state-dependent (Aldrich, 1981). Although the latter can be modelled with modified Hodgkin-Huxley kinetics (Marom and Abbott, 1994), these phenomena are best described using Markov models, a formalism more appropriate to describe single channels.

Second, in some cases, only a small number of channels contribute to an ionic current, so that the variability from single channel noise cannot be ignored. This may occur, for example, at dendritic spines. Most central excitatory synapses are made on spines and channel noise there may and have important consequences for electrical signalling (Clay and DeFelice, 1983; Strassberg and DeFelice, 1993). Markov models would be more appropriate starting point to describe such small structures as spines.

2.4.2 Markov models of ion channels

Finite-state models of ion channels rely in general on the assumption that the configuration of a channel protein in the membrane can be operationally grouped into a set of distinct states separated by large energy barriers (Hille, 1992). Because the flux of ions though single channels can be directly measured, it has been possible to observe directly the predicted rapid and stochastic transitions between conducting *open* and non-conducting *closed* states (Neher, 1992). Channels can be treated as finite-state Markov systems if one further assumption is made, namely that the probability of state transitions is dependent only of the presently occupied state.

The assumption that channels function through a succession of conformational changes does not necessarily imply a finite-state Markov description and alternative models have been proposed. Diffusional (Millhauser et al., 1988) or *continuum* gating models (Levitt, 1989), are Markovian but posit an infinite number of states. Fractal (Liebovitch and Sullivan, 1987) or deterministically chaotic models (Liebovitch and Toth, 1991) assume a finite number of states, but allow time-dependent transition rates. Differentiation between discrete multistate Markov models and any of these alternatives hinges on fast time-resolution studies of channel openings. Analysis of single-channel openings and closings has been consistent with the suitability of finite-state Markov models for many channels (McManus et al., 1988; Sansom et al., 1989).

Markov kinetic descriptions exist for the kinetics of many voltage-gated channels, including sodium channels (Aldrich et al., 1983; Chabala, 1984; Horn and Vandenberg, 1984; Keller et al., 1986; Aldrich and Stevens, 1987; Clay, 1989; Vandenberg and Bezanilla, 1991), potassium channels (Labarca et al., 1985; Hoshi and Aldrich, 1988; Perozo and Bezanilla, 1990), calcium channels (Chen and Hess, 1990), chloride channels (Labarca et al., 1980) and voltage-dependent gap junctions (Harris et al., 1981; Chanson et al., 1993). The Markov kinetic formalism includes the Hodgkin-Huxley framework as a special case and is more flexible and extensible. Within the kinetic framework it is possible to include more states, making the model more biophysically accurate, or fewer states, leading to simplified models more appropriate for network simulations (see Chapter 5).

2.4.3 Applications to model single-cell and network behaviour

We have compared here different models for voltage-dependent currents and delineated some of the differences between them. In the case of sodium channels, models of increasing complexity, from simplified two-state representations to multistate Markov diagrams, can capture many of the features of sodium channels and action potentials (Destexhe et al., 1994d). The model of choice depends on the type of experimental data available and its level of precision. It is clear that a two-state scheme cannot capture the features of single-channel recordings, which would require Markov models of sufficient complexity to account for the data. On the other hand, even simplified two- or three-state models can account for many properties of action potentials (Fig. 2.18). If the main goal is to generate action potentials, it is therefore not necessary to include all the complexity of the most sophisticated single-channel Markov models. Simplified representations, such as the Hodgkin-Huxley (1952) model, may be sufficient for modelling the macroscopic behaviour of ionic currents in a single-cell, as investigated in Chapters 3 and 4, or in networks involving a large number of neurons (see Chapters 6-8). In these cases, computational efficiency is a more important concern than reproducing all the microscopic features of the channels.

Different models for the T-type Ca^{2+} current illustrate their strengths and weaknesses (Destexhe and Huguenard, 2000b). A variety of formalisms, such as empirical Hodgkin-Huxley type models, thermodynamic models and Markov models, can capture the behaviour of the T-current in voltage-clamp and generate low-threshold spikes. Markov models are more accurate because they also account for single-channel recordings, while Hodgkin-Huxley type models do not. However, if the goal is to model the macroscopic behaviour of the T-current, such as the genesis of low-threshold spikes, Hodgkin-Huxley type models are a better alternative since they are easier to fit to voltage-clamp experiments, they are computationally more efficient and they reproduce well the physiological properties at the single-cell level. For these reasons, we chose this formalism to describe voltage-dependent currents for investigating the cellular behaviour of thalamic neurons in Chapters 3 and 4.

2.5 Summary

In this chapter we have reviewed the ionic bases of membrane excitability and the different formalisms that have been used to model the underlying mechanisms. Although the Hodgkin-Huxley model was introduced almost fifty years ago, it remains the most widely used description of ionic currents. This model has captured the essential features of many macroscopic currents in neurons and is used to model the currents found in thalamic and cortical neurons in the following chapters. The Hodgkin-Huxley model is not appropriate for modelling single-channels, which can be more precisely described by Markov models. Both types of models belong to the general family of kinetic models for which state diagrams represent the transitions between different states of the ion channel. In Chapter 5, these models will be applied to synaptic currents.

Electrophysiological properties of thalamic relay neurons

The focus of this chapter is on thalamocortical (TC) neurons, the only neurons whose axons project from the thalamus to the cortex. TC cells have intrinsic properties that allow them to actively participate in oscillations that occur in the isolated thalamus as well as in the thalamocortical system, as illustrated in Chapters 6–8. The basic anatomical and electrophysiological features of TC cells are reviewed in the first section; subsequent sections present both simple and detailed models of the ionic mechanisms responsible for their complex electrophysiological properties.

3.1 The bursting properties of thalamic relay neurons

3.1.1 Experimental characterization of the rebound burst

Sensory inputs from visual, auditory and somatosensory receptors do not reach the cerebral cortex directly, but synapse first on relay cells in specific regions of the thalamus. These relay cells in turn project to their respective area in primary sensory cortex. These topographically organized forward projections are matched by feedback projections to the corresponding afferent thalamic nucleus (Jones, 1985), as illustrated in Fig. 3.1.

In addition to relaying sensory input to the cortex, TC neurons also have intrinsic properties that allow them to generate activity. These cells can under some circumstances produce bursts of action potentials following inhibition, called a 'low-threshold spike' (LTS) or 'post-inhibitory rebound'. The importance of the rebound response of TC cells was first established by Andersen and Eccles (1962), who referred called it 'post-anodal exaltation'. It was later characterized *in vitro* by Jahnsen and Llinás (1982) and *in vivo* by Deschênes et al. (1984) and has become generally known as the 'rebound burst' or LTS.

Thalamocortical Assemblies: Sleep Spindles, Slow Waves and Epileptic Discharges. Second Edition. Alain Destexhe and Terrence J. Sejnowski, Oxford University Press. © Oxford University Press 2023.
DOI: 10.1093/oso/9780198864998.003.0003

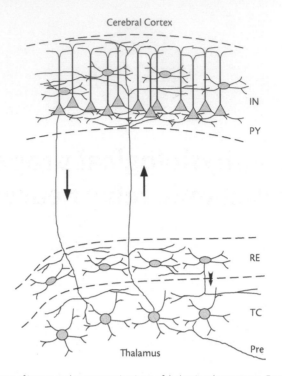

Fig. 3.1: Arrangement of inputs and output projections of thalamic relay neurons. Four cell types and their connectivity are shown: thalamocortical (TC) relay cells, thalamic reticular (RE) neuron, cortical pyramidal cells (PY) and interneurons (IN). TC cells receive prethalamic (Pre) afferent connections, which may be sensory afferents in the case of specific thalamic nuclei involved in vision, audition and somatosensory modalities. This information is relayed to the corresponding area of cerebral cortex through ascending thalamocortical fibres (upward arrow). These axons have collaterals that contact the RE nucleus on the way to the cerebral cortex, where they arborize in superficial layers I and II, layer IV and layer VI. Corticothalamic feedback is mediated primarily by a population of layer VI PY neurons that project to thalamus. The corticothalamic fibres (downward arrow) also leave collaterals within the RE nucleus and dorsal thalamus. RE cells thus form an inhibitory network that surrounds the thalamus, receive a copy of nearly all thalamocortical and corticothalamic activity, and project inhibitory connections solely to neurons in the thalamic relay nuclei. Projections between TC, RE and PY cells are usually organized topographically such that each cortical column is associated with a given sector of thalamic TC and RE cells. Modified from Destexhe et al., 1998a.

In their seminal intracellular study of thalamic neurons *in vivo*, Andersen and Eccles showed that TC cells display bursts of action potentials (Fig. 3.2A–D) at the offset of rhythmic inhibitory postsynaptic potentials (IPSPs) (Fig. 3.2E–J). To explain the genesis of oscillations, they proposed a mechanism based on interconnected thalamic cells and inhibitory interneurons (Fig. 3.2K). This mechanism is investigated in more detail in Chapter 6.

TC cells possess two different firing modes (Llinás and Jahnsen, 1982). In the 'tonic' mode, near the resting membrane potential (approximately $-60\ mV$), the relay neuron fires trains of action potentials at a frequency proportional to the amplitude of the injected current (upper curve in Fig. 3.3). This is similar to the response of many other neurons and is explained by the voltage-dependent Na^+ and K^+ currents that generate action potentials (see Fig. 2.7). In

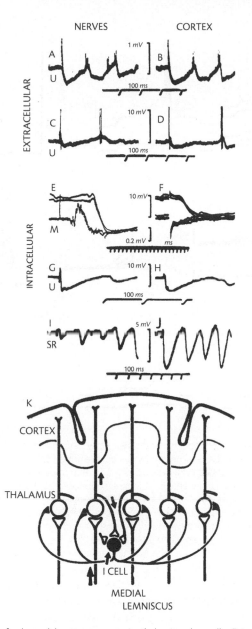

Fig. 3.2: Initial report of rebound burst responses in thalamic relay cells. Extracellular and intracellular recordings from thalamic cells of the ventrobasal complex following stimulation of contralateral foreleg nerves (NERVES) or ipsilateral somatosensory cortex (CORTEX). A–D. Extracellular recordings. A. Initial discharge followed by a large positive wave (P-wave) followed by two rhythmic burst responses. B. Same responses following cortical stimulation (thalamic neurons were antidromically activated). C and D show the response of another thalamic cell to the same stimuli as in A and B. E–J. Intracellular recordings from a thalamic relay cell (electrodes filled with potassium citrate). E–F: large IPSP in response to stimulation of median nerve and cortex, respectively. G–J. Large rhythmic IPSPs induced by stimulation. K. Mechanism proposed to account for rhythmic oscillations based on post-inhibitory rebound in the thalamus. Reproduced from Andersen and Eccles, 1962.

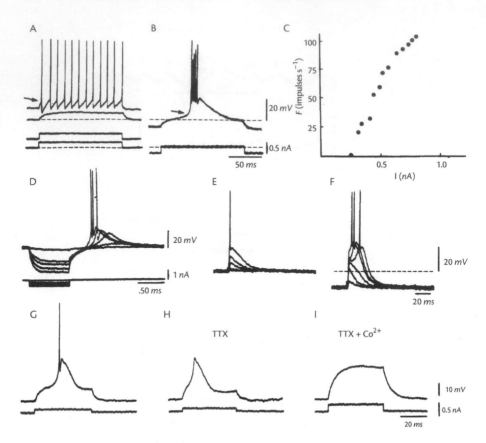

Fig. 3.3: Electrophysiological properties of thalamic relay neurons *in vitro*. A. A depolarization that is subthreshold at resting level (dashed line) produces repetitive firing if delivered at a depolarized DC level. B. When the same stimulus was given at a hyperpolarized DC level, the cell produced high-frequency bursts of action potentials. C. Graph representing the instantaneous adapted frequency as a function of injected current with protocols similar to A. D. Rebound bursts following hyperpolarizing current pulses of different amplitudes. E. Response to EPSPs of increasing amplitudes. F. EPSPs as in E but from a hyperpolarized level produce all-or-none bursts. G. Burst generated from a similar protocol as in B. H. Same cell as in G after TTX. Fast Na-dependent spikes were blocked but the slow response remained unchanged. I. Addition of $CoCl_2$ to the bath abolished the slow response. Figure reproduced from Llinás and Jahnsen, 1982.

contrast, at hyperpolarized membrane potentials, thalamic neurons can enter a 'burst mode' (lower curve in Fig. 3.3), firing high-frequency bursts of action potentials (~300 Hz) at the offset of hyperpolarizing current injection (Fig. 3.3). A burst can also occur following a strong IPSP, which provides hyperpolarization and return-to-rest similar to current injection. The response of a neuron to a depolarizing current injection depends on it previous state, producing a steady low frequency firing rate if injected at a depolarized level, but eliciting a burst followed by a long afterhyperpolarization if injected from a sufficiently hyperpolarized level (Fig. 3.3).

The ionic mechanisms underlying this 'low-threshold' behaviour was first investigated *in vitro* (Jahnsen and Llinás, 1984a, 1984b). These studies showed that the rebound responses of thalamic relay cells arose from a low-threshold Ca^{2+} current, called the T-current. Hyperpolarization of the cell deinactivated the T-current; when the cell subsequently returned to its resting level this current activates faster than it inactivates. The activation and inactivation of the T-current

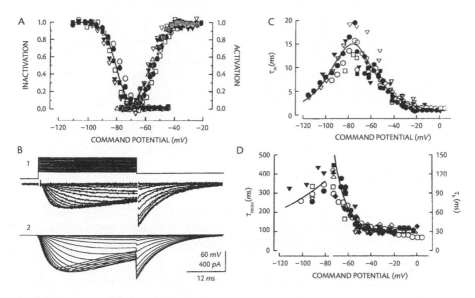

Fig. 3.4: Properties of the low-threshold calcium current revealed with voltage-clamp. A. Activation and inactivation curves obtained from six thalamic relay neurons acutely isolated from rat ventrobasal nucleus. The straight line shows the best fit using the Boltzmann equation. B. Tail current protocols to determine activation of I_T. 30 ms depolarizing current steps to various voltages were delivered before stepping the neuron back to -80 mV. The amplitude of the tail current yielded the activation curves in A. B1: experiments, B2: Hodgkin-Huxley model of I_T (see text). C. Activation kinetics of I_T. D. Inactivation kinetics. Points > -75 mV were obtained by fitting single-exponentials to inactivating currents, whereas the slower time constants reflect the recovery from inactivation. Figure reproduced from Huguenard and McCormick, 1992.

current at hyperpolarized membrane potentials has relatively slow kinetics compared with the fast sodium spikes that ride on top of the broad calcium spike.

The low-threshold Ca^{2+} current of thalamic relay neurons was characterized by voltage-clamp methods (Coulter et al., 1989; Crunelli et al., 1989; Hernandez-Cruz and Pape, 1989; Suzuki and Rogawski, 1989). This current was identified with the T-type Ca^{2+} channels characterized by Carbone and Lux (1984a, 1984b) in chick ganglion cells. Like the Na^+ current described by Hodgkin and Huxley (1952), the T-current of thalamic neurons is transient and shows activation followed by inactivation (Fig. 3.4). However, the voltage range over which I_T activates is close to the resting potential, in contrast to the Na^+ current, which activates at more depolarized voltages. The kinetics of I_T is also considerably slower than the Na^+ current. Following a voltage-clamp step from -80 mV to more depolarized values, the current displays a transient peak followed by a slow inactivation (Fig. 3.4B1). A voltage-clamp characterization of the T-current in thalamic cells was performed in dissociated TC cells by Huguenard and Prince (1992), providing quantitative data on the kinetics of activation and inactivation of this current, which will be used in computational models (see below).

3.1.2 Computational models of burst responses in TC cells

Hodgkin-Huxley type models of TC neurons were first introduced by McMullen and Ly (1988) and Rose and Hindmarsh (1989) based on the experiments of Jahnsen and Llinás (1984a).

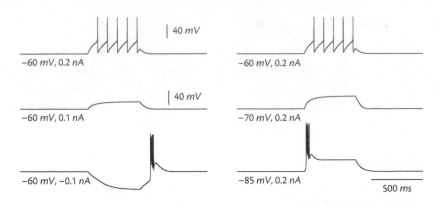

Fig. 3.5: Simulated rebound bursts in thalamic relay cells. The model contained a leakage current, the low-threshold Ca^{2+} current I_T and the fast Na^+/K^+ currents responsible for action potentials. In the left panel, three types of responses are shown for different amplitudes of injected current at the same membrane potential: tonic firing for 0.2 nA, passive response for 0.1 nA and burst firing following injection of a hyperpolarizing current of −0.1 nA. In the right panel, the same depolarizing current pulse (0.2 nA) led to three different responses: tonic firing for −60 mV, passive response at −70 mV and burst firing at −85 mV.

The more recent characterization of the T-current by voltage-clamp methods (see above) provide precise measurements for the time constants and steady-state values of activation and inactivation processes. Following the approach introduced by Hodgkin and Huxley (1952; see Section 2.1.3), these data can be incorporated into models where activation and inactivation are described by independent gates. Several Hodgkin-Huxley type models based on voltage-clamp data were introduced to account for the rebound-burst properties of TC cells (Huguenard and McCormick, 1992; Lytton and Sejnowski, 1992; McCormick and Huguenard, 1992; Toth and Crunelli, 1992; Destexhe and Babloyantz, 1993; Destexhe et al., 1993a; Wang and Rinzel, 1993; Wang, 1994; Wallenstein, 1996; Destexhe et al., 1996a, 1998c). A Hodgkin-Huxley-type model of the T-current that was directly fit to the voltage-clamp data is illustrated in Fig. 3.4.

A model derived from the measurements of the kinetic properties of a current in voltage-clamp can be tested by examining its behaviour in current-clamp. Hodgkin-Huxley type models of the T-current based on activation and inactivation kinetics estimated from voltage-clamp experiments was able to account for the rebound burst properties of TC cells (Fig. 3.5). Thus, the most salient features of the rebound burst can be reproduced by a single-compartment model containing Na^+, K^+ and T-currents described by Hodgkin-Huxley type kinetics. However, as shown in Section 3.4, reproducing all the features of the rebound burst in TC cells requires an additional constraint: the T-current must be concentrated in the dendrites.

3.2   Oscillatory properties of thalamic relay cells

3.2.1   Experimental characterization of intrinsic oscillations

In addition to the rebound burst, TC cells can also generate sustained oscillations. In experiments performed in cats *in vivo*, TC cells generated rhythmicities in the delta frequency range (0.5–4 *Hz*) after removal of the cortex (Curró Dossi et al., 1992). Oscillations in the same frequency range were also observed in TC cells *in vitro* (Leresche et al., 1990; McCormick and Pape, 1990a; Leresche et al., 1991). These intrinsic slow oscillations consisted of rebound bursts recurring periodically, and have been also called 'pacemaker oscillations' (Leresche et al., 1990, 1991). These slow oscillations were resistant to tetrodotoxin, suggesting that they were generated by mechanisms intrinsic to the TC cell.

The intrinsic delta oscillations depend on the membrane potential (McCormick and Pape, 1990a). Oscillations were only possible if TC cells were maintained at relatively hyperpolarized potentials, within the burst mode, suggesting that the T-current actively participated in its generation. Another property, illustrated in Fig. 3.6, is that these oscillations disappeared following blockade of another current, called I_h (McCormick and Pape, 1990a) with Cs$^+$. I_h is a mixed Na$^+$/K$^+$ cation current responsible for anomalous rectification in TC cells (Pollard and Crunelli, 1988). In voltage-clamp, I_h is activated by hyperpolarization in the subthreshold

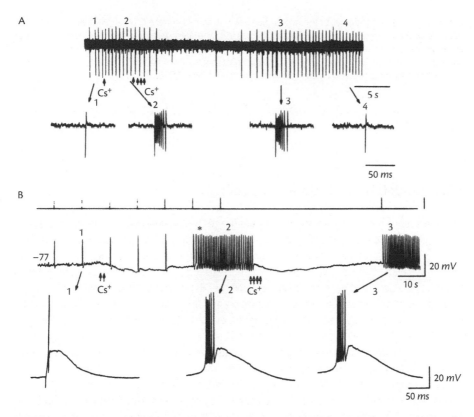

Fig. 3.6: Dependence of intrinsic slow delta-like oscillations on the presence of the hyperpolarization-activated current I_h. A. Extracellular recording of a spontaneously oscillating thalamic cell in the lateral geniculate nucleus of the ferret. Extracellularly Applied Cs$^+$ enhanced rhythmic activity from (1) single spikes and (2) bursts of up to eight action potentials. Further application of Cs$^+$ initially slowed the rate of repetitive bursting and finally abolished rhythmic activity. Spontaneous oscillations reappeared with (3) strong rhythmic bursts, which gradually weakened to (4) single action potentials. B. Intracellular recording of a thalamic cell in the lateral geniculate nucleus (−77 mV) of the ferret. Injection of short (3 ms) depolarizing current pulses evoked bursts crowned by (1) a single action potential. Application of Cs$^+$ hyperpolarized the cell and a single depolarizing pulse (*) led to (2) rhythmic bursts of six action potentials. Further application of Cs$^+$ reversibly abolished rhythmic activity. Figure from McCormick and Pape, 1990a.

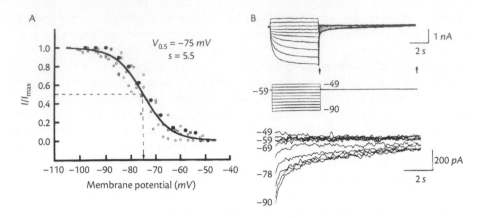

Fig. 3.7: Activation properties of the hyperpolarization-activated current I_h under voltage-clamp. A. Plot of the activation curve of I_h for seven different neurons from the lateral geniculate nucleus of the ferret (filled circles correspond to the neuron shown in B and C; the straight line represents the best fit obtained with the model. B. Voltage-clamp recording of I_h (voltage protocol is shown below). C. Expansion of tail current between arrows indicated in B. Figure from McCormick and Pape, 1990a.

range of potentials (Fig. 3.7; McCormick and Pape, 1990a; Soltesz et al., 1991). Taken together, these data indicate that intrinsic oscillations in TC cells are generated by an interplay between I_T and I_h.

3.2.2 Models of oscillations in TC cells

The interactions between I_h and I_T, which gives rise to the low-frequency oscillatory behaviour of TC cells, can be explored with a dynamical model. This requires a model for I_h. As shown in Fig. 3.8, I_h activates very slowly, with a time constant greater than 1 s at 36 °C (McCormick and Pape, 1990a; Soltesz et al., 1991). The time course of I_h activation may be quite different from the time course of deactivation at the same membrane potential. Currents similar to I_h in other preparations also show very slow activation and, in some cases, a faster time course for deactivation (see Galligan et al., 1990; Uchimura et al., 1990; Kamondi and Reiner, 1991; van Ginneken and Giles, 1991; Erickson et al., 1993). Despite the different time constants for activation and deactivation, I_h follows a single-exponential time course, which would suggest a model with first-order kinetics. However, in a simple first-order kinetic scheme, the time constant of activation is always identical to that of deactivation, so the model cannot be this simple.

A kinetic scheme that accounts for these apparently conflicting experimental data on I_h was proposed by Destexhe and Babloyantz (1993) (Fig. 3.9). It has two activation variables with different kinetics and accurately accounts for all the voltage-clamp data. Although Markov models have been developed for modelling a current similar to I_h in sino-atrial cells (Difrancesco, 1985), the Hodgkin-Huxley-type model presented here is relatively simple and explains how slow activation can coexist with faster deactivation.

The model assumes that the permeability of I_h channels depends on two independent gates (S for slow activation and F for fast activation), which must be opened simultaneously, according to the following scheme (Destexhe and Babloyantz, 1993):

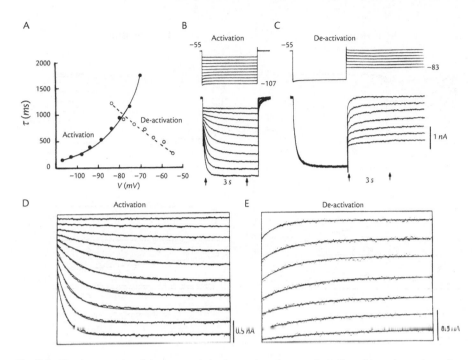

Fig. 3.8: Kinetic properties of the hyperpolarization-activated current I_h. A. Plot of the voltage-dependent activation and deactivation time constants of I_h estimated from B–E. B. Activation of I_h. The voltage was stepped to various hyperpolarized levels from a holding potential of -55 mV. C. Deactivation of I_h. The current was first activated with a voltage jump to a standard hyperpolarized level and then the cell was depolarized to various voltages. D. First 3 seconds of activating I_h (arrows in B). E. First 3 seconds of deactivating I_h (see arrows in C). Data in D–E were fit with single exponentials. Figure reproduced from McCormick and Pape, 1990a.

$$S_{closed} \underset{\beta_S}{\overset{\alpha_S}{\rightleftarrows}} S_{open} \qquad F_{closed} \underset{\beta_F}{\overset{\alpha_F}{\rightleftarrows}} F_{open} \qquad (3.1)$$

where S_{closed} and F_{closed} represent the closed states of the slow and fast activation gates of I_h, S_{open} and F_{open} represent the open states of these gates, and α_S, β_S, α_F and β_F are voltage-dependent rate constants (see below).

The corresponding kinetic equations are:

$$I_h = \bar{g}_h \, S_1 \, F_1 \, (V - E_h)$$

$$\dot{S}_1 = \alpha_S(V) \, (1 - S_1) - \beta_S(V) \, S_1 \qquad (3.2)$$
$$\dot{F}_1 = \alpha_F(V) \, (1 - F_1) - \beta_F(V) \, F_1$$

where $\bar{g}_h$ is the maximal conductance of I_h (in mS/cm^2), $E_h = -43$ mV is the reversal potential of I_h (McCormick and Pape, 1990a), S_1 and F_1 represent the fraction of activation gates in the open state. The current is proportional to the product $S_1 F_1$, reflecting the condition that both gates must be open to allow the channel to conduct ions.

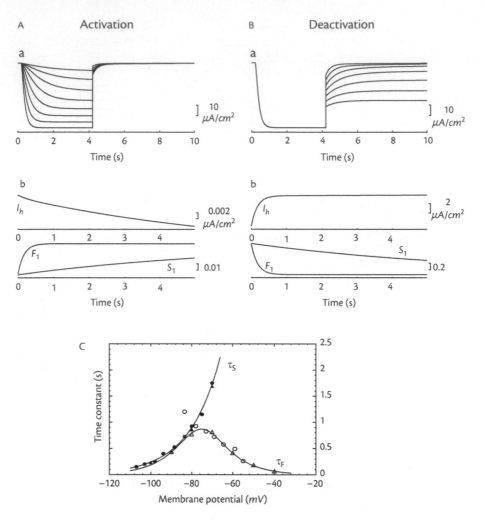

Fig. 3.9: Activation and deactivation kinetics of I_h. A. Simulation of voltage-clamp protocols of activation of I_h. (a) From an initial holding value of −55 mV, the voltage was clamped to levels from −105 mV to −70 mV for 4 s, then clamped again to −55 mV. (b) Time course of the current compared to the gating variables. During this activation protocol (initial voltage was −30 mV, current recorded after clamping to −50 mV at $t = 0$) the current followed the time course of the slow variable S_1. A time constant of approximately 3 s was estimated from fitting a single exponential to the current trace. B. Simulation of protocols for deactivation of I_h. (a) The voltage was clamped at −105 mV for 4 s, then clamped to various levels from −85 mV to −55 mV. (b) Time course of the current and gating variables. Although the voltage was clamped to the same value as in Ab (initial voltage of −110 mV and clamp to −50 mV at $t = 0$), the current followed the time course of the variable F_1 and a smaller time constant of approximately 180 ms was measured. C. Time constants for activation and deactivation of I_h as a function of the membrane potential. The time constants obtained by single-exponential fitting of the currents illustrated above for activation (filled triangles) and deactivation (open triangles) were compared to the measurements obtained by McCormick and Pape (1990a) (Fig. 3.8) during activation (filled circles) and deactivation (open circles). Solid lines represent the functions fit to these data. Reproduced from Destexhe et al., 1993a.

The rate constants are related to the activation function $H_\infty(V)$ and the time constants $\tau_S(V)$ and $\tau_F(V)$ by the following relations: $\alpha_S = H_\infty/\tau_S$, $\beta_S = (1 - H_\infty)/\tau_S$, $\alpha_F = H_\infty/\tau_F$ and $\beta_F = (1 - H_\infty)/\tau_F$. The activation function

$$H_\infty(V) = 1/(1 + \exp[(V + 68.9)/6.5])$$

was chosen so that H_∞^2 fit the data of McCormick and Pape (1990a; see Fig. 3.7).

The time constants

$$\tau_F(V) = \exp[(V + 158.6)/11.2]/(1 + \exp[(V + 75)/5.5])$$
$$\tau_S(V) = \exp[(V + 183.6)/15.24]$$

were estimated from numerical simulations of voltage-clamp protocols (Destexhe et al., 1993a).

This model exhibits two time constants. Following a depolarizing voltage jump, the two gates S and F, which are initially closed, begin to activate: the fast variable F_1 rapidly increases to its equilibrium value whereas S_1 reaches equilibrium more slowly. Since I_h is proportional to the product $S_1 F_1$, the time course of the measured current reflects the activation kinetics of the slow variable S_1 (Fig. 3.9A). The opposite occurs upon a hyperpolarizing voltage jump from a depolarized level where both gates were initially open: F_1 rapidly closes while S_1 closes more slowly. Since the decrease of F_1 immediately decreases I_h, the time course of deactivation follows the kinetics of the fast variable (Fig. 3.9B).

If the difference between activation and inactivation time constants is ignored, then I_h can be modelled by a simpler first-order activation scheme:

$$C \underset{\beta(V)}{\overset{\alpha(V)}{\rightleftarrows}} O \qquad (3.3)$$

which represents the voltage-dependent transitions of I_h channels between closed (C) and open (O) forms, with α and β as transition rates. The current is then proportional to the relative concentration of open channels:

$$I_h = \breve{g}_h [O] (V - E_h) \qquad (3.4)$$

with a maximal conductance of $\bar{g}_h = 0.02\ mS/cm^2$ and a reversal potential of $E_h = -40\ mV$.

The oscillatory behaviour in TC cells has been modelled as an interplay of I_T and I_h, based on either a double-activation model of I_h (Eq. 3.1; Destexhe and Babloyantz, 1993; Destexhe et al., 1993a), or a simplified first-order model of I_h similar to that above (Lytton and Sejnowski, 1992; McCormick and Huguenard, 1992; Toth and Crunelli, 1992; Wang, 1994; Destexhe et al., 1996a; Lytton et al., 1996). Either of these types of models for I_h is adequate for generating regular oscillations, but the double-activation model may be needed for generating waxing-and-waning oscillations (see below). Fig. 3.10 illustrates the oscillations generated by a single-compartment model of TC cells comprising I_T and I_h currents, as well as I_{Na}/I_K responsible for action potentials.

The genesis of oscillations arising from interactions between I_T and I_h is illustrated in Fig. 3.11. The activation of I_h depolarizes the membrane slowly until a low-threshold spike (LTS) is generated by activation of I_T. During the depolarization provided by the LTS, I_h deactivates,

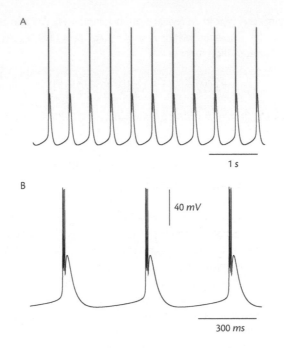

Fig. 3.10: Model of intrinsic slow delta-like oscillations in thalamic relay cells at two different time resolutions, A and B. The single-compartment model included I_T and I_h currents, as well as I_{Na}/I_K responsible for action potentials and leak current. Oscillations appeared when the resting membrane potential was set to hyperpolarized values (approximately -75 mV), and when the I_T and I_h conductances are sufficiently large. Conductance values were $\bar{g}_T = 2$ mS/cm^2, $\bar{g}_h = 0.015$ mS/cm^2, $\bar{g}_{Na} = 90$ mS/cm^2, $\bar{g}_K = 10$ mS/cm^2 (other parameters described in Destexhe et al., 1996a).

and together with the termination of the LTS the membrane becomes hyperpolarized. This hyperpolarization allows I_T to deinactivate in preparation for the next LTS; as I_h slowly activates, the cycle restarts. The same mechanism has been explored, with minor variations, in several modelling studies that used models of I_T and I_h that differed in detail (Lytton and Sejnowski, 1992; McCormick and Huguenard, 1992; Toth and Crunelli, 1992; Destexhe and Babloyantz, 1993; Destexhe et al., 1993a; Wang, 1994; Destexhe et al., 1996a; Lytton et al., 1996). The fact that these different studies reached the same conclusion indicates that the interplay between I_T and I_h is a highly robust mechanism to generate oscillations.

3.3 Intrinsic waxing-and-waning oscillations in thalamic relay cells

3.3.1 Experimental characterization of waxing-and-waning oscillations

In addition to the rebound bursts and intrinsic oscillations in the delta frequency range in TC cells, a slow modulation of these intrinsic oscillations may also occur. In cat TC cells studied in a

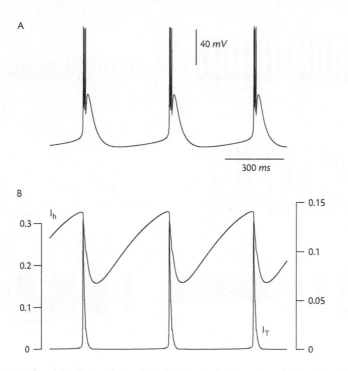

Fig. 3.11: Intrinsic slow delta-like oscillations in thalamocortical cells produced by the interaction between two voltage-dependent currents. A. Oscillations in the same TC cell model as in Fig. 3.10. B. Normalized conductance (g/g_{max}) of I_T and I_h during oscillations. Rhythmic oscillations arose from the interaction between I_T and I_h. Note that the T-current activation preceded the h-current deactivation.

low-Mg^{2+} medium *in vitro*, a few cells display waxing-and-waning slow-frequency oscillations (Fig. 3.12; Leresche et al., 1990, 1991). These 0.5–3.2 *Hz* oscillations were composed of waxing-and-waning oscillations of duration 1 to 28 *s* and silent phases 5 to 25 *s* long. The waxing-and-waning of the oscillation was resistant to tetrodotoxin, suggesting mechanisms intrinsic to the TC neuron. By analogy with the waxing-and-waning of *in vivo* spindles, they have also been called 'spindle-like oscillations' (Leresche et al., 1990, 1991). However, *in vivo* spindles occur at a much higher intraburst frequency (7–14 *Hz*) and depend on interactions with neurons of the thalamic reticular nucleus (see Chapter 6), which distinguishes them from the waxing-and-waning oscillations intrinsic to TC cells.

The ionic mechanisms underlying intrinsic waxing-and-waning oscillations were investigated by Soltesz et al. (1991), who found that the they were dependent on I_h. Slow delta-like oscillations and waxing-and-waning oscillations can be observed in the same TC cell by altering the h-current (Fig. 3.13; Soltesz et al., 1991). Increasing the amplitude of I_h by noradrenaline can transform delta-like oscillations into waxing-and-waning oscillations; application of Cs^+, an I_h blocker, has the opposite effect (Soltesz et al., 1991). In addition, the intrinsic waxing-and-waning oscillations display a characteristic hyperpolarization during the silent phase and can be transformed into sustained slow delta-like oscillations by applying a depolarizing current step (Fig. 3.12; Leresche et al., 1990, 1991).

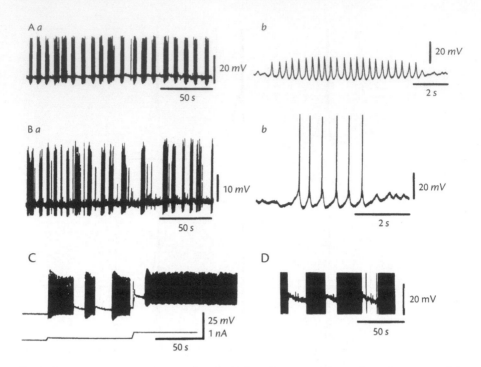

Fig. 3.12: Intrinsic waxing-and-waning ('spindle-like') oscillations in thalamic relay cells. Aa. Intracellular recording of a cell from cat lateral geniculate nucleus showing spontaneous waxing-and-waning oscillations. Oscillatory sequences repeated at a frequency of 0.11 ± 0.02 Hz and lasted for 4.7 ± 0.3 s. Ab. Same cell as in Aa after perfusion of tetrodotoxin and bicuculline. B. Spontaneous activity in another cell shown on slow (a) and fast (b) time bases. Note that the membrane is more depolarized after the oscillatory sequence in Bb. C. Transformation from waxing-and-waning to slow delta-like oscillations by depolarizing the cell (from -75 to -60 mV). D. Same cell as in C with higher amplitude magnification to show the progressive hyperpolarization between oscillatory sequences. Figure reproduced from Leresche et al., 1991.

3.3.2 Models of waxing-and-waning oscillations from Ca^{2+} regulation of I_h

Two plausible ionic mechanisms for generating waxing-and-waning oscillations have been suggested. The first model (Destexhe et al., 1993a) was inspired by experiments on the I_h current in heart cells and is based on the regulation of I_h by intracellular Ca^{2+} (Fig. 3.14; Hagiwara and Irisawa, 1989; McCormick, 1992; Toth and Crunelli, 1992). A second possible ionic mechanism depends on the interaction between I_T, I_h and a slow potassium current (Destexhe and Babloyantz, 1993).

Voltage-clamp experiments on sino-atrial node cells have suggested that the activation curve of I_h is dependent on the intracellular Ca^{2+} concentration, shifting towards more positive membrane potentials with increasing intracellular Ca^{2+} concentration ($[Ca]_i$) (Fig. 3.14; Hagiwara and Irisawa, 1989). Because calmodulin and protein kinase C were not involved in the Ca^{2+} modulation of I_h, Ca^{2+} ions may affect the I_h channels directly (Hagiwara and Irisawa, 1989), with the binding of Ca^{2+} increasing the conductance of I_h (Fig. 3.14), or indirectly through the production of cAMP (Luthi and McCormick, 1999).

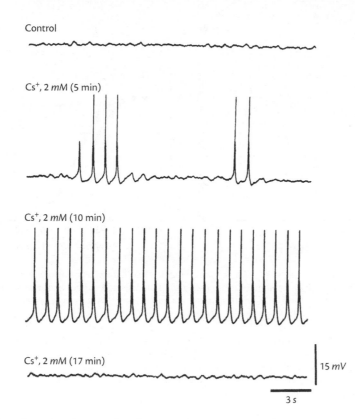

Fig. 3.13: I_h determines the type of oscillation in thalamic relay cells. Cs^+ was applied extracellularly to a cell that had no spontaneous oscillations (Control). Four minutes after the application of Cs^+, spontaneous waxing-and-waning oscillations began (silent periods of 4–9 s and oscillatory sequences lasting for 2–6 s). After an additional 4 min, the oscillations became sustained (frequency of 1–2 Hz) and persisted for approximately 6 min before all activity ceased. Figure reproduced from Soltesz et al., 1991.

A kinetic model was developed for intracellular calcium binding to the open channels of I_h that is consistent with these data (Destexhe et al., 1993a). The open state gates of scheme 3.1, S_{open} and F_{open}, were assumed to have n binding sites for Ca_i^{2+} which, when occupied, lead to the open forms S_{bound} and F_{bound} according to:

$$S_{open} + n\,Ca_i^{2+} \quad \underset{k_2}{\overset{k_1}{\rightleftharpoons}} \quad S_{bound}$$

$$\tag{3.5}$$

$$F_{open} + n\,Ca_i^{2+} \quad \underset{k_2}{\overset{k_1}{\rightleftharpoons}} \quad F_{bound}$$

where k_1 and k_2 are the forward and backward rate constants for Ca_i^{2+} binding.

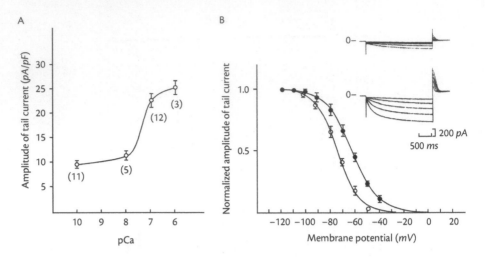

Fig. 3.14: Experiment first showing that I_h may be regulated by intracellular calcium. A. Dose-response relationship between the amplitude of the hyperpolarization-activated current and pCa = $-\log([Ca_i])$ in sino-atrial node cells from rabbit hearts. The tail current amplitudes (after stepping the voltage to +20 mV from a holding value of −100 mV) are plotted against the Ca^{2+} concentration in the patch-clamp pipette for 28 cells. The total amplitude of the current increased more than two-fold at high Ca^{2+} concentration. B. Activation curve of the hyperpolarization-activated current at low (pCa = 10, dashed line) and high (pCa = 7, continuous line) Ca^{2+} concentrations. The results from 21 cells are shown. Ca^{2+} induced a shift of approximately 13 mV in the depolarized direction. Insets shows the current recorded by stepping the voltage from −40 to −80 mV for low (pCa=10, above) and high calcium concentration (pCa=7, below). Figure reproduced from Hagiwara and Irisawa, 1989.

If S_2 and F_2 represent the fraction of gates bound to calcium, then, combining Eqs. 3.1 and 3.5, one obtains the following kinetic equations for I_h:

$$I_h = \bar{g}_h \, (S_1 + S_2) \, (F_1 + F_2) \, (V - E_h)$$

$$\dot{S}_1 = \alpha_S(V) \, (1 - S_1) - \beta_S(V) \, S_1 + k_2 \, [S_2 - C\,S_1\,]$$
$$\dot{F}_1 = \alpha_F(V) \, (1 - F_1) - \beta_F(V) \, F_1 + k_2 \, [F_2 - C\,S_1\,] \qquad (3.6)$$

$$\dot{S}_2 = -k_2 \, [S_2 - C\,S_1\,]$$
$$\dot{F}_2 = -k_2 \, [F_2 - C\,F_1\,]$$

where $C = ([Ca]_i/Ca_{crit})^n$ and α_S, β_S, α_F and β_F were obtained from H_∞ and τ_S as above. The number of binding sites was $n = 2$ in all simulations. It was assumed that $k_1 = k_2/Ca_{crit}^n$, where $Ca_{crit} = 5 \; 10^{-4} \; mM$ is the critical value of $[Ca]_i$ at which Ca^{2+} binding on I_h channels is half-activated (if $[Ca]_i \ll Ca_{crit}$, the effect of Ca_i^{2+} is negligible). The inverse of the time constant of Ca_i^{2+} binding to I_h channels is $k_2 = 4 \; 10^{-4} \; ms^{-1}$. These values were chosen to match the slow time course with which I_h is modulated by intracellular Ca^{2+}.

The dynamics of intracellular Ca^{2+} was modelled in a thin shell beneath the membrane and calcium efflux was modelled by an active Ca^{2+} transporter, as described in Chapter 2 (see section 2.2.1):

(i) the influx of Ca^{2+} due to I_T.

$$[\dot{Ca}]_i = -\frac{k}{2Fd} I_T \tag{3.7}$$

where $F = 96489\ C\ mol^{-1}$ is the Faraday constant, $d = 1\ \mu m$ is the depth of the shell beneath the membrane and the unit conversion constant is $k = 0.1$ for I_T in $\mu A/cm^2$ and $[Ca]_i$ in mM.

(ii) the efflux of Ca^{2+} due to an active transporter.

$$[\dot{Ca}]_i = -\frac{K_T\ [Ca]_i}{[Ca_i] + K_d} \tag{3.8}$$

where $K_T = 10^{-4}\ mMms^{-1}$ is the product of c_3 with the total concentration of P, and $K_d = c_2/c_1 = 10^{-4}\ mM$ is the dissociation constant, which can be interpreted here as the value of $[Ca]_i$ at which the transporter is half activated (if $[Ca]_i \ll K_d$ then the efflux is negligible).

The activation function of I_h at equilibrium as a function of the membrane potential and the intracellular Ca^{2+} concentration is, from Eq. 3.6:

$$H_\infty(V, [Ca]_i) = [(S_1 + S_2)\ (F_1 + F_2)]_{eq}$$

$$= \left(\frac{1+C}{H_\infty(V)^{-1} + C}\right)^2 \tag{3.9}$$

where $C = ([Ca]_i/Ca_{crit})^n$, $H_\infty(V)^2 = H_\infty(V, [Ca]_i = 0)$. The activation function $H_\infty(V, [Ca]_i)$ was determined from voltage-clamp measurements of TC neurons (Fig. 3.7; McCormick and Pape, 1990a) and the parameters $H_\infty(V)$ were chosen to fit these data as closely as possible (Fig. 3.15B, solid line).

Whole-cell voltage-clamp experiments (Hagiwara and Irisawa, 1989) on sino-atrial node cells have shown that increasing intracellular Ca^{2+} produces a shift of the activation function of I_h towards more positive membrane potentials (Fig. 3.14). Using patch pipettes containing various concentrations of Ca^{2+}, the shift was approximately 13 mV at the highest concentrations in Ca^{2+}. These data could be accounted for by a kinetic scheme where intracellular Ca^{2+} directly binds to the I_h channels (see Fig. 3.15A). The activation function $H_\infty(V, [Ca]_i)$ shifts towards positive membrane potentials as the value of C increases (Fig. 3.15B).

The shift at half-activation of I_h is obtained by substituting $H_\infty(V, [Ca]_i) = 0.5$ into Eq. 3.9, to obtain:

$$V_{\frac{1}{2}} = -68.9 + 6.5 \left[\log(\sqrt{2} - 1)\ \log(C + 1)\right]$$

$$\simeq -75 + 6.5\ \log\left(\left(\frac{[Ca]_i}{Ca_{crit}}\right)^n + 1\right). \tag{3.10}$$

The shift of the I_h activation is logarithmic in $[Ca]_i$ and a shift of 13 mV is obtained for $C = 6.4$.

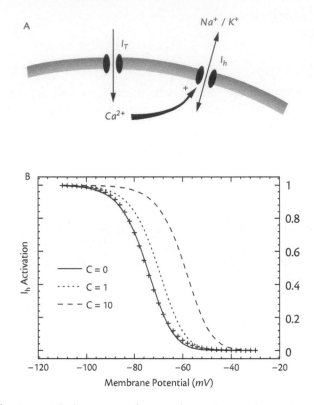

Fig. 3.15: Ca^{2+}-induced shift of the activation function of I_h. A. Schematic diagram illustrating the currents in the model. The low-threshold Ca^{2+} current (I_T) allows Ca^{2+} ions to enter the cell; these ions bind to the open form of the mixed Na^+/K^+ conductance channel I_h and modify its voltage-dependent properties. B. Direct binding of intracellular Ca^{2+} to the open form of I_h channels shifts the voltage-dependence of the current towards more positive membrane potentials. $H_\infty(V, [Ca_i])$ is represented as a function of the membrane potential V for different values of $[Ca]_i$. The activation function at the resting level of $[Ca]_i$ (solid line: $C = 0$) was estimated from voltage-clamp experiments (Fig. 3.7; McCormick and Pape, 1990a) on TC cells (+ symbols). For higher concentrations of intracellular Ca^{2+}, the activation function shifts towards a more positive membrane potential (dashed lines, $C = 1$ and $C = 10$; $C = ([Ca]_i/Ca_{crit})^2$) (modified from Destexhe et al., 1993a).

The shift should be negligible ($C < 1$) at the resting level, $[Ca]_i \sim 2 \ 10^{-4} \ mM$, which yields a lower bound: $Ca_{crit} > 2 \ 10^{-4} \ mM$. During activation of I_T, the value of $[Ca]_i$ just beneath the membrane increases to approximately 10^{-2}–$10^{-3} \ mM$ and shifts I_h by a few millivolts ($C > 1$), which gives an upper bound: $Ca_{crit} < 10^{-2}$–$10^{-3} \ mM$. The simulations shown here were obtained with $n = 2$ and $Ca_{crit} = 5 10^{-4} mM$.

3.3.3 Oscillatory behaviour from Ca^{2+}-regulated I_h

The double-activation model of I_h combined with I_T can produce a variety of resting states and slow oscillations. These patterns were obtained for different values of the maximal conductance

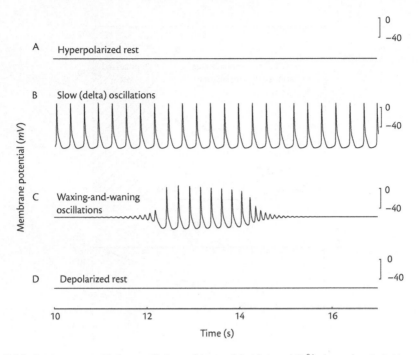

Fig. 3.16: Resting states and slow oscillations of the model with I_T and Ca^{2+}-dependent I_h. A. Hyperpolarized resting state close to -84 mV for $\bar{g}_h = 0$. B. Slow delta-like oscillations of approximately 3.5 Hz for $\bar{g}_h = 0.01$ mS/cm^2. C. Waxing-and-waning ('spindle-like') oscillations of approximately 4–8 Hz for $\bar{g}_h = 0.04$ mS/cm^2. D. Depolarized resting state approximately -58 mV for $\bar{g}_h = 0.11$ mS/cm^2. The maximum conductance of I_T was kept fixed at $\bar{g}_{Ca} = 1.75$ mS/cm^2 (modified from Destexhe et al., 1993a).

$\bar{g}_h$ of I_h (Fig. 3.16). For the lowest values of $\bar{g}_h$ (<0.01 mS/cm^2), the model remained in a hyperpolarized resting state at approximately -84 mV (Fig. 3.16A). A similar hyperpolarized resting state has been observed *in vitro* (McCormick and Pape, 1990a; Soltesz et al., 1991) after blockage of I_h. The characteristic hyperpolarization during the silent phase of the waxing-and-waning oscillations, as well as their transformation into sustained oscillations, were also observed in the model (Destexhe et al., 1993a).

The period of the Ca^{2+}-dependent waxing-and-waning oscillations was found to depend on the maximal conductance of I_h and the time constant of Ca^{2+} binding to I_h channels (Fig. 3.17). This figure also shows that the four states of Fig. 3.16 can be obtained at four different levels of conductance of I_h, in agreement with the transformation between resting and oscillatory states obtained by an increasing blockade of I_h following the application of cesium to the bathing solution (Soltesz et al., 1991).

3.3.4 K$^+$-dependent waxing-and-waning oscillations

Another possible ionic mechanism for generating waxing-and-waning oscillations (Destexhe and Babloyantz, 1993; Destexhe et al., 1993a) is based on the interaction between I_T, I_h and a slow potassium current. Several types of K$^+$ currents have been identified in TC cells (Huguenard and Prince, 1991; McCormick, 1991; Budde et al., 1992). Among these, a slowly inactivating K$^+$

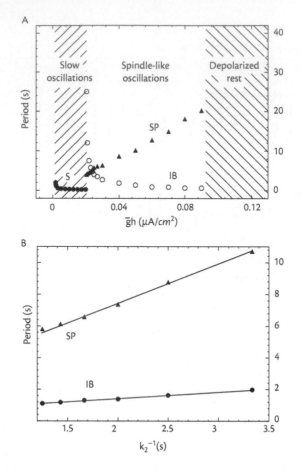

Fig. 3.17: The period of waxing-and-waning oscillations in a model TC cell depends on I_h and its regulation by Ca^{2+}. A. The length of the silent phase (SP) and the intrinsically bursting phase (IB) are shown as a function of the maximal conductance of I_h ($\bar{g}_h$). The range of values of $\bar{g}_h$ corresponding to slow delta-like oscillations (period labeled by S), waxing-and-waning oscillations and depolarized resting state are also indicated. B. Period as a function of the time constant (k_2^{-1}) of intracellular Ca^{2+} binding on I_h channels (modified from Destexhe et al., 1993a).

current activated by depolarization was characterized and termed I_{K2} by Huguenard and Prince (1991). They reported that I_{K2} inactivates slowly with two time constants (approximately 250 ms and 3 s). A similar current was found in TC cells in the lateral geniculate nucleus (McCormick, 1991). A kinetic model for this current was proposed by Huguenard and McCormick (1992):

$$I_{K2} = \bar{g}_{K2}\, m_2(0.6\, h_1 + 0.4\, h_2)\, (V - E_K)$$

$$\dot{m}_2 = -\frac{1}{\tau_{m2}(V)}\, (m_2 - m_{2\infty(V)})$$

$$\dot{h}_1 = -\frac{1}{\tau_{h1}(V)}\, (h_1 - h_{2\infty(V)})$$ (3.11)

$$\dot{h}_2 = -\frac{1}{\tau_{h2}(V)}\, (h_2 - h_{2\infty(V)})$$

where $\bar{g}_{K2}$ is the maximum value of I_{K2} conductance and $E_K = -90\ mV$ is the reversal potential for K$^+$ ions. The activation function and the time constant of the activation variables m_2, h_1 and h_2 are:

$$m_{2\infty(V)} = 1/(1 + \exp{-[(V + 43)/17]}). \tag{3.12}$$

$$\tau_{m2}(V) = 2.86 + 0.29/(\exp{[(V - 81)/25.6]} + \exp{[-(V + 132)/18]}) \tag{3.13}$$

$$h_{2\infty}(V) = 1/(1 + \exp{[(V + 58)/10.6]}) \tag{3.14}$$

$$\tau_{h1}(V) = 34.65 + 0.29/(\exp{[(V - 1329)/200]} + \exp{[-(V + 130)/7.1]}) \tag{3.15}$$

$$\tau_{h2}(V) = \tau_{h1}(V) \quad \text{for} \quad V \geq -70\ mV \tag{3.16}$$

$$= 2570\ ms \quad \text{for} \quad V < -70\ mV \tag{3.17}$$

The double-activation model of I_h when combined with I_T and I_{K2} shows the same sequence of oscillatory states observed *in vitro* (Fig. 3.18; Destexhe and Babloyantz, 1993). Characteristic properties of waxing-and-waning oscillations, such as the progressive hyperpolarization during the silent phase and its transformation into slow delta-like oscillations by a depolarizing current step, were also found in this model (Destexhe et al., 1993a). Thus, the calcium regulation of I_h is not the only way to produce waxing-and-waning oscillations with I_T and I_h.

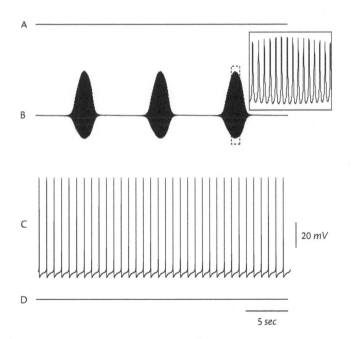

Fig. 3.18: Resting states and slow oscillations of a model that included I_T, I_h and I_{K2}. For fixed values of the conductances of I_T and I_{K2} ($\bar{g}_T = 1.75\ mS/cm^2$, $\bar{g}_{K2} = 3\ mS/cm^2$), different modes were seen the maximal conductance of I_h was decreased. A. depolarized resting state approximately $-57\ mV$ for $\bar{g}_h = 1\ mS/cm^2$. B. waxing-and-waning ('spindle- approximately like') oscillations for $\bar{g}_h = 0.4\ mS/cm^2$. C. intrinsic slow (delta) oscillations for $\bar{g}_h = 0.002\ mS/cm^2$. D. hyperpolarized resting state approximately $-82\ mV$ for $\bar{g}_h = 0$. Inset in (b): oscillations delimited by the dashed line are shown at higher temporal resolution (magnification of 10) (modified from Destexhe and Babloyantz, 1993).

The dynamical interaction between the ionic currents in the model is described by a set of differential equations that can be mathematically analysed to understand how the qualitative properties of the oscillations arise. A singular approximation analysis of the waxing-and-waning oscillations showed that these oscillations occur as an alternation between two dynamical states, a hyperpolarized stationary phase and a limit-cycle oscillatory phase (Destexhe et al., 1993a). The transitions between these two states occur via a subcritical Hopf bifurcation, similar to bursting oscillations in another model (Rinzel, 1987). Remarkably, the same dynamical mechanism underlies both the Ca^{2+}-dependent and the I_{K2}-dependent models, despite their different ionic mechanisms (for more details, see Destexhe et al., 1993a).

3.3.5 Indirect modulation of I_h

The regulation of I_h by intracellular Ca^{2+} was based on voltage-clamp data from the hyperpolarization-activated current in sino-atrial cells. This current, which is similar to I_h in thalamic neurons, has a strong Ca^{2+}-dependence (Hagiwara and Irisawa, 1989), but calcium may not bind directly to the channel (DiFrancesco and Totora, 1991; Zaza et al., 1991; Luthi and McCormick, 1998). Moreover, the Ca^{2+}-induced shift of the activation curve is unbounded (Eq. 3.10), which is inconsistent with the bounded shift found experimentally (DiFrancesco and Totora, 1991). A new model was therefore designed to incorporate the indirect modulation of I_h by Ca^{2+} (Destexhe et al., 1996a). This model assumed that the voltage-dependence and conductance is influenced by Ca^{2+} indirectly, through the binding of Ca^{2+} to an intermediate messenger (P), which itself binds to the open form of the channel, and blocks its transition to the closed form. This leads to an effective shift of the voltage-dependence to more depolarized values (Fig 3.19), as observed experimentally (Fig. 3.14; Hagiwara and Irisawa, 1989).

The full kinetic scheme is:

$$C \underset{\beta(V)}{\overset{\alpha(V)}{\rightleftharpoons}} O \tag{3.18}$$

$$P_0 + 4\,Ca^{2+} \underset{k_2}{\overset{k_1}{\rightleftharpoons}} P_1 \tag{3.19}$$

$$O + P_1 \underset{k_4}{\overset{k_3}{\rightleftharpoons}} O_L \tag{3.20}$$

where the first reaction represents the voltage-dependent transitions of I_h channels between closed (C) and open (O) forms, α and β are the transition rates, with $\alpha(V) = h_\infty(V)/\tau_h(V)$, $\beta(V) = (1 - h_\infty(V))/\tau_h(V)$, and

$$h_\infty(V) = \frac{1}{1 + \exp\left[(V + 75)/5.5\right]}$$

$$\tau_h(V) = \tau_{min} + \frac{1000}{\exp[(V + 71.5)/14.2] + \exp[-(V + 89)/11.6]} \tag{3.21}$$

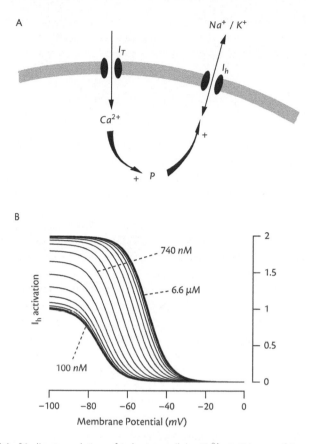

Fig. 3.19: Model of indirect regulation of I_h by intracellular Ca^{2+}. A. Scheme of I_h regulation through an intermediate messenger (P) activated by intracellular calcium. In this model, P has four binding sites for Ca^{2+} and regulates the channel by binding to its open form. B. Voltage-dependent activation of I_h for different intracellular Ca^{2+} concentrations. The successive traces from bottom to top are the activation curves ($[O] + g_{inc}[O_L]$) for different intracellular Ca^{2+} increasing from 100 nM to 6.6 μM by a multiplicative factor of 1.2. The binding of P to the open form of the I_h channel results in a shift of the voltage-dependence of the current as well as an increase of its total conductance (modified from Destexhe et al., 1996a).

which were both fit to the voltage-clamp data of McCormick and Pape (1990a) (see Figs. 3.7–3.8).

The second reaction (Eq. 3.19) represents the binding of intracellular Ca^{2+} ions to an intermediate messenger (P_0 for unbound and P_1 for bound) with four binding sites for calcium and rates of $k_1 = 2.5 \times 10^7$ mM^{-4} ms^{-1} and $k_2 = 4 \times 10^{-4}$ ms^{-1} (half-activation of 0.002 mM Ca^{2+}). The calcium-bound form, P_1, associates with the open form of the channel, leading to a 'locked' open form, O_L, with rates of $k_3 = 0.1$ ms^{-1} and $k_4 = 0.001$ ms^{-1}. The slow unbinding rate k_4 produces a slow modulation of I_h that accounts for the waxing-and-waning of the oscillations (see below). A kinetic analysis of the slow inactivation of NMDA receptors by intracellular Ca^{2+} (Legendre et al., 1993) also reported a relatively fast binding and a much slower unbinding rate with a time constant of approximately 5 seconds.

The current is proportional to the relative concentration of open channels:

$$I_h = \bar{g}_h \left([O] + g_{inc}[O_L]\right)(V - E_h) \qquad (3.22)$$

with a maximal conductance of $\bar{g}_h = 0.02$ mS/cm^2 and a reversal potential of $E_h = -40$ mV. Because of the factor $g_{inc} = 2$, the conductance of the calcium-bound open state of I_h channels is twice that of the unbound open state. This produces an augmentation of conductance following the binding of Ca^{2+}, as observed in sino-atrial cells (Hagiwara and Irisawa, 1989).

The model based on indirect modulation of I_h by Ca^{2+} is similar in spirit to the direct-binding model (Section 3.3.2; Destexhe et al., 1993a), but has important differences. First, Ca^{2+} ions do not bind directly to the channel but through a calcium-activated intermediate P, which produces shift of the activation curve (Fig 3.19B) that is limited because P has a maximum concentration inside the cell. Second, the maximal conductance of the channel increases following the binding of P (for $g_{inc} > 1$). These two properties were observed in sino-atrial cells (Hagiwara and Irisawa, 1989). The present kinetic scheme for voltage-dependence is simpler than that of the previous model of Section 3.3.2, which contained a combination of fast and slow activation gates (Destexhe and Babloyantz, 1993; Destexhe et al., 1993a). These improvements make the waxing-and-waning of the TC cell even more robust.

The behaviour of the new model of the TC cell is similar to that of the previous one: when the TC cell bursts repetitively, the enhancement of I_h depolarizes the cell, eventually terminating the bursting. The TC cell model has waxing-and-waning slow oscillations as observed *in vitro* (Leresche et al., 1991). The dynamics of the TC cell also exhibits different regimes depending on the balance between I_T and I_h conductances (Fig. 3.20A). For fixed $\bar{g}_T$, increasing $\bar{g}_h$ led successively to slow oscillations in the delta range (1–4 Hz), then to waxing-and-waning slow oscillations and, finally, to the relay resting state, consistent with *in vitro* studies (Soltesz et al., 1991).

During waxing-and-waning oscillations, calcium enters through I_T channels on each burst, resulting in an increase of calcium-bound messenger (P_1) and a gradual increase of I_h channels in the open state (O_L). This results in a progressive afterdepolarization (ADP) following each burst until the cell ceases to oscillate (Fig. 3.20B). This ADP has been observed during waxing-and-waning oscillations in TC cells maintained in low magnesium *in vitro* (Leresche et al., 1991).

In thalamic slices containing both TC and RE cells, intracellularly recorded TC cells exhibit a small (2–5 mV) ADP following repetitive burst discharges (Fig. 3.21; Bal and McCormick, 1996). This has also been seen in TC cells during waxing-and-waning oscillations (see progressive hyperpolarization in Fig. 3.12; Leresche et al., 1991). The ADP occurs after spindle oscillations (Bal and McCormick, 1996), after bicuculline-induced oscillations (Fig. 3.21A) and after a sequence of hyperpolarizing current pulses (Fig. 3.21B). In the latter case, there is a marked diminution of input resistance in successive responses (Fig. 3.21E, F). This corroborates

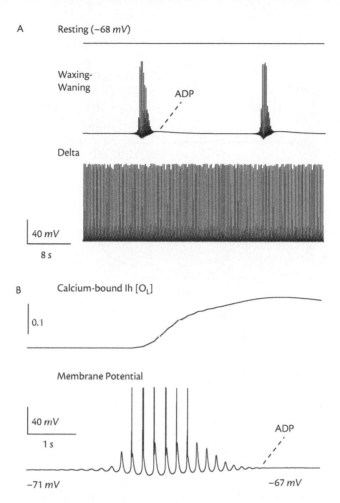

Fig. 3.20: Oscillatory properties of a model TC cell with indirect Ca^{2+}-mediated regulation of I_h. A. Three different modes with different conductance values of I_h. From top to bottom: relay state ($\bar{g}_h = 0.025$ mS/cm^2), slow waxing-and-waning ('spindle-like') oscillations ($\bar{g}_h = 0.02$ mS/cm^2) and delta oscillations ($\bar{g}_h = 0.005$ mS/cm^2). B. Intrinsic waxing-and-waning oscillation at higher time resolution. The top trace shows the fraction of channels in the calcium-bound open state (O_L) and the membrane potential is shown at bottom. Leak potassium conductance was of $g_{KL} = 5$ nS for all simulations (from Destexhe et al., 1996a).

intracellular recordings of TC cells *in vivo* showing a diminution of input resistance at the end of spindle sequences (Nuñez et al., 1992).

   In the model these two properties follow from the upregulation of I_h. In Fig. 3.22 a current-induced oscillation is compared in a model TC cell with upregulated I_h and in control conditions with unregulated I_h. Successive hyperpolarizing pulses at 4 Hz evoked rebound bursts with

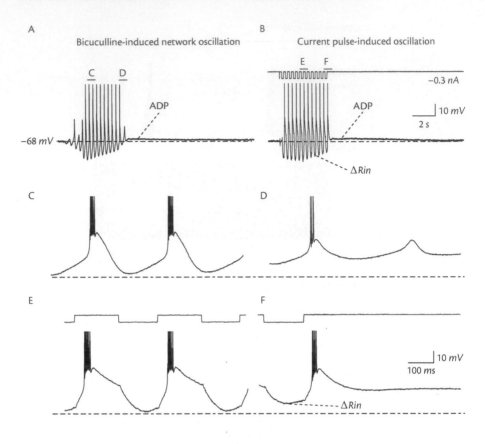

Fig. 3.21: Afterdepolarization and diminished input resistance follow a TC cell burst. ADP following a spontaneously occurring oscillation in a TC cell from the lateral geniculate nucleus *in vitro*. B. Same cell, but a sequence of rebound bursts was induced by injecting hyperpolarizing current pulses at 3 *Hz* (current shown in top trace). Examples of rebound bursts at the beginning (C) and the end (D) of the spontaneous oscillation show that IPSPs have diminished and the bursts gradually become weaker. The same phenomenon seen during injection of current pulses (E–F). During the sequence of injected pulses, there was a progressive decrease of input resistance (Δ*Rin*) (from Destexhe et al., 1996a).

a progressive diminution of input resistance with a time constant of approximately 700 *ms* (Fig. 3.22A). This phenomenon did not occur using unregulated I_h even with a stronger I_h (Fig. 3.22B). The ADP was present in both spontaneous oscillations (Fig. 3.20B) and after a sequence of rebound bursts evoked by current injection (Fig. 3.22A). The depolarization occurs because of a progressive enhancement of the I_h current.

In Figs. 3.21 and 3.22, there is a reduced tendency for further rebound bursting in the TC cell after the first few bursts. The progressive augmentation of the I_h conductance depolarizes the membrane, counteracts the deinactivation of I_T, and weakens the rebound response. This

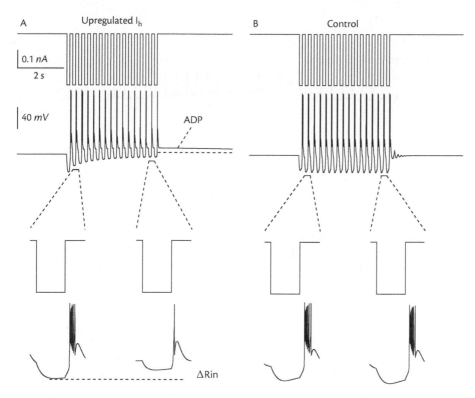

Fig. 3.22: Diminution of input resistance and after-depolarization in model TC cells. Hyperpolarizing current pulses were injected at 4 *Hz* evoking a sequence of rebound bursts (injected current shown in top trace). A. During the activity-dependent upregulation of I_h, the voltage responses gradually decreased, indicating a diminution of input resistance (ΔRin in insets). Successive bursts became progressively less powerful. At the end of the sequence, there was an after-depolarization (ADP). B. Without the upregulation of I_h, none of these phenomena was present, even when a stronger I_h conductance was used. Bottom traces show a blowup of the second and 15th pulses and responses; $\bar{g}_h = 0.025\ mS/cm^2$, $g_{KL} = 5\ nS$ in A and $\bar{g}_h = 0.08\ mS/cm^2$, $g_{Kl} = 5\ nS$ in B (from Destexhe et al., 1996a).

phenomenon was not seen in the absence of upregulation of I_h (Fig. 3.22B). This property has important consequences at the network level, as shown in Chapters 6 and 7.

Finally, recent experiments provide direct evidence for the indirect model. First, Ca^{2+} does not directly modulate I_h channels in thalamic neurons (Budde et al., 1997). Second, experiments with caged Ca^{2+} in thalamic neurons demonstrate an indirect calcium-dependent modulation of I_h (Luthi and McCormick, 1998). Fig. 3.23 illustrates the enhancement of I_h obtained by flash-photolysis of caged Ca^{2+}, which is quite similar to the prediction of the model.

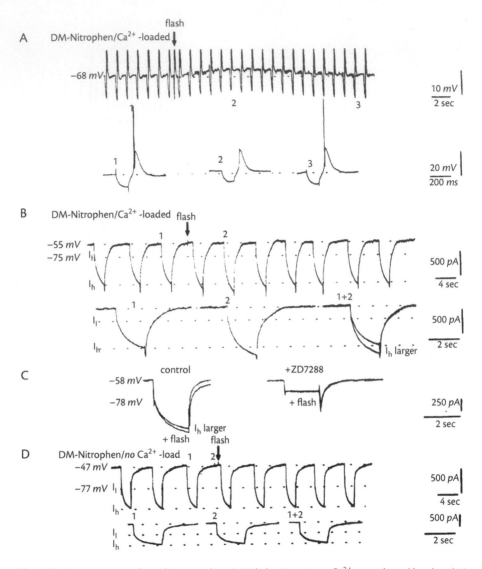

Fig. 3.23: Demonstration that calcium regulates I_h in thalamic neurons. Ca^{2+} was released by photolysis of a caged-calcium compound (DM-Nitrophen) injected in a ferret thalamic neuron from lateral geniculate nucleus. A. Injection of hyperpolarizing current pulses during photolysis of caged Ca^{2+} (arrow). Intracellular release of Ca^{2+} was accompanied by a slow depolarization of the cell and the burst responses were weaker (compare 2 to 1). Normal responses reappeared within a few seconds (3). In voltage-clamp, the released Ca^{2+} induced an increase of the inward current I_h. C. When I_h was blocked by ZD7288, the release of intracellular Ca^{2+} had no effect (right panel). E. Same experiment as in B, but DM-Nitrophen was not loaded with Ca^{2+}. In this case, no effect on I_h was detected. Figure reproduced from Luthi and McCormick, 1998.

3.4 Dendritic T-current in thalamic relay cells

The models thus far have only had a single-compartment. Although these models can account for the experimental data illustrated in the preceding sections, Ca²⁺ imaging experiments and voltage-clamp have localized T-current in the proximal dendrites of TC neurons (Munsch et al., 1997; Zhou et al., 1997) (Fig. 3.24). Given the large number of synaptic terminals on the dendrites of a TC cell (Jones, 1985; Liu et al., 1995), dendritic T-current could profoundly influence a cell's responsiveness. In order to investigate the possible electrophysiological

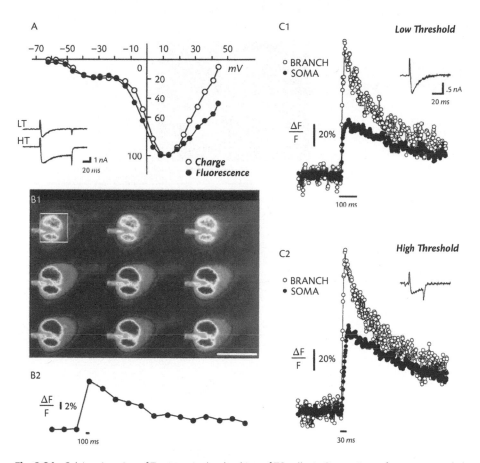

Fig. 3.24: Calcium imaging of T-current in the dendrites of TC cells. A. Comparison of currents recorded with voltage-clamp and fluorescence signals in a thalamic neuron from a rat thalamic neuron from ventrobasal nucleus. The total charge influx (current integrated over time during the voltage step) is compared to the relative fluorescence of the signal obtained in the presence of calcium green dextran, a calcium-sensitive marker. B1. Calcium signal obtained during activation of the T-current (voltage step from −110 to −35 *mV* before frame 4). B2. Time course of the total fluorescence signal in box shown in B1. C. Comparison of the signal from T-type and L-type calcium currents in soma and dendrites. C1. Signal from the T-current recorded under voltage-clamp (step from −110 to −40 *mV*; inset). C2. Signal obtained from a protocol activating the L-type current (step from −50 to 0 *mV*; inset). Figure reproduced from Zhou et al., 1997.

consequences of dendritic I_T, a computational model of T-current in the dendrites of TC cells was examined based on *in vitro* recordings and computational modelling techniques (Destexhe et al., 1998c).

3.4.1 Morphology

A TC neuron was recorded in intact slice preparation and stained with biocytin. The cell is shown in Fig. 3.25A (see also Huguenard and Prince, 1992). The morphology of that cell was reconstructed from serial sections of 80 μm, using a computerized tracing system. The

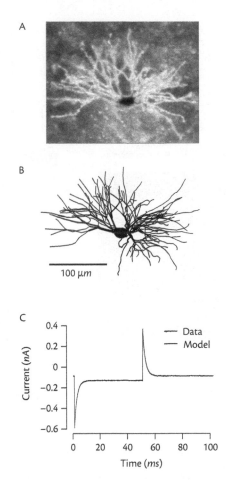

Fig. 3.25: Recording, staining, reconstruction and simulation based on of the same cellular geometry. A. Thalamic relay cell from rat ventrobasal nucleus, intracellularly recorded in slices (Huguenard and Prince, 1992). B. Three-dimensional reconstruction of the same cell. The complete dendritic arbor (of which only part appears in A) was reconstructed from thin serial sections. C. Computational model of the reconstructed cell. The simulation (continuous line) is compared to passive responses obtained from recordings of that cell (noisy trace). The adjustment of the model to the experimental response by a simplex fitting procedure gives estimates of passive parameters (from Destexhe et al., 1998c).

reconstructed TC neuron is shown in Fig. 3.25B. There were eleven primary dendrites, with a total length of 7,095 μm; the total membrane area of the cell was 23,980.5 μm^2, including 2,625 μm^2 for the soma, which had a diameter of 20–25 μm (An error of 0.1 μm in the diameter leads to approximately $\pm 9\%$ error of the total membrane area). The dendritic arborizations tended to be organized in a bush-like structure, similar to previous morphological observations (Jones, 1985).

3.4.2 Passive properties and electrotonic structure

Fig. 3.25A shows voltage-clamp recordings of passive responses obtained in the reconstructed TC cell. These recordings were used to estimate the passive parameters by fitting the model to the data (Fig. 3.25C). Because the model and data had the same cellular geometry, this procedure leads to a unique set of passive parameters if they are uniform (Rall et al., 1992).

To estimate the passive parameters, leak currents were inserted in all compartments of the reconstructed cell model. The values of the passive parameters (leak conductance g_L, leak reversal potential E_L, axial resistivity R_a, specific membrane capacitance C_m) and the electrode series resistance (R_s) were obtained by fitting the simulations to experimental data using a simplex algorithm (Press et al., 1986). At each iteration of the simplex algorithm, the model minimized the root mean squared (RMS) error between the experimental recording and the model. This procedure was repeated from different initial conditions to avoid local minima. The values of passive parameters were assumed to be uniform over the neuron and were consistent with the values estimated from the recordings. approximately 50 to 300 iterations were required to converge to a minimum error.

The optimal set of values obtained was: $C_m = 0.878$ $\mu F/cm^2$, $R_a = 173$ Ωcm, $g_L = 0.0379$ mS/cm^2, $E_L = -69.85$ mV and $R_s = 8.1$ $M\Omega$. This optimal set was obtained from different initial conditions with ranges of values tested of 0.2–2 $\mu F/cm^2$, 50 to 500 Ωcm, 0.02 to 0.2 mS/cm^2, −70 to −80 mV and 1 to 40 $M\Omega$, respectively. The range of parameter values yielding similar fitting errors (within a 2 pA maximal RMS error) were: $C_m = 0.856$ to 0.899 $\mu F/cm^2$, $R_a = 147$ to 200 Ωcm, $g_L = 0.0376$ to 0.0383 mS/cm^2, $E_L = -69.72$ to −69.97 mV and $R_s = 7.0$ to 9.5 $M\Omega$. These ranges were obtained by varying each parameter individually around the optimal fit.

Because the window current of I_t may affect the resting membrane potential of the cell, the same fitting was also performed with a somatic-dendritic distribution of the T-current (see below for equations of the model). In this case, the values obtained were close (within standard error) to those of the passive fitting, except for the leak reversal potential, which was readjusted to more negative values to compensate for the window current ($E_L = -70.1$ to −73.4 mV for dendritic T-current densities of 1.7 10^{-5} to 12.5 10^{-5} cm/s; see below).

The electrotonic length of the longest dendrite was 0.34 space constants using the optimal set of passive parameters given above. The attenuation characteristics of the cell were estimated by injecting a current pulse of 10 ms and 0.1 nA into the middle of a representative terminal branch in the intact cell model with passive currents. The attenuation, approximately a ten-fold decrease (0.098), was calculated by measuring the maximal voltage deflections evoked by this current injection at the site of injection compared with the soma. When the current was injected in the soma the ratio of voltage deflections measured in the soma compared to the dendrites was close to unity (0.91).

This ventrobasal TC cell is therefore relatively electrotonically compact, consistent with conclusions of previous studies (Bloomfield et al., 1987; Crunelli et al., 1987). Note that an RE cell reconstructed from the same animal was found to be highly non-compact using similar methods (see Section 4.3 in Chapter 4; Destexhe et al., 1996b). Further comparison with other studies is difficult because different animals were used, using different types of recording electrodes, different passive parameters and different methods to estimate the electrotonic length.

3.4.3 Model of dendritic I_T

A Hodgkin-Huxley model for the T-current was used, similar to the model illustrated in Fig. 3.4. In contrast to the previous model, which was based on the Nernst equation, a formalism was used that captures more accurately the nonlinear and behaviour of calcium currents under conditions that are far from equilibrium. The calcium current was described by Goldman-Hodgkin-Katz constant-field equations (see Section 2.2.2):

$$I_T = \bar{P}_{Ca}\, m^2 h\, G(V, Ca_o, Ca_i)$$

$$\dot{m} = -\frac{1}{\tau_m(V)}\,(m - m_\infty(V)) \tag{3.23}$$

$$\dot{h} = -\frac{1}{\tau_h(V)}\,(h - h_\infty(V))$$

where $\bar{P}_{Ca}$ (in cm/s) is the maximum permeability of the membrane to Ca^{2+} ions, m and h are, respectively, the activation and inactivation variables. $G(V, Ca_o, Ca_i)$ is a nonlinear function of voltage and ionic concentrations:

$$G(V, Ca_o, Ca_i) = Z^2 F^2 V/RT\, \frac{Ca_i - Ca_o \exp(-ZFV/RT)}{1 - \exp(-ZFV/RT)} \tag{3.24}$$

where $Z=2$ is the valence of calcium ions, F is the Faraday constant, R is the gas constant and T is the temperature in Kelvin. Ca_i and Ca_o are the intracellular and extracellular Ca^{2+} concentrations (in M), respectively.

Expressions for the steady-state activation and inactivation functions were obtained from voltage-clamp experiments on dissociated TC cells (Huguenard and Prince, 1992). The activation function was empirically corrected in order to account for the contamination of inactivation (Huguenard and McCormick, 1992). An overall hyperpolarizing shift of 2 mV was applied to compensate for screening charge (voltage-clamp experiments were performed with 3 mM extracellular Ca^{2+} compared with 1.5–2 mM under physiological conditions). In addition, an overall depolarizing shift of 3 mV was needed to reproduce current-clamp simulations of TC cells. The functions optimized for both voltage-clamp and current-clamp data on TC cells were:

$$m_\infty(V) = 1\, /\, (1 + \exp[-(V + 56)/6.2])$$
$$h_\infty(V) = 1\, /\, (1 + \exp[(V + 80)/4]).$$

The voltage-dependent time constant for activation was:

$$\tau_m(V) = 0.204 + 0.333 / (\exp[-(V + 131)/16.7] + \exp[(V + 15.8)/18.2]) \tag{3.25}$$

and for inactivation:

$$\tau_h(V) = \begin{array}{ll} 0.333\ \exp[(V + 466)/66.6] & \text{for } V < -81\ mV \\ 9.32 + 0.333\ \exp[-(V + 21)/10.5] & \text{for } V >= -81\ mV. \end{array} \tag{3.26}$$

Ca^{2+} concentration of 2 mM at a temperature of 36 °C. All voltage-clamp simulations were done at 24 °C assuming Q_{10} values of 2.5 for both m and h; current-clamp behaviour was simulated at 34 °C.

Calcium handling was modelled by a first-order system representing Ca^{2+} transporters and buffers, as in Eq. 2.29 in Chapter 2, with a time constant of decay of Ca^{2+} of 5 ms. At equilibrium, the free intracellular Ca^{2+} concentration was 240 nM and the extracellular Ca^{2+} concentration was 2 mM, corresponding to a reversal potential of approximately +120 mV.

Na^+ and K^+ currents responsible for fast action potentials were included in the soma with kinetics from a model of hippocampal pyramidal cells (Traub and Miles, 1991), as given in Table 2.1, with maximal conductances of $\bar{g}_{Na} = 100\ mS/cm^2$ and $\bar{g}_K = 100\ mS/cm^2$, and reversal potentials of $E_{Na} = 50\ mV$ and $E_K = -100\ mV$. This model exhibits repetitive firing within bursts of action potentials (Traub and Miles, 1991; Destexhe et al., 1996a, 1996b; see above).

3.4.4 Low density of T-current in soma and proximal dendrites

The T-current in TC cells was characterized previously in acutely dissociated neurons (Huguenard and Prince, 1992). This preparation is quite useful because the dissociation procedure removes most of the dendritic arbor, leaving the soma intact with proximal bits of dendrites (see Fig. 3.26A1). This preparation therefore leads to highly compact cells, in which voltage-clamp recordings can be made with minimal space clamp errors. The kinetics of the T-current used here were obtained from such recordings (see Huguenard and McCormick, 1992; Huguenard and Prince, 1992).

Another advantage of the dissociated-cell preparation is that it allows a direct estimate of the T-current density in the perisomatic region of the cell. This estimate was made by matching a dissociated-cell model to voltage-clamp recordings of the T-current in dissociated TC cells. The model in Fig. 3.26B1 was obtained by truncating the dendrites of the original cell (Fig. 3.26B1). This was compared with the model in Fig. 3.26B1, which was obtained by keeping the soma and proximal bits of dendrites of the reconstructed cell based on the morphology of dissociated TC cells (Fig 3.26B2; see also Huguenard and Prince, 1992) and the ratio of input capacitance measured experimentally (113 pF for the intact cell and 16.7 pF on average for dissociated TC cells; see Huguenard and Prince, 1992), leading to an area of approximately 3500 μm^2. Distal dendrites were removed from the reconstructed cell until the model matched this area. The

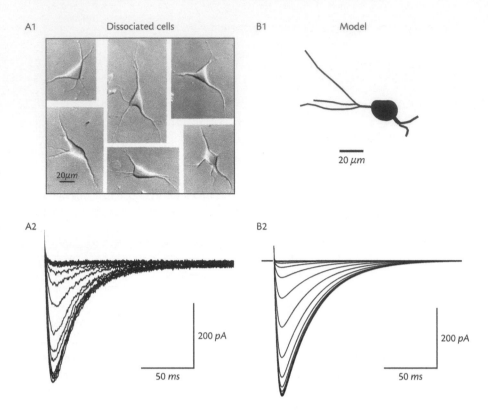

Fig. 3.26: Low amplitude T-currents in dissociated TC cells under voltage-clamp. A1. Typical structure of dissociated TC cells, with most of the dendrites removed by the dissociation procedure, leaving only the soma and proximal dendrites. A2. Voltage-clamp recordings of the T-current in a dissociated TC cell. The voltage-clamp protocol consisted in conditioning the cell at various voltage levels (from −125 to −60 mV) for 1 second, then stepping the voltage to −30 mV, revealing the transient activation of the current. The peak current was approximately 400 pA (different cell than that shown in A1). B1. Model of a dissociated TC cell, consisting of the soma with proximal bits of dendrites, adjusted from the input capacitance of the model. B2. Same voltage-clamp protocol as in A2, simulated with the dissociated-cell model. The model reproduced the peak amplitude of the T-current in dissociated cells with a moderate density of T-channels (permeability of $1.7\ 10^{-5}$ cm/s). This procedure provided an estimate of the perisomatic T-current density in TC cells. All experiments and simulations at 24°C (from Destexhe et al., 1998c).

resulting dissociated-cell model had only the soma and two proximal dendritic branches, with a total membrane area of 3430 μm^2 (Fig. 3.26B1).

The peak T-current amplitude was relatively low in dissociated TC cells, approximately 400 pA (Fig. 3.26A2). Similar T-current peak amplitudes were obtained assuming a uniform density of T-current in the dissociated-cell model ($1.7\ 10^{-5}$ cm/s; see Fig. 3.26B2). A range of T-current densities of 0.5–3.0 10^{-5} cm/s reproduced the range of T-current amplitudes measured in dissociated TC cells under voltage-clamp (350 ± 27 pA, n = 49 from Coulter et al., 1989, and 280 ± 23 pA, n = 26 from Huguenard and Prince, 1992).

3.4.5 Increased density of T-current in more distal dendrites

In previous studies, the T-current was found in the dendrites of TC cells using optical imaging techniques (Munsch et al., 1997; Zhou et al., 1997). Given that intact TC cells have considerably more extended dendritic area than dissociated TC cells, this would predict a higher T-current amplitude in intact TC cells than shown in Fig. 3.26A2 for dissociated cells. This was demonstrated by voltage-clamp recordings of intact TC cells (Fig. 3.27). The maximal T-current amplitude ranged from approximately 2 *nA* to 8.3 *nA* in different intact cells (5.8 ± 1.7 *nA*, *n* = 7), which is on average approximately 16 to 21 times larger than in dissociated cells.

The T-current densities in the dendrites were then estimated by matching the reconstructed TC cell model with the range of T-current amplitude found experimentally in intact TC cells. As

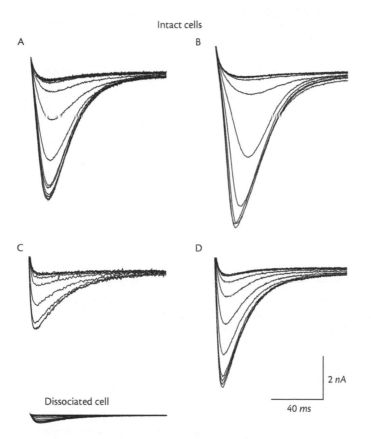

Intact cells

A

B

C

D

2 *nA*

40 *ms*

Dissociated cell

Fig. 3.27: High amplitude T-currents in intact TC cells under voltage-clamp. Inactivation protocol in four TC cells (A–D) recorded in thalamic slices of the ventrobasal nucleus. For each cell, the voltage-clamp protocol giving rise to the largest peak current is shown. The inactivation protocol consisted in conditioning the cell at various voltage levels (from −105 to −40 *mV*) for 1 second, then stepping the voltage to a fixed voltage value (−55 *mV* in A, −65 *mV* in B, −60 *mV* in C and −45 *mV* in D). For comparison, a similar protocol in a dissociated cell is shown in the bottom at the same calibration. All cells were the same age (rat p12); recording temperature was 24 °C (from Destexhe et al., 1998c).

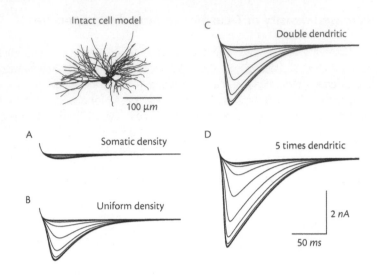

Fig. 3.28: High densities of dendritic T-current are needed to match the T-current amplitude recorded in intact TC cells. The model shows the same inactivation protocol using different distributions of T-current. In all cases, the perisomatic I_T density was compatible with recordings in dissociated cells (permeability of $1.7\ 10^{-5}$ cm/s in soma and proximal dendrites), while the density of I_T in more distal dendrites was varied. A. No dendritic T-current: I_T was limited to the perisomatic region. B. Uniform T-current: I_T had the same density throughout the neuron. C. Twice dendritic: I_T density was twice in dendrites compared to the perisomatic density. D. five times dendritic: dendrites had five times as much T-current density as in the perisomatic region. The latter produced peak I_T amplitudes comparable to the average value of intact TC cells recorded in slices (Fig. 3.27). All simulations at 24 °C (from Destexhe et al., 1998c).

shown in Fig. 3.28, the density of T-current estimated from dissociated cells (1.5–$2.0\ 10^{-5}$ cm/s) was insufficient to reproduce T-current amplitudes comparable to intact cells (Fig. 3.28A), even when the same density was extended to the entire dendritic tree (Fig. 3.28B). In order to obtain T-current peak amplitudes of approximately 2 nA and greater, a higher density of T-current had to be assumed in dendrites (Fig. 3.28C–D). To reproduce peak amplitude in the range of 2 to 7 nA observed in Fig. 3.27, the range of dendritic densities needed in the model were 1.7–$6.5\ 10^{-5}$ cm/s for a low series resistance ($R_s = 0.01\ M\Omega$) to 2.5–$50\ 10^{-5}$ cm/s for a high series resistance ($R_s = 12\ M\Omega$), up to twenty-nine times higher than the density in the soma.

3.4.6 Dendritic T-currents affect current-voltage relations

Further evidence for a higher density of T-current in dendrites was found in the current-voltage (I/V) relations. The I/V curves obtained from voltage-clamp recordings of dissociated TC cells (Fig. 3.29A) were markedly different than those obtained from intact cells (Fig. 3.29B). In addition to the larger T-currents, shown in previous section, the shape of the I/V curve is different: the peak occurs at approximately −40 to −30 mV in dissociated TC cells, while in intact cells, the I/V curve was significantly shifted, peaking at −70 to −60 mV.

The simulated I/V curves of the dissociated-cell model matched the experimental curves (compare Fig. 3.29A and C). In intact cells, however, the behaviour depended markedly on

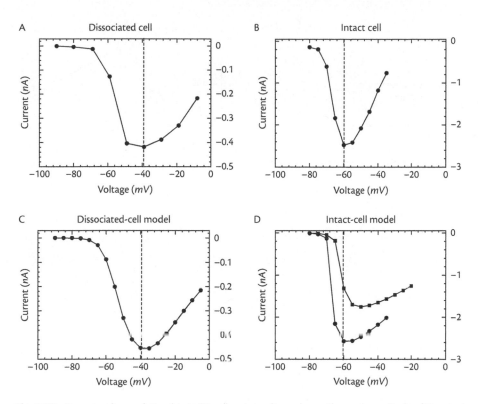

Fig. 3.29: Current-voltage relationship in TC cells using voltage-clamp. The peak amplitude of T-current obtained during activation protocols under voltage-clamp; each protocol consisted in conditioning the cell at −105 mV for 1 second and stepping to other voltage values (shown in abscissa). A. Current-voltage (I/V) relation for the T-current in a dissociated TC cell. The peak current was approximately 0.4 nA and occurred at −40 mV. B. I/V curve in an intact TC cell (same cell as in Fig. 3.27C). The peak T-current was approximately 2.5 nA and occurred at −60 mV. The steep decline of the I/V curve above −60 mV is probably due to incomplete block of outward currents by cesium. C. Dissociated-cell model. A similar I/V curve as in A could be reproduced with moderate T-current density (permeability of $1.7 \ 10^{-5} \ cm/s$). D. Intact-cell model. In this case, the I/V curve of the intact cell shown in B could be reproduced using a larger T-current density in dendrites ($2.5 \ 10^{-5} \ cm/s$) compared to the perisomatic region ($1.7 \ 10^{-5} \ cm/s$), and a series resistance of $R_s = 12 \ M\Omega$ (circles). This I/V curve is compared to the same simulation with uniform T-current density of $1.7 \ 10^{-5} \ cm/s$ (squares). Experiments and simulations were at 24°C and the peak currents shown were leak-subtracted (from Destexhe et al., 1998c).

the density of the T-current used in the model. With higher T-current densities in dendrites, the I/V curve was comparable to those from experimental recordings in intact cells (compare Fig. 3.29B and D, circles). In contrast, with uniform T-current densities, the I/V curves were not consistent with experimental observations, fitting neither the peak amplitude, which was too low, nor the position of the peak current, which was too depolarized (Fig. 3.29D, squares). Over a wide range of parameters (see Destexhe et al., 1998c), the I/V curve shifts were consistent with those observed in experiments only when the density of T-current was significantly higher in the dendrites.

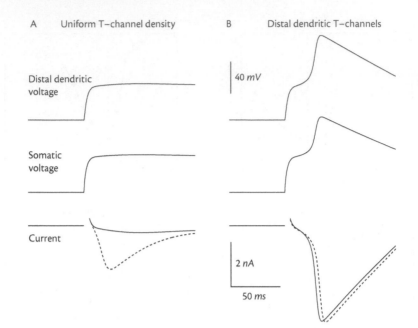

Fig. 3.30: Poor space clamp together with dendritic T-currents alters current-voltage relationship. Voltage-clamp simulation of the intact TC cell model consisting in conditioning the cell at −115 *mV* for 1 second and stepping the voltage to −65 *mV*. The voltage and current shown in soma and dendrite. The maximal current that could be evoked (peak of the I/V curve) is shown by dotted lines for comparison. A. uniform T-channel density of $1.7 \ 10^{-5}$ *cm/s* shows little current activation at −65*mV* (0.187 *nA*), representing approximately 10% of the total T-current. B. Same simulation with high densities of T-current in dendrites ($1.7 \ 10^{-5}$ *cm/s* perisomatic and $8 \ 10^{-5}$ *cm/s* in dendrites). In this case, large uncontrolled voltage transients occurred in both soma and dendrites, leading to a large peak current (4.09 *nA*), representing 98% of the total available T-current. Series resistance was 12 *M*Ω and temperature was 24°C in both cases (from Destexhe et al., 1998c).

The I/V curve shift of intact TC cells compared to dissociated cells also could be explained by poor voltage-clamp of the intact cell, due to space clamp and/or series resistance artefacts. During a somatic voltage-clamp, the voltage at the dendritic T-channels was not easily controlled. Thus, the same phenomenon observed in thalamic reticular cells was also found in TC cells (Fig. 3.30) (Section 4.3 in Chapter 4; Destexhe et al., 1996b). From a holding potential of −115 *mV*, stepping the voltage to a hyperpolarized value (−65 *mV*), where only few T-channels should open, activated approximately 10% of the total T-current available for a uniform density of $1.7 \ 10^{-5}$ *cm/s* (Fig. 3.30A). On the other hand, with high dendritic densities of T-current, the same protocol led to activation of almost all of the available T-current (Fig. 3.30B). In this case, a low-threshold spike was elicited in the dendrites and therefore the recorded current in the soma was anomalously large. The inflexion of the current trace in Fig. 3.30B is indeed indicative of poor voltage control (see Huguenard et al., 1988). Consequently, the I/V curve's peak occurred at more hyperpolarized values due to the poor control over the dendritic

T-channels. This is consistent with a previous modelling study showing that imperfect space clamp results in alterations of I/V curves (Müller and Lux, 1993).

3.4.7 Burst generation in TC cells with dendritic T-current

Additional evidence for dendritic T-current was obtained from current-clamp simulations. Responses to depolarizing current pulses from rest (-73 to -74 mV) during current-clamp recordings (same cell as the reconstructed neuron) are shown in Fig. 3.25A. Current pulses of 50 pA and 75 pA amplitude elicited low-threshold spikes with one and two action potentials, respectively (Fig. 3.31A). When a uniform density of T-current was used in the model based on the value obtained from voltage-clamp recordings in dissociated TC cells, the same current-clamp protocol did not give rise to a low-threshold spike (LTS) (Fig. 3.31B), over a wide range of parameters values including the resting level of the cell and the kinetics of I_T.

Generation of an LTS similar to that in an intact cell could be produced using high densities of T-current in dendrites (Fig. 3.31C). One- and two-spike bursts were generated by 50 pA and 75 pA current injection, respectively. Although the one-spike burst had a longer latency, it did not last as long as in experiments. The latency could be increased by using a more negative resting membrane potential (not shown), but then the resting values did not agree with the data. The longer latency of the first burst may therefore depend on the presence of other currents that were not included, such as I_h, which is active at rest and therefore likely to affect the resting level of the cell (Pape, 1996).

Dendritic T-current densities in only a narrow range yielded the correct bursting behaviour. In comparing burst responses containing one and two spikes (as in Fig. 3.31A), too low a density of the T-current in dendrites failed to produce an LTS, whereas too high a density gave rise to a correct burst for 50 pA current injection, but generated a far too powerful burst at 75 pA (not shown). Therefore, the current-clamp recordings of the reconstructed TC cell provides a strong constraint on the total amount of T-current in the cell. A uniform T-current based on that estimated from dissociated cells ($1.7 \ 10^{-5}$ cm/s) (Fig. 3.31B) also failed to reproduce a burst. Correct burst behaviour could be obtained using uniformly high densities of T-current (approximately $7 \ 10^{-5}$ cm/s), but that model was inconsistent with dissociated cells. Using the experimentally observed perisomatic density of T-current, a relatively narrow range of T-current density in dendrites gave rise to burst generation consistent with experimental data. Simulations with this type of T-channel distribution were in agreement with all voltage-clamp and current-clamp data. Assuming a ± 3 mV error on all voltages (the voltage of the experimental data of Fig. 3.31A as well as the voltage of the kinetics of I_T), gave a possible range of dendritic densities of T-current between $7.6 \ 10^{-5}$ and $12.5 \ 10^{-5}$ cm/s, which is approximately 4.5 to 7.6 times the density in the soma.

The above dendritic T-current permeabilities were converted into conductances by simulating the calcium current using the Nernst relation (Hille, 1992). The range of dendritic T-current conductance obtained was approximately 0.8 to 1.4 mS/cm^2 (8–14 pS/μm^2). This estimated conductance range is slightly larger, but close to the average T-current densities of 7 to 10 pS/μm^2 measured in the dendrites of hippocampal pyramidal cells (Magee and Johnston, 1995a).

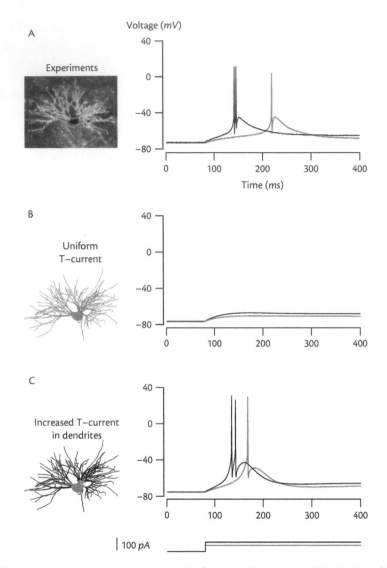

Fig. 3.31: Low-threshold spike generation in model of an intact TC cell requires high densities of dendritic T-current. Current-clamp recordings of low-threshold spikes in the recorded intact TC cell are compared with simulations based on the same cellular geometry. A. Experimental recordings of low-threshold spikes in the intact TC cell at rest using two different amplitudes of injected depolarizing current: 50 *pA* (thin trace) and 75 *pA* (thick trace). B. Simulations of the same current injection did not generate low-threshold spikes using a uniform density of T-current based on dissociated cells ($1.7 \, 10^{-5}$ *cm/s*). C. Successful low-threshold spike generation with increased density of T-current in dendrites. In B and C, the grey levels indicate the density of T-current in different regions of the cell: $1.7 \, 10^{-5}$ *cm/s* (light grey) and $8.5 \, 10^{-5}$ *cm/s* (black). All experiments and simulations at 34 °C (from Destexhe et al., 1998c).

3.4.8 T-channels can be controlled more efficiently if they are dendritic

What are the electrophysiological consequences of dendritic T-currents? To answer this question, the reconstructed TC cell was simulated using two different distributions of T-current with same total number of T-channels ('Somatic & dendritic' and 'Somatic only' in Fig. 3.32).

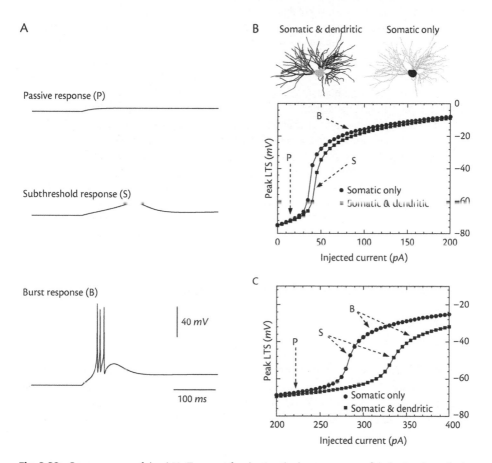

Fig. 3.32: Consequences of dendritic T-current for shaping the burst response of thalamic relay cells. A. Three representative types of response to depolarizing current injection from rest: passive (P), subthreshold (S) and burst (B) responses. B. LTS peak amplitude represented as a function of the amplitude of injected current in the soma. For this simulation, fast Na+ and K+ currents underlying action potentials were not included and the reconstructed TC cell had two different somato-dendritic distributions of T-current: 'Somatic & dendritic' (1.7 10^{-5} cm/s perisomatic and 8.5 10^{-5} cm/s dendritic) and 'Soma only' (density of 56.53 10^{-5} cm/s in soma with none in dendrites; adjusted such that the total number of T-channels was the same as for 'Somatic & Dendritic'). These two curves do not overlap, reflecting the effects of channel segregation on burst generation in normal conditions. C. LTS peak amplitudes in the presence of dendritic shunt conductances. With a dendritic shunt conductance of $g_L = 0.15$ mS/cm^2, the burst response generated when the T-current was located in the dendrites differed significantly from the response when the T-current was located exclusively in the soma (from Destexhe et al., 1998c).

As the stimulation intensity of injected current increased, the cell first generated passive responses, then a subthreshold LTS and, finally, full-blown bursts with sodium spikes (Fig. 3.32A). The threshold for these LTS responses depended on whether the T-current was somatic or somato-dendritic, with an LTS evoked by weaker current injections for somatic-only distributions (Fig. 3.32B). This property is already apparent in the I/V curves, where somatically localized T-channels led to higher peak T-current amplitudes compared to T-channels distributed in dendrites (see Fig. 7D in Destexhe et al., 1998c). Therefore, localizing T-channels in the dendrites decreased the excitability of the LTS.

The model was used next to investigate how burst properties can be modulated by other currents in the dendrites. The dendrites of TC cells are densely covered by synaptic terminals from various excitatory and inhibitory neurons. During the tonic activity that characterizes active states (Steriade and McCarley, 1990) TC cells are bombarded by mixed excitatory and inhibitory inputs, which should produce a significant dendritic shunt. The effect of dendritic shunt conductances on burst generation is shown in Fig. 3.32C. The differences between 'Somatic only' and 'Somatic & dendritic' T-channel localization were greatly enhanced (Fig. 3.32C). In addition, the LTS response was less steep, leading to more 'graded' bursting behaviour in the presence of shunt conductances. More importantly, a dendritic shunt shifted the LTS response curve further for dendritic T-channels (compare 'Somatic & dendritic' curves between Fig. 3.32B and 3.32C) than for somatic T-current (compare 'Somatic only' curves). These simulations therefore show that the same amount of T-channels can be controlled differently if they are exclusively somatic or distributed throughout the dendrites. Possible functional consequences of this property for the behaviour of TC cells *in vivo* are discussed in section 3.6.

3.4.9 Simplified three-compartment models of TC cells

To provide a simplified model for a TC cell, the dendritic arbor was collapsed into fewer compartments based on the conservation of axial resistance (modified from Bush and Sejnowski, 1993). This method consists in merging dendritic branches into equivalent cylinders that preserve the axial resistance of the original branches. If the cross-sectional area of the equivalent cylinder equals the sum of each individual cross-sectional areas, this is equivalent to summing parallel resistances since $1/R = \sum_j 1/R(j)$, where $R(j)$ are the axial resistances of the collapsed branches. The radius (r) of the equivalent cylinder is then given by:

$$r = \sqrt{\sum_i r_i^2} \qquad (3.27)$$

where r_i are the radii of the collapsed branches.

The length (l) of the equivalent cylinder is taken as an average of the lengths of the collapsed branches (l_i), weighted by their respective diameters (r_i), such as:

$$l = \frac{\sum_i l_i\, r_i}{\sum_i r_i} \qquad (3.28)$$

This extension of the Bush-Sejnowski method accommodates the merging of branches of different length, which is often encountered while reducing dendritic morphologies, as in the reconstructed TC cell studied here.

Since the total membrane area is not conserved by this method, the reduced model may not have a correct input resistance. This is compensated by introducing in each equivalent cylinder a dendritic correction factor (C_d) that rescales the values of conductances (g_i) and membrane capacitance (C_m) in the dendrites such that

$$g'_i = C_d\, g_i, \qquad C'_m = C_d\, C_m. \tag{3.29}$$

C_d is chosen so that the reduced model has the correct input resistance and time constant (Bush and Sejnowski, 1993).

This algorithm was used to reduce the TC cell into three compartments, corresponding to the three regions considered previously: (a) the soma; (b) the proximal dendrites (corresponding to those of dissociated cells); and (c) the remaining dendrites. The resulting model is shown in Fig. 3.33A1. The three compartments had the following lengths (l) and diameters (*diam*): $l = 38.4\ \mu m$ and *diam* = 26 μm for the soma (area of 2624 μm^2); $l = 12.5\ \mu m$ and *diam* = 10.3 μm for the proximal segment (area of 403 μm^2); $l = 84.7\ \mu m$ and *diam* = 8.5 μm for the distal segment (area of 2261 μm^2). The total area was 5,289 μm^2. The dendritic correction factor was $C_d \sim 8.02$ as calculated from the ratio between the total surface area of the dendritic segments and their equivalent cylinders. A more accurate estimation of $C_d = 7.95$ was obtained by fitting simulations to voltage-clamp recordings until the input resistance and other passive properties of the three-compartment model perfectly matched the values obtained for the reconstructed cell (Fig. 3.33A2).

Compared to the detailed model with reconstructed geometry, the three-compartment model is approximately sixty-six times faster to simulate.

The next step was to obtain passive properties consistent with experimental data. The simplified model was given with the same passive parameters as the detailed model and a dendritic correction factor (C_d) was applied to the dendrites in order to compensate for their reduced membrane area (see above). The value of dendritic correction was adjusted by fitting the passive responses of the model to that obtained during voltage-clamp recordings (Fig. 3.33A2). With $C_d = 7.95$ (range 7.92 to 7.97) the model fitted the data remarkably well (Fig. 3.33A2). This value was close to the reduction ratio of 8.02 for the dendritic membrane.

The simplified model generated voltage-clamp behaviour close to the detailed model when T-channels were inserted using somatic and dendritic densities similar to the detailed mode. The peak T-current amplitudes were comparable in both models (compare triangles with squares in Fig. 3.33A3). Using a slightly elevated density of T-current in the dendrites, the simplified model closely matched the complete I/V curve of the detailed model (compare circles with squares in Fig. 3.33A3).

The behaviour of the simplified model was then examined with a current-clamp protocol. First, the somato-dendritic density that matched closely the I/V curve generated low-threshold responses similar to those obtained by current injection in the intact cell and in the detailed model (Fig. 3.33A4; compare with Fig. 3.31). Second, the genesis of low-threshold bursts required increased densities of dendritic currents in the simplified model as in the detailed model (Fig. 3.34A). This suggests that high densities of dendritic calcium currents are required in

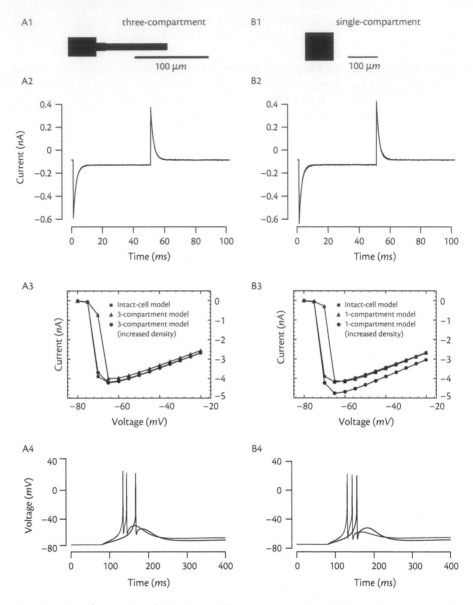

Fig. 3.33: Simplified models of TC cells. A1. Three-compartment model obtained by collapsing the dendritic morphology of the intact cell. A2. Dendritic correction of the model adjusted to obtain passive properties consistent with experimental recordings. A3. Voltage-clamp I/V curves, comparing the intact-cell model (squares) with the simplified model with the same density of T-current (triangles; same densities as in Fig. 3.31C). A slightly increased dendritic density matched the I/V curve of the intact model (circles; dendritic density of 9.5 10^{-5} cm/s). A4. Low-threshold bursts with the simplified model (9.5 10^{-5} cm/s). B1. Single-compartment model of the TC cell. B2. Adjustment of the membrane area such that the model fits experimental passive responses. B3. Voltage-clamp I/V curves, with intact-cell model (squares) and its best match with the single-compartment model (triangles; density of 6 10^{-5} cm/s). The circles are from the intact model and were the target current-clamp behaviour (density of 8 10^{-5} cm/s). B4. Low-threshold bursts with the single-compartment model (density of 8 10^{-5} cm/s) (from Destexhe et al., 1998c).

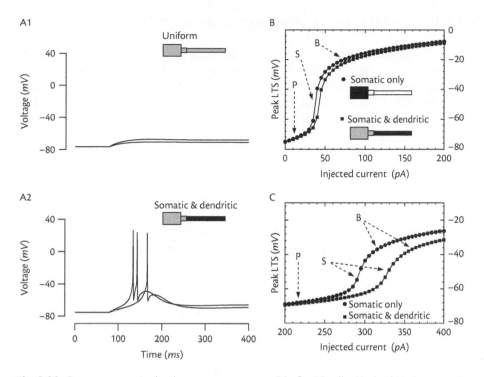

Fig. 3.34: Burst responses in the three-compartment model of a TC cell with dendritic T-current. A. Current-clamp simulations of LTS generation in the three-compartment TC cell model using two different amplitudes of injected depolarizing current (50 pA and 75 pA from rest; see Fig. 3.31B–C). A1: No LTS could be generated with uniform T-current density based on dissociated-cell recordings ($1.7\ 10^{-5}\ cm/s$, indicated by grey shades in the scheme). A2: Successful LTS generation in the simplified model with high densities of T-current in dendrites ($9.5\ 10^{-5}\ cm/s$, indicated by black shades in the scheme). B. LTS peak amplitude represented as a function of the amplitude of injected current in the soma (similar description as in Fig. 3.32). Somato-dendritic T-current distributions were 'Somatic & dendritic' (density of $1.7\ 10^{-5}\ cm/s$ perisomatic and $9.5\ 10^{-5}\ cm/s$ dendritic) and 'Somatic only' (density of $56.36\ 10^{-5}\ cm/s$ exclusively in the soma). Burst responses and the differences due to channel segregation in the cell were similar as the reconstructed TC cell model (Fig. 3.32B). C. LTS peak amplitudes in the presence of dendritic shunt conductances ($g_L = 0.15\ mS/cm^2$). The enhanced differences between the two curves is comparable to the reconstructed TC cell model (Fig. 3.32C) (from Destexhe et al., 1998c).

order to be consistent with all the data, and that this conclusion is independent of the particular morphological details of the model.

The effect of the electrical separation of currents in the cell were tested with the three-compartment model (Fig. 3.34B). As in the detailed model, localizing T-channels in the dendrites diminished the excitability of the cell for LTS generation. These differences were markedly enhanced in the presence of dendritic shunt conductances (Fig. 3.34C), similar to the detailed model (compare with Fig. 3.32B–C). This demonstrates that somato-dendritic interactions underlying LTS generation critically depend on the pattern of distribution of T-current in dendrites.

3.4.10 Simplified single-compartment model of TC cells

Finally, a single-compartment model of the TC cell was generated for comparison. When the dendrites were removed from the detailed model, the isolated soma did not generate bursts of action potentials and had an incorrect input resistance. When the membrane area was increased until the model matched the input resistance of the intact cell, the passive properties were similar to those of the detailed model under voltage-clamp conditions (Fig. 3.33B2). The optimal model had length and diameter of 100 μm and 76.6 μm, respectively.

The density of T-current was adjusted so that the total amount of T-channels was identical to that in the detailed model. This made the peak amplitude of the T-current similar to the detailed model, but the I/V curves were different (compare triangles with squares in Fig. 3.33B3). Unlike for the three-compartment model, it was not possible to match these I/V curves. Under these conditions, the model did not generate low-threshold burst responses consistent with experimental data (not shown). However, if the T-current density was increased, such that the shape of the I/V curve was similar to the detailed model (compare circles with squares in Fig. 3.33B3), then the single-compartment model could generate low-threshold burst responses consistent with experiments (Fig. 3.33B4).

This single compartment model is of course too simple to account for data from intact and dissociated TC cells at the same time, but it still generates LTS consistent with intact TC cells in current-clamp. Single-compartment models might therefore be appropriate for building network simulations in which the central aim is to reproduce the rebound burst of TC cells (see Chapters 6–8; Destexhe et al., 1996a, 1998a).

3.5 Further findings on thalamocortical neurons

Important experimental and theoretical results were obtained about thalamocortical neurons in the last years and are worth being mentioned.

3.5.1 Detailed models

Models of thalamocortical (TC) relay cells have greatly improved over the last decade. A model proposed by Rhodes and Llinás (2005), accounted for fast oscillations in addition to the classic dual bursting and tonic modes of TC cells. A compartmental model of TC cells proposed by Zomorrodi et al. (2008) was used to study the impact of T-type current density in different TC cell morphologies. Birska et al. (2003) explored models of relay cells in the lateral geniculate (LGN) nucleus of the thalamus, investigating the transmission of signals through TC dendrites, based on reconstructed morphologies. Finally, the dendritic integrative properties of a detailed TC cell model was investigated, and in particular the impact of inhibitory inputs and the clustering of excitatory synapses (Neubig and Destexhe, 2001).

3.5.2 Simplified models

In addition to these detailed biophysical models, several simplified models of TC cell were also introduced. The 'integrate, fire and burst' model, proposed by Smith and Sherman (2002), consisted of an integrate-and-fire model augmented with a variable to account for bursting. An augmented integrate-and-fire model, called the adaptive exponential (AdEx model (Brette and Gerstner, 2005)), was used to model the bursting activity of thalamic neurons (Destexhe, 2009). In both cases, the simplified models reproduced both the burst and tonic modes in TC cells. These simplified models are particularly relevant for large-scale network simulations.

3.5.3 Impact of synaptic background activity

Wolfart et al. (2005) investigated how the tonic/burst modes are affected by the synaptic background activity observed *in vivo*. They showed that in the presence of this 'synaptic noise' TC cells do not display the classic dual firing modes, but instead, bursts and single spikes are mixed at all levels of the membrane potential, *Vm*, with an increased propensity for bursting at hyperpolarized *Vm* (Fig. 3.35). These findings explain the occurrence of low-threshold bursts at depolarized levels where the tonic firing mode prevails. This study also showed that the relay properties of the cells take on a different character in the presence of noise: a higher propensity for bursts at hyperpolarized levels compensates for fewer tonic spikes and insures that an approximately constant number of spikes are transmitted to cortex, independently of the *Vm* (Fig. 3.36). This remarkable property is only possible if the T type current is combined with synaptic noise in the membrane. It also suggests that the transfer properties of thalamic neurons depend not only on their intrinsic neuronal properties, but also on the level of synaptic activity.

3.5.4 Calcium channels and sleep oscillations

Recent advances in our ability to identify the subunits of the ion channels, and to knock-out the corresponding genes in mice, has made it possible to study the molecular mechanisms underlying sleep oscillations. In a knock-out study of the Cav3.1 isoform of the T-type calcium channel, the subunit mostly found in TC cells, there was no change in low-frequency slow-wave activity, but a significant decrease in frequencies (8–10 *Hz*) of sleep spindles (Lee et al., 2004).

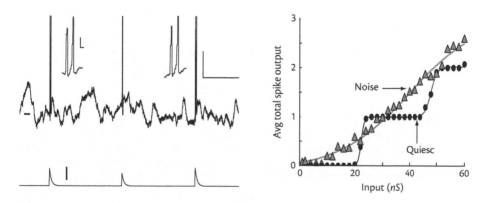

Fig. 3.35: Noise increases occurrence of bursts at resting potential. (a) With noise, high-frequency bursts of action potentials occurred at resting potential (left). Plotting the average total number of spikes per burst response against the input shows that noise linearized the staircase-like transfer function across the whole input range (right). Adapted from Wolfart et al. (2005).

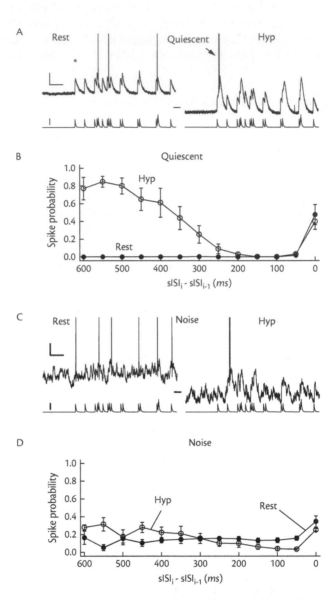

Fig. 3.36: Spike transfer becomes independent of membrane potential in the presence of synaptic noise. A TC cell was injected with physiologically realistic Poisson-distributed inputs, mimicking retinal input. At resting potential without synaptic noise (a), the spike transfer (probability to evoke at least one spike) was highly dependent on the membrane potential (b). In the presence of synaptic noise (c), the spike transfer tended to be independent of membrane potential (d). Modified from Wolfart et al. (2005).

Despite the absence of LTS bursts in TC cells, spindles of reduced amplitude were still present in this Cav3.1 knock-out mouse. It was later shown that Cav3.1 knock-out mice have a markedly higher levels of arousal and the organization of their slow-wave sleep is highly irregular (Anderson et al., 2005). Although thalamocortical rhythms are severely perturbed in the Cav3.1 knock-out mice (Choi et al., 2015), another study confirmed that spindles can still be generated without bursts of spikes in TC cells (Lee et al., 2013).

It should be borne in mind that the results from experiments on knock-out mice may be misleading because of possible compensatory mechanisms during development. Thus it is possible that other isoforms of the T-type channel were expressed in TC cells to compensate, which may be why spindle activity was still present. Conditional knock-out experiments, which can be induced in adults, are needed to clarify these conclusions.

The initial study on Cav3.1 knock-out mice did not report any effects on slow waves (Lee et al., 2004), and similarly, nor did the Cav3.3 knock-out (Astori et al., 2011). These results suggest that the generator for delta and slow waves is cortical rather than thalamic. But these channels were also absent in the cortex of knock-out mice, which would imply that Cav3.1 and Cav3.3 T-type calcium channels are not involved in generating delta waves. Once again, this interpretation of the knock-out results must be taken under advisement because of possible compensatory mechanisms (Leresche and Crunelli, 2013).

An alternative strategy to dissecting the origin of sleep spindles and slow waves is to block T-type channels pharmacologically with TTA-P2 (Uebele et al., 2009), which is highly specific for T-type Cav channels. Unlike the knock-out mice, there is no long-term compensation during acute experiments. There is also a higher areal specificity because TTA-P2 can be infused locally in the thalamus. The perfusion of TTA-P2 in slices fully suppressed the rebound burst firing of all thalamic cells, and thalamic infusion of TTA-P2 *in vivo* strongly diminished spindle waves, both in anesthetized and naturally sleeping animals (David et al., 2013). In addition, the same thalamic infusion also altered delta waves, markedly diminishing their frequency (David et al., 2013). These experiments provide converging evidence that thalamic T-type channels are directly involved in both spindle waves and delta oscillations.

The slowing-down of slow-waves following the infusion of thalamic TTA-P2 is consistent with the original findings of Grenier et al. (1998), who reported that thalamic neurons systematically fire a burst of action potentials before spiking in cortical neurons during the transition from Down to Up states. Thus, blocking or knocking out the T-type Cav channels underlying these bursts in the thalamus could account for the observed slow down of the oscillation frequency of slow oscillations. The results of the TTA-P2 experiments therefore confirm a cortical site for the genesis of slow oscillations, consistent with the observations of these oscillations in cortical slices (Sanchez and McCormick, 2000).

3.6 Discussion

In this chapter, we introduced several models of thalamic relay neurons with varying degrees of detail, from a single-compartment model to a multicompartment models that included the dendritic morphology of a reconstructed thalamic relay cell. Computational models have made a number of contributions to understanding the properties of thalamic relay neurons. We discuss here these contributions and the application of these models to investigate network behaviour.

3.6.1 The interplay of ionic currents in thalamic relay cells

3.6.1.1 Rebound burst generation by I_T

The rebound burst properties of TC cells were identified first by Andersen and Eccles (1962; Fig. 3.2) and later characterized *in vitro* by Llinás and Jahnsen (1982; Fig. 3.3). The same authors also showed that a low-threshold Ca^{2+} current I_T underlies rebound burst generation in TC cells (Jahnsen and Llinás, 1984b). The characterization of I_T by voltage-clamp methods (Coulter et al., 1989; Crunelli et al., 1989; Hernandez-Cruz and Pape, 1989; Suzuki and Rogawski, 1989; Huguenard and McCormick, 1992; Huguenard and Prince, 1992) showed that this current is unusual in that the voltage ranges of activation and inactivation are relatively hyperpolarized, in the range of -80 to -50 *mV* (Fig. 3.4). As a consequence, I_T could in principle be activated below the threshold for action potential generation in TC neurons, and therefore explain the genesis of the rebound burst.

Many models have been proposed for rebound burst generation in TC cells (Rose and Hindmarsh, 1985, 1989; Huguenard and McCormick, 1992; Lytton and Sejnowski, 1992; McCormick and Huguenard, 1992; Toth and Crunelli, 1992; Destexhe and Babloyantz, 1993; Destexhe et al., 1993a; Wang and Rinzel, 1993; Wang, 1994; Wallenstein, 1996; Destexhe et al., 1996a, 1998c). Similar models for thalamic reticular cells have also been examined (Destexhe and Babloyantz, 1992; Destexhe et al., 1993b; Wang and Rinzel, 1993; Destexhe et al., 1994a; Golomb et al., 1994; Wallenstein, 1994a; Destexhe et al., 1996a, 1996b). Despite their diversity, all of these models have concluded that the voltage-dependent properties of I_T, determined by voltage-clamp experiments, are sufficient to account for the rebound burst response observed in these cells (Fig. 3.5).

3.6.1.2 Oscillations from I_T/I_h interactions

In addition to having a prominent rebound-burst property, TC cells also have the intrinsic ability to generate sustained oscillations. Experiments performed in cats *in vivo* showed that TC cells can generate clock-like rhythmicities in the delta frequency range (0.5–4 *Hz*) after removal of the cortex (Curró Dossi et al., 1992). Oscillations in the same frequency range were also observed in TC cells *in vitro* (Fig. 3.6; McCormick and Pape, 1990a; Leresche et al., 1990, 1991). These low frequency oscillations, which consisted of rebound bursts recurring periodically, have also been called 'pacemaker oscillations' (Leresche et al., 1990, 1991). *In vitro* experiments have determined that intrinsic oscillations are dependent on the hyperpolarization-activated current I_h (Fig. 3.6; McCormick and Pape, 1990a; Soltesz et al., 1991). The hypothesis adopted here is that the interplay between I_T and I_h produce these oscillations.

This hypothesis has been investigated and confirmed by a number of modelling studies (Lytton and Sejnowski, 1992; McCormick and Huguenard, 1992; Toth and Crunelli, 1992; Destexhe and Babloyantz, 1993; Destexhe et al., 1993a; Wang, 1994; Lytton et al., 1996). The mechanism underlying the oscillation is illustrated in Fig. 3.11: the depolarization during the burst deactivates I_h relatively rapidly; the ensuing hyperpolarization slowly reactivates, resulting in a slow rise of membrane potential, until a new burst occurs, and the same cycle repeats.

Models have shown that the biophysical properties of these currents, such as their typical rise and decay kinetics and their voltage-dependence, are sufficient to reproduce the main characteristics of oscillations such as their low frequency and dependence on membrane voltage.

3.6.1.3 Waxing and waning oscillations: the Ca^{2+} regulation of I_h

In TC cells from cat studied in low-Mg^{2+} medium *in vitro*, a few cells displayed intrinsic waxing-and-waning oscillations that were resistant to tetrodotoxin, suggesting mechanisms intrinsic to the TC neuron. Furthermore, intrinsic slow delta-like oscillations and waxing-waning oscillations could be observed in the same TC cell by altering the h-current (Fig. 3.13; Soltesz et al., 1991). Increasing the amplitude of I_h by noradrenaline could transform these slow oscillations into waxing-and-waning oscillations; application of Cs^+, an I_h blocker, had the opposite effect (Soltesz et al., 1991). It was concluded that interactions between I_T and I_h were sufficient to produce intrinsic waxing-and-waning oscillations (Soltesz et al., 1991).

One explanation for the intrinsic waxing-and-waning oscillations in TC cells arises from experiments on the I_h current in heart cells (Fig. 3.14; Hagiwara and Irisawa, 1989). These experiments showed that I_h can be upregulated by intracellular Ca^{2+}, suggesting that a similar mechanism may also be present in TC neurons (McCormick, 1992; Toth and Crunelli, 1992; Destexhe et al., 1993a). In a model of I_h channels (Destexhe et al., 1993a) it was assumed that the direct binding of Ca^{2+} ions to the open form of the channel produced a positive shift of the activation curve by calcium (Fig. 3.15), as observed experimentally (Fig. 3.14; Hagiwara and Irisawa, 1989). The model generated waxing-and-waning oscillations and their coexistence with other oscillatory and resting states in TC cells (Fig. 3.16) (Fig. 3.13; Soltesz et al., 1991). This model therefore provided a biophysically plausible mechanism that confirmed the suggestion that slow delta and waxing-and-waning oscillations observed *in vitro* correspond to two different equilibria between I_T and I_h (Soltesz et al., 1991).

However, subsequent experiments showed that Ca^{2+} ions do not bind directly to I_h channels (DiFrancesco and Totora, 1991; Zaza et al., 1991; Budde et al., 1997; Luthi and McCormick, 1998), but may modulate the current only indirectly. A second model based on indirect regulation of I_h by intracellular Ca^{2+} (Fig. 3.19; Destexhe et al., 1996a) also produced oscillatory properties consistent with experiments (Fig. 3.20), including the prominent ADP identified in TC neurons (Fig. 3.21; Bal and McCormick, 1996).

Other models have been proposed to account for the waxing-and-waning oscillations in TC cells. A different calcium-dependent mechanism for I_h regulation has been proposed (Crunelli et al., 1993). Wallenstein (1996) has suggested self-regulation of I_h by adenosine released in an activity-dependent manner. However, the adenosine-dependence of I_h in TC cells (Pape, 1992) produces a progressive depolarization during the interspindle silent period, which is inconsistent with the progressive hyperpolarization observed experimentally (Leresche et al., 1991; Bal and McCormick, 1996).

Models for waxing and waning oscillations were also proposed based on the interaction between I_T, I_h and a slow potassium current I_{K2} (Destexhe and Babloyantz, 1993), and on the interactions between I_T, I_h and a Ca^{2+}-activated potassium current $I_{K[Ca]}$ (Destexhe and Babloyantz, 1993; Hindmarsh and Rose, 1994b). Although these models were based on currents identified in TC cells, and produced a progressive hyperpolarization during the inter-spindle period, the frequency exhibited was approximately 10 Hz, which is much higher than the 0.5–4 Hz frequency range observed experimentally (Leresche et al., 1991).

The recent demonstration that I_h is upregulated by photolysis of caged Ca^{2+} (Fig. 3.23; Luthi and McCormick, 1998) provides strong evidence that Ca^{2+} is on the pathway to modulation of I_h. Cyclic AMP was shown to be implicated in the modulation of I_h by intracellular Ca^{2+} (Luthi and McCormick, 1999), in agreement with the sensitivity of I_h to cyclic AMP (McCormick and Pape, 1990b; Akasu and Tokimasa, 1992; DiFrancesco and Mangoni, 1994). Since several types of adenylate cyclase are activated by intracellular Ca^{2+} (reviewed in MacNeil et al., 1985), they could provide the missing link between Ca^{2+} and I_h (Luthi and McCormick, 1999).

In conclusion, models suggest that interactions between I_T, I_h and intracellular calcium allow the coexistence of tonic firing, slow delta-like oscillations and waxing-and-waning oscillations in TC cells. The consequences of these single-cell properties at the network level are investigated in Chapters 6, 7 and 8.

3.6.2 Dendritic T-current in thalamic relay cells

In addition to possessing several types of intrinsic currents, TC cells also have voltage-dependent currents localized in their dendrites, as observed by calcium imaging (Munsch et al., 1997; Zhou et al., 1997), and more recently by direct measurements of channel activity in dendrites (Williams and Stuart, 1999, 2000). The dendritic localization of voltage-dependent currents may have important consequences for their electrophysiological behaviour. Investigating these consequences is a task for which computational models are particularly useful.

Models were tightly constrained by *in vitro* recordings of calcium currents in various preparations (with and without dendrites). These measurements were used to estimate the density of T-current in the dendrites of TC cells using the model. Not only was dendritic T-current needed to account for the data, which was not a surprising result in light of optical imaging evidence (Munsch et al., 1997; Zhou et al., 1997), but the density needed to reproduce electrophysiological data was approximately 4.5 to 7.6 times that of the somatic region. This surprising finding is summarized below, as well as its possible implications for the physiology of TC cells.

3.6.2.1 Dendritic T-current in TC cells

The T-current density in the dendrites of TC cells was estimated here by combining electrophysiological measurements with computational models: (a) The peak amplitude of T-current in dissociated TC cells was approximately 400 *pA*. As these cells have lost most of the dendritic tree, matching a dissociated-cell model to those recordings provides estimates of the perisomatic density of T-current. (b) The model showed that the peak T-current amplitude in intact TC cells was fairly large (2–8.3 *nA*), and simulations of a reconstructed TC cell showed that not only must the T-current be present in dendrites, but its dendritic density must be increased several fold in order to reproduce the experimental measurements. (c) I/V curves of intact TC cells under voltage-clamp were significantly shifted compared to those from dissociated TC cells. This phenomenon could be replicated in the model and was accounted for by the poor voltage control over dendritic T-channels. (d) The current-clamp behaviour of the reconstructed TC cell was compared to the traces obtained in the same cell during recording. These date highly constrained the model, as burst behaviour required a more narrow range of T-current densities in dendrites, approximately 4.5 to 7.6 times higher than in the soma. (e) Similar conclusions were reached using a simplified model with only three compartments, provided that dendrites had high densities of T-channels.

This combination of computational models with *in vitro* recordings points to the conclusion that TC cells have a somatodendritic distribution of T-channels with most channels concentrated in dendrites (> 11 μm from soma), similar to the conclusions reached for thalamic RE cells (Section 4.3 in Chapter 4; Destexhe et al., 1996b) as well as for hippocampal pyramidal cells (Karst et al., 1993).

3.6.2.2 Possible functional consequences of dendritic T-current in TC cells

One consequence of the dendritic T-current in TC cells is that it is likely to enhance the rebound burst responses following IPSPs. The dendrites of TC cells are densely covered by inhibitory synaptic terminals from RE cells (Jones, 1985), which evoke powerful IPSPs that can trigger rebound bursts (Steriade and Deschênes, 1984; Bal et al., 1995a, 1995b; Kim et al., 1997). A high density of T-current at the same site as these terminals would be most likely facilitate the genesis of rebound bursts to dendritic IPSPs. On the other hand, RE cells receive excitatory synaptic contacts from TC cells and have a more hyperpolarized resting state, where the T-current is deinactivated (Contreras et al., 1993; Bal et al., 1995b; Destexhe et al., 1996b; see Chapter 4). Because they may also have dendritic T-current (see Section 4.3 in Chapter 4; Destexhe et al., 1996b), RE cells seem well designed to produce bursts in response to EPSPs. Taken together, these data suggest that the interconnected TC-RE structure with dendritic T-currents makes a highly efficient and robust oscillator, in which cells very powerfully elicit bursts in each other. Oscillations based on TC-RE interactions will be considered in Chapter 6.

Another possible consequence of dendritic T-current in TC cells is to provide optimal conditions for the modulation of burst responses by corticothalamic synapses. The evidence for high densities of T-current in dendrites is paralleled by morphological data showing that TC cell's dendrites receive a considerable amount of corticothalamic synaptic terminals (Liu et al., 1995; Erisir et al., 1997a, 1997b), which are more numerous than terminals from sensory afferents (Jones, 1985). In addition, the dendrites of TC cells are also densely covered by inhibitory synapses (Jones, 1985; Liu et al., 1995). Therefore, the increase of background activity of excitatory and inhibitory synapses may counteract the genesis of the rebound burst if the T-channels are dendritic, because synaptic currents would act directly on T-channels at their precise site of localization, through shunting effects. These types of dendritic interactions were simulated in the model (Fig. 3.32) and similar interactions were also investigated in a model of dendritic T-current in RE cells (see Section 4.3 in Chapter 4; Destexhe et al., 1996b).

The TC model therefore suggests that the T-current can be modulated more effectively when it is distributed in dendrites (Fig. 3.32), leading to a finer and more graded control of bursts. One reason why most of the T-channels are found in the dendrites could be that bursts are optimally tuned by synaptic currents if the T-current is dendritic. This type of dendritic interaction could provide a fast switch between the burst mode (cortical synapses silent) and the tonic mode (sustained cortical drive), as suggested previously (Destexhe et al., 1996b, 1998c). In addition to conventional neuromodulatory mechanisms, which operate over time scales of hundreds of milliseconds (McCormick, 1992), these types of local dendritic interactions give corticothalamic feedback the ability to control the state of thalamic neurons on a time scale of a few milliseconds.

3.7 Summary

In summary, we presented here several different models for the electrophysiological behaviour of thalamic relay cells, ranging from morphologically simple (single compartment) to morphologically complex models (reconstructed neuron). The latter model showed that taking into account the properties of dendrites is necessary to explain the electrophysiological features of spike bursts in these cells. The electrophysiological consequences of dendritic T-current are only beginning to be explored and future studies should address the interplay of T-current with other voltage-dependent currents in dendrites, as data become available. Models should also address dendritic integration in thalamic cells, and more generally how synaptic currents interact with voltage-dependent currents in dendrites.

Single-compartment models are useful for investigating the dynamical interactions between several voltage-dependent currents and their regulation by intracellular calcium. Although they do not include dendrites, single-compartment models nevertheless display the salient bursting, oscillatory, and resting properties of TC cells. They should be useful for investigating simulations in which the qualitative properties dominate, such as in network simulations that include a large number of thalamic cells. Such network models will be investigated in Chapters 6–8.

In Chapter 4, a similar approach will be taken to thalamic reticular neurons. A single-compartment model will be used first to explain the bursting and oscillatory behaviour of these cells based on the interplay of ionic mechanisms, then a morphologically complex model will be used to investigate the presence of T-current in dendrites and how this shapes the electrophysiological properties of the cell.

Electrophysiological properties of thalamic reticular neurons

In this chapter, we focus on neurons from the thalamic reticular (RE) nucleus. These neurons share some electrophysiological properties with TC cells, such as a rebound burst, but there are also significant differences owing to the presence of a different calcium-dependent and other voltage-dependent currents. These properties are reviewed here and are used to constrain single-compartment models of RE cells. In the last section, a more detailed multicompartment model is considered that takes into account the important influence of dendrites and dendritic currents in shaping the behaviour of these cells.

4.1 The rebound burst of thalamic reticular cells

4.1.1 Experimental characterization of rebound responses in RE cells

In common with thalamic relay cells, RE neurons display two distinct modes of firing. Typical firing patterns of RE cells observed *in vivo* are illustrated in Fig. 4.1. An RE neuron recorded in an awake animal shows a tonic firing activity (upper trace in Fig. 4.1A). During natural slow-wave sleep, the activity of thalamic cells changes to more rhythmic firing with bursts of action potentials (Steriade et al., 1986). A typical burst of firing in an RE cell during natural sleep shows an *accelerando-decelerando* pattern of action potentials (Fig. 4.1A, lower). In intracellular recordings, it was possible to elicit both modes of firing in RE cells, depending on the membrane potential. A depolarizing current pulse from a level of $-68\ mV$ produced tonic firing, whereas the same pulse delivered at more hyperpolarized levels elicited a burst (Fig. 4.1B). In this cell, the burst showed a slowly rising phase, was broader than in TC cells, and there was always an *accelerando-decelerando* pattern of sodium spikes, typical of RE cells in unanesthetized, naturally sleeping animals, as well as in animals under different anesthetics (Domich et al., 1986; Mulle et al., 1986; Steriade et al., 1986; Contreras et al., 1993).

Thalamocortical Assemblies: Sleep Spindles, Slow Waves and Epileptic Discharges. Second Edition. Alain Destexhe and Terrence J. Sejnowski, Oxford University Press. © Oxford University Press 2023.
DOI: 10.1093/oso/9780198864998.003.0004

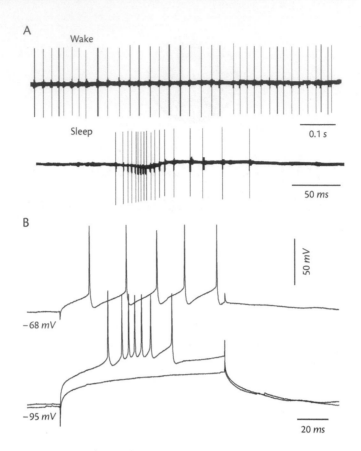

Fig. 4.1: Tonic and bursting firing in thalamic reticular neurons *in vivo*. A. RE neuron from the rostrolateral region of the RE nucleus was recorded extracellularly in an unanesthetized, chronically implanted cat. The cell discharged tonically during the waking state (Wake), with periods of sustained firing at frequencies between 20 and 60 *Hz*. During slow wave sleep (Sleep) the same neuron discharged bursts of action potentials that had a characteristic *accelerando-decelerando* pattern at the core of the burst, reaching frequencies of up to 400 *Hz*, followed by a tail of tonic discharge. This pattern is characteristic of RE cells *in vivo*. B. Intracellular recording of a RE neuron from the rostrolateral region in a urethane-anesthetized cat. At depolarized membrane potentials (−68 *mV*) the cell responded to depolarizing square current pulses with tonic discharges with frequencies proportional to the amplitude of injected current. After DC hyperpolarization to −95 *mV* the same current pulse gave rise to a passive response. Increasing the amount of injected current triggered bursts of action potentials that had a characteristic *accelerando-decelerando* pattern with a longer first interspike interval (see also Sleep burst in A) (modified from Destexhe et al., 1996b).

The ionic basis of the rebound burst in RE cells was investigated *in vitro* (Avanzini et al., 1989; Bal and McCormick, 1993). The rebound burst of RE cells, like that of TC cells, is mediated by a low-threshold Ca^{2+} current. However, the characterization of the T-type Ca^{2+} current in RE cells by voltage-clamp methods revealed marked differences with TC cells. In RE cells, the kinetics were slower and the activation was less steep and more depolarized than that found in TC cells (Huguenard and Prince, 1992). This current was named 'slow I_T', or I_{Ts}. Its characteristics are illustrated in Fig. 4.2.

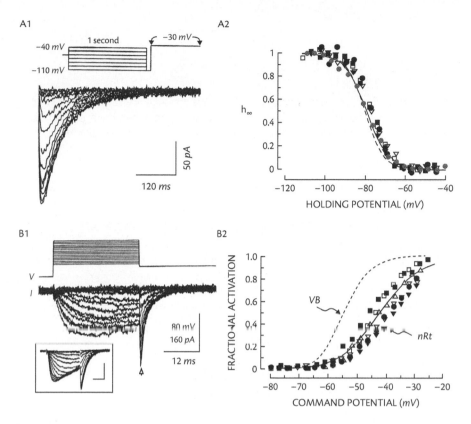

Fig. 4.2: Voltage-clamp characterization of the T-current in isolated RE neurons. A. Steady-state inactivation of I_{Ts}. A1. Voltage-clamp protocol consisting in stepping the voltage to various holding potentials, then clamping the cell to -30 mV. A2. Steady-state inactivation of I_{Ts}. The normalized current amplitude is plotted against holding voltage for six different RE cells (symbols). The lines show the best fits obtained with the Boltzmann function $(1/(1 + \exp[(V - V_h)/K_h])$; see text for values) for RE neurons (continuous line) and TC neurons (dashed line). B. Steady-state activation of I_{Ts}. B1. Voltage-clamp protocol consisted in applying various depolarizing holding currents, then stepping the membrane to -80 mV and measuring the amplitude of tail currents (arrow). The inset shows the same protocol applied to a TC cell. B2. Normalized tail current amplitude for five different RE cells (symbols). The best fit is shown by a continuous line for RE cells and is compared to the activation curve of TC cells (dashed line). All recordings were obtained in acutely dissociated cells from the ventrobasal thalamus of rats. Figure modified from Huguenard and Prince, 1992.

4.1.2 Model of the T-current and rebound burst in RE cells

Early models of RE cells (Destexhe and Babloyantz, 1992; Wang and Rinzel, 1993) used a T-current similar to that found in TC cells, based on the data available at that time (Llinás and Geijo-Barrientos, 1988; Avanzini et al., 1989; McCormick and Wang, 1991). More recent characterization of the kinetics of the T-current in RE cells (Huguenard and Prince, 1992) has led to a more realistic model of I_{Ts} in these cells (Destexhe et al., 1994a):

$$I_{Ts} = \bar{g}_{Ca}\ m^2 h\ (V - E_{Ca})$$

$$\dot{m} = -\frac{1}{\tau_m(V)}(m - m_\infty(V)) \tag{4.1}$$

$$\dot{h} = -\frac{1}{\tau_h(V)}(h - h_\infty(V))$$

where $\bar{g}_{Ca} = 1.75\ mS/cm^2$ is the maximum value of the conductance of the Ca^{2+} current, E_{Ca} is the Ca^{2+} reversal potential, given by the Nernst relation, and m and h are, respectively, the activation and inactivation variables. The activation function and time constants were:

$$m_\infty(V) = 1 / (1 + \exp[-(V + 52)/7.4])$$
$$\tau_m(V) = 0.44 + 0.15 / (\exp[(V + 27)/10] + \exp[-(V + 102)/15])$$
$$h_\infty(V) = 1 / (1 + \exp[(V + 80)/5])$$
$$\tau_h(V) = 22.7 + 0.27 / (\exp[(V + 48)/4] + \exp[-(V + 407)/50]).$$

These values correspond to a temperature of 36°C, assuming Q_{10} values of 5 and 3 (Coulter et al., 1989) and an extracellular Ca^{2+} concentration of 2 mM.

The low-threshold behaviour generated by this model is illustrated in Fig. 4.3. The three upper curves of Fig. 4.3 show a protocol similar to the experiment shown in Fig. 4.1B. Depolarizing

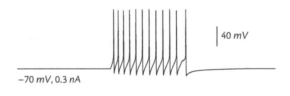

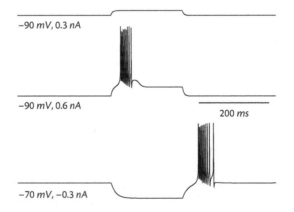

Fig. 4.3: Simulated burst responses of thalamic reticular neurons. The model contained a leakage current, the low-threshold Ca^{2+} current I_{Ts} and the fast Na^+/K^+ currents responsible for action potentials. The four curves show the result of injecting of a 200 ms current pulse at different membrane potentials. At −70 mV rest, injection of 0.3 nA current resulted in tonic firing. At −90 mV injection of the same current led to a passive response but increasing the current amplitude led to a burst of action potentials. A hyperpolarizing current injection starting from −70 mV (−0.3 nA) generated a rebound burst (bottom curve). The burst generated by I_{Ts} was broader and had a slower rising phase compared to bursts generated by I_T from TC cells.

current injection generated tonic firing from −70 mV rest, and generated burst firing at more negative membrane potentials (−90 mV). Hyperpolarizing current injection from −70 mV generated a rebound burst (Fig. 4.3, bottom).

4.2 Intrinsic oscillations in RE cells

4.2.1 Experimental characterization of repetitive bursting

In intracellular recordings from cat RE cells *in vivo* (Mulle et al., 1986; Contreras et al., 1993), rhythmic bursting activity at a frequency of 8-12 *Hz* occurred either spontaneously, or following internal capsule or thalamic stimulation (Fig. 4.4). Typically, a depolarizing envelope

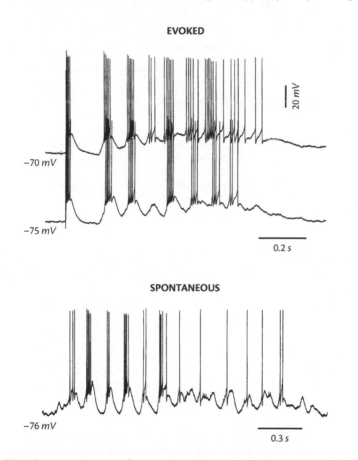

Fig. 4.4: 8-12 *Hz* bursting properties of cat reticular thalamic neurons *in vivo*. Sequences of rhythmic bursts appear in RE cells spontaneously or evoked by electrical stimulation of the internal capsule. During spindle oscillations RE cells displayed spikes bursts at 8-12 *Hz* riding on a depolarizing envelope. Top panel (EVOKED): two different responses to electrical stimulation of the internal capsule in an RE cell from the rostrolateral region. The cell was progressively hyperpolarized by d.c. current injection. At a more depolarized membrane potential, the cell displayed a tonic tail of spikes. Bottom panel (SPONTANEOUS): spontaneous spindle sequence in a different RE cell (from Destexhe et al., 1994a).

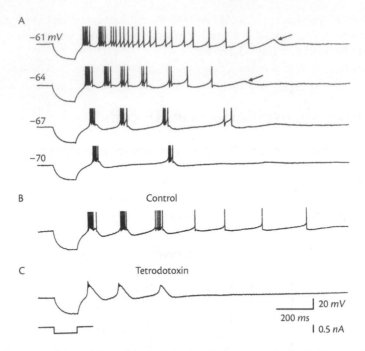

Fig. 4.5: Sequence of rhythmic burst firing in thalamic reticular neurons *in vitro*. A. Response to injection of a hyperpolarizing current pulse in a guinea-pig RE cell results in rhythmic bursts followed by a tonic tail of spikes (−61 *mV*). When the cell was held at more hyperpolarized levels (between −64 and −70 *mV*), the tonic tail disappeared and repetitive bursting slowed in frequency. B. Sequence of rhythmic bursts and tonic tail in another cell. C. Same cell as in B after application of TTX showing the rhythmic low-threshold spikes. Figure reproduced from Bal and McCormick, 1993.

accompanied this oscillatory behaviour and with a slow afterhyperpolarization (AHP) following each burst; the termination of the oscillatory sequence was followed by a tonic tail of spike activity (Domich et al., 1986).

These features were confirmed by intracellular recordings of RE cells *in vitro* (Avanzini et al., 1989; Bal and McCormick, 1993). A rebound sequence of rhythmic bursts could be elicited in RE cells following current injection (Fig. 4.5). In an *in vitro* study of rodent RE cells (Bal and McCormick, 1993) this rhythmic behaviour was resistant to tetrodotoxin and was, therefore, intrinsic to the cell. The same study also showed that blocking the AHP with apamin, which blocks a class of calcium-activated potassium current $I_{K[Ca]}$, abolished the rhythmic activity. Rhythmic oscillations in RE cells therefore seem to imply an interplay of the T-type Ca^{2+} current and $I_{K[Ca]}$.

The *in vitro* study (Bal and McCormick, 1993) also reported that the rhythmic oscillations at 7–12 *Hz* were often followed by a short tonic tail of spikes. Application of TTX revealed an afterdepolarization (ADP) mediated by a non-specific cation current, activated by intracellular calcium, called I_{CAN}. This current, encountered in many other cell types in the nervous system (Partridge and Swandulla, 1988), could underlie the tonic tail of spikes in RE cells (Bal and McCormick, 1993).

4.2.2 Models of oscillatory properties of RE cells

In RE cells, $I_{K[Ca]}$ did not show voltage-sensitivity (Bal and McCormick, 1993). Among the different classes of calcium-activated channels, the apamin-sensitive $I_{K[Ca]}$ is independent of voltage (for review, see Latorre et al., 1989; McManus, 1991). The model used for $I_{K[Ca]}$ was the one in Chapter 2 (Eq. 2.33):

$$I_{K[Ca]} = \bar{g}_{K[Ca]} \, m^2 \, (V - E_K) \tag{4.2}$$

where $\bar{g}_{K[Ca]} = 10 \, mS/cm^2$ is the maximal conductance, $E_K = -95 \, mV$ is the reversal potential, m is the fraction of open channels, and $[Ca]_i$ is the intracellular calcium concentration. The rate constants were assumed to be independent of voltage such that the activation of $I_{K[Ca]}$ depended solely on intracellular Ca^{2+}. The values $n = 2$, $\alpha = 48 \, ms^{-1} \, mM^{-2}$ and $\beta = 0.03 \, ms^{-1}$ yielded AHPs similar to those of RE cells recorded *in vivo* and *in vitro*.

The non-specific cation current I_{CAN} is also independent of voltage (Partridge and Swandulla 1988). Because data on the voltage-dependence of I_{CAN} in RE neurons are not available, we used the activation scheme above for $I_{K[Ca]}$, but with slower kinetics:

$$I_{CAN} = \bar{g}_{CAN} \, m^2 \, (V - E_{CAN}) \tag{4.3}$$

where $\bar{g}_{CAN} = 0.25 \, mS/cm^2$ is the maximal conductance and $E_{CAN} = -20 \, mV$ is the reversal potential. The values $n = 2$, $\alpha = 20 \, ms^{-1} \, mM^{-2}$ and $\beta = 0.002 \, ms^{-1}$ yielded ADPs similar to those recorded in RE cells *in vitro* (see below).

In modelling studies, the interaction between the T-current and $I_{K[Ca]}$ has been shown to robustly generate oscillations at low frequency (2–4 Hz) (Destexhe et al., 1993b, 1994a; Wang and Rinzel, 1993), as shown in Fig 4.6A. These oscillations could be elicited as a rebound rhythmic bursting activity in response to a hyperpolarizing pulse. The 2–4 Hz frequency was mainly dependent on the level of the resting potential and on the kinetics of I_{Ts} and $I_{K[Ca]}$. Despite the ease with which these low-frequency oscillations could be generated, none of the kinetic parameters tested for $I_{K[Ca]}$ were able to produce frequencies in the range 7–14 Hz.

The mechanism underlying oscillations in RE cells was similar to that suggested by current-clamp experiments (Avanzini et al., 1989). In the model, the interplay of I_{Ts} and $I_{K[Ca]}$ produced rhythmic oscillations in a way that was similar to the analysis of Chapter 2 (Fig. 2.11). This type of rhythmical activity is illustrated in Fig. 4.7: following an LTS, Ca^{2+} enters and activates $I_{K[Ca]}$, which then hyperpolarizes the membrane and deinactivates I_{Ts}. When the membrane depolarizes due to the deactivation of $I_{K[Ca]}$, a new LTS is produced and the cycle repeats. The robustness of this mechanism was confirmed in several modelling studies (Destexhe et al., 1993b; Wang and Rinzel, 1993; Destexhe et al., 1994a; Hindmarsh and Rose, 1994a; Wallenstein, 1994a).

The outward current I_{CAN}, present in RE cells (Bal and McCormick, 1993) produced a marked afterdepolarization (ADP) after application of TTX and apamin. In the model, such an ADP can be produced in the presence of I_{Ts} and I_{CAN} as a rebound following a hyperpolarizing pulse (Fig. 4.6B). The kinetic scheme used for describing I_{CAN} was not voltage-dependent (see above).

Simulations of a single compartment containing the combination of currents I_{Ts}, $I_{K[Ca]}$ and I_{CAN} are shown in Fig. 4.6C–D. The model cell produced a rebound bursting activity at a

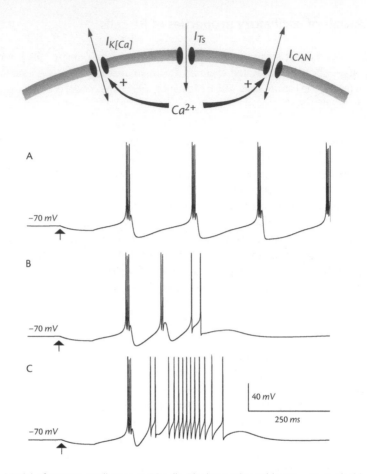

Fig. 4.6: Model of intrinsic oscillations in RE cells. Rhythmic rebound bursts were evoked by injection of a hyperpolarizing pulse (arrow). A. Rhythmic bursting following the interaction between I_{Ts} and $I_{K[Ca]}$ ($\bar{g}_{CAN} = 0$). B. ADP in the model cell with only I_{Ts} and I_{CAN} present ($\bar{g}_{Na} = 0$, $\bar{g}_{K[Ca]} = 0$, $\bar{g}_{CAN} = 0.25\ mS/cm^2$). C. Same simulation in the presence of all currents, with $I_{K[Ca]}$ dominating ($\bar{g}_{CAN} = 0.25\ mS/cm^2$). The cell produced a few bursts at higher frequency and terminated with a short tonic tail of spikes. D. Same simulation as in C, but with I_{CAN} dominating ($\bar{g}_{CAN} = 2\ mS/cm^2$). The tonic tail activity is more prominent. All hyperpolarizing pulses were 2 pA (0.2 $\mu A/cm^2$) in amplitude and 100 ms in duration. Same calibration for A, C, D; for all traces, the starting voltage was −70 mV (modified from Destexhe et al., 1994a).

frequency of 9–11 *Hz*. The activation of I_{CAN} accelerated the rising phase of the burst and increased the frequency of the rebound burst sequence. The presence of I_{CAN} also terminated the oscillatory behaviour by producing a tonic tail of spikes before the membrane returned to its resting level. Varying $\bar{g}_{K[Ca]}$ and $\bar{g}_{CAN}$ modulated both the frequency and the relative importance of rhythmic bursting relative to tonic tail activity (compare Fig. 4.6C and D). These results were confirmed in another modelling study (Wallenstein, 1994a).

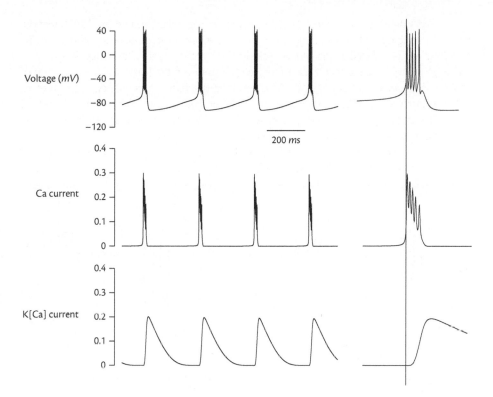

Fig. 4.7: Rhythmic oscillations in thalamic reticular cells through an interplay of I_{Ts} and $I_{K[Ca]}$. Model based on a single compartment RE cell containing I_{Ts}, $I_{K[Ca]}$, I_{CAN}, submembranal Ca^{2+} and fast Na^+/K^+ currents for generating action potentials. Oscillations were generated from the interaction between the calcium current I_{Ts} and the Ca^{2+}-dependent K^+ current $I_{K[Ca]}$ (similar to Fig. 2.11). The membrane potential (top) is represented together with the fraction of activated conductance (g/g_{max}) for I_{Ts} and $I_{K[Ca]}$. The right panel shows a burst at five times higher temporal resolution. $\bar{g}_{K[Ca]} = 1.2\ mS/cm^2$; see text for other parameters.

4.3 Dendritic T-current in thalamic reticular cells

A number of experimental studies (reviewed below) suggest that the dendrites of RE cells participate in burst generation. Moreover, a number of electrophysiological features of RE cells (also reviewed below) are fundamentally different comparing *in vivo* and *in vitro* recordings. These differences were explored in a study (Destexhe et al., 1996b) that combined *in vivo* and *in vitro* recordings with computational models to investigate the possible role of dendrites in generating bursts of action potentials in RE cells.

4.3.1 Experimental evidence for dendritic calcium currents

The bursts observed in RE cells differ from those in TC cells in several ways. First, RE cells typically produce broad bursts that develop more slowly than in TC cells (Mulle et al., 1986; Steriade et al., 1986; Llinás and Geijo-Barrientos, 1988). This is reflected in the slower activation and inactivation kinetics of the T-current in these neurons (Huguenard and Prince, 1992). Second, relatively strong current pulses need to be injected to evoke bursts in RE cells compared to TC cells (Mulle et al., 1986; Bal and McCormick, 1993; Contreras et al., 1993), suggesting that I_{Ts} might be located at a site that is electrotonically distant from the soma. Third, the pattern of sodium spikes within a burst typically increases then decreases in frequency (Domich et al., 1986; Avanzini et al., 1989; Contreras et al., 1993; Huguenard and Prince, 1994; Bal et al., 1995). This *accelerando-decelerando* pattern is routinely used as a criterion to identify RE cells in extracellular recordings (Steriade et al., 1986).

Perhaps the most distinctive feature of RE cell bursts is that they can be evoked in a graded fashion *in vivo* (Contreras et al., 1993). In intracellular recordings from three RE cells (Fig. 4.8), small amplitude current pulses elicited a passive response, whereas strong current pulses gave

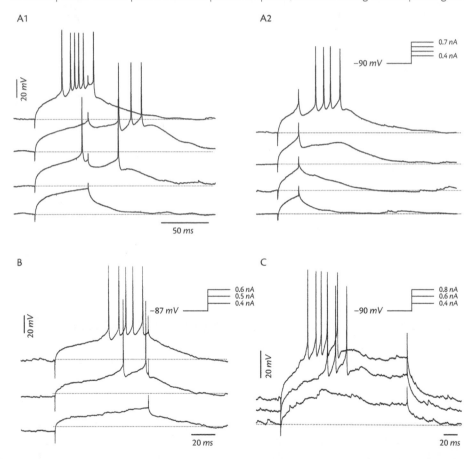

Fig. 4.8: Burst discharges may occur in a graded fashion in RE cells *in vivo*. Three different cells from the rostrolateral sector of the cat RE nucleus were recorded *in vivo* under urethane anesthesia. A. Injection of square current pulses of identical length and increasing amplitudes. Injection of long duration pulses (A1) revealed gradually stronger bursts with shorter latencies; pulses of shorter duration (A2) evoked a burst of spikes proportional to the pulse intensity. In both cases, there was a hyperpolarizing DC current bringing the membrane potential to −90 mV. B. In a different cell, depolarizing current pulses of increasing amplitudes applied at −87 mV elicited a burst with an increasing number of spikes. C. The same progressive increase in the number of spikes in a burst shown in a different cell. The pulse protocols and the membrane potential are indicated in the inset (from Destexhe et al., 1996b).

rise to a fully developed burst, as in TC cells. However, pulses of intermediate amplitude elicited active, embryonic bursts that did not develop into a full burst response, unlike the characteristic all-or-none behaviour observed in TC cells.

In RE cells studied *in vitro*, the burst also showed a slowly rising phase, a relatively broad structure, and an *accelerando-decelerando* pattern of sodium spikes, as described in previous studies (Avanzini et al., 1989; Huguenard and Prince, 1994; Bal et al., 1995). However, an important difference with *in vivo* recordings is the all-or-none nature of the burst *in vitro*. In rat thalamic slices, the burst has a clear threshold for activation and its structure is relatively stereotyped (Fig. 4.9). Both extracellular stimulation (Fig. 4.9A) and intracellular injection of

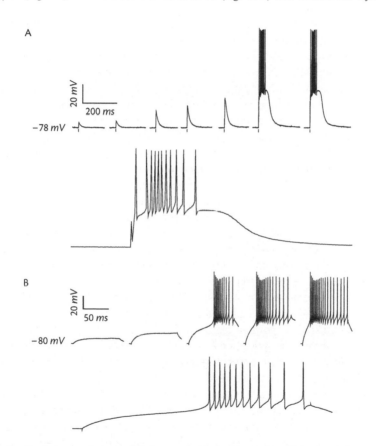

Fig. 4.9: Bursting behaviour in intracellularly recorded thalamic reticular cells *in vitro*. A. Bursts obtained from a rat RE cell following extracellular stimulation in a slice from the ventrobasal region of the RE nucleus. Successive stimuli (indicated by bars) were of 100 μA intensity and 40, 60, 80, 100, 150, 200 and 250 μs duration, respectively (from left to right). The evoked burst had a threshold that was all-or-none. The bottom trace indicates a burst on a ten times faster time scale. B. Burst obtained following intracellular injection of depolarizing current pulses. The amplitudes of the pulses were of 100, 200, 300, 400 and 500 pA, respectively (from left to right). The bursts evoked for the strongest stimuli had a clear threshold for activation, as in thalamocortical cells (Jahnsen and Llinás, 1984); they differed only in the number of spikes produced after the burst. The bottom trace shows the first burst on a five times faster time scale. In all cases the bursts of RE cells had a slow rising phase and sodium spikes within a burst showed a typical *accelerando-decelerando* pattern. All recordings were at 34°C (modified from Destexhe et al., 1996b).

current pulses (Fig. 4.9B) elicit stereotyped bursts with a slow rising phase, an *accelerando-decelerando* pattern of sodium spikes, and a sharp threshold.

These comparisons reveal important similarities as well as clear differences in the electrophysiological properties of RE cells in different preparations, and also differences between RE and TC cells. The computational model, presented below, was designed to investigate whether these properties could be accounted for by dendritic currents (Destexhe et al., 1996b).

4.3.2 Model of dendritic calcium currents in thalamic reticular cells

An intact RE neuron was recorded from the reticular region of the ventrobasal thalamus in rat and stained with biocytin. The cell, shown in Fig. 1 of Huguenard and Prince (1992), was reconstructed from serial sections of 80 μm, using a computerized tracing system (Eutectic Electronics, Raleigh, NC). Using a ×100 objective, and correcting for tissue shrinkage, the theoretical accuracy with which dendritic diameters could be measured was 0.1 μm. The morphology of the neuron is shown in Fig. 4.10A (see also stereoscopic view in Fig. 4.11). There were four primary dendrites, having a total length of 3,785 μm; the total membrane area of the cell was 15,115.5 μm^2, including 1,760 μm^2 for the soma which was approximately 20 to 25 μm diameter. The dendritic arborizations tended to spread in planes parallel to the long axis of the nucleus, as described previously by Ramón y Cajal (1909).

In acutely dissociated cells, most of the dendritic arborizations are removed by the dissociation procedure (see Fig. 2 of Huguenard and Prince, 1992). Acutely dissociated RE cells were simulated using a cable geometry obtained by truncating the dendrites of the original reconstructed cell (Fig. 4.10B). The model shown in Fig. 4.10B had only the soma and the most proximal parts of dendrites (eight compartments), with a total membrane area of 3,639 μm^2. The input capacitance of this model was similar to, but slightly greater than, that measured from dissociated RE cells (Huguenard and Prince, 1992).

Voltage-clamp recordings were obtained in the cell shown in Fig. 4.10A. These recordings were used to estimate the passive parameters by fitting the model to the data (Fig. 4.12A). It has been shown that when the cable geometry of the cell is known, this procedure should lead to a unique set of parameter values if they are uniform (Rall et al., 1992), which was the case here. A direct fit to current-clamp recording was not attempted since the model included only a subset of the currents present in these cells.

Voltage-dependent conductances were modelled using a Hodgkin-Huxley type of kinetic model (Hodgkin and Huxley, 1952). The Na$^+$ and K$^+$ currents responsible for fast action potentials were inserted in the soma and their kinetics were taken from a model of hippocampal pyramidal cells (Traub and Miles, 1991), assuming a resting potential of -67 mV, maximal conductances of $\bar{g}_{Na} = 100$ mS/cm^2 and $\bar{g}_K = 80$ mS/cm^2, and reversal potentials of $E_{Na} = 50$ mV and $E_K = -100$ mV.

The kinetics of activation and inactivation of I_{Ts} in RE cells, as well as the activation curves of I_{Ts}, obtained from voltage-clamp recordings in acutely dissociated RE cells (Huguenard and Prince, 1992), were the same as described above (Eq. 4.1), with the exception that the voltage-dependent time constants were:

$$\tau_m(V) = 1 + 0.33 / (\exp[(V+27)/10] + \exp[-(V+102)/15]) \qquad (4.4)$$

$$\tau_h(V) = 28.3 + 0.33 / (\exp[(V+48)/4] + \exp[-(V+407)/50])$$

A

Intact cell model

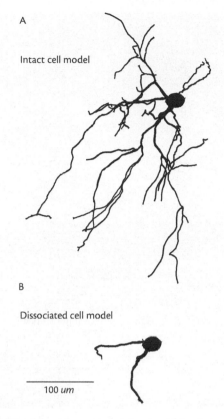

B

Dissociated cell model

100 *um*

Fig. 4.10: Reconstructed RE cell from the ventrobasal region of the thalamic reticular nucleus of the rat. A. The cell was stained with biocytin during recordings in a slice preparation (Huguenard and Prince, 1992); reconstructed using a computerized tracing system and incorporated into the NEURON simulation program. Models with either 80 or 230 compartments were used. B. Model of a dissociated RE cell, obtained by removing most of the dendrites from the intact cell (modified from Destexhe et al., 1996b).

as obtained from fitting the time constants measured from the same experiments (see below). These values correspond to a temperature of 36°C, assuming Q_{10} values of 2.5 and an extracellular Ca^{2+} concentration of 2 *mM*. Different densities of I_{Ts} were used for the soma and the dendrites.

Intracellular calcium handling was modelled by a first-order system representing Ca^{2+} pumps and buffers, as in Eq. 2.29 of Chapter 2, with a time constant of decay of Ca^{2+} of 5 *ms*. The free intracellular Ca^{2+} concentration at equilibrium was 240 *nM* and the extracellular Ca^{2+} concentration was 2 *mM*, corresponding to a reversal potential of approximately +120 *mV*. As the intracellular Ca^{2+} concentration varied, the reversal potential was calculated using the Nernst relation. This assumption was compared with constant field equations (see Section 2.2.2 in Chapter 2), which provides a better model for the nonlinear behaviour of Ca^{2+} currents (Hille, 1992). Simulations of either voltage-clamp or current-clamp experiments did not reveal any significant difference, even with high densities of Ca^{2+} channels, suggesting that details of Ca^{2+} handling had minimal effect on this model.

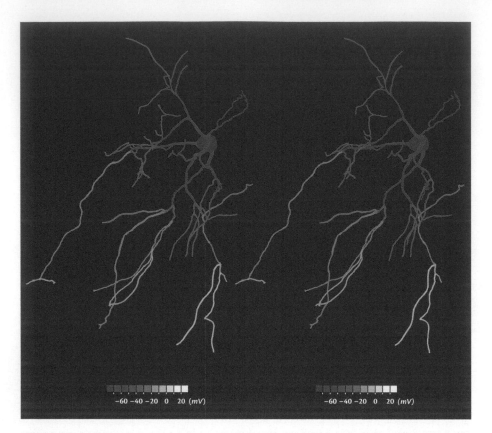

Fig. 4.11: Stereoscopic view of the reconstructed RE cell. This three-dimensional view shows that the dendrites tend to lie in three parallel planes. The distribution of membrane potential is shown by colours during a burst (colour code below; bar is 100 μm). Yellow indicates that the distal dendrites participate actively in burst generation (see text and Fig. 4.15).

4.3.3 Passive properties

Short voltage pulses were used to determine the passive properties of the cell in voltage-clamp mode. Fig. 4.12A shows the recordings of capacitive transients in the same cell that was filled with biocytin and reconstructed (Fig. 4.10A). The current transients were multiexponential, and were well fit using two exponentials:

$$I(t) = A_1 \exp(-t/\tau_1) + A_2 \exp(-t/\tau_2) \qquad (4.5)$$

where $A_1 = -11\ pA$, $A_2 = -57\ pA$, $\tau_1 = 12.9\ ms$ and $\tau_2 = 4.3\ ms$. Using Rall's expression for the electrotonic length in voltage-clamp mode (Rall, 1969, 1995):

$$L = \frac{\pi}{2} \sqrt{\frac{9\tau_2 - \tau_1}{\tau_1 - \tau_2}}, \qquad (4.6)$$

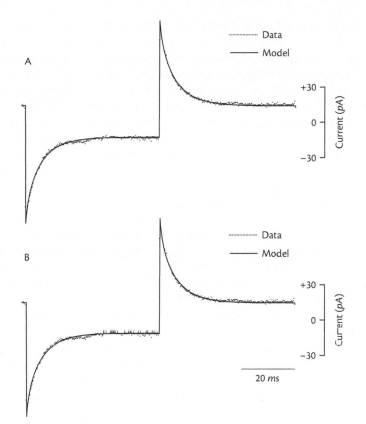

Fig. 4.12: Passive properties of thalamic reticular cells. A. Parameter fitting of the multicompartment model to voltage-clamp recordings from the same cell using the simplex method. Dots: −5 mV voltage-clamp step during 50 ms from a holding potential of −80 mV (average of eight traces). Continuous line: best simulation obtained after 450 iterations of the simplex method. On each iteration, the same voltage-clamp protocol was simulated in the model cell and the least square error was computed between the simulation and the data. The procedure was repeated until the model converged to a stable set of values of the parameters, from different initial conditions. Models with 80 and 230 compartment gave identical results. Best parameters: membrane capacitance $1.01 \pm 0.01\ \mu F/cm^2$, axial resistivity ($R_a$) $260 \pm 30\ \Omega cm$, leak conductance $0.05 \pm 0.0001\ mS/cm^2$ and resting potential $−82.844 \pm 0.002\ mV$ (mean values and standard deviation obtained from different initial conditions; ranges of values tested were 0.2 to 2 $\mu F/cm^2$, 50 to 500 Ωcm, 0.02 to 0.2 mS/cm^2 and −80 to −85 mV, respectively). B. Parameter fitting of a simplified multicompartment model to voltage-clamp recordings of a RE cell using the simplex method. In this case, the same parameters were used as in A, and only the dendritic correction factor, C_d, was fit. The optimal value obtained was $C_d = 3.69 \pm 0.01$ (range tested 1.5 to 5). The least square error was slightly higher (<1%) than that obtained using the multicompartment model (modified from Destexhe et al., 1996b).

leads to an electrotonic length of $L = 2.7$ for this neuron. This high value is consistent with the dendritic structure of this cell and the lack of control over the dendrites (see below).

The passive values measured from these capacitive transients were $R_{in} = 170\ M\Omega$ and $C = 69\ pF$. The capacitance was measured by dividing the integrated charge by the voltage step. This resulted in a specific capacitance of approximately 0.5 $\mu F/cm^2$, which is low compared with the standard value of 1 $\mu F/cm^2$ (see Rall et al., 1992). The strong non-isopotentiality of the cell (estimated $L > 2.5$) together with a high series resistance could explain this discrepancy.

In current-clamp, the time course of voltage transients in response to small current pulses was monoexponential with a membrane time constant of $\tau_m = 23 \pm 1$ ms. Based on the estimate of L and time constants of capacitive current transients in voltage-clamp (see above), the predicted value of the membrane time constant was $\tau_m = 17.3$ ms (see Rall, 1995). If this neuron is representative, then the disagreement between these two values indicates that Rall's equivalent cylinder model cannot be used to accurately describe the passive properties RE neurons. In agreement with this, measurements of diameters at branching points gave values which significantly depart from the 3/2 power rule necessary for Rall's equivalent cylinder model (A. Destexhe, unpublished data).

Passive parameters were also obtained by fitting the simulations of the multicompartment model directly to the recordings obtained in the same cell. The results are shown in Fig. 4.12A. From several initial conditions, the model always converged towards similar values, suggesting that only a single set of parameters can match this recording using the geometry of the cell. The input resistance (R_{in}) values obtained from the optimal passive parameters were 141 to 146 $M\Omega$ with axial resistivities between 200 and 300 Ωcm; the membrane time constant was 20 ms and the total capacitance (C) was approximately 151 pF.

The axial resistivity obtained here was between 200 and 300 Ωcm. This value is high compared to the values approximately 70 to 100 Ωcm measured in mammalian neurons (Barrett and Crill, 1974; Stuart and Spruston, 1998). However, the latter is probably an underestimate (see Rall et al., 1992) and the estimated value of R_a in the RE cell is consistent with the 100 to 300 Ωcm range obtained by matching detailed models to experimental data in other cell types (Stratford et al., 1989; Cauller and Connors, 1992; Major et al., 1994; Rapp et al., 1994; but see Stuart and Spruston, 1998).

The series resistance was simulated by including a resistance between the cell and the voltage source (voltage-clamp amplifier), and was included as one of the parameters of the fitting procedure. The best fit shown in Fig. 4.12A could only be obtained using rather high series resistance (20 to 50 $M\Omega$), consistent with the value measured during the recording. The model fit the non-isopotentiality of the cell as well. The values obtained using this procedure are self-consistent, and were also in agreement with the passive parameters obtained from RE cells in other preparations (Avanzini et al., 1989; Huguenard and Prince, 1992; Bal and McCormick, 1993; Huguenard and Prince, 1994; Warren et al., 1994). These values were used as the basis for the passive properties of the RE cell in the models presented here.

4.3.4 Localization of the T-current

The voltage-dependent rate constants and activation function used for modelling I_{Ts} were obtained from voltage-clamp experiments on acutely dissociated RE cells (Huguenard and Prince, 1992). The current was best described by a Hodgkin-Huxley formalism with two activation gates and one inactivation gate, for which the voltage-dependent parameters were obtained by fitting exponential expressions to the values determined experimentally (see below).

In acutely dissociated cells, the total I_{Ts} current showed a low peak amplitude, approximately 130 pA (Fig. 4.13A; Huguenard and Prince, 1992). Simulations of steady-state activation and

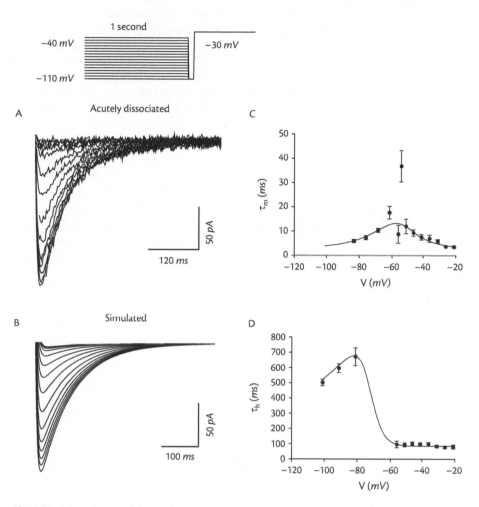

Fig. 4.13: Measurement of the steady-state inactivation parameters for I_{Ts} with voltage-clamp recordings. The voltage protocol is depicted at the top: the holding potential was 1 second, followed by a 10 ms hyperpolarization to -110 mV that preceded the depolarizing command to -30 mV. A. Inactivation protocol for the low threshold Ca^{2+} current I_{Ts} in an acutely dissociated rat RE cell (from Huguenard and Prince, 1992). The currents shown were obtained with depolarizations to -30 mV from different holding potentials. B. Simulations of the same inactivation protocols using the cell shown in Fig. 4.10B, and the density of I_{Ts} was estimated as 0.045 mS/cm^{2}; kinetics from Destexhe et al. (1994a). I_{Ts} was located uniformly in the cell and the density of channels was adjusted to match the amplitude of the currents recorded in dissociated cells. C. Fit of the voltage-dependence of time constants of activation (τ_m). D. Fit of the inactivation time constant (τ_h). For both C and D, the data were from acutely dissociated RE cells (circles are mean values and vertical bars are standard deviation from the mean) at 24°C (Huguenard and Prince, 1992). Continuous lines show the best fit obtained using exponential expressions (modified from Destexhe et al., 1996b).

inactivation protocols captured the main features of I_{Ts} in RE cells, namely (a) a relatively slow inactivation of this current compared to TC cells; (b) a nearly voltage-independent rate of inactivation; (c) an activation voltage range that was more depolarized than TC cells. Simulated steady-state inactivation of I_{Ts} is shown in Fig. 4.13B. The time constants for activation (τ_m) and inactivation (τ_h) of I_{Ts} are shown in Fig. 4.13C–D (see also Fig. 6C–D of Huguenard and Prince, 1992). The best fit of these values assumed a voltage-dependence described by bi-exponential expressions (Eq. 4.5 above) and are shown in Fig. 4.13C–D (see also Destexhe et al., 1994a).

No change in these kinetics parameters were needed to match the voltage-clamp data from dissociated RE cells using the dissociated cell model of Fig. 4.10B. The only free parameter was the density of T-channels, assuming a uniform density. The best fit was obtained with 0.045 mS/cm^2 (Fig. 4.13), independently of the particular details of the morphology chosen to represent dissociated RE cells. However, if I_{Ts} were located only in the soma using the geometry of Fig. 4.10B, the density needed to match the same recordings was 0.1 mS/cm^2.

The conductance density of I_{Ts} estimated from the dissociated cell model was then introduced in the intact cell model. Neither a uniform density of 0.045 mS/cm^2 (Fig. 4.14, left panel) nor a somatic density of 0.1 mS/cm^2 produced bursts in current-clamp mode (although they produced a total current of approximately 500 pA). The density of I_{Ts} had to be about an order of magnitude higher than these values to elicit bursts: the threshold density for burst generation was of approximately 0.3 mS/cm^2 for a uniform density, and approximately 3 mS/cm^2 for a somatic density (no I_{Ts} in the dendrites). This was the case for bursts generated using either depolarizing or hyperpolarizing current pulses. However, these densities, as well as the current amplitude in voltage-clamp, are inconsistent with the recordings from acutely dissociated cells.

A possible way to account for the differences between current-clamp in intact cells and voltage-clamp in dissociated cells is to assume a high density of I_{Ts} in the distal dendrites. A uniform density of 0.045 mS/cm^2 was considered in the soma and proximal dendrites, consistent with dissociated cells, and a higher density in distal dendrites (drawn in black in Fig. 4.14, right panel). Under these conditions, the minimal distal density for burst generation was of approximately 0.5 mS/cm^2. With distally located I_{Ts} (0.5 to 1 mS/cm^2), and in the absence of I_{Na}, the amplitude of the low-threshold spike (8 to 20 mV; not shown) was consistent with the amplitudes seen experimentally in RE neurons following application of TTX (Avanzini et al., 1989; Bal et al., 1993).

4.3.5 Properties of dendritically generated bursts

A high density of T-current in distal dendrites may have an important impact on the properties of burst generation in RE cells. The properties of these bursts were studied assuming the same I_{Ts} density as in Fig. 4.14 (left panel). Following a brief pulse of hyperpolarizing current, a broad calcium spike appeared in distal dendrites (Fig. 4.15), which then elicited sodium spikes in the soma. During the burst, as shown by the successive snapshots and dendritic potential profiles in Fig. 4.15, the membrane potential remained high in the distal part of the dendritic tree, 'feeding' the soma with current. The sodium spikes were generated in the soma and spread only into proximal dendritic segments. Due to the low-pass filtering properties of the dendritic tree, the slow calcium spike had a strong effect in the soma, whereas the backward propagation of the

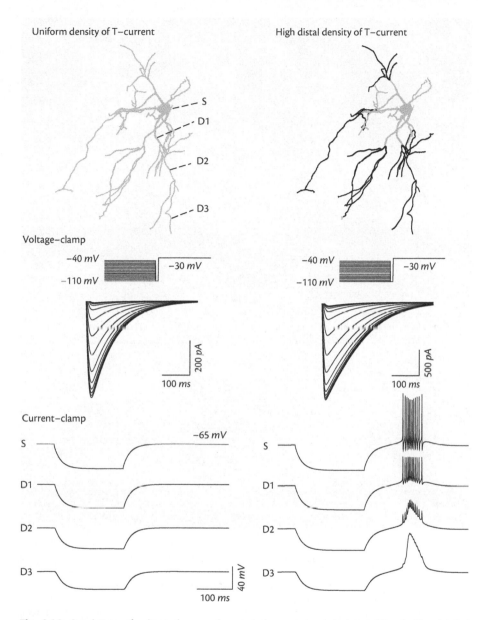

Fig. 4.14: Simulations of voltage-clamp and current-clamp protocols in intact RE cells. The detailed morphology of the RE cell was used with different densities of I_{Ts}. S, D1, D2 and D3 indicate dendritic locations of recording sites depicted below. Left: with a uniform density of 0.045 mS/cm^2, no low-threshold burst could be elicited from current injection in the soma (0.3 nA during 200 ms). Right: with a higher density of 0.6 mS/cm^2 in distal dendrites (shown in black), bursting behaviour could be generated in current-clamp. The slow rise of the burst and the *accelerando-decelerando* pattern of spikes were most prominent with dendritic I_{Ts}. In both cases shown here, the voltage-clamp behaviour was consistent with the current amplitudes in Fig. 4.13 if the dendrites were removed (modified from Destexhe et al., 1996b).

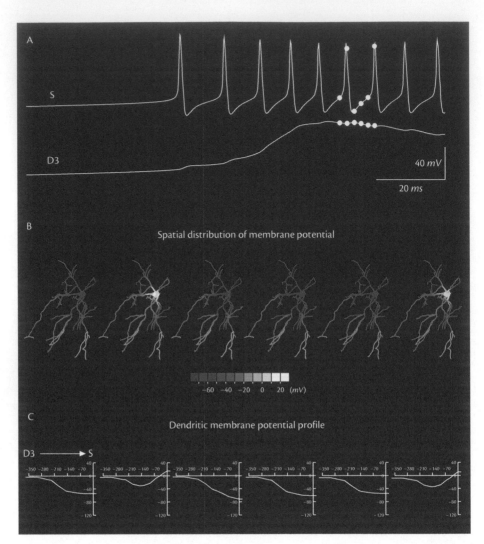

Fig. 4.15: Distribution of the membrane potential during dendritically-generated bursts. A. Expansion of the somatic and dendritic membrane potential during a burst (same simulation as in Fig. 4.14, lower right). B. Snapshots of electrical activity in the cell. The distribution of membrane potential is indicated by false colours, as shown in the colour scale. The six successive frames correspond to the six circles indicated in A. C. Profile of the membrane potential along a path from distal dendrite to the soma (from D3 to S in Fig. 4.14). The abscissa shows the distance from soma (in μm) and the ordinate shows the membrane potential along this path (in mV) (modified from Destexhe et al., 1996b).

higher frequency sodium spikes was limited. Dendritic sodium currents, not included here, are likely to affect the dendritic propagation of action potentials (Huguenard et al., 1989; Stuart and Sakmann, 1994; Mainen et al., 1995).

With distal I_{Ts}, the simulated bursts showed *accelerando-decelerando* patterns of sodium spikes, similar to bursts patterns seen in RE cells (Figs. 4.14–4.15). The same patterns also persisted after shifting the voltage-dependence of I_{Na} and I_K currents towards more depolarized or hyperpolarized levels (not shown), suggesting that these currents were not responsible for

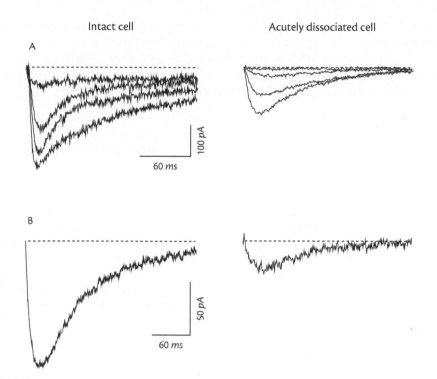

Intact cell   Acutely dissociated cell

A

100 pA

60 ms

B

50 pA

60 ms

Fig. 4.16: Comparison of voltage-clamp recordings of I_{Ts} in intact and acutely dissociated RE cells *in vitro*. A. Activation protocols for I_{Ts} in an intact cell (left) compared to an acutely dissociated cell (right). The command membrane potentials were −60, −50, −40 and −30 mV (left) and −50, −40, −30 and −20 mV (right); both were at 23°C; details of the protocols are given in Huguenard and Prince (1992). For the cell on the left, the amplitude of I_{Ts} was exceptionally low but was, however, still higher than that for dissociated cells (see also Fig. 4.13A); the time course of inactivation was slower in the intact cell. B. Voltage-clamp steps in an intact and a dissociated cell. In the intact cell (left), the holding potential was −71 mV and the command potential was −46 mV (34°C); in the dissociated cell, they were −110 mV and −40 mV, respectively (23°C). The amplitude was higher and the time course was slower in the intact cell. A was modified from Destexhe et al., 1996b; B was modified from Huguenard and Prince, 1994a.

the *accelerando-decelerando*, but that the slow rising and decaying phases of the burst were the determining factors.

The presence of I_{Ts} in the dendrites is further supported by a series of additional observations. First, recordings of intact RE cells in slices show a much higher peak amplitude of the current than in acutely dissociated cells (see Fig. 4.16). Although most of the T-channels would be inactivated at the holding potential used (−71 mV in Fig. 4.16B, left), the current evoked with steps to −46 mV was much larger than the fully inactivated current obtained with a similar command potential (−40 mV in Fig. 4.16B, right) in a dissociated neuron. In the model, a high density of I_{Ts} in distal dendrites reproduced the higher amplitude of the current seen in the soma (approximately 2 nA; Fig. 4.14) compared to dissociated cells (approximately 150 pA; Fig. 4.13B).

Second, voltage-clamp control is extremely difficult to obtain experimentally in intact RE cells. Occasionally with either appropriate voltage protocols to reduce the size of the current, or with recordings in cells that had exceptionally small I_{Ts} amplitude (as in Fig. 4.16A), reasonably good clamp was obtained as judged by the lack of notches in the inward current trace. An example from each of these two situations is shown in Fig. 4.16. In simulations of voltage-clamp in intact cells, it was virtually impossible to clamp the voltage in the dendrites from a somatic electrode (Fig. 4.17). The voltage difference between soma and distal dendrites could be as high as 60 mV during transients (50 mV was obtained using constant field equations). This poor voltage control of the dendrites was due to the presence of high densities of dendritic I_{Ts}.

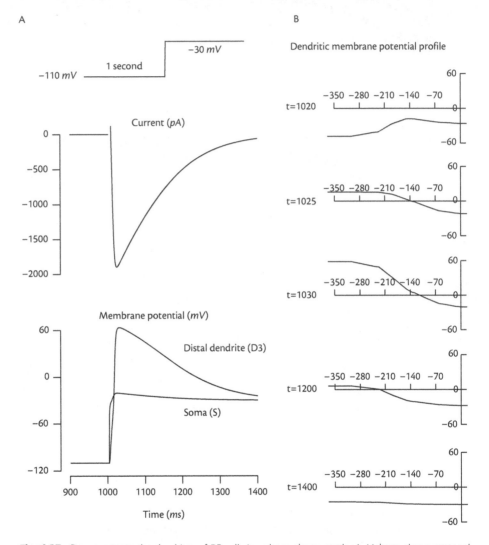

Fig. 4.17: Poor access to the dendrites of RE cells in voltage-clamp mode. A. Voltage-clamp protocol similar to Fig. 4.14 with a high density of distal T-current. In addition to the current recorded at the electrode, the somatic and distal dendritic membrane potentials are shown (S and D3 are as described in Fig. 4.14). The deviation from voltage-clamp in the dendrites was pronounced. Similar results were obtained using constant field equations for I_{Ts}. B. Dendritic membrane potential profile (defined identically as in Fig. 4.15) taken during the voltage-clamp protocol in A (same units of time as in A). From an initially good space clamp at −110 mV, the attempt to clamp the cell at −30 mV was unsuccessful because of powerful current transients in the dendrites, which could not be controlled from the soma. Due to these transients, the kinetics of the current recorded in the soma appeared to be slower than in dissociated cells (modified from Destexhe et al., 1996b).

Third, the observed kinetics of I_{Ts} were generally slower in intact RE cells compared to acutely dissociated cells (Fig. 4.16). In Fig. 4.16A, the time course of inactivation was slower in the intact cell; in Fig. 4.16B, even though the temperature in the dissociated cell experiment was much cooler than the intact cell experiment (23 °C versus 34 °C), the kinetics were similar. The current in the intact neuron should be much slower at the lower temperature used in dissociated cell experiments. In the model, this property arose only when the intact cell had a high density of dendritic I_{Ts}. As a consequence of the poor voltage-clamp of the dendrites, dendritic current transients occurred and added a slower component to the current decay seen in the soma (Fig. 4.17). It could, however, be argued that the slower T-current kinetics seen in RE cells compared to TC cells (Huguenard and Prince, 1992) could be due to poor voltage-clamp in dendrites. Indeed, the T-current of RE cells does not always show a slower time course (Tsakiridou et al., 1995). Simulations clearly indicate that the slower kinetics of I_{Ts} cannot be accounted for by distal dendritic localization (Fig. 4.18) but reflects the channel properties. Moreover, the space-clamp was almost perfect in the dissociated cell model (not shown), suggesting that in this preparation the estimates of the kinetics of the T-current were reliable.

A fourth argument for the presence of T-current in dendrites is the similarity between the model and recordings from the dendrites of RE cells. A presumed dendritic impalement of an RE cell *in vivo* shows a burst structure dominated by a broad spike with small-amplitude spikes (Fig. 4.19). The broad spike was presumably a calcium spike, whereas the small-amplitude spikes

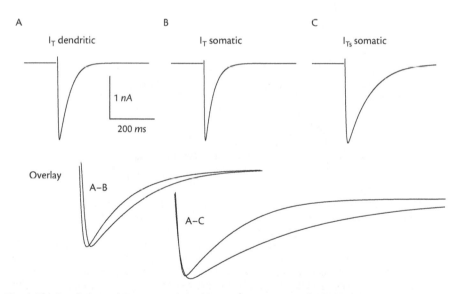

Fig. 4.18: Distally located T-current with fast kinetics cannot account for the slow T-current kinetics in RE cells. The detailed morphology of the RE cell was used with different models of the T-current. A. TC-cell like T-current located with highest densities in distal dendrites (I_T dendritic; 0.6 mS/cm^2 in distal dendrites and 0.045 mS/cm^2 elsewhere). B. TC-cell like T-current located exclusively in soma (I_T somatic; density of 0.845 mS/cm^2). C. RE-cell-like T-current located exclusively in soma (I_{Ts} somatic; density of 0.845 mS/cm^2). For A–C, the current trace is shown for a voltage-clamp protocol consisting of a conditioning pulse of 1 sec at −110 mV, followed by a voltage jump to −30 mV. The bottom traces (Overlay) compare the current traces of I_T dendritic and somatic (A–B), showing the slow component due to distal T-current. The comparison between dendritic I_T and somatic I_{Ts} in A–C shows that the slow kinetics of an RE cell's T-current cannot be accounted for by a TC-cell-like T-current located in distal dendrites.

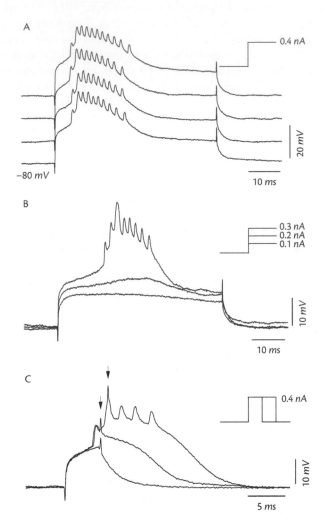

Fig. 4.19: Presumed dendritic recording in a cell from the rostrolateral region of the RE nucleus *in vivo*. A. Intracellular recording in a urethane-anesthetized cat *in vivo*. Depolarizing current pulses of constant intensity applied at a hyperpolarized membrane potential elicited an active response that was crowned by small spike potentials (presumably electrotonically attenuated somatic spikes) with a stereotyped bursting pattern in which the first interval was slightly longer with the frequency increasing then decreasing during the response. Four different responses, displaced vertically for clarity, display the stereotyped burst response. B. In response to depolarizing pulses of increasing amplitudes, first a passive response was obtained, then an active response of small amplitude, and finally a fully developed burst. Increasing the amount of current injected (as in A) only decreased the latency to the burst but did not change its stereotyped shape. C. Increasing the duration of a constant amplitude depolarizing current pulse also elicited a burst. The shorter pulse occasionally elicited active responses that did not develop into a burst. The arrows indicate the end of the current pulses. All the pulses in the figure were applied at the same membrane potential (indicated in A) (modified from Destexhe et al., 1996b).

were presumed sodium spikes, as they showed the typical *accelerando-decelerando* structure of RE cells. Sodium spikes with increasing frequency were correlated with the rising phase of the calcium spike, whereas spikes with decreasing frequencies were associated with the decaying phase of the calcium spike. A strikingly similar behaviour was observed in the dendritic membrane potential of the model (Fig. 4.14, right, D2). These observations, together with the stability of the recording and with the negative value of the resting potential, suggest that the impalement was dendritic. These patterns of bursting in the dendrites were seen in the model only when the distal dendritic density of I_{Ts} was high.

Finally, another consequence of the possible dendritic localization of I_{Ts} is that other currents that depend on the entry of Ca^{2+} through T-channels might also be localized in the dendrites. For example, a calcium-dependent potassium current, $I_{K[Ca]}$, is involved in the repetitive bursting of RE cells (Avanzini et al., 1989; Bal and McCormick, 1993). This current was not included model, which also did not generate rhythmic bursting from steady inputs. Determining the properties of $I_{K[Ca]}$ is itself a difficult problem. This current activates relatively slowly, as indicated by the broadness of the burst in RE cells, but it must also deactivate relatively quickly to allow repetitive bursting to occur at frequencies approximately 10 Hz (Avanzini et al., 1989; Bal and McCormick, 1993). More data are needed before precise models can be designed to investigate the somatic vs. dendritic localization for $I_{K[Ca]}$.

4.3.6 Properties of RE bursts *in vivo*

A major consequence of the dendritic T-current is that dendritic synaptic currents can potentially exert a powerful control over the bursting properties of RE cells. With dendritic T-current, the model shown above reproduced many features of RE cell bursts seen in slices, but no explanation has yet been proposed to account for the graded properties of the bursts in these cells *in vivo* (see Fig. 4.8). The possibility that the presence of synaptic background activity in the dendrites could lead to graded bursting behaviour is investigated below.

Neurons *in vivo* are under constant synaptic bombardment due to the spontaneous activity of the network. This is particularly true of RE cells, which receive collaterals from the majority of thalamocortical and corticothalamic fibres. A series of previous computational studies have considered the effects of synaptic bombardment on the passive and integrative properties in other types of neurons (Barrett, 1975; Holmes and Woody, 1989; Bernander et al., 1991; Rapp et al., 1992; De Schutter and Bower, 1994; Destexhe and Paré, 1999; Hô and Destexhe, 2000). These studies found that synaptic bombardment significantly decreases the input resistance and the time constant, as well as influence the effectiveness of postsynaptic potentials to discharge the cell (Hô and Destexhe, 2000), but they did not investigate the effects on intrinsic properties. A modelling study of Purkinje cells examined the effects of dendritic synaptic currents on the firing patterns of the cell and was able to reproduce *in vivo* recordings in the presence of synaptic bombardment (De Schutter and Bower, 1994). The impact of synaptic currents on the properties of the bursts in RE cells was simulated by adding sustained depolarizing currents distributed uniformly on the dendritic tree.

In the absence of synaptic background activity, the intact cell model normally generated all-or-none burst responses (Fig. 4.20A), in agreement with recordings of RE cells *in vitro* (Fig. 4.9; see also Llinás and Geijo-Barrientos, 1988). For depolarizing pulses of progressively larger amplitudes, the bursts in the model appeared fully developed, and the pattern of spikes

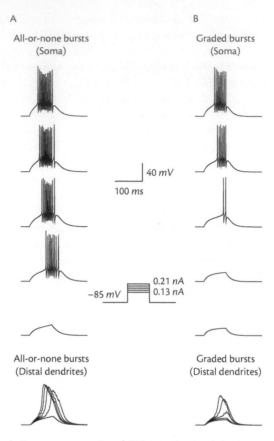

Fig. 4.20: Graded and all-or-none properties of RE bursts. Bursting behaviour was elicited by injecting depolarizing current pulses from $-85\,mV$ in the simulated intact RE cell. The time courses of the membrane potential in the soma (top panels) and in the distal dendrites (superimposed in bottom panels) are indicated. A. All-or-none bursts in the intact RE cell with high density of I_{TS} in distal dendrites and no additional dendritic current (same parameters as in Fig. 4.14, right). The membrane potential in the distal dendrite (bottom) shows that there is a threshold for bursting behaviour. B. Graded bursts in the presence of a sustained depolarizing current uniformly distributed in the dendrites (same simulation as in A, but with an additional current of 0.02 mS/cm^2 density and $-20\,mV$ reversal potential). In this case, the number of action potentials in the burst was proportional to the amplitude of the current pulse (modified from Destexhe et al., 1996b).

remained similar for stronger current pulses. It was possible to obtain intermediate patterns, but only by carefully tuning the current amplitude, since these patterns only occurred within a narrow range of values (not shown).

In contrast, a graded burst response was observed in RE cells *in vivo* (Fig. 4.8; see also Fig. 5 in Contreras et al., 1993). We explored the hypothesis that this discrepancy could be explained by the presence of synaptic background activity (Destexhe et al., 1996b). RE cells are particularly subject to background activity because they receive collaterals of thalamocortical as well as corticothalamic axons (Jones, 1985). These cells therefore receive a copy of the traffic between thalamus and cortex, under the form of EPSPs. We represented here synaptic background activity either by simulating randomly occurring EPSPs distributed in dendrites, or by an equivalent depolarizing leak current.

In the presence of synaptic background activity, the simulated RE cell generated bursts in a graded fashion, with either leak-current (Fig. 4.20B) or explicit representations of background activity (Fig. 4.21). These graded burst responses agreed with *in vivo* recordings and were due here to the depolarizing action of background activity which counteracted the activation of the T-current in the dendrites.

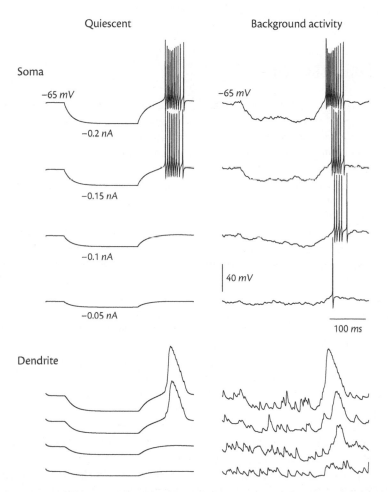

Fig. 4.21: Burst generation becomes graded in RE cells in the presence of synaptic background activity. Bursting behaviour was elicited by a current injection protocol similar to that in Fig. 4.20. In this case, synaptic background activity was explicitly simulated using 645 AMPA synapses uniformly distributed in dendrites (quantal conductance of 1200 *pS*; random release frequency of 2.5 *Hz* per synapse). A. All-or-none bursts in the absence of synaptic background activity (simulation similar to Fig. 4.20A). B. Graded burst responses in the presence of synaptic background activity.

Because synaptic background activity has not been measured in RE cells, a wide range of possible excitatory current densities and reversal potentials were tested. The tendency to generate graded bursting behaviour depended on the strength of the currents underlying synaptic background activity. Typical models with moderate strength are shown in Figs. 4.20B and 4.21B. For background activity with higher intensities, it became increasingly difficult to generate a burst, which suggests that tonic bombardment of the dendritic tree could also be a decisive factor in controlling burst generation itself.

The possibility of generating graded bursting behaviour from a purely somatic current (0 mV reversal), representing for example the leak caused by electrode impalement, was also tested. However, graded burst responses could not be obtained based on this factor alone. In the model, bursting was generated in the dendrites and the only way to alter its properties was to change the balance of currents in the dendrites.

4.3.7 Reduced models

Models that incorporate morphological details of the dendritic tree are certainly relevant for studying single-cell behaviour. However, in simulating large-scale networks, simplified models are much more computationally efficient. Another application of reduced models is hybrid recordings, in which a simulated cell is connected artificially to a recording from a real cell (Yarom, 1991; LeMasson et al., 1992; Renaud-Le Masson et al., 1993). This hybrid method requires models that can be simulated in real time, while still conserving a sufficient degree of realism. One- and three-compartment models derived from the detailed morphological model of the RE cell are investigated below.

The dendritic morphology was reduced to three compartments using a method based on the conservation of the axial resistance (Bush and Sejnowski, 1993) rather than conserving the membrane area (see Rall, 1995), as summarized in Chapter 3. Because the total membrane area is not conserved by this method, the reduced model may not have a correct input resistance. This was compensated by introducing in each equivalent cylinder a dendritic correction factor (C_d), that rescales the values of conductances (g_i) and membrane capacitance (C_m) in the dendrites. If C_d is estimated correctly, the reduced model will have correct input resistance and time constant (Bush and Sejnowski, 1993).

In principle, C_d is the ratio between the total surface area of the dendritic segments and their equivalent cylinders (which was of approximately 2.5 here). A more accurate estimation of C_d was obtained by fitting simulations of the reduced model directly to the voltage-clamp recordings obtained in the cell (Fig. 4.12B). The other passive parameters were the same as those obtained from fitting the detailed model to voltage-clamp recordings.

The resulting model had three compartments with the following lengths (l) and diameters (diam): $l = 34.546$ μm and $diam = 14.075$ μm for the soma (area of 1527.55 μm^2); $l = 103.24$ μm and $diam = 5.56$ μm for the proximal segment (area of 1803.32 μm^2); $l = 190.69$ μm and $diam = 3.06$ μm for the distal segment (area of 3636.48 μm^2).

The kinetics and localization of I_{Ts} were investigated in this reduced model. Voltage-clamp traces obtained with a uniform density of I_{Ts} of 0.045 mS/cm^2 were similar to those in the intact cell model with the same density. As in the more detailed model, the current-clamp behaviour did not show bursting activity (Fig. 4.22, left panel) and the uniform density of I_{Ts} had to be increased by about an order of magnitude to observe bursts (not shown).

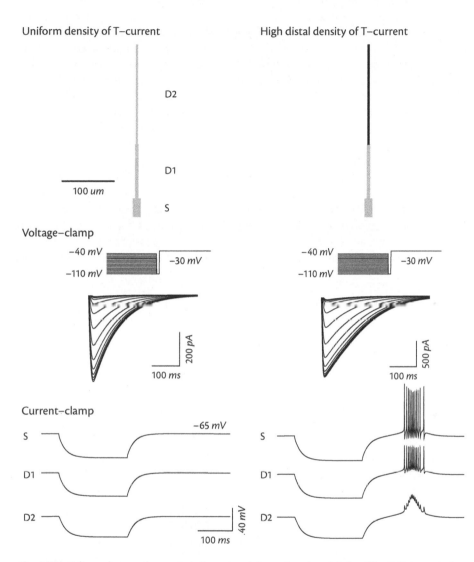

Fig. 4.22: Voltage-clamp and current-clamp protocols in a reduced model of an RE cell. The morphology of the cell shown in Fig. 4.10A, was reduced to three compartments: the soma (S), proximal (D1) and distal (D2) dendritic compartments (corresponding approximately to S, D1, D2 in Fig. 4.14). Right: uniform density of I_{T_s} of 0.045 mS/cm^2. In current-clamp, a low-threshold burst could not be elicited from current injection in the soma (0.3 nA during 200 ms). Right: higher density of I_{T_s} in distal dendrites (0.68 mS/cm^2). In this case, similar bursts with a slow rise and an *accelerando-decelerando* pattern of spikes are seen, as in Fig. 4.14. The simulation of this model was approximately twenty-five times faster than the detailed morphological model. In both cases shown here, the voltage-clamp behaviour was consistent with Fig. 4.13 in the absence of distal dendrites (modified from Destexhe et al., 1996b).

With a high density of I_{Ts} in the distal dendritic compartment, the reduced model showed bursting behaviour that was remarkably similar to that in the detailed morphological model (compare Fig. 4.14 with Fig. 4.22). However, the threshold for bursting behaviour was slightly different than that of the intact cell model. The properties of the bursts generated by distal I_{Ts} were the same as described above for the intact cell: namely, they had a slow rise and decay, they needed strong current pulses to be evoked, and they possessed the typical *accelerando-decelerando* pattern of sodium spikes (Fig. 4.22, right panel). Graded burst responses could also be simulated using the three-compartment model, using similar densities of currents as in the detailed model (not shown).

Finally, a single-compartment model of the RE cell was investigated. The size of the compartment was determined by fitting the passive properties to voltage-clamp recordings of a RE cell (Fig. 4.23A). The single compartment model could not capture all the details of the capacitive transients, but nevertheless provided an acceptable fit. Using a density of I_{Ts} close to threshold for bursting behaviour, the single-compartment model generated bursts in response to injection of current pulses (Fig. 4.23B). The burst still had a relatively broad structure, although narrower than in the presence of dendrites. However, it was not possible to obtain the typical *accelerando-decelerando* structure of sodium spikes. Nonetheless the first spike interval

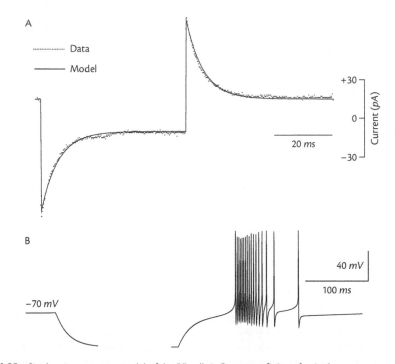

Fig. 4.23: Single-compartment model of the RE cell. A. Parameter fitting of a single-compartment model to estimate the optimal size for the cell using the simplex method. The fit gave an equivalent cylinder of $l = 64.52 \pm 0.01$ *μm* with a diameter of 70 *μm*. Other parameters were identical as in Fig. 4.12A; the least square error was, however, larger in this case (by approximately 3%). B. Low-threshold burst obtained in this model following current injection (0.3 *nA* for 200 *ms*; density of I_{Ts} was 3 *mS/cm*^{2}). In this case, there was no *accelerando-decelerando* pattern of spikes, and the cell could not produce graded bursting behaviour (modified from Destexhe et al., 1996b).

of the burst shown in Fig. 4.23B was longer than for the subsequent ones. More importantly, depolarizing currents were ineffective in obtaining the graded burst response found in models having dendritic I_{Ts}.

4.4 Further findings on thalamic reticular cells

Several key experimental and theoretical results were obtained for thalamic reticular neurons in the last years and are worth being mentioned.

4.4.1 Calcium channels and sleep oscillations

Calcium and voltage-sensitive imaging studies of thalamic reticular (RE) cells revealed that calcium-dependent bursts are present in dendrites (Bal et al., 2012; Chausson et al., 2013), as we had predicted based on computational models (Destexhe et al., 1996). In addition to the calcium currents, Na+-dependent action potentials also propagated in the dendrites of RE neurons.

The Cav3.3 isoform of the T-type channels is mostly present in the RE nucleus, and its deletion in Cav3.3 knockout mice resulted in the total abolition of bursts in the RE nucleus, as well as a strongly reduced T-type current in voltage-clamp recordings (Astori et al., 2011). Cav3.3 knock-out mice displayed diminished EEG power in the spindle frequency band (10–12 Hz), especially before the transition from SWS to rapid eye movement (REM) sleep when many sleep spindles normally occur (Astori et al., 2011). As in the Cav3.1 knockout, oscillations in the spindle frequency range were reduced but not abolished. Thus, both the Cav3.1 and Cav3.3 knockout mice have globally reduced—but not totally suppressed—sleep spindle activity. However, since both TC-RE loops and isolated RE oscillations require bursts in RE cells, these experiments do not fully support either of the two mechanisms for spindle generation.

4.4.2 Simplified models

Simplified models of RE cells consisting of an integrate-and-fire mechanism augmented to allow the model to produce bursts of action potentials reproduced the typically prolonged bursts in RE cells, as well as inhibitory-rebound burst sequences from a depolarized Vm level (Destexhe 2009). Such models will be further described in the next chapters, focusing the on network level.

4.5 Discussion

In this chapter, we reviewed several models for thalamic reticular neurons, from a single-compartment to compartmental models that included the dendritic morphology of RE cells. The models have made several important contributions to understand the intrinsic electrophysiological properties of RE cells, as summarized here.

4.5.1 The interplay of ionic currents in thalamic reticular cells

4.5.1.1 Rebound burst generation by I_{Ts}

RE cells, Like TC cells, can generate low-threshold spikes (LTS) following a strong hyperpolarization. The presence of this rebound burst response in RE cells was first demonstrated by Mulle et al. (1986) *in vivo* and by Llinás and Geijo-Barrientos (1988) *in vitro*. However, the bursts generated by RE cells are broader than that of TC cells and require more current to be elicited (Mulle et al., 1986). Voltage-clamp experiments in these cells revealed the presence of a T-type Ca^{2+} current, with properties that are significantly different from those found in TC cells, including a slower kinetics and activation over a more depolarized range of membrane potentials (Huguenard and Prince, 1992; but see Tsakiridou et al., 1995).

Computational models were designed to investigate burst generation in RE cells (Destexhe and Babloyantz, 1992; Destexhe et al., 1993b; Wang and Rinzel, 1993; Destexhe et al., 1994a; Golomb et al., 1994; Wallenstein, 1994a; Destexhe et al., 1996a, 1996b). The models established that the differences of activation of I_{Ts} in RE cells (compared to I_T in TC cells) could account for the broader bursts displayed by these cells. Following the approach taken in Chapter 3, the models established that the ion channel properties characterized in voltage-clamp experiments are sufficient to account for features of RE cell's bursts.

However, a number of features could not be reproduced, such as the graded nature of the burst *in vivo* (Contreras et al., 1993), the relative difficulty of evoking LTS in RE cells (Mulle et al., 1986) and the typical *accelerando-decelerando* patterns of spikes within RE bursts (Domich et al., 1986). To account for these properties, it was necessary to take into account dendritic morphology and the presence of T-current in the dendrites of RE cells (Destexhe et al., 1996b; see below).

4.5.1.2 Oscillations from the interplay of I_{Ts} and $I_{K[Ca]}$

RE neurons, like TC cells, are characterized by rhythmic firing properties, but in contrast to TC cells, they rarely display sustained oscillations. Typically, short sequences of rhythmic bursts at 7 to 12 Hz can be evoked by injection of current pulses (Avanzini et al., 1989; Bal and McCormick, 1993), or by stimulating excitatory afferents (Contreras et al., 1993; Huguenard and Prince, 1994b). The resistance of these oscillations to tetrodotoxin (Bal and McCormick, 1993) indicates that they are due to mechanisms intrinsic to the cell. Recordings from RE neurons *in vitro* revealed, in addition to I_{Ts}, the presence of a Ca^{2+}-activated current $I_{K[Ca]}$, responsible for an after-hyperpolarization (AHP) following a burst (Avanzini et al., 1989; Bal and McCormick, 1993). These *in vitro* studies suggested that rhythmic oscillations could occur through interactions between I_{Ts} and $I_{K[Ca]}$.

Models have confirmed that the combination of these two currents can produce rhythmic bursting similar to that found in RE cells (Fig. 4.7; see also Fig. 2.11 in Chapter 2). These models took advantage of the characterization of I_{Ts} and $I_{K[Ca]}$, and established that $I_{K[Ca]}$ can be activated in the same range of subthreshold membrane potentials within which Ca^{2+} enters through I_{Ts} channels. Thus, the following sequence of events likely occur: Ca^{2+} enters through T-channels during an LTS; Ca^{2+} then activates $I_{K[Ca]}$, which then hyperpolarizes the membrane and deinactivates I_{Ts}; when the membrane depolarizes due to the deactivation of $I_{K[Ca]}$, a new LTS is produced and the cycle repeats. The robustness of this mechanism has been confirmed in several modelling studies of thalamic neurons (Destexhe et al., 1993b; Wang and Rinzel, 1993; Destexhe et al., 1994a; Hindmarsh and Rose, 1994a; Wallenstein, 1994a).

4.5.1.3 Oscillatory behaviour from interactions between I_{Ts}, $I_{K[Ca]}$, and I_{CAN}

Another characteristic of RE cells is that a sequence of bursts is often terminated by a tonic tail of spikes, *in vivo* (Domich et al., 1986; Contreras et al., 1993) and *in vitro* (Avanzini et al., 1989; Bal and McCormick, 1993). Bal and McCormick (1993) reported the presence of a nonspecific cation current, I_{CAN}, in RE cells and suggested that the slow activation of this current could explain the genesis of a tonic tail of spikes.

In the model, the activation of I_{CAN} accelerated the rising phase of the burst and increased the frequency of the rebound burst sequence to approximately 10 Hz, the frequency observed experimentally (Bal and McCormick, 1993). The presence of I_{CAN} also terminated the oscillatory behaviour by producing a tonic tail of spikes before the membrane returned to its resting level. The relative values of $\bar{g}_{K[Ca]}$ and $\bar{g}_{CAN}$ modulated both the frequency and the relative importance of rhythmic bursting relative to tonic tail activity (Fig. 4.6; Destexhe et al., 1993b, 1994a). These results were confirmed in another modelling study (Wallenstein, 1994a).

In conclusion, single-compartment models suggest that the interaction between three ionic currents, I_{Ts}, $I_{K[Ca]}$ and I_{CAN}, accounts for the qualitative features of the intrinsic bursting and oscillatory properties of RE cells. Such simple models will be used to investigate network behaviour in Chapters 6, 7 and 8.

4.5.2 The role of the dendrites in the electrophysiology of thalamic reticular neurons

4.5.2.1 Model of dendritic T-current in thalamic reticular cells

As outlined above, not all properties of RE cells could be reproduced by single-compartment models. These properties include the relative difficulty of evoking a rebound response in RE cells (Mulle et al., 1986; Contreras et al., 1993), the typical *accelerando-decelerando* pattern of spikes consistently seen in RE cell bursts (Fig. 4.1A; Domich et al., 1986; Avanzini et al., 1989; Contreras et al., 1993; Huguenard and Prince, 1994; Bal et al., 1995), and the graded nature of the burst in RE cells recorded *in vivo* (Fig. 4.8; Contreras et al., 1993), which contrasts with the all-or-none properties seen *in vitro* (Fig. 4.9).

We investigated the hypothesis that these features could be explained if the T-current is located in the dendrites of RE cells (Section 4.3; Destexhe et al., 1996b). A computational model that included the dendritic morphology of RE cells was used and was constrained by voltage-clamp recordings with and without dendrites to evaluate the density of T-channels located in dendrites and soma, respectively. We concluded that the major portion of T-channels is located in the distal dendrites of RE cells. This pattern of localization not only accounted for voltage-clamp recordings, but also reproduced all the aforementioned properties of the burst in RE cells. In particular, models with distal dendritic T-current accounted for the graded properties of the bursts in the presence of synaptic bombardment in dendrites (Fig. 4.21), thus providing a plausible explanation for the differences seen between *in vivo* and *in vitro* recordings.

4.5.2.2 Physiological consequences of dendritic T-current in thalamic reticular cells

The dendritic localization of I_{Ts} in RE cells has consequences for their physiological properties. First, a rebound burst could be evoked locally in dendrites by IPSPs from neighbouring RE cells.

RE neurons indeed contact their neighbours through GABAergic axon collaterals (Jones, 1985; Bal et al., 1995; Liu et al., 1995) or dendro-dendritic GABAergic synapses (Deschênes et al., 1985; Yen et al., 1985; Pinault et al., 1997). Intracellular recordings revealed GABAergic IPSPs in RE cells (Destexhe et al., 1994a; Bal et al., 1995; Ulrich and Huguenard, 1996; Sanchez-Vives, 1997a, 1997b; Zhang et al., 1997) but they are of relatively low amplitude and might not be sufficient to elicit a rebound burst (Ulrich and Huguenard, 1997b). However, if the T-current is dendritic, IPSPs arising from neighbouring RE cells might initiate rebound bursts in localized dendritic regions, without necessarily any evidence of IPSP in somatic recordings. The possibility that mutual inhibitory interactions between RE cells generate rebound bursts and oscillatory behaviour will be considered in Chapter 6.

GABAergic IPSPs could also shunt the burst discharges of RE cells and counteract their initiation. In this case, mutual inhibition between RE cells would be a factor that protects RE cells from generating large burst discharges, which may have important implications for the genesis of epileptic discharges (see details in Chapter 8). It thus seems that the interactions between RE cells will critically depend on the membrane potential and the reversal potential of IPSPs. The range of possible interactions, and their impact at the network level, will be considered in Chapters 6 and 8.

Second, if dendritic branches containing a high density of I_{Ts} are hyperpolarized, the T-current will be deinactivated and bursts could be evoked by depolarizing inputs. Dendritic T-current therefore could make RE cells highly sensitive to incoming EPSPs. This hypothesis was investigated recently using computational models (Destexhe, 2001). This possibility is also supported by in vivo experiments showing that corticothalamic excitatory volleys can trigger bursts in RE cells and promote spindle oscillations (Contreras et al., 1993; Contreras and Steriade, 1996). The sensitivity of RE cells to cortical EPSPs is also the source for the 'inhibitory dominance' hypothesis (Destexhe et al., 1998a). This hypothesis is critical for understanding the large-scale synchrony of oscillations, which will be elaborated in Chapters 6 and 8.

Finally, the dendritic origin of burst discharges suggests a new mechanism by which RE cells could participate in controlling the transition of the thalamocortical system between wakefulness and sleep. In the model, the presence of additional depolarizing currents in the dendrites counteracts the genesis of the burst. This behaviour could not be observed in a single compartment model, suggesting that the soma with sodium spikes and the dendrites with calcium spikes constitute a device that functions according to the level of synaptic background activity. The dendrites of RE cells would 'sample' the overall synaptic activity between the thalamus and the cortex, and tune the responsiveness of the RE nucleus according to these inputs. For high levels of background activity, which occurs during tonic activity in the thalamus and cortex, the RE cell does not have the tendency to fire bursts. For lower levels of background activity, or more phasic inputs such as during synchronized sleep, the dendrites would no longer be bombarded in a sustained manner, and bursting behaviour would be enhanced. We propose that this type of interaction between currents in the dendrites acts in concert with neuromodulation in switching the thalamus between tonic and bursting modes.

4.6 Summary

In this chapter two kinds of models for thalamic reticular cells were reviewed. The first, a morphologically simple model consisting of a single isopotential compartment, was used to investigate the interaction between several intrinsic ionic currents such as I_{Ts}, $I_{K[Ca]}$ and I_{CAN}. The problem here was how to account for the patterns of rhythmic bursting, oscillations and plateau potentials (ADP) recorded in these cells in current-clamp. The models suggest that the kinetics of ionic currents as found in RE cells can account qualitatively for their electrophysiological features, including oscillations and bursting behaviour.

Second, a RE cell model that incorporated their profuse dendritic morphology, was used to investigate the electrophysiological consequences of localizing the T-current in dendrites. This issue can at present only be addressed through computational studies. Models of reconstructed RE cells indicate that dendritic T-current are necessary to account for the typical features that distinguish the morphology of the burst in RE cells compared to bursts in other cell types. More importantly, the model provides an explanation for the contrasting experimental observations that RE bursts are all-or-none *in vitro* but are graded *in vivo*. In the model, all-or-none bursts become graded if background synaptic activity is included in dendrites, as observed *in vivo*, but the presence of dendritic T-current is needed.

In conclusion, not only has it been possible to reasonably accurately model the intrinsic properties of RE cells, but models have also provided an explanation for contrasting experimental observations. The same model can display *in vivo* or *in vitro* type behaviour depending on the presence of synaptic background activity. These morphologically complex models could be used to investigate further questions regarding the physiology of the thalamic reticular nucleus, such as examination of the effect of dendro-dendritic GABAergic synapses combined with dendritic T-current, the modulation of bursts by corticothalamic synapses, and others. Single-compartment models may nonetheless provide the basic units for large-scale network simulations that include thalamic reticular cells, as illustrated in Chapters 6–8.

Biophysical models of synaptic interactions

Neurons communicate primarily through specialized junctions called synapses. The purpose of this chapter is to provide a variety of models of synaptic transmission for use in detailed compartmental models of single neurons and in simulations of large-scale networks of neurons. We first consider relatively simple models of neurotransmitter release (Section 5.1), and then combine these with Markov models of postsynaptic receptors (Section 5.2). These simplified models capture the most salient features of synaptic transmission at a variety of synapses. We next consider more sophisticated models that include extracellular diffusion of neurotransmitter (Section 5.3) to evaluate the precision required to capture the dynamics of inhibitory transmission in the thalamus and cerebral cortex. The models introduced here provide the foundation for subsequent large-scale network simulations (see Chapters 6–8).

5.1 Transmitter release

5.1.1 Experimental characterization of neurotransmitter release

In contrast to the intracellular spread of electrical currents, intercellular communication is most often mediated by the release of chemical substances from a presynaptic neurons and their binding to receptors on a postsynaptic neuron. There are many advantages to chemical neurotransmission. First, with direct electrical coupling, the cells are biased to synchronize their dynamics on a fast time scale, whereas with chemical transmission, a brief event in the presynaptic cell can trigger long-term events in the postsynaptic cell. Second, chemical transmission can mediate inhibition as well as excitation of the postsynaptic cell. A single chemical released from a presynaptic terminal can even have multiple effects on the postsynaptic cell through different

Thalamocortical Assemblies: Sleep Spindles, Slow Waves and Epileptic Discharges. Second Edition. Alain Destexhe and Terrence J. Sejnowski, Oxford University Press. © Oxford University Press 2023.
DOI: 10.1093/oso/9780198864998.003.0005

receptors, which may have different binding affinities and couple into different effectors. Finally, chemical transmission can be modulated on a wide range of time scales postsynaptically, by changing the properties or the nature of postsynaptic receptors, which can affect the kinetics and amplitude of postsynaptic events, and presynaptically, by regulating the release of neurotransmitter, which can produce presynaptic reduction or presynaptic enhancement of transmission.

The detailed mechanisms underlying synaptic release were first investigated at the neuromuscular junction, where the release of acetylcholine triggers muscle contraction. Central synapses are qualitatively similar, though quantitatively quite different than the neuromuscular junction (Katz, 1966; Sakmann and Neher, 1995). We focus here on relatively simple models of the presynaptic processes leading to neurotransmitter release.

When an action potential invades a synaptic terminal, it induces Ca^{2+} entry through voltage-dependent Ca^{2+} channels; Ca^{2+} then triggers a cascade of events leading to the rapid release of neurotransmitter into the synaptic cleft. Neurotransmitter molecules then bind to receptors and affect the voltage or the intracellular metabolism of the target cells. The detailed mechanisms whereby Ca^{2+} enters the presynaptic terminal, the specific proteins with which Ca^{2+} interacts, and the molecular mechanisms leading to exocytosis are under active investigation (Neher, 1998; Jahn and Sudhof, 1999; Llinás, 1999). Most of the proteins that mediate these processes have been identified. One of the most important properties of the Ca^{2+}-induced release of neurotransmitter is that Ca^{2+} ions act cooperatively (Fig. 5.1). Cooperativity provides the nonlinearity needed to produce a sharp threshold for release, which is typically an all-or-none (or quantal) event. At some specialized 'drip' synapses, release of neurotransmitter is graded and the transmission is continuous.

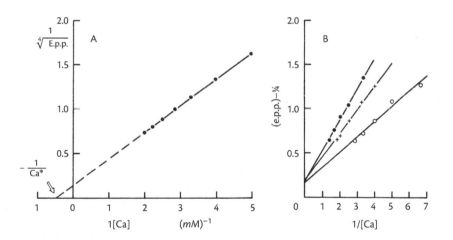

Fig. 5.1: Cooperativity of calcium-dependent mechanisms underlying neurotransmitter release. The dependence of neurotransmitter release on presynaptic calcium is shown here for the frog neuromuscular junction. A. Graph representing the fourth root of EPSP amplitude versus the inverse of calcium concentration (1/[Ca]). Circles indicate experimental recordings and the dashed line is the best fit. B. EPSP amplitudes recorded using different Mg^{2+} concentrations (0.5, 2.0 and 4.0 mM Mg^{2+} for open circles, + and filled circles, respectively). Reproduced from Dodge and Rahamimoff, 1967.

5.1.2 Kinetic models of neurotransmitter release

The mechanisms underlying synaptic release include clustering of calcium channels, Ca^{2+} diffusion and gradients, Ca^{2+}-dependent enzymatic reactions involved in exocytosis, and the particular properties of the diffusion of neurotransmitter across the fusion pore and synaptic cleft. An accurate model of these processes would require simulating the shape of the presynaptic terminal and synaptic cleft, as done in Monte Carlo simulations of neuro-transmitter release (Bartol et al., 1991; Stiles et al., 1996; Stiles et al., 2000). The kinetic model of calcium-induced release presented here was inspired by Yamada and Zucker (1992). This model assumed that: (a) upon invasion by an action potential, Ca^{2+} enters the presynaptic terminal through a high-threshold Ca^{2+} channel; (b) Ca^{2+} activates a calcium-binding protein that promotes release by interacting with the neurotransmitter-containing vesicles; (c) an inexhaustible supply of 'docked' vesicles are available in the presynaptic terminal, ready to be released; (d) the binding of the activated calcium-binding protein to the docked vesicles leads to the release of n molecules of neurotransmitter in the synaptic cleft. The release is modelled here as a first-order process with a stoichiometry coefficient of 4 (see details in Destexhe et al., 1994d).

The calcium-induced cascade leading to the release of neurotransmitter is described here by the following kinetic scheme:

$$4\ Ca^{2+} + X \underset{k_u}{\overset{k_b}{\rightleftharpoons}} X^* \tag{5.1}$$

$$X^* + V_e \underset{k_2}{\overset{k_1}{\rightleftharpoons}} V_e^* \tag{5.2}$$

$$V_e^* \overset{k_3}{\longrightarrow} n\,T \tag{5.3}$$

$$T \overset{k_c}{\longrightarrow} \dots \tag{5.4}$$

where calcium ions bind to a calcium-binding protein, X, with a cooperativity factor of 4 (see Fig. 5.1; Dodge and Rahamimoff, 1967), leading to an activated calcium-binding protein, X^* (Eq. 5.1). The associated forward and backward rate constants are k_b and k_u. X^* then reversibly binds to neurotransmitter-containing vesicles, V_e, with corresponding rate constants k_1 and k_2 (Eq. 5.2). Exocytosis is modelled by an irreversible reaction (Eq. 5.3) where activated vesicles, V_e^*, release n molecules of neurotransmitter, T, into the synaptic cleft with a rate constant k_3. The values of the parameters of these reactions were based on previous models and measurements (see Yamada and Zucker, 1992).

The concentration of the liberated neurotransmitter in the synaptic cleft, $[T]$, was assumed to be uniform in the cleft and cleared by processes of diffusion outside the cleft (to the extra-junctional extracellular space), uptake or degradation. These contributions were collectively modelled by a first-order reaction (Eq. 5.4), where k_c is the rate constant for clearance of T. The values of the parameters are given in Destexhe et al. (1994d).

Fig. 5.2A shows a simulation of this model of neurotransmitter release. The presynaptic terminal had a single compartment containing mechanisms for an action potential, a high-threshold calcium current and calcium dynamics (see Destexhe et al., 1994d for details). Injection of a short current pulse into the presynaptic terminal elicited a single action potential. The depolarization of the terminal by the action potential activated high-threshold calcium channels, producing a rapid influx of calcium. The elevation of intracellular $[Ca^{2+}]$ was transient due to clearance by an active pump. The time-course of activated calcium-binding proteins and vesicles followed closely the time-course of the transient calcium rise in the presynaptic terminal. This resulted in a brief ($\approx$ 1 ms) rise in neurotransmitter concentration the synaptic cleft (Fig. 5.2A, bottom curve). The rate of neurotransmitter clearance was adjusted to match the time course of neurotransmitter release estimated from patch clamp experiments (Clements et al., 1992; Clements, 1996) as well as for detailed simulations of extracellular diffusion of neurotransmitter (Bartol et al., 1991; Destexhe and Sejnowski, 1995; see below).

5.1.3 Simplified models of the release process

The release model given above is computationally expensive and would preclude simulations of networks with thousands of synapses. A simpler model of the release process is needed that preserves the essential features. One alternative is to use a continuous function to transform the presynaptic voltage directly into neurotransmitter concentration. This function can be derived assuming that all intervening reactions in the release process are relatively fast and can be considered at steady state (Destexhe et al., 1994d). The stationary relationship between the neurotransmitter concentration $[T]$ and presynaptic voltage was well fit by:

$$[T] = \frac{T_{max}}{1 + \exp[-(V_{pre} - V_p)/K_p]} \tag{5.5}$$

where T_{max} is the maximal concentration of neurotransmitter in the synaptic cleft, V_{pre} is the presynaptic voltage, $K_p = 5$ mV gives the steepness and $V_p = 2$ mV sets the value at which the function is half-activated (Destexhe et al., 1994d). One of the principal advantages of using Eq. 5.5 is that it provides a fair approximation of the neurotransmitter concentration using a simple and smooth transformation between presynaptic voltage and neurotransmitter concentration (Fig. 5.2B). This form, in conjunction with simple kinetic models of postsynaptic channels, has the virtue of modelling synaptic interaction solely with autonomous differential equations having only one or two variables (see also Wang and Rinzel, 1992).

An alternative model assumes that the change in the neurotransmitter concentration occurs as a brief pulse. At many synapses the exact time course of neurotransmitter in the synaptic cleft does not, under physiological conditions, determine the time course of postsynaptic responses (e.g. Magleby and Stevens, 1972; Lester et al., 1990; Colquhoun et al., 1992; Clements et al., 1992). Indeed, techniques for rapidly applying neurotransmitter have shown that brief pulses of glutamate reproduced PSCs in membrane patches that were quite similar to synaptic responses recorded at an intact synapse (Colquhoun et al., 1992; Hestrin, 1992; Standley et al., 1993; Sakmann and Neher, 1995).

This type of experiments is illustrated in Fig. 5.3. From a whole-cell configuration in brain slices (Fig. 5.3A1), an outside-out patch is excised (Fig. 5.3A2) and placed into the flow path

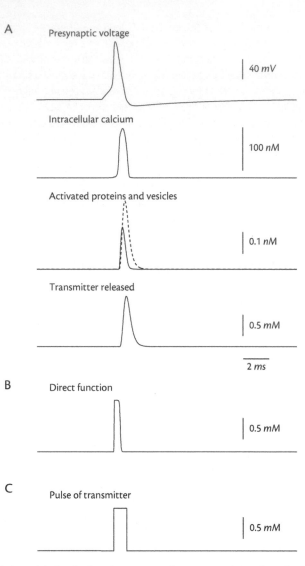

Fig. 5.2: Simplified models for the fast time course of neurotransmitter release in the synaptic cleft. A. Kinetic model of presynaptic processes leading to neurotransmitter release. A presynaptic action potential elicited by injection of a 0.1 nA current pulse lasting 2 ms in the presynaptic terminal (top curve). The intracellular Ca^{2+} concentration in the presynaptic terminal increased (second curve) due to the presence of a high-threshold calcium current that provided a transient calcium influx during the action potential. Calcium was removed by an active calcium pump. The relative concentration of activated calcium-binding protein X^* (third curve; solid line) and vesicles V_e^* (third curve; dotted line) also increased transiently, as did the concentration of neurotransmitter in the synaptic cleft (bottom curve). B. Approximation of the neurotransmitter time course using a direct sigmoid function of the presynaptic voltage. C. Pulse of 1 ms and 1 mM shown for comparison. Panel A was modified from Destexhe et al. (1994d), which contains all parameters of the model.

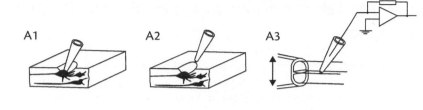

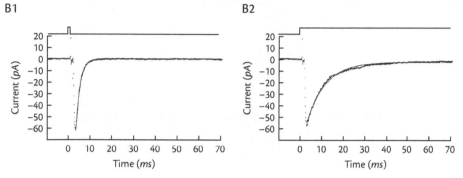

Fig. 5.3: Brief square pulses of glutamate reproduce postsynaptic currents of similar time course as those recorded in the intact synapse. A. Schematic drawing of fast application of glutamate to outside-out patches isolated from neurons in brain slices. A1. Whole-cell configuration obtained after the cleaning procedure. A2. Excision of an outside-out patch. A3. Fast agonist application to the excised patch using a piezo-driven double-barrel pipette containing the transmitter at two different concentrations. B. Fast application of glutamate in outside-out patches excised from hippocampal CA3 pyramidal neurons. B1. Current evoked by a 1 *ms* pulse of 1 *mM* glutamate. The current evoked had a decay time constant of about 1.6 *ms*, similar to the intact synapse. B2. Current evoked by a pulse of 100 *ms* in the same experiment. In this case, the decay time constant was of about 8.3 *ms*. Figure modified from Colquhoun et al., 1992.

of a double-barrel perfusion pipette (Fig. 5.3A3). The patch pipette can be moved extremely rapidly between the two flows using a piezo-electric system, producing a step change in the concentration of transmitter if the two flows contain different glutamate concentrations, and a 'pulse' if moved back to the first pipette. This type of experiment has been used to show that 1 *ms* pulses of 1 *mM* glutamate produce synaptic currents with very similar time course to those observed in an intact synapse (Fig. 5.3B1). The application of long pulses of transmitter reveals the desensitization of the receptor (Fig. 5.3B2), showing that the decay of the synaptic current is not governed by densensitization.

Thus, the stereotyped pulse of 1 *ms* and 1 *mM* shown in Fig. 5.2C provides the simplest approximation to the time course of the neurotransmitter release. Using biophysical models of synaptic transmission, simulating the release by a kinetic model or by a pulse of neurotransmitter had a barely detectable influence on the time course of the synaptic current (see below; see also Destexhe et al., 1994d, 1998b). This simplification has the further advantage that pulse-based Markov models can be solved analytically (see Appendix B), leading to fast algorithms for simulating synaptic currents (Destexhe et al., 1994c, 1994d, 1998b). Pulse-based models are considered in more detail below.

5.2 Models for different types of postsynaptic receptors

Conventional synaptic transmission in the central nervous system is principally mediated by the excitatory and inhibitory amino acid neurotransmitters, glutamate and γ-aminobutyric acid (GABA), respectively. Glutamate activates α-amino-3-hydroxy-5-methyl-4-isoxazolepropionic acid (AMPA) and kainate receptors, responsible for most fast excitatory transmission, as well as N-methyl-D-aspartate (NMDA) receptors, whose activation is much slower than that of AMPA/kainate receptors and is highly voltage dependent. Glutamate also activates metabotropic receptors, which may trigger long-term changes in cellular metabolism through second messengers. GABA activates two main classes of receptors, $GABA_A$ receptors, which have relatively fast kinetics and $GABA_B$ receptors, which are much slower and involve second messengers.

Within these different classes of receptors, there exists a wide range of physiological subtypes that arise from combining different receptor subunits that vary in their molecular composition. The properties of a receptor depend on the particular subunits that make up the receptor. For example, the NMDA receptor subunit NR2 has a variety of subtypes (A, B, C and D), that confer differences to the Mg^{2+} sensitivity and kinetics of the channel (Monyer et al., 1994). Similarly, the presence of the GluR-B subunit determines the Ca^{2+} permeability of AMPA receptors (Jonas et al., 1994), while the GluR-B and GluR-D subunits affect their desensitization (Mosbacher et al., 1994). Interneurons and principal cells express AMPA receptor channels with distinct subunit composition and hence distinct properties—in particular, interneurons express AMPA receptors that are faster and more Ca^{2+} permeable (Geiger et al., 1995). The subunit composition of receptors in different cell types and brain regions is currently the subject of intense study (see reviews by Huntley et al., 1994; Molinoff et al., 1994; Zukin and Bennett, 1995; McKernan and Whiting, 1996). Although the results of these molecular studies will undoubtedly continue to refine our understanding, in this chapter we focus on the most frequently encountered types of glutamatergic and GABAergic receptors and their prototypical properties. These models are sufficiently flexible to fit a wide range of subtypes.

Central synapses are difficult to study owing to their inaccessibility, rapid kinetics, the difficulty of measuring or controlling the time course of neurotransmitter, and their electrotonically remote location from somatic recording sites. Nevertheless, progress in understanding the gating of these synaptic receptors has been made through the use of excised membrane patches containing receptors and rapid perfusion of neurotransmitter and other molecules (Franke et al., 1987). Using these and other methods, it has been shown that the time course of neurotransmitter in the synaptic cleft is much briefer than the kinetics of the postsynaptic receptors, which are responsible for the prolonged time courses of the synaptic currents Lester et al., 1990; (Clements et al., 1992; Clements, 1996).

Three important aspects of receptor gating kinetics must be included in models. (a) *Activation/binding*. The time course of the rising phase of the synaptic current can be determined either by the rate of opening after neurotransmitter is bound to the receptor, or, at low concentrations, the probability of neurotransmitter binding to the receptor. The rising phase can be delayed (made more sigmoidal) by requiring more than one neurotransmitter molecule to be bound before the receptor is activated (analogous to the gating of the 'delayed-rectifier' potassium channel). (b) *Deactivation/unbinding*. The time course of decay can be determined by either deactivation following neurotransmitter removal or desensitization (see below). The rate of deactivation is limited either by the closing rate of the receptor, or more typically, by the rate at which neurotransmitter unbinds from the receptor. (c) *Desensitization*. Synaptic receptor-gated channels can enter a so-called 'desensitized' state analogous to the 'inactivated' states of voltage-gated channels. Desensitization decreases the fraction of channels that can be opened during a synaptic response and can affect the synaptic time course in several ways, including prolonging the decay time and shortening the rise time.

Models should also account for the temporal interactions between successive synaptic release events. Several types of interactions can occur between events occurring in rapid succession. (a) *Priming*. Due to slow activation kinetics, a pulse of neurotransmitter may bind to a channel without opening it; this can prime the receptor to respond to a subsequent pulse. For GABA$_B$ responses, this priming can occur through residual G-proteins on the K$^+$ channels (see Section 5.2.6; Destexhe and Sejnowski, 1995). (b) *Desensitization*. A response that leads to significant desensitization may leave many receptors unable to open in response to a release of neurotransmitter shortly afterwards, causing a progressive decline in responsivity. (c) *Saturation*. When a large fraction of receptors are bound by an initial pulse of neurotrans-mitter, subsequent pulses can produce greatly diminished incremental responses since most channels are already open.

Thus, receptor kinetics are important not only in determining the time course of individual synaptic events, but for the temporal integration of signals during a sequence of synaptic events. Kinetic models for the main receptor types mediating synaptic transmission in the central nervous system are reviewed below.

5.2.1 Markov models of neurotransmitter-gated channels

In general, for a ligand-gated channel, transition rates between the unbound and bound states of the channel depend on the binding of a ligand:

$$T + S_i \underset{r_{ji}}{\overset{r_{ij}}{\rightleftharpoons}} S_j \, , \tag{5.6}$$

where T is the ligand, S_i is the unbound state, S_j is the bound state (sometimes written S_j T), r_{ij} and r_{ji} are rate constants as defined before.

The same reaction can be rewritten as:

$$S_i \underset{r_{ji}}{\overset{r_{ij}([T])}{\rightleftharpoons}} S_j \, , \tag{5.7}$$

where $r_{ij}([T]) = [T] \, r_{ij}$ and $[T]$ is the concentration of ligand. Written in this form, Eq. 5.7 is equivalent to Eq. 2.40. Ligand-gating schemes are generally equivalent to voltage-gating schemes, although the functional dependence of the rate on $[T]$ is simpler than the voltage-dependence discussed in Chapter 2.

5.2.2 Simplified models of neurotransmitter-gated channels

The principle upon which simplified models of neurotransmitter-gated channels are designed is illustrated in Fig. 5.4. The first step is to approximate the release process that determines the neurotransmitter concentration T by a brief pulse (see Section 5.1.3). Voltage-clamp

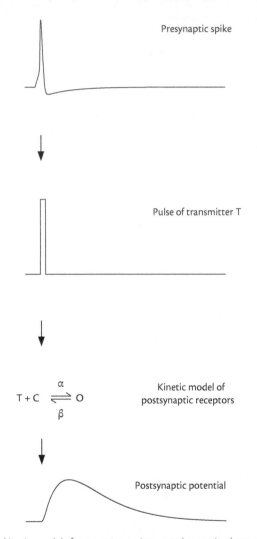

Fig. 5.4: Pulse-based kinetic models for neurotransmitter-gated synaptic channels. When a presynaptic spike occurs (top trace), a pulse of neurotransmitter T (1 mM amplitude, 1 ms duration) is triggered and is used in a simple two-state kinetic scheme to calculate the synaptic current and postsynaptic voltage (lower trace). This type of representation has a number of advantages. First, by adjusting the parameters (α, β, E_{syn}), the model can fit typical time course (rise time and decay) of postsynaptic currents mediated by GABAergic and glutamatergic ionotropic receptors. Second, pulse-based kinetic models are analytically solvable, leading to fast algorithms for simulating synaptic interactions. Third, this type of model automatically handles the summation of responses to a rapid sequence of neurotransmitter pulses (see text for further explanation).

recordings in excised membrane patches have indeed shown that 1 ms pulses of 1 mM glutamate reproduced PSCs that were quite similar to those recorded in the intact synapse (Colquhoun et al., 1992; Hestrin, 1992; Standley et al., 1993). It was therefore assumed that the neurotransmitter (here either glutamate or GABA) is released as a brief pulse when an action potential invades the presynaptic terminal (Destexhe et al., 1994c).

The second step is to approximate the kinetics of the postsynaptic receptors by a simple two-state (open/closed) kinetic scheme. When combined with pulses of neurotransmitter, this produces a simple scheme that can be solved analytically (see Appendix B). The same approach applied to three-state and higher-order models also yields analytical solutions, although they are more complex (Destexhe et al., 1994d). These extremely fast algorithms (see Appendix B) can be used to simulate most types of synaptic receptor dynamics (Destexhe et al., 1994c; Lytton, 1996).

The most elementary state diagram for a ligand-gated channel is:

$$C \underset{r_2}{\overset{r_1([T])}{\rightleftharpoons}} O \qquad (5.8)$$

where C and O represent the closed and open states of the channel, $r_1([T])$ and r_2 are the associated rate constants and $[T]$ is the transmitter time course, represented by a pulse.

In addition to allowing synaptic interactions to be simulated with fast algorithms, the model has a number of other advantages. First, these models can accurately fit the time course of the main ionotropic receptor types (AMPA, NMDA, GABA$_A$; see below). Second, common phenomena such as summation and saturation are naturally taken into account by this class of models. Third, the biophysical interpretation of the equations is similar to the Hodgkin-Huxley (1952) type of model for voltage-dependent currents, which allows all the dynamical elements in the modelled neuron to be described with the same formalism (Destexhe et al., 1994d). These issues are explored further for each receptor type in the following sections.

5.2.3 AMPA/kainate receptors

The prototypical fast excitatory synaptic interactions in the central nervous system are mediated by AMPA/kainate receptors. In specialized auditory nuclei AMPA/ kainate receptor kinetics may be extremely rapid with rise and decay time constants in the submillisecond range (Raman et al., 1994). In the neocortex and hippocampus, responses are somewhat slower (e.g. Hestrin et al., 1990). The 10–90% rise time of the fastest currents measured at the soma (representing those with least cable filtering) is 0.4 to 0.8 ms in cortical pyramidal neurons, while the decay time constant is about 5 ms (e.g. Hestrin, 1993). It may be worth noting that inhibitory interneurons express AMPA receptors with significantly different properties. First, they are about twice as fast in rise and decay time as those on pyramidal neurons (Hestrin, 1993) and second, they have a significant Ca^{2+} permeability (Koh et al., 1995). The latter property appears to be conferred by the lack of the GluR-B subunit in these receptors.

The fast kinetics of AMPA/kainate responses is thought to be due to a combination of rapid clearance of neurotransmitter and rapid channel closure (Hestrin, 1992). Desensitization of these receptors does occur but is somewhat slower than deactivation. The physiological signifi-cance of AMPA receptor desensitization has not been well established. Although desensitization may contribute to the fast synaptic depression observed at neocortical synapses (Thomson and Deuchars, 1994; Markram and Tsodyks, 1996), a study of paired-pulse facilitation in the hippocampus suggested that the contribution of desensitization was minimal even at 7 *ms* intervals between stimuli (Stevens and Wang, 1995).

A Markov kinetic model that accounts for these properties was introduced by Patneau and Mayer (1991) (see also Jonas et al., 1993) and had the following state diagram:

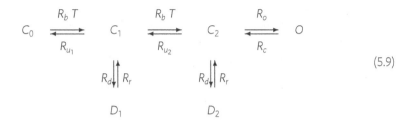

$$(5.9)$$

where the unbound form of the receptor C_0 binds to one molecule of neurotransmitter T, leading to the singly bound form C_1, which itself can bind another molecule of T leading to the doubly bound form C_2. R_b is the binding rate and R_{u_1} and R_{u_2} are unbinding rates. Both C_1 and C_2 can desensitize, leading to D_1 and D_2 with rates R_d and R_r for desensitization and resensitization respectively. Finally, the doubly bound receptor C_2 can open, leading to the open form O, with opening and closure rates of R_o and R_c respectively. This model was simulated with the mechanism for neurotransmitter release described in Section 5.1.2 and its parameters were optimized by directly fitting the full model to whole-cell recorded AMPA currents (see Destexhe et al., 1998b). The fitting procedure gave the following values for the rate constants (Fig. 5.5A): $R_b = 13 \times 10^6 \, M^{-1} \, s^{-1}$, $R_{u_1} = 5.9 \, s^{-1}$, $R_{u_2} = 8.6 \times 10^4 \, s^{-1}$, $R_d = 900 \, s^{-1}$, $R_r = 64 \, s^{-1}$, $R_o = 2.7 \times 10^3 \, s^{-1}$ and $R_c = 200 \, s^{-1}$.

The AMPA current is then given by:

$$I_{AMPA} = \bar{g}_{AMPA} \, [O] \, (V - E_{AMPA}) \qquad (5.10)$$

where $\bar{g}_{AMPA}$ is the maximal conductance, $[O]$ is the fraction of receptors in the open state, V is the postsynaptic voltage and $E_{AMPA} = 0 \, mV$ is the reversal potential. In neocortical and hippocampal pyramidal cells, measurements of miniature synaptic currents (10–30 *pA* amplitude; see McBain and Dingledine, 1992; Burgard and Hablitz, 1993) and quantal analysis (e.g. Stricker et al., 1996) lead to estimates of maximal conductances around 0.35 to 1.0 *nS* for AMPA-mediated currents at a single synapse.

The simplest model that approximates the kinetics of the fast AMPA/kainate type of glutamate receptors can be represented by the two-state diagram:

$$C + T \underset{\beta}{\overset{\alpha}{\rightleftharpoons}} O \qquad (5.11)$$

Glutamate AMPA receptors

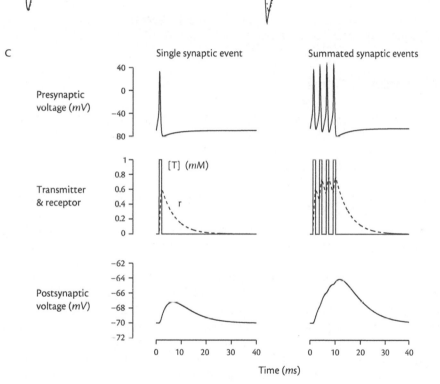

Fig. 5.5: Fast excitatory synaptic interactions mediated by glutamate AMPA receptors. A. AMPA/kainate-mediated currents (noisy trace; obtained from Xiang et al., 1992; recorded at 31°C) compared to Markov kinetic model of AMPA receptors (continuous trace; Eq. 5.9 with neurotransmitter modelled as in Fig. 5.2A). B. Same data fitted by a simplified two-state model for AMPA receptors (Eq. 5.12) with neurotransmitter represented as a pulse (Fig. 5.2C). C. Summation of EPSPs using the model shown in B. A single compartment model was simulated (10 μm diameter, 10 μm length, 0.2 mS/cm^2 leak conductance and −70 mV leak reversal) containing a synaptic conductance of 0.1 nS. Left panel: behaviour with a single presynaptic spike (top graph), leading to a single EPSP (bottom graph; neurotransmitter time course and fraction of open receptors are shown in the middle graph). Right panel: Summated synaptic events. The same model is shown following a burst of four presynaptic spikes at high frequency (300–400 Hz; top graph), leading to summated EPSPs (bottom graph). Figure modified from Destexhe et al., 1998b.

where α and β are voltage-independent forward and backward rate constants. If r is defined as the fraction of the receptors in the open state, the following first-order equation describes the kinetics:

$$\frac{dr}{dt} = \alpha \,[T]\,(1-r) - \beta r \qquad (5.12)$$

and the postsynaptic current I_{AMPA} is given by:

$$I_{AMPA} = \bar{g}_{AMPA}\, r \,(V - E_{AMPA}) \qquad (5.13)$$

where $\bar{g}_{AMPA}$ is the maximal conductance, E_{AMPA} is the reversal potential and V is the postsynaptic membrane potential.

The best fit of this kinetic scheme to whole-cell recorded AMPA/kainate currents (Fig. 5.5B) gave $\alpha = 1.1 \times 10^6\ M^{-1}s^{-1}$ and $\beta = 190\ s^{-1}$ with $E_{AMPA} = 0\ mV$.

5.2.4 NMDA receptors

NMDA receptors are also activated by glutamate, but they mediate synaptic currents that are substantially slower than AMPA-kainate currents, with a rise time of about 20 ms and decay time constants of around 25 to 125 ms at 32°C (Hestrin et al., 1990). The kinetics of activation are slow because two agonist molecules must bind to open the receptor, and also because the channel opening rate of bound receptors is relatively slow (Clements and Westbrook, 1991). The slow decay is due primarily to slow unbinding of glutamate from the receptor (Lester and Jahr, 1992; Bartol and Sejnowski, 1993). The open probability of an NMDA channel at the peak of a synaptic response has been estimated to be as high as 0.3 (Jahr, 1992), raising the possibility that significant saturation of synaptic NMDA receptors may occur during high-frequency stimulus trains (but see Mainen et al., 1999).

Another fundamental difference between the NMDA and AMPA-kainate receptors is that NMDA receptors are voltage-dependent. Under physiological conditions, the activation of NMDA-receptor-mediated currents require not only the binding of glutamate but also a sufficient depolarization of the postsynaptic membrane. The simultaneous presence of presynaptic neurotransmitter and postsynaptic depolarization are needed to open the NMDA receptor, making it a molecular coincidence detector. Furthermore, NMDA currents are carried partly by Ca^{2+} ions, which have a prominent role in triggering many intracellular biochemical cascades. Together, these properties suggest that the NMDA receptor is important for gating synaptic plasticity (Bliss and Collingridge, 1993) and for activity-dependent development (Constantine-Paton et al., 1990).

The voltage-dependence of NMDA receptors arises from a block of the ion channel by physiological concentrations of extracellular Mg^{2+} (Nowak et al., 1984; Jahr and Stevens, 1990a, 1990b). The Mg^{2+} block is voltage-dependent, allowing NMDA receptor channels to conduct ions only when the membrane is depolarized.

Several kinetic schemes have been proposed for the NMDA receptor (Clements and Westbrook, 1991; Clements et al., 1992; Edmonds and Colquhoun, 1992; Lester and Jahr, 1992; Hessler et al., 1993) based on essentially the same state diagram:

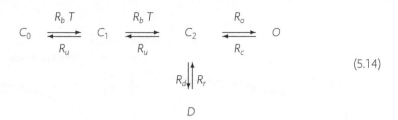

$$(5.14)$$

This kinetic scheme is similar to that of AMPA receptors (Eq. 5.9), with only one desensitized form of the receptor (D) and a single unbinding rate R_u. Direct fitting of this model to whole-cell recorded NMDA currents (in free Mg^{2+}; see below) yielded the following values for the rate constants (Fig 5.6A): $R_b = 5 \times 10^6 \ M^{-1} \ s^{-1}$, $R_u = 12.9 \ s^{-1}$, $R_d = 8.4 \ s^{-1}$, $R_r = 6.8 \ s^{-1}$, $R_o = 46.5 \ s^{-1}$ and $R_c = 73.8 \ s^{-1}$.

The NMDA current can be described by:

$$I_{NMDA} = \bar{g}_{NMDA} \ B(V) \ [O] \ (V - E_{NMDA}) \qquad (5.15)$$

where $\bar{g}_{NMDA}$ is the maximal conductance, $B(V)$ is the magnesium block (see below), $[O]$ is the fraction of receptors in the open state, V is the postsynaptic voltage and $E_{NMDA} = 0 \ mV$ is the reversal potential.

Miniature excitatory synaptic currents also have an NMDA-mediated component (McBain and Dingledine, 1992; Burgard and Hablitz, 1993) and the conductance of dendritic NMDA channels have been reported to be between 3% and 62% of the AMPA channels (Zhang and Trussell, 1994; Spruston et al., 1995), leading to estimates for the maximal conductance of NMDA-mediated currents at a single synapse of around $\bar{g}_{NMDA} = 0.01 - 0.6 \ nS$.

The magnesium block of the NMDA receptor channel is extremely fast compared to the other time constants of the receptor (Jahr and Stevens, 1990a, 1990b). The block can therefore be accurately modelled as an instantaneous function of voltage (Jahr and Stevens, 1990b):

$$B(V) = \frac{1}{1 + \exp(-0.062 \ V) \ [Mg^{2+}]_o / 3.57} \qquad (5.16)$$

where $[Mg^{2+}]_o$ is the external magnesium concentration (1 to 2 mM in physiological conditions).

The simplest model for the kinetics of the slow NMDA-mediated EPSPs is a two-state kinetic scheme

$$C + T \underset{\beta}{\overset{\alpha}{\rightleftharpoons}} O \qquad (5.17)$$

where α and β are voltage-independent forward and backward rate constants. As with AMPA receptors, r is defined as the fraction of receptors in the open state, and is described by the following first-order equation:

$$\frac{dr}{dt} = \alpha \ [T] \ (1 - r) - \beta r. \qquad (5.18)$$

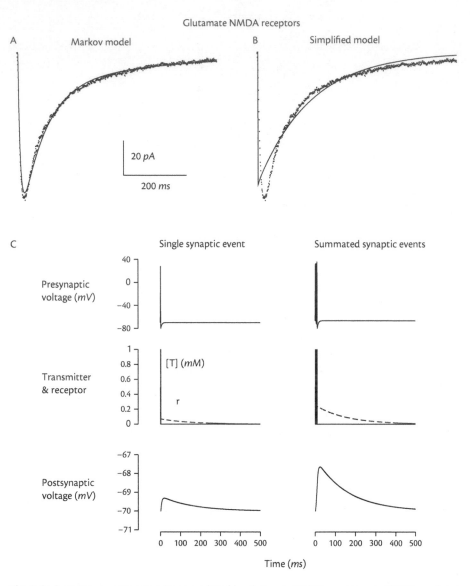

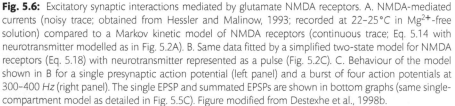

Fig. 5.6: Excitatory synaptic interactions mediated by glutamate NMDA receptors. A. NMDA-mediated currents (noisy trace; obtained from Hessler and Malinow, 1993; recorded at 22–25°C in Mg^{2+}-free solution) compared to a Markov kinetic model of NMDA receptors (continuous trace; Eq. 5.14 with neurotransmitter modelled as in Fig. 5.2A). B. Same data fitted by a simplified two-state model for NMDA receptors (Eq. 5.18) with neurotransmitter represented as a pulse (Fig. 5.2C). C. Behaviour of the model shown in B for a single presynaptic action potential (left panel) and a burst of four action potentials at 300–400 *Hz* (right panel). The single EPSP and summated EPSPs are shown in bottom graphs (same single-compartment model as detailed in Fig. 5.5C). Figure modified from Destexhe et al., 1998b.

The magnesium block is represented by a voltage-dependent term similar to that above, leading to the following expression for the postsynaptic current:

$$I_{NMDA} = \bar{g}_{NMDA} \, B(V) \, r \, (V - E_{NMDA}) \tag{5.19}$$

where $\bar{g}_{NMDA}$ is the maximal conductance, E_{NMDA} is the reversal potential and $B(V)$ represents the magnesium block (same equation as Eq. 5.16).

The best fit of the parameters for this kinetic scheme to whole-cell recorded NMDA currents (Fig 5.6B) was $\alpha = 7.2 \times 10^4 \, M^{-1}s^{-1}$ and $\beta = 6.6 \, s^{-1}$ with $E_{NMDA} = 0 \, mV$.

The voltage-dependent properties of the simplified model of NMDA receptors are illustrated in Fig. 5.7. Under physiological conditions ($[Mg^{2+}]_o = 1 \, mM$), the EPSP had a small amplitude at $-70 \, mV$ that increased over three-fold at $-30 \, mV$ (Fig. 5.7, continuous curves). Upon removal of Mg^{2+} ($[Mg^{2+}]_o = 0$), the full amplitude was revealed at both membrane potentials (dashed curves). In this case, the amplitude difference was due to the different driving forces from the reversal potential. This simulation thus shows one of the most salient property of NMDA receptors: they have a small contribution at rest, but may become important if the postsynaptic membrane potential is depolarized.

5.2.5 GABA$_A$ receptors

GABA$_A$ receptors mediate most fast inhibitory postsynaptic potentials (IPSPs) in the central nervous system. GABA$_A$-mediated IPSPs are elicited following minimal stimulation, in contrast

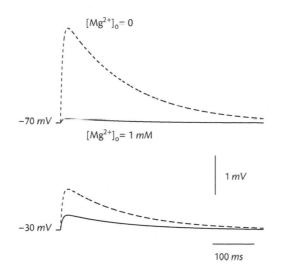

Fig. 5.7: Voltage-dependence of simulated NMDA responses. An NMDA-mediated EPSP was simulated at two different voltages (-70 and $-30 \, mV$) and two different Mg^{2+} concentrations (continuous lines: $[Mg^{2+}]_o = 1 \, mM$; dashed lines: $[Mg^{2+}]_o = 0$). The EPSP was voltage dependent only in the presence of Mg^{2+}. These simulations were done using a simplified two-state model of NMDA receptors (Eq. 5.17–5.18; same model as in Fig. 5.6C). The voltage dependence was described by the empirical equation of Jahr and Stevens (1990b) describing the magnesium block of NMDA receptors (Eq. 5.16).

to GABA$_B$ responses which require strong stimuli (see Section 5.2.6). GABA$_A$ receptors have a high affinity for GABA and are believed to be saturated by release of a single vesicle of neurotransmitter (see Mody et al., 1994; Thompson, 1994). GABA$_A$ receptors have at least two binding sites for GABA and show a weak desensitization (Busch and Sakmann, 1990; Celentano and Wong, 1994). However, blocking uptake of GABA reveals prolonged GABA$_A$ currents that last for more than a second (Thompson and Gähwiler, 1992; Isaacson et al., 1993), suggesting that, as with AMPA/kainate receptors, deactivation following neurotransmitter removal is the main determinant of the decay time.

The kinetic model introduced by Busch and Sakmann (1990) was used for GABA$_A$ receptors based on the following state diagram:

$$
C_0 \underset{R_{u_1}}{\overset{R_{b_1} T}{\rightleftharpoons}} C_1 \underset{R_{u_2}}{\overset{R_{b_2} T}{\rightleftharpoons}} C_2
$$

$$
R_{o_1} \updownarrow R_{c_1} \qquad\qquad R_{o_2} \updownarrow R_{c_2}
$$

$$
O_1 \qquad\qquad\qquad O_2
$$

(5.20)

Here, the neurotransmitter GABA (T) binds to the unbound form C_0, leading to singly bound C_1 and doubly bound form C_2, with binding and unbinding rates R_{b_1}, R_{u_1}, R_{b_2} and R_{u_2} respectively. Both singly and doubly bound forms can open, leading to O_1 and O_2 forms with opening and closure rates of R_{o_1}, R_{c_1}, R_{o_2} and R_{c_2} respectively. Directly fitting this model to whole-cell recorded GABA$_A$ currents gave the following values for the rate constants (Fig 5.8A): $R_{b_1} = 20 \times 10^6$ M^{-1} s^{-1}, $R_{u_1} = 4.6 \times 10^3$ s^{-1}, $R_{b_2} = 10 \times 10^6$ M^{-1} s^{-1}, $R_{u_2} = 9.2 \times 10^3$ s^{-1}, $R_{o_1} = 3.3 \times 10^3$ s^{-1}, $R_{c_1} = 9.8 \times 10^3$ s^{-1}, $R_{o_2} = 10.6 \times 10^3$ s^{-1} and $R_{c_2} = 410$ s^{-1}.

The current is then given by:

$$
I_{GABA_A} = \bar{g}_{GABA_A} ([O_1] + [O_2]) (V - E_{Cl})
$$

(5.21)

where $\bar{g}_{GABA_A}$ is the maximal conductance, $[O_1]$ and $[O_2]$ are the fractions of receptors in the open states, and $E_{Cl} = -70$ mV is the chloride reversal potential. Estimates of the maximal conductance of a single GABAergic synapse from miniature GABA$_A$-mediated currents (Ropert et al., 1990; De Koninck and Mody, 1994) leads to $\bar{g}_{GABA_A} = 0.25 - 1.2$ nS.

The simplest kinetic model for GABA$_A$ receptors is based on the two-state diagram:

$$
C + T \underset{\beta}{\overset{\alpha}{\rightleftharpoons}} O
$$

(5.22)

where α and β are voltage-independent forward and backward rate constants. The fraction of the receptors in the open state, r, is described by the following first-order equation:

$$
\frac{dr}{dt} = \alpha [T] (1 - r) - \beta r
$$

(5.23)

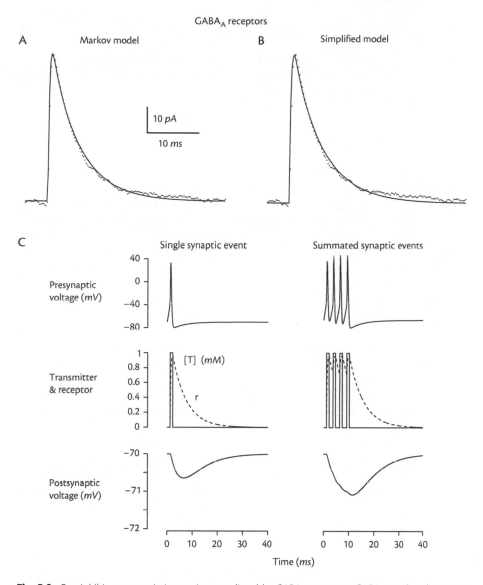

Fig. 5.8: Fast inhibitory synaptic interactions mediated by GABA$_A$ receptors. GABA$_A$-mediated currents (noisy trace; obtained from Otis and Mody, 1992; recorded at 33–35°C) compared to a Markov kinetic model of GABA$_A$ receptors (continuous trace; Eq. 5.20 with neurotransmitter modelled as in Fig. 5.2A). B. Same data fitted by a simplified two-state model for GABA$_A$ receptors (Eq. 5.23) with neurotransmitter represented as a pulse (Fig. 5.2C). C. Behaviour of the model shown in B for a single presynaptic action potential (left panel) and a burst of four action potentials at 300–400 Hz (right panel). The single IPSP and summated IPSPs are shown in bottom graphs (same single-compartment model as detailed in Fig. 5.5C). Figure modified from Destexhe et al., 1998b.

and the postsynaptic current is given by:

$$I_{GABA_A} = \bar{g}_{GABA_A}\, r\,(V - E_{Cl}) \tag{5.24}$$

where $\bar{g}_{GABA_A}$ is the maximal conductance and E_{Cl} is the chloride reversal potential.

The best fit of this kinetic scheme to whole-cell recorded $GABA_A$ currents (Fig 5.8B) was $\alpha = 5 \times 10^6\ M^{-1}s^{-1}$ and $\beta = 180\ s^{-1}$ with $E_{GABA_A} = -80\ mV$.

5.2.6 GABA$_B$ receptors

In contrast to ionotropic receptors, for which the receptor and ion channel are both part of the same protein complex, metabotropic synaptic responses are mediated by an ion channel that is independent of the receptor. In this case, the binding of the neurotransmitter to the receptor induces the formation of an intracellular second-messenger, which in turn activates (or inactivates) separate ion channels. One advantage of this type of neurotransmission is that a single activated receptor can lead to the formation of thousands of second-messenger molecules, an efficient form of amplification. The time course of the response is relatively slow compared to fast synaptic transmission.

These so-called *metabotropic* receptors not only act on ion channels but may also influence several key metabolic pathways in the cell through second-messengers such as G-proteins or cyclic nucleotides. Kinetic models for synaptic interactions acting through second-messengers are reviewed below. This type of interaction is illustrated here by the GABA$_B$ receptor, whose response is mediated by K$^+$ channels through the activation of G-proteins (Andrade et al., 1986; Dutar and Nicoll, 1988).

In contrast to GABA$_A$-mediated responses which are observable following weak stimuli, GABA$_B$ responses require high levels of presynaptic activity (Dutar and Nicoll, 1988; Davies et al., 1990; Huguenard and Prince, 1994; Kim et al., 1997; Thomson and Destexhe, 1999). This property might be due to extrasynaptic localization of GABA$_B$ receptors (Mody et al., 1994), but a detailed model of synaptic transmission on GABAergic receptors suggests that this effect could also be due to cooperativity in the activation kinetics of GABA$_B$ responses (Destexhe and Sejnowski, 1995; Thomson and Destexhe, 1999; see 'Priming' in Section 5.2). Typical properties of GABA$_B$-mediated responses in hippocampal, thalamic and cortical slices can be reproduced assuming that several G-proteins bind to the associated K$^+$ channels (Destexhe and Sejnowski, 1995; Thomson and Destexhe, 1999), based on the following scheme:

$$R_0 + T \;\rightleftharpoons\; R \;\rightleftharpoons\; D \tag{5.25}$$

$$R + G_0 \;\rightleftharpoons\; RG \;\longrightarrow\; R + G \tag{5.26}$$

$$G \;\longrightarrow\; G_0 \tag{5.27}$$

$$C_1 + n\,G \;\rightleftharpoons\; O \tag{5.28}$$

Here the neurotransmitter, T, binds to the receptor, R_0, leading to its activated form, R, and desensitized form, D. The G-protein is transformed from an inactive (GDP-bound) form, G_0, to an activated form, G, catalyzed by R. Finally, G binds to open the K$^+$ channel, with n independent binding sites. If quasi-stationarity is assumed in Eqs. 5.26 and 5.28, G_0 is considered in excess, then the kinetic equations for this system are:

$$\frac{d[R]}{dt} = K_1 [T] (1 - [R] - [D]) - K_2 [R] + K_3 [D]$$

$$\frac{d[D]}{dt} = K_4 [R] - K_3 [D] \tag{5.29}$$

$$\frac{d[G]}{dt} = K_5 [R] - K_6 [G]$$

$$I_{GABA_B} = \bar{g}_{GABA_B} \frac{[G]^n}{[G]^n + K_d} (V - E_K)$$

where $[R]$ and $[D]$ are respectively the fraction of activated and desensitized receptor, $[G]$ (in μM) is the concentration of activated G-protein, $\bar{g}_{GABA_B} = 1$ nS is the maximal conductance of K$^+$ channels, $E_K = -95$ mV is the potassium reversal potential, and K_d is the dissociation constant of the binding of G on the K$^+$ channels. This model accounted accurately for both the time course and other properties of GABA$_B$ responses. Directly fitting the model to whole-cell recorded GABA$_B$ currents gave (Fig 5.9A): $K_d = 100$ μM^4, $K_1 = 6.6 \times 10^5$ M^{-1} s^{-1}, $K_2 = 20$ s^{-1}, $K_3 = 5.3$ s^{-1}, $K_4 = 17$ s^{-1}, $K_5 = 8.3 \times 10^{-5} M$ s^{-1} and $K_6 = 7.9$ s^{-1} with $n = 4$ binding sites.

It is difficult to estimate the conductance of a GABA$_B$-mediated IPSPs at a single synapse. As discussed above, evoking a GABA$_B$-mediated response typically requires high stimulus intensities. Consistent with this observation, miniature GABAergic synaptic currents never contain a GABA$_B$-mediated component (Otis and Mody, 1992; Thompson and Gähwiler, 1992; Thompson, 1994). As a consequence, GABA$_B$-mediated unitary IPSPs are difficult to obtain experimentally. However, it is possible to obtain unitary GABA$_B$-mediated IPSPs in dual impalements if the presynaptic interneuron fired high-frequency bursts of action potentials (Thomson and Destexhe, 1999). In this case, the maximal conductance of GABA$_B$ IPSPs could be estimated by matching computational models with the intracellular recordings, and was of about 0.03–0.04 nS. A peak GABA$_B$ conductance of around 0.06 nS was also reported using release evoked by local application of sucrose (Otis et al., 1992).

This model of GABA$_B$-mediated synaptic transmission, which requires the binding of multiple G-proteins to activate K$^+$ channels, is consistent with other models of the these K$^+$ channels. Following Hodgkin and Huxley (1952), a tetrameric structure has been established for most K$^+$ channels and this has been shown to account for the voltage-dependent and calcium-dependent properties of various K$^+$ channel subtypes (Hille, 1992). This is consistent with the present model, in which the K$^+$ channel associated with GABA$_B$ receptors might be described as a multimer composed of four identical subunits, with one G-protein binding site on each subunit. The model therefore predicts that each subunit must be bound to a G-protein before the K$^+$ channel can open, leading to a nonlinear stimulus/response relationship and slow rise time typical of GABA$_B$ responses.

The simplest model that reproduced the stimulus dependencies of GABA$_B$ responses had two variables and was obtained from Eq. 5.29:

$$\frac{dr}{dt} = K_1 [T] (1 - r) - K_2 r$$

$$\frac{ds}{dt} = K_3 r - K_4 s \tag{5.30}$$

$$I_{GABA_B} = \bar{g}_{GABA_B} \frac{s^n}{s^n + K_d} (V - E_K)$$

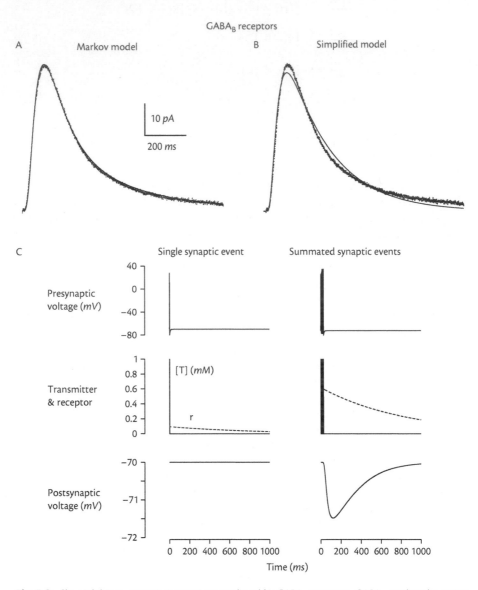

Fig. 5.9: Slow inhibitory synaptic interactions mediated by GABA$_B$ receptors. GABA$_B$-mediated currents (noisy trace; obtained from Otis et al., 1993; recorded at 33–35°C) compared to a kinetic model of GABA$_B$ responses (continuous trace; Eq. 5.29 with neurotransmitter modelled as in Fig. 5.2A). B. Same data fitted by a simplified model for GABA$_B$ receptors (Eq. 5.30) with neurotransmitter represented as a pulse (Fig. 5.2C). C. Behaviour of the model shown in B for a single presynaptic action potential (left panel) and a burst of ten action potentials at 300–400 Hz (right panel). A single action potential did not generate a detectable response (bottom-left) but a burst of high-frequency spikes generated a slow IPSP (bottom-right; same single-compartment model as detailed in Fig. 5.5C). Figure modified from Destexhe et al., 1998b.

where all symbols have the same meaning as in Eq. 5.29 with $r = [R]$ and $s = [G]$. The best fit of this model to whole-cell recorded $GABA_B$ currents (Fig 5.9B) was: $K_d = 100\ \mu M^4$, $K_1 = 9 \times 10^4\ M^{-1}\ s^{-1}$, $K_2 = 1.2\ s^{-1}$, $K_3 = 180\ s^{-1}$ and $K_4 = 34\ s^{-1}$ with $n = 4$ binding sites.

The main difference between this model and Eq. 5.29 is the absence of a desensitized state for the receptor. It was found that the desensitized state was necessary to account accurately for the time course of $GABA_B$ currents (Fig 5.9A), but had little influence on the properties of $GABA_B$ responses at the network level (see Section 6.5 in Chapter 6; Destexhe et al., 1996a).

5.2.7 Simplified models of second-messenger gated channels

A series of neuromodulatory pathways, acting on metabotropic receptors, innervate thalamic nuclei and cerebral cortex. These neuromodulators include acetylcholine (ACh), noradrenaline (NE), serotonin (5HT) and other substances such as dopamine, histamine, opioids, glutamate and GABA. These neuromodulators evoke slow intracellular responses in cortical and thalamic cells, mostly mediated by K^+ conductances (reviewed in McCormick, 1992). This neuromodulatory control of K^+ currents is also seen in cortical pyramidal neurons and is a general mechanism by which ascending neuromodulatory systems, originating in brain stem and basal forebrain, control the state of thalamic and cortical neurons (Steriade and McCarley, 1990; McCormick, 1992; Steriade et al., 1997). In addition to affecting conductances, neuromodulators also affect various metabolic pathways in the cell, up to gene regulation (Berridge and Irvine, 1989; Tang and Gilman, 1991; Gutkind, 1998; Selbie and Hill, 1998).

The second-messengers involved in these metabotropic responses are G-proteins, most probably through direct activation/deactivation of the target K^+ channel by the G-protein itself (Brown, 1990; Brown and Birnbaumer, 1990). Therefore, as for $GABA_B$ receptors, the stimulation of the receptor induces the formation of activated G-proteins, which bind to K^+ channels and affect their gating properties. The model given above for $GABA_B$ responses may therefore be valid for modelling other metabotropic receptors, with rate constants adjusted to fit the time courses reported for each particular response. For example, combined patch-clamp recordings and modelling of the K^+ channels affected by muscarinic receptors activated by ACh indicated that the channel activates according to an allosteric model with four G-protein binding sites (Yamada et al., 1993; Hosoya et al., 1996). This is similar to the four binding sites predicted in the model for $GABA_B$ responses above, suggesting that this model may be valid for describing these types of muscarinic responses as well.

5.2.8 Noradrenergic and serotonergic receptors

Simpler two-state models can also be designed to represent the action of metabotropic receptors. This is particularly relevant when there are no data available that would permit detailed models of these responses. Such a model was used to describe the action of noradrenaline (NE) and serotonin (5HT) on thalamic neurons (Destexhe et al., 1994b; see also Section 6.4 in Chapter 6).

NE and 5HT depolarize thalamic cells by blocking a leak K^+ current through G-proteins (McCormick and Wang, 1991). This action was represented by a simple two-state kinetic model in which the leak K^+ current was given by:

$$I_{KL} = \bar{g}_{KL}\, m\, (V - E_K) \tag{5.31}$$

where $\bar{g}_{KL} = 1\,nS$ is the maximal leak K^+ conductance, m is the fraction of open K^+ channels, V is the postsynaptic membrane potential and E_K is the potassium reversal potential. The activation of K^+ channels was assumed to follow a simple open/closed reaction represented by a first-order kinetic equation:

$$\frac{dm}{dt} = -\alpha\, [G]\, m + \beta\, (1 - m) \tag{5.32}$$

where $[G]$ is the intracellular G-protein concentration, and α and β are respectively the binding and unbinding rate constants. K^+ channels are open at rest ($m = 1$ if $[G] = 0$) and close after G proteins bind to them. To model the opposite effect of G-proteins on K^+ channel, the following equation can be used:

$$\frac{dm}{dt} = \alpha\, [G]\, (1 - m) - \beta\, m \tag{5.33}$$

Here, G binds on the closed form to open the channel, and α and β are respectively the binding and unbinding rate constants.

These models assumed that the presynaptic spike induces a long pulse of $[G]$ (80 to 100 ms), which leads to a prolonged effect of the channels. This procedure also leads to an analytically solvable model, similar to those investigated in Section 5.2.2.

5.2.9 Synaptic summation

The summation of postsynaptic potentials (PSPs) and postsynaptic currents (PSCs) is an important part of synaptic signalling. An additional complexity arises because of short-term synaptic facilitation and depression; impact of a particular synaptic event on the postsynaptic cell will depend on the previous history of a spike train. A sufficiently flexible model for synaptic transmission should be capable of representing all of these effects.

Perhaps the most commonly used model for postsynaptic currents is a template function, such as the *Alpha function*:

$$r(t - t_0) = \frac{(t - t_0)}{\tau_1}\, \exp[-(t - t_0)/\tau_1] \tag{5.34}$$

This template was originally introduced to fit a single PSP (Rall, 1967). Modelling the interaction between successive synaptic events using template functions requires the summation overlapping waveforms, which may be cumbersome since each event in time must be stored in a queue for the duration of the waveform and necessitates the calculation of several waveforms during this period (see Srinivasan and Chiel, 1993). In addition, there is no natural provision for saturation of the conductance, which may lead to inaccuracies.

The kinetic models used here provide a natural way to handle summation because receptor kinetics by their nature correctly integrate the effects of successive releases of neurotransmitter. The summation behaviour of a simple kinetic model for AMPA receptors is shown in Fig. 5.5C.

A burst of high-frequency presynaptic spikes leads to successive synaptic events that naturally summate within the kinetic formalism of the AMPA receptor model (Fig. 5.5C, right panel). A similar protocol is also shown for NMDA (Fig. 5.6C) and GABA$_A$ receptors (Fig. 5.8C). Because the membrane potential was sufficiently far from the reversal potential for these currents, the peak amplitudes of these summated PSPs varied approximately linearly with the number of presynaptic spikes. Thus the simplified two-state kinetic schemes account for the most basic properties of synaptic summation.

Although the models for AMPA, NMDA and GABA$_A$ receptors showed linear summation for voltages far from the reversal potential, the situation was radically different for GABA$_B$ receptors: isolated release events must not evoke a detectable response, since miniature and unitary GABAergic events do not contain a GABA$_B$ component (Miles and Wong, 1984; Otis and Mody, 1992; Thompson and Gahwiler, 1992; Thompson, 1994). On the other hand, GABA$_B$-mediated currents are reliably evoked when a burst of high-frequency presynaptic spikes occurs (Kim et al., 1997; Thomson and Destexhe, 1999). This nonlinear stimulus dependence typical of GABA$_B$ receptors can be accounted for by a model in which several G-protein bindings are needed to activate the K$^+$ channels associated to the receptors (Section 5.2.6; Destexhe and Sejnowski, 1995). This behaviour is illustrated in Fig. 5.9C. A single action potential did not generate a detectable response (Fig. 5.9C, left panel) but a burst of high-frequency spikes generated a slow IPSP (Fig. 5.9C, right panel). This property will be analysed in more detail in Section 5.3 and its consequences at the network level will be investigated in Chapter 8.

5.3 Models of synaptic transmission including extracellular diffusion of transmitter

Postsynaptic receptors are located outside the synaptic cleft and may be activated by neurotransmitter diffusion outside the active zone of the synapse, a phenomenon called *neurotransmitter spillover*. In order to investigate the conditions when spillover may occur, the extracellular diffusion of neurotransmitter must be taken into account. This was investigated by Destexhe and Sejnowski (1995) for GABAergic transmission.

5.3.1 Model of neurotransmitter spillover

The release, diffusion and uptake of GABA in the synaptic cleft was described by the following reaction-diffusion equations:

$$\frac{\partial T(\bar{x},t)}{\partial t} = f_{release}(\bar{x},t) - \frac{V_{max}\, T(\bar{x},t)}{T(\bar{x},t)+K_m} + D\,\nabla^2 T(\bar{x},t) \qquad (5.35)$$

where $T(\bar{x},t)$ is the concentration of GABA at point $\bar{x}$ and time t, and the three terms on the right side of Eq. 5.35 represent, respectively, the release, uptake and diffusion of GABA. The diffusion coefficient was $D = 8 \times 10^{-6}\, cm^2/s$, based on values of compounds of similar molecular weight (Atkins, 1986).

The extracellular space was simulated using a two-dimensional array of square compartments (0.5 $\mu m \times$ 0.5 μm) (Fig. 5.10A, B), representing the thin extracellular space between the postsynaptic neuron and processes emanating from other cells, including other neurons and astrocytes. The area of each compartment was that of a typical single synaptic terminal

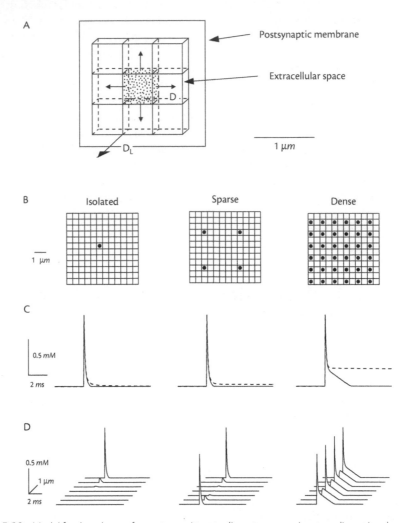

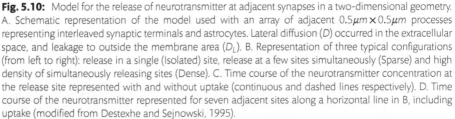

Fig. 5.10: Model for the release of neurotransmitter at adjacent synapses in a two-dimensional geometry. A. Schematic representation of the model used with an array of adjacent 0.5$\mu m \times$ 0.5μm processes representing interleaved synaptic terminals and astrocytes. Lateral diffusion (*D*) occurred in the extracellular space, and leakage to outside the membrane area (*D$_L$*). B. Representation of three typical configurations (from left to right): release in a single (Isolated) site, release at a few sites simultaneously (Sparse) and high density of simultaneously releasing sites (Dense). C. Time course of the neurotransmitter concentration at the release site represented with and without uptake (continuous and dashed lines respectively). D. Time course of the neurotransmitter represented for seven adjacent sites along a horizontal line in B, including uptake (modified from Destexhe and Sejnowski, 1995).

(Harris and Landis, 1986); the array therefore represented many interleaved synaptic and glial terminals in a volume of neuropil. The underlying assumptions were that (a) the width of the synaptic cleft (about 200 Å in Harris and Landis, 1986) and extracellular space is less than the typical size of the synaptic terminal, allowing a two-dimensional approximation; (b) the diffusion outside the area of terminals is negligible; (c) the diffusion is instantaneous inside each compartment.

When the presynaptic voltage crossed a threshold value of 0 *mV*, the release of GABA was simulated by transiently increasing the concentration of GABA by 1 *mM* in the corresponding compartment. For a cleft width of 200–500 Å, a peak neurotransmitter concentration of around 1 *mM* (Clements et al., 1992) would correspond to 3,000 to 7,500 molecules of released neurotransmitter (Bartol et al., 1993).

The GABA transporter is present in both the presynaptic terminals of interneurons and in astrocytes (Hertz, 1979). It was therefore included in all extracellular compartments. Uptake was modelled by a standard Michaelis-Menten equation, with a value of K_m of 4 μM, estimated from kinetic studies of GABA transporters (Clark and Amara, 1994). The value of V_{max} could only be roughly estimated from the literature, and was taken to be $V_{max} = 0.1\ M\ s^{-1}$ unless uptake was blocked.

In the absence of uptake, the slow decay of GABA was modelled in two ways: first, by simulating a large patch of postsynaptic membrane (900 μm^2) from which the neurotransmitter leaked out only through the borders, neglecting diffusion in the third dimension; second, in an alternative model, a leak was introduced in each compartment with a smaller diffusion coefficient ($D_L = 10^{-8} cm^2/s$; see Fig. 5.10A). Both methods gave slow decay times comparable to that estimated from experiments (Thomson and Gahwiler, 1992; Isaacson et al., 1993) but the latter model was more convenient.

The reaction-diffusion equations (Eq. 5.35) were integrated using a first-order explicit integration method with a discretization step of $\Delta x = 0.5\ \mu m$. The von Neumann criterion (see Press et al., 1986) gives a minimal time step of $\Delta t = \Delta x^2/2D \simeq 150\ \mu s$ for numerical stability, so $\Delta t = 10$ to $100 \mu s$ was used.

GABA$_A$ receptors have at least two binding sites for GABA and show desensitization (Busch and Sakmann, 1990; Celentano and Wong, 1994). However, blocking uptake revealed prolonged GABA$_A$ currents (Thomson and Gahwiler, 1992; Isaacson et al., 1993), suggesting that desensitization was minimal. Desensitization was neglected and these receptors were modelled using a simple first-order kinetic scheme similar to Eqs. 5.23–5.24:

$$\frac{dr}{dt} = \alpha\ [T]^2\ (1-r) - \beta\ r \qquad (5.36)$$

$$I_{GABA_A} = \bar{g}_{GABA_A}\ r\ (V - E_{Cl})$$

where the binding of two molecules of neurotransmitter, T, leads to the opening of the channel with rate constants of $\alpha = 2 \times 10^{10}\ M^{-2}\ s^{-1}$ and $\beta = 162\ s^{-1}$ (obtained by fitting the model to whole-cell recorded GABA$_A$ current; Fig. 5.11, top-left); the maximal conductance is $\bar{g}_{GABA_A} = 1\ nS$, r is the fraction of receptor in the open state, and $E_{Cl} = -80\ mV$ is the chloride reversal potential.

The model for GABA$_B$ receptors was identical to the detailed model given in Eq. 5.29.

5.3.2 Time course of GABA in the synaptic cleft

The three typical release configurations considered are shown in Fig. 5.10B–D. In the first configuration, release occurred at an isolated site and GABA was present in the cleft extremely briefly (Fig. 5.10C, left), consistent with other models (Busch and Sakmann, 1990; Bartol, 1991; Clements et al., 1992). GABA was practically undetectable 2 μm away from the release site (Fig. 5.10D, left). The decay of neurotransmitter was biphasic with a fast initial decay governed by lateral diffusion (initial time constant of $\Delta x^2/4D \simeq 80\ \mu s$) and a second slower component of low amplitude. The decay of the second component depended on the capacity (V_{max}) of GABA uptake and its time constant was of about 1.2 ms in the absence of uptake.

In the second configuration, GABA was released from sparsely spaced co-releasing sites and the time course of GABA in the cleft was almost as brief as an isolated site (Fig. 5.10C–D, middle). In the absence of uptake, the initial fast decay dominated by lateral diffusion was unchanged, but the slow component of decay was more prominent than at an isolated site.

In the third configuration, releasing GABAergic terminals were densely packed (Fig. 5.10B, right) and although there was still a fast decaying phase due to lateral diffusion, the presence of neurotransmitter was prolonged. In the absence of uptake, the concentration of GABA stayed relatively high and decayed slowly everywhere (Fig. 5.10C, right).

For intermediate configurations, a similar behaviour was obtained over a wide range of geometries, values of the diffusion coefficient and efficiency of uptake, provided the density of co-releasing terminals was adjusted accordingly. In particular, using reduced diffusion coefficients to account for tortuosity of extracellular space (Perez-Pinzon et al., 1995) led to equivalent results.

These simulations therefore show that the dwell time for GABA in the synaptic cleft is very brief if one or few synapses are activated. Lateral diffusion accounts primarily for the fast initial decay, with a time constant of about 80 μs, and a much lower amplitude, slower decaying component. However, when a high density of adjacent sites release GABA at the same time, the peak concentration was no greater than during isolated release, but its second decaying component was significantly enhanced, depending on the capacity (V_{max}) of the uptake mechanism. This suggests (a) that uptake is essential for maintaining a brief time course of GABA concentration when many adjacent sites are activated, as proposed earlier (Isaacson et al., 1993), and (b), that there is a prolonged presence of neurotransmitter when many adjacent sites co-release neurotransmitter.

5.3.3 Time course of GABAergic currents

For GABA$_A$ receptors, the model was first adjusted to reproduce whole-cell recorded GABA$_A$ currents (obtained from Otis and Mody, 1992). Using a single release site, the kinetic model of GABA$_A$ receptors gave an excellent fit to GABA$_A$ currents recorded in hippocampal cells (Fig. 5.11, top-left). For these values, release saturated the GABA$_A$ receptors (see Mody et al., 1994; Thompson, 1994).

If there is more than one G-protein binding site, the activation of GABA$_B$-mediated currents is sigmoidal and shows multiple exponential decay, as seen experimentally (Otis et al., 1993). Excellent fits to whole-cell recorded GABA$_B$ currents in hippocampal cells were obtained for both $n = 2$ and $n = 4$ binding sites for G-proteins (Fig. 5.11, top-right).

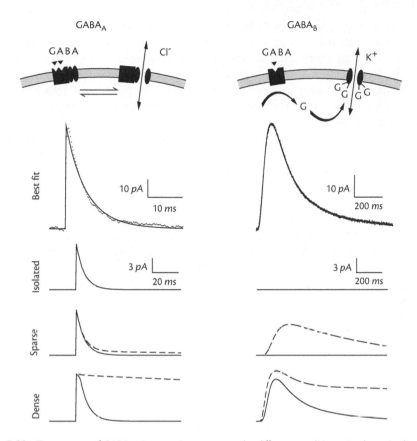

Fig. 5.11: Time course of GABAergic synaptic currents under different conditions. A schematic diagram is shown (top) For GABA$_A$ (left panel) and GABA$_B$ (right panel) receptors, as well as the time course of the current under different conditions. Best fit: the traces indicate the best fit obtained after running a simplex procedure to optimize the parameters (continuous traces). Whole-cell recorded GABAergic IPSCs were obtained from granule cells of the dentate gyrus (noisy traces; provided by T. Otis, Y. Dekoninck and I. Mody; see Otis and Mody, 1992; Otis et al., 1993). The traces below show GABAergic IPSCs at a single synapse for the three densities indicated in Fig. 5.10. The model IPSCs are shown in the presence (continuous trace) and in the absence of uptake (dashed lines) (modified from Destexhe and Sejnowski, 1995).

These kinetic models were tested by using different densities of co-releasing terminals. For release from isolated GABA terminals, the GABA$_A$ current was insensitive to uptake and no GABA$_B$ current was evoked, even when uptake was blocked (Fig. 5.11). For simultaneous release from adjacent terminals at a low density, the time course of both GABA$_A$ and GABA$_B$ IPSCs were indistinguishable from isolated release when uptake was present ('Sparse' in Fig. 5.11). However, blocking uptake evoked a prolonged tail in the GABA$_A$ current, and a GABA$_B$ response occurred for a relatively narrow range of densities of releasing terminals. Finally, for high densities of simultaneously releasing sites, both GABA$_A$ and GABA$_B$ IPSCs occurred and their time courses were prolonged in the absence of uptake ('Dense' in Fig. 5.11).

Because of receptor saturation, GABA$_A$-mediated currents were relatively insensitive to the density of terminals and the exact time course of GABA; the decay was dominated by the low value of the unbinding constant, β. In comparison, the amplitude of GABA$_B$-mediated currents was highly sensitive to the time course of GABA in the cleft, as explored in more detail below.

These simulations show that the multiple G-protein binding sites on K$^+$ channels confer to the GABA$_B$ response the cooperativity needed to account for its observed activation characteristics. A sufficient level of G-protein must be activated intracellularly in order to produce a detectable K$^+$ current. This occurs only when GABA$_B$ receptors are exposed to prolonged stimulation by GABA; this is consistent with the finding that GABA$_B$ currents can be evoked when neurotransmitter release is facilitated using sucrose (Otis et al., 1992).

Multiple G-protein binding sites also imply that there should be no GABA$_B$ component when GABA release is not prolonged. This is consistent with recordings of miniature IPSCs, which do not show any GABA$_B$ component and are unaffected by uptake blockers (Otis and Mody, 1992; Thompson and Gahwiler, 1992). Similar conclusions apply also to unitary IPSPs recorded from dual impalements (Miles and Wong, 1984) or IPSPs obtained from minimal stimulation (Isaacson et al., 1993). The present model suggests that in these experiments, release occurred at single—or distantly located—sites, and that the intracellular level of activated G-protein was insufficient to open the K$^+$ channels.

An extrasynaptic location of GABA$_B$ receptors was postulated to account for their properties (Dutar and Nicoll, 1988; Mody et al., 1994). This 'spillover hypothesis' proposes that multiple GABA release would result in extracellular accumulation of GABA. This extrajunctional GABA would reach levels sufficient to activate extrasynaptically located GABA$_B$ receptors only when multiple presynaptic spikes occur at high frequency or when GABA uptake is blocked. As an alternative to this 'spillover hypothesis', the 'G-protein hypothesis' (Destexhe and Sejnowski, 1995) reviewed here proposes that the nonlinear activation properties arise from the intra-cellular mechanisms involving G-protein cascades that underlie K$^+$ channel activation. In the latter case, the G-protein must reach a sufficiently high concentration before detectable IPSPs are activated. The need for activated G-protein to build up would also explain the minimal 10-20 ms onset latency, the sigmoidal rise and the multiexponential decay of GABA$_B$ IPSPs (observed experimentally by Otis et al., 1993). The model predicts that the prolonged time course of GABA with densely packed releasing sites should also activate extrajunctional GABA$_B$ receptors (Fig. 5.10D, dense).

5.3.4 Stimulus intensity dependence of GABAergic currents

In hippocampal slices, only strong stimulation in stratum radiatum evoked GABA$_B$ responses, and the ratio between GABA$_B$ and GABA$_A$ components varied with the stimulus intensity (Dutar and Nicoll, 1988; Davies et al., 1990). The model can account for these observations assuming that the intensity of stimulation affects the number of adjacent synaptic terminals releasing simultaneously. The amplitude of the GABA$_B$ current evoked under normal conditions depended on the density of releasing sites (Fig. 5.11). This effect is shown quantitatively in Fig. 5.12A where a single release event was simulated while varying the number of release sites. The total GABA$_A$ current increased linearly with the number of release sites, as predicted from Fig. 5.11. On the other hand, because of G-protein cooperativity, GABA$_B$ responses only appeared for the strongest stimuli, corresponding to the highest densities of terminals.

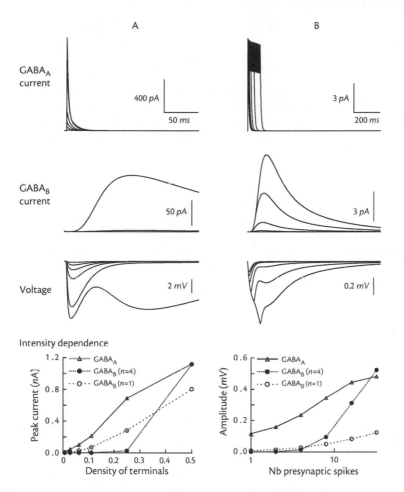

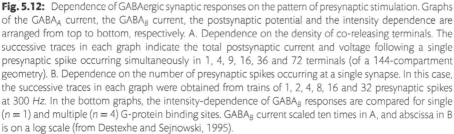

Fig. 5.12: Dependence of GABAergic synaptic responses on the pattern of presynaptic stimulation. Graphs of the GABA$_A$ current, the GABA$_B$ current, the postsynaptic potential and the intensity dependence are arranged from top to bottom, respectively. A. Dependence on the density of co-releasing terminals. The successive traces in each graph indicate the total postsynaptic current and voltage following a single presynaptic spike occurring simultaneously in 1, 4, 9, 16, 36 and 72 terminals (of a 144-compartment geometry). B. Dependence on the number of presynaptic spikes occurring at a single synapse. In this case, the successive traces in each graph were obtained from trains of 1, 2, 4, 8, 16 and 32 presynaptic spikes at 300 Hz. In the bottom graphs, the intensity-dependence of GABA$_B$ responses are compared for single ($n = 1$) and multiple ($n = 4$) G-protein binding sites. GABA$_B$ current scaled ten times in A, and abscissa in B is on a log scale (from Destexhe and Sejnowski, 1995).

In thalamic slices, RE neurons produce stereotyped high-frequency bursts of action potentials (Huguenard and Prince, 1994b; Kim et al., 1997). In the model, the GABA$_A$ and GABA$_B$ responses also depended on the presynaptic pattern of activity. High-frequency trains of presynaptic action potentials (300 Hz) were used to mimic the frequency of bursting neurons in the thalamus. During high-frequency release from a single terminal, the time course of GABA during each individual release event was identical to that for an isolated release site (as in Fig. 5.10C), with the same pattern repeated at 300 Hz. When increasingly long presynaptic bursts were delivered, the amplitude of the GABA$_A$ response increased proportionally (Fig. 5.12B), but the GABA$_B$ response only occurred for longer bursts (Fig. 5.12B). G-protein cooperativity was responsible for this effect.

The intensity dependence was highly influenced by the number of G-protein binding sites, n. A model with no cooperativity ($n = 1$) was optimized using the simplex procedure as described above. The simulations are shown in Fig. 5.12. For $n = 1$ the GABA$_B$ response was proportional to the stimulus, in contrast to the strongly nonlinear response with $n = 4$ (compare filled with open circles in Fig. 5.12, lower panels).

These simulations therefore show two different ways of activating GABA$_B$ responses. First, if GABA is released from a large number of terminals sufficiently close to each other, GABA spillover can occur and eventually provides the prolonged activation of GABA$_B$ receptors needed to evoke a detectable current (Fig. 5.12A). This situation is probably more typical of the hippocampus, where GABAergic terminals are relatively dense on the dendrites of hippocampal cells (Babb et al., 1988). Second, a prolonged activation of GABA$_B$ receptors can also occur at single terminals under high-frequency release conditions. This situation is probably more relevant for thalamic circuits, because the inhibitory neurons from the thalamic reticular nucleus generate bursts of high-frequency action potentials (see Chapter 4).

5.3.5 Slow IPSPs in neocortex

A recent combination of dual intracellular recordings and computational models (Thomson and Destexhe, 1999) was used to investigate the sensitivity of slow IPSPs to the number of spikes arising from a single interneuron. In three of eighty-five dual recordings involving an interneuron, no response was elicited by single presynaptic spikes, but a long latency ($\sim$20 ms) IPSP was elicited by a train of three or more interneuronal spikes at frequencies of 50 to 200 Hz. This slow IPSP was insensitive to bicuculline (Fig. 5.13A) and had an extrapolated reversal potential close to -90 mV. The amplitude of the slow IPSP varied nonlinearly with the number of presynaptic spikes (Fig. 5.13B1). No response was present for less than three to four spikes; the response increased sharply from four to ten spikes and showed a tendency to saturate for more than ten spikes. These data therefore show that the GABA$_B$ response is a highly nonlinear function of the number of presynaptic spikes, as predicted by the model (Destexhe and Sejnowski, 1995), and that the GABA released by a single interneuron can saturate the GABA$_B$ receptor mechanism(s) accessible to it.

A computational model of GABA release, diffusion and uptake similar to that in Section 5.3 was investigated based on data from the neocortex (Thomson and Destexhe, 1999). As in the thalamic model, extracellular accumulation of GABA alone could not account for the nonlinear relationship between spike number and IPSP amplitude (see details in Thomson and Destexhe, 1999). Different kinetic models were considered for how G-proteins activate K$^+$ channels, including a model similar to Eq. 5.30 (Fig. 5.13B2) and an allosteric model (Fig. 5.13B3). All models fit to experimental data predicted an optimum of $n = 4$ G-protein binding sites on K$^+$ channels, consistent with the tetrameric structure of K$^+$ channels (Hille, 1992).

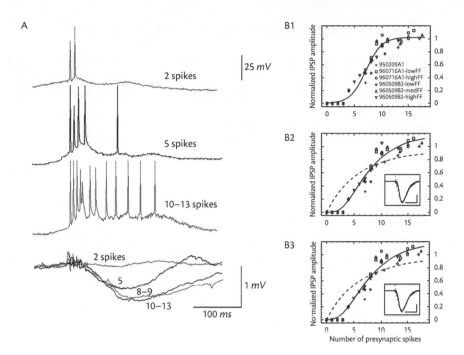

Fig. 5.13: Slow IPSPs in cerebral cortex are dependent on the number of presynaptic spikes. A. Dual intracellular recordings in rat cerebral cortex *in vitro*. The presynaptic cell elicited slow bicuculline-resistant IPSPs in the postsynaptic cell. The top traces show different patterns of presynaptic activity evoked by current injection and the bottom traces show the simultaneously recorded (presumably $GABA_B$-mediated) slow IPSPs in the postsynaptic cell. B. Nonlinear dependence of slow IPSP amplitude on the number of presynaptic spikes. The symbols indicate the normalized IPSP amplitudes for different experiments involving the same postsynaptic membrane potential and the same presynaptic firing frequency. The smooth curve is the best sigmoidal curve ($1/(1 + \exp[-(x-x_0)/K])$ fit to the spike number/IPSP amplitude plot and yielded an optimal fit for $x_0 = 7.1$ and $K = 1.4$ (units of number of spikes). B2. Same data shown with a model of $GABA_B$ responses involving multiple G-protein binding sites on K^+ channels (Eq. 5.30; continuous line, $n = 4$ binding sites; dashed line, $n = 1$ binding site). The inset shows the best fit of the $n = 4$ model to a slow IPSP obtained with ten presynaptic spikes (similar to Fig. 5.9C, right panel). B3. Same fit as in B2 but using an allosteric model for the K^+ channel regulation by G proteins. In this case, $n = 4$ binding sites also gave the best fits to the experimental data. Figure modified from Thomson and Destexhe, 1999.

In conclusion, models incorporating the diffusion of GABA in extracellular space were used to investigate the possible role of extrajunctional GABAergic receptors and the neurotransmitter spillover needed to activate them. The models reviewed here show that the cooperativity arising from multiple G-protein binding sites on K^+ channels can account for the properties of $GABA_B$ responses in hippocampus, thalamus and neocortex. Spillover of GABA may occur at densely packed releasing sites and may explain the hippocampal results. Differences in the high-frequency release conditions at single terminals may account for all the data from thalamic and neocortical slices. In all cases, $GABA_B$-mediated currents show a highly nonlinear stimulus/response relationship, which may have important consequences in generating epileptic discharges, as investigated in detail in Chapter 8.

5.4 Further developments on synaptic models

The biophysical models of synapses reviewed in the previous sections had simplified release mechanisms in the presynaptic terminal and macroscopic kinetic models for the postsynaptic responses. These models did not include synaptic plasticity, which ranges from short-term that last seconds, to long-term that can last many days.

A new class of particle-based models based on Monte Carlo techniques can model the diffusion and interactions of molecules in three-dimensional spatial compartments. In this chapter, the diffusion of neurotransmitter molecules in extracellular space was modelled by filling space with finite elements and updating concentrations with continuous reaction-diffusion equations. Another approach, simulated by the MCell environment, diffusion is modelled by the random walk of molecules in a neuropil reconstructed from electron microscopic cross sections (Kinney et al., 2013). This can be directly compared to experimental measurements of tortuosity and volume fraction. Synaptic conductances are modelled in MCell by the probabilistic binding of molecules to receptors, and by Markov transitions between open and closed states (http://mcell.org/) (Kerr et al., 2008).

MCell is computation intensive compared with traditional finite element techniques for large-scale problems, but gives more accurate results in the limit of small volumes and molecule numbers. This holds for synapses, where the number of receptors at many central synapses is less than a hundred and the number of free calcium ions inside a pyramidal spine head at resting levels is less than ten. The high degree of stochasticity of responses observed at central synapses is reflected in the stochastic simulations.

Using MCell, it is possible to develop highly realistic models of presynaptic release that account for paired-pulse facilitation from calcium accumulation and synaptic depression from vesicle depletion (Nadkarni et al., 2012). The regulation of long-term synaptic strengths can be modelled following the biochemical reactions triggered by the entry of individual calcium ions into a postsynaptic spine (Bartol et al., 2015). The results of these Monte Carlo simulations using accurate spatial geometries can then be used to constrain simplified models of synapses that retain the temporal complexities of the particle-based simulations (Garcia, 2017).

5.5 Discussion

In this chapter, we explored models of synaptic transmission, including simplified models of the calcium-dependent enzyme cascade implicated in neurotransmitter release, the gating of channels by neurotransmitter, and the G-protein cascade underlying the metabotropic action of GABA on K^+ channels via $GABA_B$ receptors. We have illustrated that the same formalism can be used to describe all these mechanisms and also generate simplified models. We also discuss the assumptions, advantages and drawbacks of this approach in the context of modelling large-scale networks, as considered in Chapters 6–8.

5.5.1 Modelling synaptic interactions

In Section 5.2, we have presented simple and detailed models for various types of synaptic receptors (AMPA, NMDA, $GABA_A$, $GABA_B$ and neuromodulators). We discuss below the advantages and drawbacks of the different models for a given receptor type.

5.5.1.1 Which is the best model?

Modelling chemical synaptic transmission began with studies of endplate currents at the neuromuscular junction (Katz, 1966). Early quantitative descriptions of these currents made use of simple kinetic models (e.g. Katz, 1966; Magleby and Stevens, 1972). Single-channel recordings led to the development of more sophisticated Markov models to describe the behaviour of individual ligand-gated channels (reviewed in Sakmann and Neher, 1995). As improved physiological, pharmacological, and molecular techniques have revealed additional complexities, the models have increased in sophistication accordingly. The most rigorous quantitative accounts of single-channel recordings now have Markov models with over a dozen states.

The most accurate synaptic model is not always the best model if the goal is to explore the effects of synaptic interactions in large-scale network models. Simplified models are needed that can faithfully reproduce the responses of the synapses over a wide range of conditions and be computationally efficient when simulating thousands or more synapses. We have found that even the simplest one- or two-variable Markov models are remarkably good at capturing the most salient properties of synaptic interactions, such as the rise and decay of single events and the summation of successive events (Section 5.2; Destexhe et al., 1994c, 1994d).

In the simplified model the time-course of neurotransmitter in the synaptic cleft was a brief pulse (Destexhe et al., 1994c; see also Staubli et al., 1992). This pulse waveform approximated the brief presence of neurotransmitter at high concentration in the synaptic cleft (Clements et al., 1992; Clements, 1996). The pulse waveform also allowed us to derive analytic expressions for the postsynaptic currents (Destexhe et al., 1994c, 1994d; Appendix B). Because no equations must be solved, this type of model is computationally highly efficient and is faster than other algorithms in simulating a large numbers of convergent synapses (Lytton, 1996; see Appendix B).

5.5.1.2 Approximations of kinetic models

Kinetic equations involving concentrations, fractions of channels, and rate constants rely on ensemble averages over a large populations of channels or other proteins. These assumptions break down when a single synapse is considered. First, the release of neurotransmitter-containing vesicles in response to depolarization following spike invasion is a probabilistic process (Katz, 1966). Whereas a kinetic model gives the same, reliable neurotransmitter release following each action potential, actual release at terminals is all-or-none and the probabilities of release are less than 0.1 at some central synapses (Hessler et al., 1993; Rosenmund et al., 1993; Murthy et al., 1997). One could take into account this stochasticity by making the release probabilistic, i.e. by triggering a pulse of neurotransmitter following a presynaptic action potential conditional on a random variable.

A second problem in the application of kinetic equations to individual synapses is that there may be a relatively small number of postsynaptic receptors at a release site. The liberation of a quantum of neurotransmitter at central synapses may open only about 5–50 postsynaptic receptors (Edwards et al., 1990; Hestrin, 1992; Traynelis et al., 1993). Consequently, individual synaptic currents have significant variability due to probabilistic channel openings (see e.g. Silver et al., 1992). When this variability is important, stochastic rather than deterministic models would be appropriate. A model incorporating Markov kinetics could incorporate Monte-Carlo methods in order to simulate the behaviour of individual channels rather than population averages (see e.g. Wathey et al., 1979; Bartol et al., 1991; Stiles et al., 2000).

Finally, the representation of synaptic currents by models involving only two states of the channel neglects interactions such as receptor desensitization. For AMPA receptors, desensitization may contribute to the fast synaptic depression observed at neocortical synapses (Thomson and Deuchars, 1994; Markram and Tsodyks, 1996), but its contribution was minimal in the hippocampus even for high-frequency stimulation (Stevens and Wang, 1995). Desensitization is also minimal for $GABA_A$ receptors (Busch and Sakmann, 1990; Celentano and Wong, 1994), and was also a minor effect for $GABA_B$ receptors (see Section 5.2.6). Two-state models neglect desensitization but reproduce the correct rise and decay time of the currents, as well as their nonlinear summation.

5.5.1.3 Metabotropic responses and link to biochemistry

The kinetic formalism also applies to synaptic interactions through metabotropic receptors, which involve mechanisms for intracellular biochemical transduction. The case of $GABA_B$ receptors considered here is representative of a wide range of physiological responses in which G-proteins link receptor stimulation to ion channel activation. Here again, an accurate biophysical model of these processes would require many variables to represent faithfully all biochemical operations involved. However, it was shown that a simplified kinetic model involving only two variables captured both kinetic properties and summation behaviour of $GABA_B$ IPSPs. These simplified models will be used in subsequent chapters to represent synaptic interactions in networks involving up to several hundred neurons (Chapters 6–8).

Kinetic models therefore allow processes to be described in a common formalism that are as diverse as voltage-dependent channels, calcium-activated channels, neurotransmitter-gated channels and channels gated by G-proteins (Destexhe et al., 1994d; Destexhe, 2000). Kinetic models are compatible with molecular and biochemical descriptions, which makes it possible to consider ion channels in a molecular and biochemical context as well as an electrical one. The intricate web of second-messengers, protein phosphorylation cascades, and gene regulation is just beginning to be integrated into computational models of neural activity. As these biochemical pathways are worked out, appropriate models can be used to understand the range of complex interactions that may occur. Kinetic models should be particularly useful because they naturally integrate electrophysiology with cellular biochemistry.

5.5.1.4 Application to interacting neurons

An important theoretical application of the kinetic description presented here is to the analysis of neural interactions. The general kinetic approach allows an entire network of neurons to be described by autonomous equations, even if individual cells contain voltage-dependent currents or if they interact via second-messenger pathways. The coupling between pre- and post-synaptic cells can be accomplished either by kinetic models, such as the presynaptic release mechanism described here, or more simply through functions that approximate the release process, such as Eq. 5.5. Although the set of equations describing a network of neurons might be extremely complex, it may still be possible to calculate the stability of steady state solutions, determine the existence of periodic solutions, analyse bifurcations, etc.

The models permit the assessment of how specific intrinsic properties of neurons contribute to the organization of the collective dynamics in neural populations. In central neurons, the presence of intrinsic currents can confer to the cell extremely complex properties (Llinás, 1988). These complex neurons interact through many different types of receptors, making the dynamics of interconnected populations of such neurons extremely difficult to understand intuitively. Investigating the behaviour of thalamocortical populations by computational models is one of the leading aims of this book. Chapters 6–8 illustrate how kinetic models of synapses can be applied to study network behaviour.

5.5.2 Model of synaptic responses including extracellular diffusion

Section 5.3 was devoted to more detailed models of synaptic interactions that included the extracellular diffusion of neurotransmitter and the possible 'spillover' to adjacent sites. This model was proposed (Destexhe and Sejnowski, 1995) to explain the properties of $GABA_B$ responses, which typically require high-intensity inputs to evoke release (Dutar and Nicoll, 1988; Davies et al., 1990; Huguenard and Prince, 1994; Kim et al., 1997; Thomson and Destexhe, 1999). Several hypotheses have been proposed for explaining the properties of $GABA_B$ responses (reviewed in Benardo, 1994; Mody et al., 1994; Thompson, 1994): (a) A co-released factor is needed to activate $GABA_B$ receptors; (b) $GABA_B$ receptors are located extrajunctionally and are activated by spillover; (c) different populations of interneurons mediate $GABA_A$ and $GABA_B$ responses; and (d) intrinsic sensitivity of $GABA_B$ responses to the number of presynaptic spikes (Destexhe and Sejnowski, 1995). The latter hypothesis was explored here using a model of the G protein transduction mechanism underlying $GABA_B$ responses, and the contribution of spillover was evaluated by modelling the extracellular diffusion of GABA.

5.5.2.1 The time course of GABA in the synaptic cleft

The model assumed that the time course of the neurotransmitter in the cleft is brief if only one or a few neighbouring synapses are activated. This is consistent with experimental data and other models (Busch and Sakmann, 1990; Bartol et al., 1991; Clements et al., 1992). Lateral diffusion accounts primarily for the fast initial decay, with a time constant of about 80 μs, and a more slowly decaying component of much lower amplitude. When a high density of adjacent sites released GABA at the same time in the model, the peak concentration was no greater than during isolated release, but the slowly decaying component was significantly prolonged, the duration depending on the capacity (V_{max}) of the uptake mechanism. This suggests that (a) uptake is essential for rapidly reducing the GABA concentration when many adjacent sites are activated, as proposed earlier (Isaacson et al., 1993), and (b), the presence of neurotransmitter may be prolonged when many adjacent sites co-release.

The extracellular space in this model was approximated by a two-dimensional surface, which is equivalent to assume that the distance between cells is small. Thus, the two-dimensional approximation neglected diffusion to neighbouring synapses through extracellular space in the third dimension. The relatively sharp spatial decay of the neurotransmitter concentration due to diffusion and uptake suggests that the effects of this approximation should be minimal, and that spillover should be limited to nearby terminals. A more accurate three-dimensional model of diffusion is being developed to investigate this point in more detail (Stiles et al., 2000).

When uptake was blocked, the effects on an isolated release site was minimal, consistent with the absence of effects of uptake blockers on miniature events (Thompson and Gahwiler, 1992). However, when GABA was released from many adjacent sites simultaneously, the concentration remained elevated throughout the area surrounding the terminals. In this case, the amplitude of GABA depended on the density of release sites and its slow decay depended on the geometry of the re-uptake system and the leak of neurotransmitter to extracellular space outside this region. These predictions are consistent with the markedly prolonged GABAergic currents observed following application of uptake blockers (Thomson and Gahwiler, 1992; Isaacson et al., 1993).

Note that this type of model neglected the possible effect of depletion of extracellular calcium when many synapses are simultaneously activated. Because the extracellular space is so reduced, this effect may be significant in some cases, such as for example when both release events and backpropagating action potentials occur within a small volume of neuropil (Egelman and Montague, 1999).

5.5.2.2 The intrinsic properties of GABA$_B$ responses

The central assumption—and prediction—of this model is that the characteristic properties of GABA$_B$ responses are a consequence of mechanisms intrinsic to single terminals containing these receptors, while interactions between different release sites, such as spillover, are insufficient to fully account for GABA$_B$ interactions. The model thus predicts that GABA$_B$ responses should be intrinsically sensitive to the number and frequency of presynaptic spikes, a prediction that is confirmed by the sensitivity to number of spikes observed in dual impalements (Kim et al., 1997; Thomson and Destexhe, 1999).

The model assumed that these intrinsic properties of GABA$_B$ responses occurred as a consequence of the need of multiple independent G-protein bindings to activate the K$^+$ channels associated with GABA$_B$ receptors. The multiplicity of G-protein binding sites had previously been suggested as an explanation for the multiexponential time course of the GABA$_B$ current (Otis et al., 1993). This hypothesis was taken a step further here by exploring the possible effects of multiple binding sites in generating the dynamics of the response patterns of GABA$_B$-mediated synaptic interactions.

With several G-protein binding sites, a sufficient level of G-protein must be activated intracellularly in order to produce a detectable K$^+$ current. This implies that a prolonged activation of the receptors must occur to evoke GABA$_B$ responses. This property can account for the following observations: (a) GABA$_B$ currents can be revealed by facilitating the release of neurotransmitter with sucrose (Otis et al., 1992b). (b) There is no GABA$_B$ component in miniature IPSCs (Otis and Mody, 1992a; Thompson and Gahwiler, 1992), in unitary IPSPs recorded from dual impalements (Miles and Wong, 1984) or in IPSPs obtained from very weak stimulation (Isaacson et al., 1993). In the model this occurred when the release was at single or distantly-located sites. (c) GABA$_B$ currents show multiexponential decay, a 10–20 *ms* delay of onset and a sigmoidal rising phase (Otis et al., 1993; see Fig. 5.11B). The delay occurred in the model because the active G-protein needed time to build up and reach a level sufficient to activate the K^+ channels. Other potential mechanisms may also contribute to the sigmoidal rising phase (see Hille, 1992).

Several different model reached the same conclusion that cooperativity in the activation properties of $GABA_B$ responses is critical to explain experimental measurements. These included a detailed model of $GABA_B$ transduction (Destexhe et al., 1994d), simplified models with pulses of neurotransmitter (Eq. 5.30; Destexhe et al., 1996a, 1998b), as well as allosteric and Hodgkin-Huxley type models (Thomson and Destexhe, 1999). These models produced similar results, but only if multiple G-protein binding sites were included. The need for cooperativity was also predicted in another independent modelling study of thalamic interactions (Golomb et al., 1996).

The predicted multiplicity of G-protein binding sites could be tested experimentally by application of activated G-proteins on membrane patches (see VanDongen et al., 1988), or by voltage-clamp experiments. Application of G-proteins on membrane patches was performed for muscarinic K^+ channels and a nonlinear response was observed with a Hill coefficient greater than 3 (Ito et al., 1992); thus, at least in this system, several G-protein bindings are needed to activate the K^+ channels. Other kinetic studies have also suggested that G proteins act on ion channels at multiple binding sites (see Yamada et al., 1993 for the muscarinic current; Boland and Bean, 1993; Golard and Siegelbaum, 1993 for Ca^{2+} currents in sympathetic neurons).

Finally, it is worth noting that the optimal $n = 4$ binding sites found in the model is consistent with the subunit composition of K^+ channels. K^+ channels are tetramers with four identical subunits (Hille, 1992), which implies that these channels have four identical sets of binding sites for G-proteins, but whether one or several bindings are necessary to activate the channel is unknown. One can speculate that, if K^+ channels are gated in a manner that is similar to that in the model of Hodgkin and Huxley (1952)[1], then a plausible mechanism for activation of K^+ channels by G-proteins is that (a) each subunit has one G-protein binding site; (b) G-protein binding induces a conformational change setting the subunit in an activated state; (c) all four subunits must be activated for the channel to conduct K^+ ions. This mechanism would result in a fourth-power nonlinearity, similar to K^+-channel activation in the Hodgkin-Huxley model (see Eq. 2.14), as predicted by the present model for $GABA_B$ responses. In the model we assumed that the binding sites are independent of each other, but the same results can be obtained if they are concerted and influence each other.

5.5.2.3 Possible physiological consequences

One consequence is that the activation of $GABA_B$ responses is sensitive to the number of presynaptic spikes. This predicted property (Destexhe and Sejnowski, 1995) and has since been confirmed experimentally using dual impalements (Kim et al., 1997; Thomson and Destexhe, 1999; see Fig. 5.13). Thus, a single axon can induce $GABA_B$ responses[2] and can even saturate

---

[1] All four n-gates of the K^+ channel must be open for the channel to conduct ions.

[2] In contrast to the spillover activation of extrasynaptic $GABA_B$ receptors, which would require the synchronized release at many terminals.

the response (Fig. 5.13B1). However, detectable $GABA_B$ responses were seen only when several (>3) presynaptic spikes occurred at a high frequency (> 50–100 Hz).

The sensitivity of $GABA_B$ responses to the number of presynaptic spikes has the properties of a 'high-pass filter'. $GABA_B$ responses would not occur in response to tonic spike activity at low-frequency (10–40 Hz), but would be fully activated by high-frequency bursts of action potentials (100–300 Hz). Tonic activity is common in cortical and thalamic neurons during states of vigilance, whereas the same neurons show bursts of action potentials during sleep (Hubel, 1959; Evarts, 1964; Steriade, 1978). The model therefore predicts that $GABA_B$ responses would not be activated when neurons display tonic activity (during waking or REM sleep), but could be activated when neurons generate bursts of action potentials (during sleep or epileptic seizures). In agreement with this filtering effect, the slow component of IPSPs disappears in the thalamus after brain activation (Steriade and Deschênes, 1984).

Another consequence of these properties is that $GABA_B$ responses seem to act as an 'emergency brake', which would only be invoked when interneurons produce excessive discharges. $GABA_B$ inhibition may therefore be activated in states where excessive discharges are produced, such as typically during epileptic seizures. Indeed, $GABA_B$ receptors have been shown to play a critical role in some forms of experimental seizure activity (Hosford et al., 1992; Snead, 1992). The sensitivity of $GABA_B$ responses to the number of spikes is central to our explanation for these types of epileptic patterns (see details in Chapter 8).

5.6 Summary

This chapter has reviewed kinetic models for synaptic interactions. As in Chapter 2, the general class of Markov kinetic models accurately reproduced many of the properties at the single-channels associated with the main receptor types, including ionotropic and metabotropic receptors. Such models have been used to account for the fine details in experimental data at the single-channel and the single-synapse level.

The same class of Markov models also allowed us to generate simplified models to represent synaptic interactions in network simulations. These simplified models must be computationally efficient when simulating thousands of synapses, but at the same time, they must capture the most salient properties of synaptic interactions. For ionotropic receptors (AMPA, NMDA, $GABA_A$), a simple two-state kinetic model (with an additional voltage-dependent term for NMDA receptors) fits these currents remarkably well. For metabotropic receptors, such as $GABA_B$ receptors, the response depends on G-protein transduction, which links receptor stimulation to channel activation. In this case, the minimal model that captured both kinetic properties and summation behaviour had two variables. These simplified models will be used in subsequent chapters to represent synaptic interactions in networks involving several hundred neurons (Chapters 6–8).

Neighbouring synaptic terminals may interact via 'spillover' of neurotransmitter. A two-dimensional model of extracellular space was used to investigate the extracellular diffusion of neurotransmitter (Section 5.3). The neurotransmitter time course depended on several processes: release, lateral diffusion, uptake and binding to postsynaptic receptors. The purpose of this model was to evaluate the extent to which 'spillover' occurs between neighbouring terminals. Simulations support the prediction that hippocampal GABAergic interactions have significant spillover effects. However, spillover need not be invoked to explain data from thalamus and neocortex, where the observed $GABA_B$-mediated interactions can be explained in a model where four G-proteins are needed to activate a K^+ channel, consistent with its tetrameric structure. This model predicts that $GABA_B$-mediated synaptic interactions are intrinsically sensitive to the number of presynaptic spikes, a property that may have important consequences for the genesis of epileptic discharges, as explored in Chapter 8.

Spindle oscillations in thalamic circuits

In the preceding chapters, we presented models of thalamic relays cells, reticular thalamic cells and synaptic interactions. The stage is now set to explore the genesis of thalamic spindle oscillations. This has become possible because of the extensive experimental data available for modelling thalamic oscillations, from *in vivo* recordings of large-scale networks to *in vitro* recordings from synaptic and voltage-dependent ion channels in single cells. Some of these experimental results are apparently inconsistent. We show here that large-scale network modelling can provide a single coherent framework that unifies these results.

6.1 Experimental characterization of sleep spindle oscillations

Spindle waves are a hallmark of slow-wave sleep. In the humans electroencephalogram (EEG), which is recorded from the scalp, these oscillations are grouped in short 1–3 s periods of 7–14 *Hz* oscillations, organized within a waxing-and-waning envelope, that recur periodically every 10–20 seconds. These oscillations typically appear during the initial stages of slow-wave sleep (stage II). In cats and rodents, spindle waves of similar characteristics appear during slow-wave sleep, and are typically more prominent at sleep onset. They are enhanced by some anesthetics, such as barbiturates, which, when administered at an appropriate dose, generate an EEG dominated by spindles (see Andersen and Andersson, 1968). In ferret thalamic slices, spindle oscillations occur spontaneously (von Krosigk et al., 1993). This *in vitro* model of spindle waves has made it possible to precisely characterize the ionic mechanisms underlying spindle oscillations, as shown below.

The first hint that spindles could originate outside the cerebral cortex was from Bishop (1936), who showed that rhythmical activity was suppressed in cerebral cortex following destruction

Thalamocortical Assemblies: Sleep Spindles, Slow Waves and Epileptic Discharges. Second Edition. Alain Destexhe and Terrence J. Sejnowski, Oxford University Press. © Oxford University Press 2023.
DOI: 10.1093/oso/9780198864998.003.0006

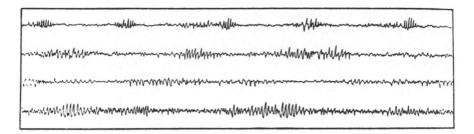

Fig. 6.1: Persistence of spindles in the thalamus following decortication. Field potential recordings in the thalamus of a bilaterally decorticated cat with optic nerves divided and brainstem transected at the intercollicular level. Traces from top to bottom show recordings obtained after 2, 8.45, 19 and 72 hours after operation. Figure reproduced from Morison and Bassett, 1945.

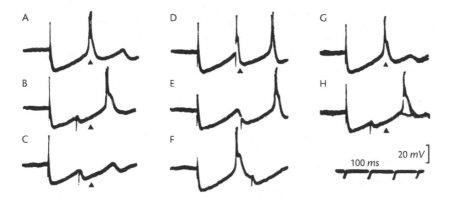

Fig. 6.2: Importance of the inhibition in shaping thalamic oscillations. Intracellular recording from a thalamic cell from the VPL nucleus eight days after removing sensorimotor cortex (resulting in the degeneration of all cortico-thalamic fibres). A. Stimulation of the internal capsule antidromically activated the TC cell, which was immediately followed by an IPSP and a rebound burst (arrow). B–H show the effect of a second stimulation at various latencies. The second stimulus delayed (B, H) or abolished the rebound (C, G). When the stimulus was coincident with the rebound, the rhythmic activity was reinforced (D). Figure reproduced from Andersen and Sears, 1964.

of its connections with the thalamus. Bremer (1938) showed that rhythmical activity is still present in the white matter following destruction of the cortical mantle. Later, Adrian (1941) and Morison and Bassett (1945) observed that spindle oscillations persist in the thalamus upon removal of the cortex, providing firm experimental evidence for the genesis of these oscillations in the thalamus (Fig. 6.1). These experiments led to the development of the 'thalamic pacemaker' hypothesis (Andersen and Andersson, 1968; Steriade and Deschênes, 1984), according to which rhythmic activity is generated in the thalamus and communicated to the cortex, where it entrains cortical neurons and is responsible for the rhythmical activity observed in the EEG.

Andersen and Eccles (1962) were the first to propose a cellular mechanism for the genesis of spindle oscillations. In intracellular recordings from thalamocortical (TC) relay neurons during spindles, they reported that TC cells fired bursts of action potentials interleaved with inhibitory postsynaptic potentials (IPSPs) (Fig. 6.2). They suggested that TC cells fire in response to IPSPs (post-inhibitory rebound), which was later demonstrated to be a characteristic electrophys-iological feature of thalamic cells (see Fig. 3.2 in Chapter 3). In particular, they suggested that the oscillations arose from the reciprocal interactions between TC cells and inhibitory local-circuit interneurons. This mechanism was later incorporated into a computational model that provided a phenomenological description of the inhibitory rebound (Andersen and Rutjord, 1964).

The mechanism proposed by Andersen and Eccles (1962) was almost correct. Reciprocal connections between TC cells and thalamic interneurons have not been observed in anatom-ical studies. Subsequently, intrathalamic loops of varying complexity have been found with inhibitory neurons of the thalamic reticular (RE) nucleus, which receive collaterals from cor-ticothalamic and thalamocortical fibres and project to specific and nonspecific thalamic nuclei (Scheibel and Scheibel, 1966a). That 'TC-RE' loops could underlie recruitment phenomena and spindle oscillations was suggested by Scheibel and Scheibel (1966a, 1966b, 1967), shifting the role of the interneuron in the 'TC-interneuron' loops of Andersen and Eccles to the reticular thalamic neurons. In particular, they predicted that the output of the RE nucleus should be inhibitory (Scheibel and Scheibel, 1967) and that the inhibitory feedback from RE cells onto TC cells should be critical for the genesis of thalamic rhythmicity. This hypothesis was supported by the observation that the pattern of firing of RE neurons was tightly correlated with IPSPs in TC neurons (Schlag and Waszak, 1971; Yingling and Skinner, 1977; Steriade and Deschênes, 1984).

Several critical experiments firmly established the involvement of the RE nucleus in the generation of spindles in cats in vivo (Steriade et al., 1985, 1987). First, cortically projecting thalamic nuclei lose their ability to generate spindle oscillations if deprived of input from the RE nucleus (Fig. 6.3; Steriade et al., 1985). Second, the isolated RE nucleus can itself generate rhythmicity in the spindle frequency range (Fig. 6.4; Steriade et al., 1987). In these experiments, the RE nucleus in the rostral pole, the thickest region of the RE nucleus, was surgically isolated from dorsal thalamic and cortical afferents. This isolation created an isolated 'island' of RE cells (Fig. 6.4A), where blood supply was preserved and in which the only remaining afferents were fibres from the brainstem and basal forebrain. Extracellular field potentials from the isolated RE nucleus showed rhythmicity in the same frequency range as in the intact thalamus (Fig. 6.4B; Steriade et al., 1987). This observation suggested that the RE nucleus is the pacemaker of spindle activity and that oscillation in TC cells were entrained by rhythmic IPSPs from RE cells.

Thus, thalamic rhythmicity could be explained by three different mechanisms: the 'TC-interneuron' loops of Andersen and Eccles, the 'TC-RE' loops of Scheibel and Scheibel, and the 'RE pacemaker' hypothesis of Steriade. The recent introduction of an in vitro model of spindle waves in young ferrets (von Krosigk et al., 1993) supported the second of these mechanisms. Slices of the visual thalamus that contain the dorsal (lateral geniculate nucleus or LGN) and reticular nuclei (perigeniculate nucleus or PGN) as well as the interconnections between them generated spindles spontaneously (Fig. 6.5; von Krosigk et al., 1993), confirming earlier experimental evidence (Adrian, 1941; Morison and Bassett, 1945) for the genesis of spindles in the thalamus.

The in vitro preparation allowed detailed pharmacological investigation of the ionic currents and synaptic receptors underlying the spindle oscillations. In particular, the spindle waves disappeared after the connections between TC and RE cells were severed (Fig. 6.6), consistent with the mechanism based on intrathalamic TC-RE loops proposed by Scheibel and Scheibel

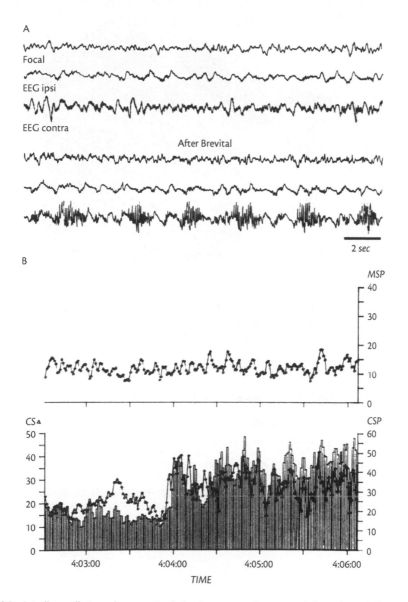

Fig. 6.3: Spindle oscillations disappear in thalamic neurons disconnected from the reticular nucleus. A. Field potentials recorded in the dorsal thalamus (CL nucleus) after surgical transection of the RE nucleus in one hemisphere. Spindle oscillations, either occurring spontaneously (top traces) or potentiated by brevital (bottom traces), were present only in the intact hemisphere. B. Graphs of the total power in the 7–14 *Hz* frequency range in ipsilateral (top) and contralateral (bottom) hemispheres. Brevital was given at 4:03:35 and potentiated spindles only in the intact hemisphere (bar graphs indicate power in 0.5–4.0 *Hz* range). Figure reproduced from Steriade et al., 1985.

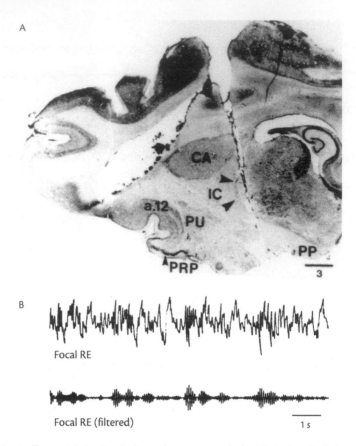

Fig. 6.4: The deafferented thalamic reticular nucleus generates rhythmicity in the spindle frequency range. A. Saggital section stained with cresyl violet. The rostral pole of the RE nucleus is indicated by arrows. Two surgical cuts were performed to isolate the RE nucleus from dorsal thalamus and from cerebral cortex. CA, caudate nucleus; IC, internal capsule; PP, pes pedunculi; PRP, prepyriform cortex; PU, putamen; a12, orbital area 12. B. Field potential recording in the isolated RE nucleus showing rhythmicity in the spindle frequency range (second trace, filtered between 7–14 *Hz*). Figure modified from Steriade et al., 1987.

(1996a, 1966b, 1967). This *in vitro* experiment also confirmed the observation that the input from RE neurons is necessary to generate spindles (Fig. 6.3; Steriade et al., 1985). However, the RE nucleus maintained *in vitro* did not generate oscillations without connections from TC cells (Fig. 6.6; von Krosigk et al., 1993), in contrast with the observation of spindle rhythmicity in the isolated RE nucleus *in vivo* (Fig. 6.4; Steriade et al., 1987).

Thus the experiments performed *in vitro* appear to agree with a mechanism by which oscillations are generated by the TC-RE loop, in contrast with the 'RE pacemaker' hypothesis. We show here how computational models suggested a way to reconcile these apparently contradictory experimental observations.

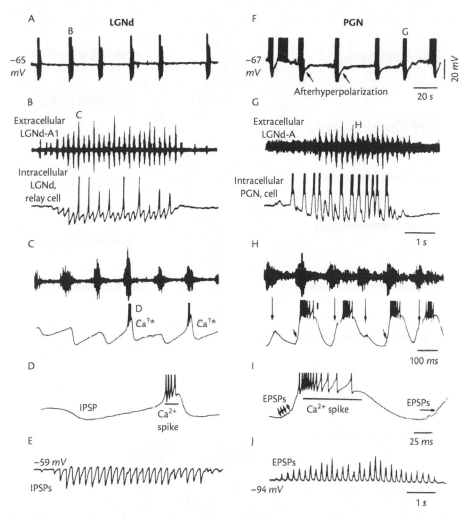

Fig. 6.5: Spindle waves *in vitro*. A. Intracellular recording from a neuron in the dorsal lateral geniculate nucleus (LGNd) revealed the occurrence of spindle waves approximately once every 20 s. B–C. Details of one spindle sequence recorded both intracellularly and extracellularly in LGNd cells. D. Detail of a burst discharge in a LGNd cell, which occurred following an IPSP. E. Injection of depolarizing current to maintain the cell at −59 mV inactivated the burst responses and revealed barrages of IPSPs during the spindle wave. F. Intracellular recording from a neuron in the perigeniculate (PGN) sector of the reticular nucleus. G. Expansion of one spindle wave recorded intracellularly in the PGN and extracellularly in LGNd. Detail of the simultaneous recording from G reveals that bursts in LGNd are associated with barrages of EPSPs in PGN (arrows). I. In some cases, groups of 3–5 EPSPs occurred at the same frequency as LGNd bursts (compare D and I). J. Maintaining the cell hyperpolarized at −94 mV revealed rhythmic barrages of EPSPs during a spindle. Figure reproduced from von Krosigk et al., 1993.

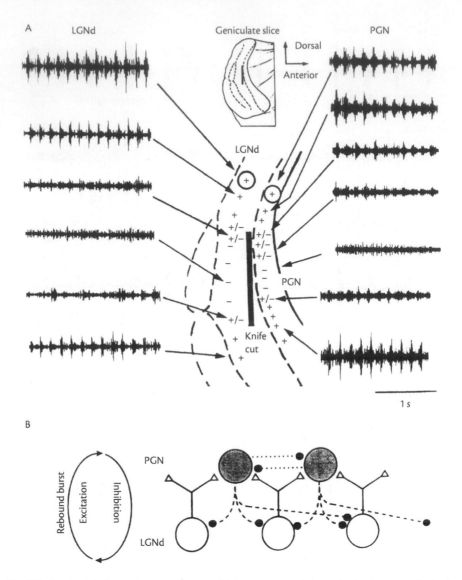

Fig. 6.6: *In vitro* spindle waves require functional interconnections between thalamic relay and reticular neurons. A. A small knife cut (1 *mm*) was performed between the LGNd and PGN in a thalamic slice from a ferret. Extracellular recordings at various locations of the LGNd and PGN revealed robust spindling in locations away from the cut (+), and the absence of spindling (−) in regions anterior and posterior to the centre of the cut. B. Mechanism proposed based on *in vitro* observations: the oscillations are generated by a loop involving interconnected PGN and LGNd neurons, with AMPA-mediated excitation (LGNd→PGN) and GABA$_A$-mediated inhibition (PGN→LGNd; PGN→PGN). Figure reproduced from von Krosigk et al., 1993.

6.2 Models of rhythmicity in the isolated reticular nucleus

That the thalamic reticular nucleus can generate spontaneous oscillations might not be expected, since all of the neurons in the RE are inhibitory and are interconnected. However, the discovery that thalamic RE neurons were capable of a rebound burst (Llinás and Geijo-Barrientos, 1988; Avanzini et al., 1989), similar to TC cells, suggests that these cells may sustain oscillations through sequences of reciprocal inhibitory rebounds. The genesis of oscillations in networks of inhibitory neurons possessing rebound burst properties was investigated earlier (Perkel and Mulloney, 1974), but in this type of model, the neurons typically oscillate anti-phase, inconsistent with the synchronized oscillations observed in the RE nucleus.

During the last few years, a series of computational models were designed to explain the genesis of synchronized oscillations in the RE nucleus (Destexhe and Babloyantz, 1992; Wang and Rinzel, 1992, 1993; Golomb and Rinzel, 1993, 1994; Destexhe et al., 1993c, 1994a, 1994b; Destexhe and Sejnowski, 1997; Bazhenov et al., 1999). These models are reviewed here with a view toward the more general problem of dynamics in inhibitory networks.

Two different hypotheses to explain the genesis of synchronized oscillations in the RE nucleus were proposed at about same time. Wang and Rinzel (1992, 1993) investigated the 'slow-inhibition hypothesis' postulating that networks of inhibitory neurons can generate synchronized oscillations if they interact through slow (presumably $GABA_B$ mediated) inhibition. In contrast, other investigators investigated synchronized oscillations based on fast (presumably $GABA_A$-mediated) type of inhibition (Destexhe and Babloyantz, 1992; Destexhe et al., 1994a), the 'fast-inhibition hypothesis'. In the Wang and Rinzel (1993) model, the oscillatory behaviour of two coupled RE neurons depended on the decay rate of inhibition: anti-phase oscillations typically arose for fast inhibition (Fig. 6.7A), whereas slow inhibition led to in-phase oscillations (Fig. 6.7B). These two different mechanisms for producing oscillations can be explained most clearly with phase plane diagrams (see Wang and Rinzel, 1992). One problem with the 'slow-inhibition hypothesis' was that it generated synchronized oscillations slower than the ~10 *Hz* frequency observed in the thalamic RE nucleus *in vivo* (Steriade et al., 1987).

In a model of the 'fast-inhibition hypothesis' for generating synchronized oscillations in the RE nucleus (Destexhe and Babloyantz, 1992) the RE cells were more densely connected and interacted through fast inhibitory synaptic interactions. The rebound burst properties of RE cells were based on the data available at that time (Llinás and Geijo-Barrientos, 1988; Avanzini et al., 1989; McCormick and Wang, 1991). Two-dimensional networks of RE cells were investigated, in which each RE neuron densely connected its neighbours within some diameter (see Fig. 6.8A). The connectivity was based on anatomical studies showing that RE cells are connected through GABAergic dendro-dendritic synapses (Deschênes et al., 1985; Yen et al., 1985). In addition, the axons emanating from RE cells give rise to collaterals in the RE nucleus before projecting to relay nuclei (Jones, 1985; Yen et al., 1985; Pinault et al., 1997). As these collaterals probably extend over larger distances than the size of the dendritic tree, a given RE cell should receive inputs from cells outside its immediate neighbourhood. Under these conditions, the network generated synchronized oscillations at a frequency of about 7 *Hz* (Fig. 6.8B). Unlike the model of Wang and Rinzel (1992, 1993), this model did not generate fully synchronized oscillations. The neurons in the network exhibited phases distributed around zero and the average membrane potential displayed oscillations reflecting the synchrony of the population (Fig. 6.8B).

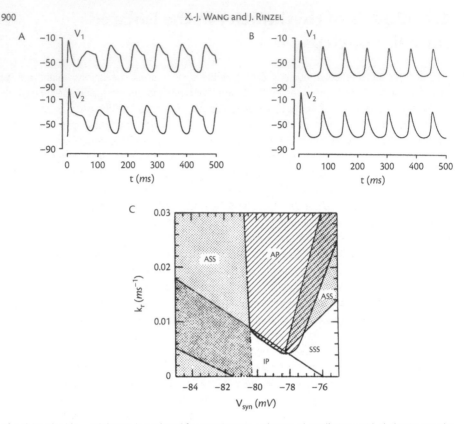

Fig. 6.7: The 'slow-inhibition hypothesis' for generating synchronized oscillations with thalamic reticular neurons. A. Anti-phase oscillation with fast synaptic decay ($k_r = 0.5\ ms^{-1}$). B. In-phase oscillation with a fast-rising and slow-decaying synaptic conductance ($k_r = 0.005\ ms^{-1}$). C. State diagram indicating the behaviour of the model as a function of the synaptic current decay (k_r) and reversal potential (V_{syn}). SSS, symmetric steady-state (blank); ASS asymmetric steady-state (stipled); IP, in-phase oscillation (shaded) as in B; AP, anti-phase oscillation (striped) as in A. The synchronous rhythmic behaviour was possible only for sufficiently slow inhibition (small k_r) and negative V_{syn}. Reproduced from Wang and Rinzel, 1993.

More recent biophysical measurements of the T-current (Huguenard and Prince, 1992) and other currents (Bal and McCormick, 1993) in RE cells allowed more realistic models of the RE nucleus to be developed. These were included in a model of the RE nucleus (Destexhe et al., 1994a), which was used to investigate the influence of connectivity and synaptic kinetics on the frequency of oscillation and conditions under which in-phase and anti-phase oscillations occur, as presented in the next section.

6.2.1 Model networks of RE cells

The model of an RE neuron in the network model used a single-compartment, as in Section 4.2.2. The network of N RE cells was given by:

A

B

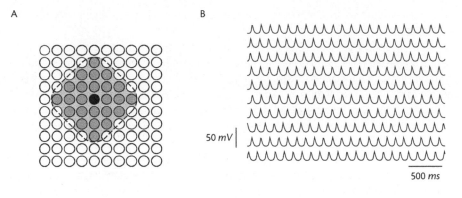

50 mV

500 ms

20 mV

Fig. 6.8: The 'fast-inhibition hypothesis' for generating rhythmicity in the thalamic reticular nucleus. A two-dimensional network of 100 thalamic RE neurons was simulated based on Hodgkin-Huxley kinetics for describing the rebound burst in RE cells. A. Scheme of network connectivity. Every cell contacts all its near neighbours within some radius. B. Oscillatory behaviour in networks of RE neurons interconnected with fast inhibitory synapses. Ten individual RE neurons are shown along the diagonal of a network of 100 RE neurons. The bottom trace in B displays the average membrane potential (calculated over the 100 RE cells), showing a synchronized oscillation around 7 Hz. Modified from Destexhe and Babloyantz, 1992.

$$C_m \dot{V}_i = -g_L (V_i - E_L) - f_I(V_i) - \sum_j I_{GABA}(V_i, V_j) \qquad (6.1)$$

where V_i is the membrane potential of neuron i ($i = 1 \ldots N$), the sum runs over all neurons j sending a connection to neuron i, $f_I(V_i)$ represents the intrinsic currents to cell i ($f_I = I_T + I_{K[Ca]} + I_{CAN} + I_{Na} + I_K$). The synaptic current is given by:

$$I_{GABA}(V_i, V_j) = \bar{g}_{GABA} \, r_{ji} (V_i - E_{GABA})$$

where the variable r_{ji} is the fraction of open receptors in the synapse from neuron j to neuron i, and depends on presynaptic activity as described in Chapter 5:

$$\frac{dr_{ji}}{dt} = \alpha \, [T_{ji}] \, (1 - r_{ji}) - \beta r_{ji} \qquad (6.2)$$

where r_{ji} is the fraction of open postsynaptic receptors in neuron i, $[T_{ji}]$ is the concentration of transmitter in the synaptic cleft released by neuron j and α and β are rate constants. When the presynaptic neuron j generates a spike ($V_j > 0$), $[T_{ji}]$ is set to 1 mM for 1 ms.

Neurons were arranged on a $\sqrt{N} \times \sqrt{N}$, two-dimensional square lattice with uniform connectivity: the same pattern of connectivity was used for each neuron in the network and all values of synaptic conductances were identical. Several types of connectivity were considered. The simplest case was the 'nearest-neighbour' connectivity, in which every cell connects to its four immediate neighbours. The standard connectivity was 'dense proximal' coupling in which every neuron connected to all other neurons within a fixed radius. Finally, 'fully connected' networks were studied, in which every cell connected 'all-to-all' to the whole array of cells in the network.

'Mirror' boundary conditions were used, in which connections near the boundaries of the square array were reflected back into the array. That is, a connection to a neuron outside the array was instead connected to the mirror image of that neuron in the array. This allowed every neuron in the network to receive the same number of synapses, therefore minimizing the effects of having a finite size. By increasing the size of the array it is possible to check whether these boundary effects are important, since their impact should decrease as the size of the array increases.

All neurons were initially set at their resting membrane potential, approximately −70 mV. Because RE neurons only display bursts in response to an external stimulus, a network of RE cells will remain silent unless an external perturbation occurs. Oscillatory behaviour was elicited by injecting a random hyperpolarizing pulse into each neuron during the first 2 seconds of the simulation. These pulses had 100 ms duration, amplitudes uniformly distributed between 0 and 0.05 nA and occurred at latencies uniformly distributed between 0 and 2 seconds. Following this initial perturbation during the first 2 seconds, the network developed self-sustained oscillatory activity (see below).

The average membrane potential was evaluated using:

$$< V > = \frac{1}{N'} \sum_i V_i \qquad (6.3)$$

where the sum runs over a subset of N' cells taken in the network ($N' \leq N$). For $N = 100$, all cells were averaged ($N' = N$), but for larger networks, a local average was performed by choosing a disk of $N' = 113$ cells in the centre of the network.

The variable $<V>$ represents the 'spatial' average value of the membrane potentials in the network and is a useful indicator of synchronized activity.[1] This variable can be compared to field potentials recorded in the isolated RE nucleus. Although the origin of field potentials is still not clearly established, it is likely that they result from a spatial summation of current sources from synapses and dendrites near the electrode (Nunez, 1981). In a network of single compartment neurons, estimates of the field potentials derived from current sources and from the average membrane potential have similar time courses (Destexhe and Babloyantz, 1992).

6.2.2 Oscillatory behaviour in simple circuits with GABA_A synapses

RE cells maintained *in vitro* are sensitive to GABA$_A$ agonists (McCormick and Prince, 1986; Spreafico et al., 1988; Bal and McCormick, 1993; Huguenard and Prince, 1994). Intracellular recordings *in vivo* reveal the presence of fast IPSPs in RE cells (Fig. 6.9; Destexhe et al., 1994a). Taken together, these data strongly suggested that fast IPSPs in RE cells are mediated by

---

[1] A 'synchronous' state is defined here as a state in which some neurons, but not necessarily all, oscillate in-phase. Note that neurons involved in a synchronized oscillation need not produce spikes since subthreshold oscillations of membrane potentials also contribute to $<V>$. Systems in which all cells oscillate in-phase will be described as 'fully synchronized'.

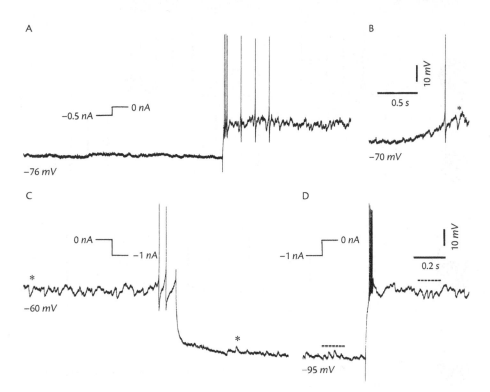

Fig. 6.9: Fast inhibitory synaptic potentials in RE cells *in vivo*. RE cells occasionally display short lasting IPSPs. A–B. An intracellular recording from a RE cell from the rostrolateral sector shows spontaneous short lasting IPSPs (*), lasting 20–40 *ms*. C–D. The IPSPs were inverted by injection of hyperpolarizing current (indicated by * and dashed line). Reproduced from Destexhe et al., 1994a.

GABA$_A$ receptors. *In vitro* experiments subsequently demonstrated that the strong inhibitory interactions between RE neurons are indeed mediated by GABA$_A$ conductances (Bal et al., 1995b; Ulrich and Huguenard, 1996; Sanchez-Vives et al., 1997a, 1997b; Zhang et al., 1997) although weak GABA$_B$ conductances are also present (Ulrich and Huguenard, 1996).

The properties of RE neurons interconnected with GABA$_A$ synapses was examined first with two interconnected neurons (Fig. 6.10). The strengths of the mutual inhibition was adjusted to produce sufficiently strong IPSPs to produce oscillations. Two types of oscillatory behaviour were observed depending on the pattern of connectivity of the two cells. In the absence of inhibitory self-connections or autapses (Fig. 6.10A), the two model RE cells produced alternating multiple bursts. The first cell (cell 1) produced a series of bursts in the range 9–11 *Hz* that induced a sequence of IPSPs in the companion cell (cell 2). These IPSPs deinactivated I_T in cell 2 and, when the bursts in cell 1 stopped, the release of the inhibition induced a series of bursts in cell 2. This cycle continued indefinitely with a repetition rate of about 1 *Hz*. Between periods of rhythmic burst firing, the membrane was hyperpolarized below the range of I_T activation, preventing I_{AHP} and I_{CAN} from activating because Ca^{2+} entry could not occur. In the following, 'Type 1 oscillations' will be defined as an oscillation in which at least two groups

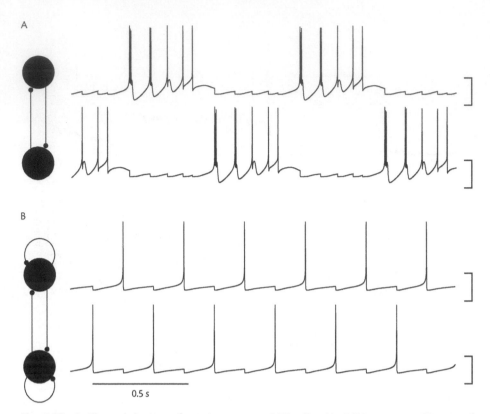

Fig. 6.10: Oscillatory behaviour of two interconnected RE cells with GABA$_A$ synapses. Diagram of connectivity is shown on the left and traces from simulations of the two neurons are shown on the right. A. Reciprocal interactions produce Type 1 oscillation. B. The addition of self connections produces Type 2 oscillation. $\bar{g}_{GABA} = 1$ μS (A) and 0.5 μS (B). Vertical calibration bars are from -100 to -50 mV (from Destexhe et al., 1994a).

of cells produce sequences of bursts in anti-phase alternation, separated by periods where the membrane is hyperpolarized. This is a common type of oscillatory mechanism for neurons coupled with mutual inhibition (Perkel and Mulloney, 1974). A similar oscillatory behaviour was also described by Golomb and Rinzel (1994) for mutually inhibiting cells.

Another type of oscillatory behaviour occurred when self-inhibitory were included (Fig. 6.10B). Cell 1 produced a single burst that terminated the bursting in cell 1 and inhibited cell 2. A rebound burst was produced in cell 2 and the cycle repeated at a frequency of around 6.5 Hz. During the interburst period, the hyperpolarized membrane potential prevented I_T, I_{AHP} and I_{CAN} from activating. Single cells with an auto-GABA$_A$ synapse do not oscillate (not shown). Although many combinations of parameter values were tested, synchronous bursting activity was never observed with GABA$_A$ synapses, unless their decay was slowed down significantly (see Section 8.2.1 in Chapter 8). In the following 'Type 2 oscillations' is defined as an oscillation in which the whole system shows a phase-locked subthreshold oscillations, but individual cells fire bursts only occasionally. In the case shown in Fig. 6.10B, the Type 2 oscillations are periodic and each of the two cells fires a single burst of spikes in alternation.

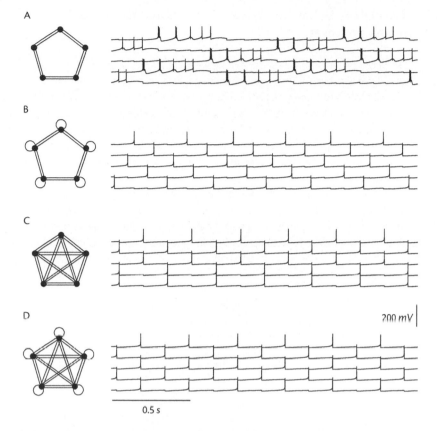

Fig. 6.11: Oscillatory behaviour of systems of five RE neurons with GABA$_A$ synapses. The patterns of connectivity are shown on the left and the corresponding traces from the five neurons is shown on the right. In A and B the connectivity is to nearest neighbours along a circle, and in C and D, full reciprocal connectivity was used. The neurons in B and D have self connections. Vertical calibration bars are 100 mV. $\bar{g}_{GABA} = 0.5\ \mu S$ (A), 0.33 μS (B), 0.25 μS (C) and 0.2 μS (D) (from Destexhe et al., 1994a).

Similar oscillations were found in more complex circuits. The oscillatory behaviour in a system of five RE neurons interconnected with GABA$_A$ synapses are shown in Fig. 6.11. With nearest neighbour connections, an alternating multiple burst activity was seen similar to the Type 1 oscillations (Fig. 6.11A). However, the five neurons did not alternate in two populations, but burst sequentially in an ordered pattern. In particular, neurons that were separated from each other by one intervening neuron tended to fire in pairs, with one following the other after a single burst. The configurations with nearest-neighbour with auto-synapses (Fig. 6.11B) and full connectivity (Fig. 6.11C–D) produced an oscillatory mode that was similar to the Type 2 oscillations seen in the two-cell model. There was a marked synchrony of the membrane potential of the five cells at around 6.5 Hz, although they never fired in unison.

6.2.3 Oscillatory behaviour in two-dimensional networks with GABA$_A$ synapses

Similar oscillatory patterns were also found in networks of 100 RE neurons whose connectivity was organized in a two-dimensional array, which better approximates the organization of the RE than full connectivity. The nearest-neighbour configuration is shown in Fig. 6.12A. This topology of the GABA$_A$ synapses favoured alternating multiple burst activity in neighbouring neurons (Fig. 6.12B) that was similar to the Type 1 oscillations described in the two-neuron model. In addition, there was a tendency for waves of bursting activity to sweep across the network (Fig. 6.12B). The average membrane potential of the network did not show any evidence of synchrony (Fig. 6.12C).

A more realistic pattern of connectivity for RE neurons is 'dense proximal connectivity', in which a given neuron connects to all other neurons within some radius, as shown in Fig. 6.13A. With connections from twenty-five neighbouring neurons including an auto-synapse, the activity of the network showed a complex, oscillatory dynamics at a frequency of 6.5–9 Hz (Fig. 6.13B). There was little evidence of alternating multiple burst activity and the dynamics of

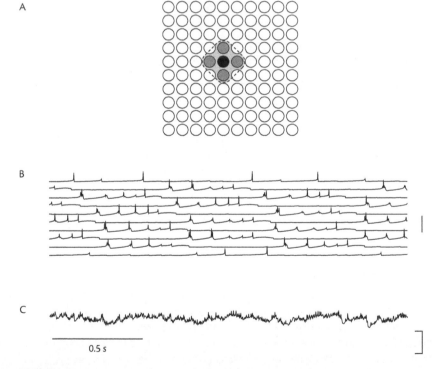

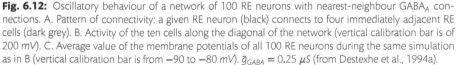

0.5 s

Fig. 6.12: Oscillatory behaviour of a network of 100 RE neurons with nearest-neighbour GABA$_A$ connections. A. Pattern of connectivity: a given RE neuron (black) connects to four immediately adjacent RE cells (dark grey). B. Activity of the ten cells along the diagonal of the network (vertical calibration bar is of 200 mV). C. Average value of the membrane potentials of all 100 RE neurons during the same simulation as in B (vertical calibration bar is from −90 to −80 mV). $\bar{g}_{GABA} = 0.25$ μS (from Destexhe et al., 1994a).

A

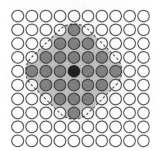

B

C

2 s

Fig. 6.13: Oscillatory behaviour of a network of 100 RE neurons with dense proximal GABA$_A$ connections. A. Pattern of connectivity: a given RE neuron (black) connects with twenty-four other RE cells (dark grey) and itself. B. Activity of the ten cells along the diagonal of the network (vertical calibration bar is of 200 mV). C. Average value of the membrane potentials of the 100 RE neurons during the same simulation as in B. Vertical calibration bar for the membrane potentials is from −90 to −80 mV). $\bar{g}_{GABA} = 0.04$ μS (from Destexhe et al., 1994a).

the oscillation was similar to the Type 2 oscillation observed in the two-neuron model. The main difference, however, was that the network with dense proximal connectivity produced a more disordered dynamics. The average membrane potential showed waxing-and-waning patterns of oscillatory activity (Fig. 6.13C) which corresponded to alternating periods of synchronization and desynchronization of the activity of the RE cells (see next section). This pattern of average activity was consistent with the waxing-and-waning patterns of field potentials recorded extracellularly in the RE nucleus after deafferentation (Fig. 6.4; Steriade et al., 1987).

Waxing-and-waning patterns of the average potential with a period of several seconds were observed for a large range of connectivity patterns; virtually all connection patterns except nearest neighbours and full connectivity gave waxing-and-waning Type 2 oscillations. The same type of oscillations was also observed in networks of various sizes (simulations were performed up to N = 1600 networks). This type of oscillation represents the prototypical behaviour of a two-dimensional network of RE neurons with GABA$_A$ synapses.

The effects of varying other parameters such as the conductances and the kinetics of synaptic interactions were also investigated. Blocking I_{CAN} greatly influences the firing patterns of the isolated RE cell and also affects Type 1 oscillations. However, this current had no detectable effect on Type 2 waxing-and-waning oscillations, which still occurred with I_{CAN} either completely blocked, or at a higher value (such as in Fig. 4.6D). Similarly, different sets of values for the kinetic constants of the IPSPs were tested (see for example values used in Destexhe et al., 1993c) and did not affect qualitatively the behaviour of the system. Finally, the conductances of the different currents, including synaptic conductances were tested in the range within approximately ± 50% of the values given here. Here again, a qualitatively equivalent behaviour was obtained.

6.2.4 Spatiotemporal dynamics of two-dimensional networks with GABA$_A$ synapses

The spatiotemporal patterns of activity underlying spindle rhythmicity were investigated in the two-dimensional network model. The dynamical behaviour of the network exhibited ordered spatial as well as temporal patterns of activity. Snapshots of the activity in the two-dimensional array were constructed by assigning a grey level to the value of the membrane potential of each RE neuron. A succession of such frames, played as a movie, allowed coherent patterns of activity to be tracked in the population of neurons.[2] Examination of Fig 6.14 for a network with 100 neurons revealed that, at any given moment, all neurons that were firing in phase form a coherent patch, or 'wave'. As time evolved, this wave of synchronous activity moved around the network, forming travelling patterns. This suggests that the spindle rhythmicity observed in the average firing activity of neurons in the network is a reflection of coherent waves of activity that recur on every cycle of the oscillation (Fig. 6.14).

Networks with 400 neurons displayed similar behaviour as networks with 100 neurons, although the patterns were more complex. Fig. 6.15 shows the spatiotemporal patterns of activity in a network of 400 RE neurons with dense proximal connectivity that produced waxing-and-waning Type 2 oscillations. The network displayed spatial patterns of propagating activity in the form of clockwise rotating waves or spiral waves (Fig. 6.15).

These travelling wave patterns were closely related to the oscillation period in the population activity. Comparison of these snapshots with the averaged voltage and single cell traces of Fig. 6.13 reveals that the periodicity of the waves of activity, 110 to 150 *ms*, corresponds closely to the oscillations observed in the averaged voltage (6.5–9 *Hz*). For $N = 400$, the period of these oscillations was 120 to 140 *ms* (assuming that the spiral-like wave is symmetric), which is in the 6.5–9 *Hz* spindle range. The travelling wave of activity involved every cell in the network, but the oscillation was subthreshold, only occasionally giving rise to a burst of spikes, as indicated by black squares in Figs. 6.14 and 6.15). Thus, although the average activity was dominated by the subthreshold 6.5–9 *Hz* oscillation, single cells fired irregularly, once every two or three cycles.

The waxing-and-waning envelope of the oscillation were long (several seconds) and corresponded in the spatiotemporal pattern to coherent waves that formed and dissolved. The activity of neurons in the network alternated between periods of coherent oscillatory activity, characterized by well-defined travelling waves, and periods of less coherent activity, in which the pattern of spatiotemporal activity was less organized (Fig. 6.16A).

---

[2] Computer-generated movies can be viewed at: http://www.cnl.salk.edu/~alain or http://cns.iaf.cnrs-gif.fr.

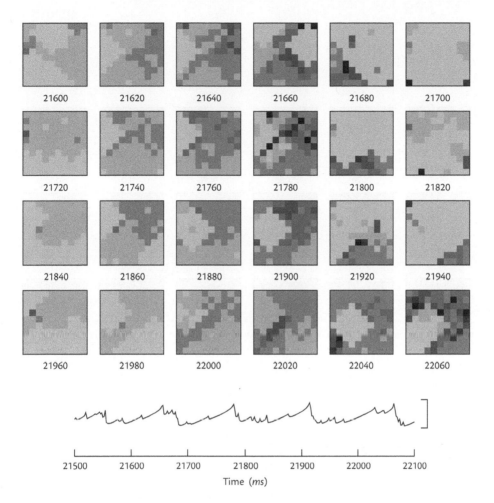

Fig. 6.14: Spatial activity of the 100 neuron network during waxing-and-waning oscillations. Same connectivity as in Fig. 6.13. Each square frame represents the activity of the 100 neurons, arranged in a 10 × 10 array. The value of the membrane potential for each neuron is shown as a grey scale ranging in 10 steps from −90 mV (white) to −60 mV (black). Membrane potentials higher than −60 mV are shown in black. Each frame is labeled by the time at which the snapshot was taken (time is in ms from the beginning of the simulation; interval between frames is 20 ms). The averaged activity of the network during the same period of time is shown at the bottom. Vertical calibration bar for the average membrane potential is from −80 to −70 mV (from Destexhe et al., 1994a).

The spatiotemporal patterns of activity became more complex when the size of the system was increased (Fig. 6.15), but travelling waves persisted with about the same wavelength. For large networks (N = 400, N = 1600), domains formed within the spatial array that oscillated independently, giving rise to an average activity over the entire network that was weaker in amplitude because the phases tended to cancel out (not shown). However, local domains of about 100 neurons in a large network showed waxing-and-waning patterns very similar to those observed in N = 100 networks (Fig. 6.16B). Thus, the waxing-and-waning patterns exhibited

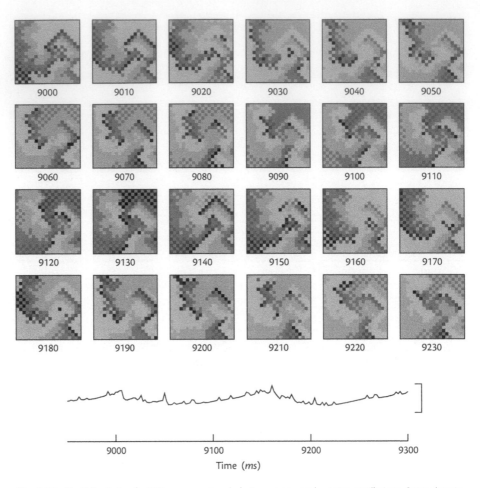

Fig. 6.15: Spatial activity of a 400 neuron network during waxing-and-waning oscillations. Same description as in Fig. 6.14 with nearest-neighbour connectivity and self-connections. In this case, a clockwise rotating spiral-like pattern of activity occurred transiently, as well as patterns similar to Fig. 6.14. Interval between frames were 10 *ms* and the averaged activity was evaluated by averaging over a disk of 113 neurons in the centre of the network. (from Destexhe et al., 1994a).

in small networks (Fig. 6.13) may be closely related to similar patterns observed within the domains of larger networks. This is important in comparing models to the much larger network of neurons found in the RE nucleus. Anatomical studies have found that the RE nucleus seems organized in tightly packed bundles, interleaved with regions with a much lower density of neurons (Scheibel and Scheibel, 1972). In agreement with this nonhomogeneous structure, Steriade et al. (1987) have reported that the field potentials from the RE recorded *in vivo* can display waxing-and-waning oscillations at one location but be silent if the electrode is slightly displaced to a neighbouring location in the RE. These data suggest that a relatively small number of RE neurons, perhaps no more than a few dozen, can generate the field potentials observed in the RE nucleus with extracellular recordings. Thus, the small sizes of the networks studied here may in fact match the relevant scale for collective phenomena occurring in the RE.

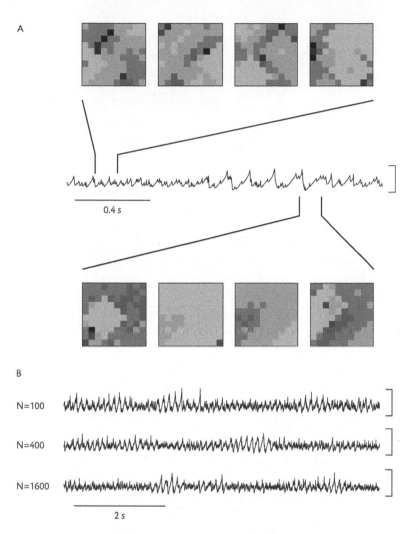

Fig. 6.16: Spatiotemporal patterns giving rise to waxing-and-waning activity. Snapshots of activity in a 100 neuron network during waxing-and-waning oscillations corresponding to the regions of the averaged membrane potential as indicated. The top series of snapshots was taken during the 'desynchronized' phase and shows highly irregular spatiotemporal behaviour. The bottom series of snapshots was taken during the 'oscillatory' phase, when the network is more synchronized and coherent oscillations were found in the averaged activity. The time interval between frames was 40 *ms*. B. Averaged membrane potentials for networks with $N = 100$, $N = 400$ and $N = 1600$ neurons. For $N = 400$ and $N = 1600$, the local average membrane potential was obtained by averaging over a disk of 113 neurons in the centre of the network. All simulations were performed under conditions identical to those in Fig. 6.13. Vertical calibration bars for the average membrane potential traces are from -80 to -70 *mV* (from Destexhe et al., 1994a).

6.2.5 Oscillatory behaviour in the presence of GABA$_B$ synapses

Application of GABA$_B$ antagonists has an insignificant effect on spindle oscillations *in vitro* (von Krosigk et al., 1993), suggesting that GABA$_B$ receptors are not essential to these oscillations.

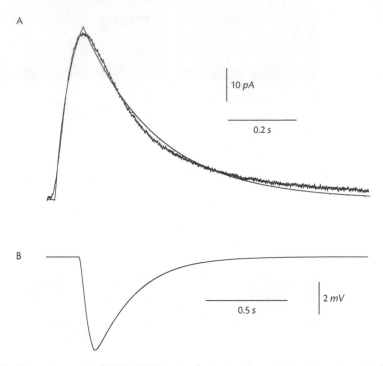

Fig. 6.17: First-order model of GABA$_B$ IPSPs. A. Best fit obtained from the kinetic-based model of synaptic currents (smooth curve), compared with GABA$_B$ currents obtained from whole-cell recordings in granule cells of rat dentate gyrus (noisy trace, from Otis et al., 1993). The model was fit to the data using a least squares procedure. B. Corresponding GABA$_B$ IPSP obtained from the model with a maximal conductance of $\bar{g}_{GABA} = 0.25$ nS (from Destexhe et al., 1994a).

However a weak postsynaptic GABA$_B$ component was seen in RE neurons (Ulrich and Huguenard, 1996). Indirect evidence also supports a prominent GABA$_A$ IPSP with a much weaker GABA$_B$ component (Huguenard and Prince, 1993). Although there is no evidence for purely GABA$_B$ synapses between RE cells, the strength of GABA$_A$ IPSP can be reduced with pharmacological blockers *in vitro* and may also be abnormally low in some pathological conditions *in vivo*. A modelling study has suggested that 7–14 Hz rhythmicity could result from slow inhibition between RE cells (Wang and Rinzel, 1992, 1993). In this section, the properties of RE neurons interacting by slow inhibition were explored with simulations (see details in Destexhe et al., 1994a).

A simple first-order kinetic model was used to represent GABA$_B$-mediated IPSPs. The current flowing through K$^+$ channels associated to GABA$_B$ receptors was given by:

$$I_{GABA_B} = \bar{g}_{GABA_B} \, m \, (V - E_K) \tag{6.4}$$

where $\bar{g}_{GABA_B}$ is their maximal conductance, m is the fraction of open K$^+$ channels, V is the postsynaptic membrane potential and E_K is the potassium reversal potential. The activation of K$^+$ channels was assumed to occur following a simple open/closed reaction represented by following first-order kinetic equation:

$$\frac{dm}{dt} = \alpha \, [G] \, (1 - m) - \beta m \tag{6.5}$$

where [G] is the intracellular G-protein concentration, and α and β are rate constants. By assuming that the occurrence of a presynaptic spike induces a long pulse of [G], an analytically solvable first-order model similar to those investigated in Section 5.2.2 can be obtained. Fitting this kinetic scheme to whole-cell currents obtained from hippocampal neurons (Otis et al., 1993) yielded a value of 1 μM for the pulse amplitude, a pulse duration of 85 ms and rate constants of $\alpha = 0.016\ ms^{-1}\mu M^{-1}$ and $\beta = 0.0047\ ms^{-1}$ (Fig. 6.17). The reversal potential was of $-95\ mV$, according to values estimated from rat TC cells ($-103 \pm 8\ mV$, J. Huguenard, personal communication) and guinea-pig RE cells (-94 to $-97\ mV$, Bal and McCormick, 1993). To yield IPSPs comparable in amplitude to the GABA$_A$-mediated IPSPs, the maximal conductance of the GABA$_B$ current was set 100-fold lower to compensate for the longer duration of the GABA$_B$ current.

Two types of oscillations were observed in a two-neuron circuit with GABA$_B$ synapses (Fig. 6.18). In the absence of GABA$_B$ self-connections (Fig. 6.18A), there was an alternating multiple burst activity, very similar to the Type 1 oscillation of Fig. 6.10A. However, the IPSP produced by the sequence of bursts was longer lasting and the period of the oscillation, around 3 seconds, was much longer. In the presence of self-inhibitory connections, a new type of oscillatory behaviour was observed (Fig. 6.18B). The two cells oscillated in full synchrony at

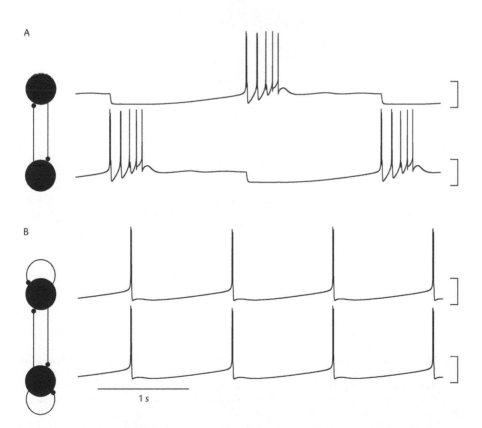

Fig. 6.18: Oscillatory behaviour of two interconnected RE cells with GABA$_B$ synapses. Identical configuration and parameters as in Fig. 6.10, but with slow inhibitory synapses. A. Reciprocal inhibitory connections produced Type 1 oscillations. B. Reciprocal and self connections produced Type 3 oscillations that were highly synchronized. $\bar{g}_{GABA} = 10\ nS$ (A) and 5 nS (B). Vertical calibration bars for the membrane potentials are from -100 to $-50\ mV$ (from Destexhe et al., 1994a).

a frequency of 0.5–1 *Hz*. Variations of the conductances values of all currents present in the system failed to produce oscillation at a frequency higher than about 2 *Hz*.

Unlike a single neuron with an auto-GABA$_A$ synapse, which does not oscillate, a single neuron with an auto-GABA$_B$ synapse displays oscillatory behaviour (not shown). This shows that a fully synchronized solution is possible in a network of RE neurons with GABA$_B$ synapses. In the following, such fully synchronized oscillation are defined as 'Type 3 oscillations'.

Circuits of 5 RE neurons (similar to Fig. 6.11) were also investigated. Type 1 oscillation only occurred for nearest-neighbour connectivity, as in Fig. 6.11A. Type 3 oscillations were only seen for full connectivity, in which each neuron connects to every other neuron (as in Fig. 6.11C–D). The other types of connectivities gave asynchronous oscillations.

In two-dimensional networks of RE neurons interacting with GABA$_B$ synapses, the dynamics appeared to be much less synchronous than with GABA$_A$ interactions. Networks with nearest-neighbour connections displayed a Type 1 oscillation (not shown) similar to those observed for GABA$_A$. The output from a network with dense proximal connections is shown in Fig. 6.19.

A

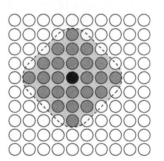

B

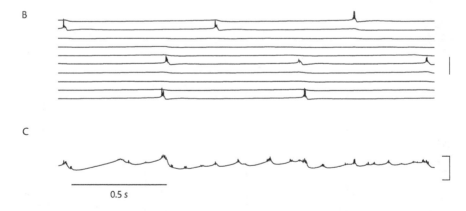

C

0.5 s

Fig. 6.19: Oscillatory behaviour of a network of 100 RE neurons with dense proximal GABA$_B$ connections. Identical configuration and parameters as Fig. 6.13, but with slow inhibitory synapses. A. Pattern of connectivity: a given RE neuron (black) connects to immediately adjacent RE cells (dark grey) including an auto-synapse. B. Activity of the ten cells along the diagonal of the network (vertical calibration bar is of 200 *mV*). C. Average value of the membrane potentials of all 100 RE neurons during the same simulation as in B. Vertical calibration bar for the membrane potential is from −90 to −80 *mV*. $\bar{g}_{GABA} = 0.4$ *nS* (from Destexhe et al., 1994a).

Unlike the same configuration with $GABA_A$ synapses (Fig. 6.13), only a few cells became synchronized, although a very slow oscillatory tendency was observed in the average value of the membrane potential (Fig. 6.19C). Typically, a majority of cells were hyperpolarized by $GABA_B$ IPSPs for a long time (several seconds), occasionally producing a rebound burst. At any given time most of the cells were silent, as reflected in the average membrane potential, which rarely went above -80 mV (by comparison, in the case of $GABA_A$ networks, the average potential oscillated around -75 mV). 'Clusters' of fully synchronous cells (2 to 5) were sometimes observed. This behaviour is similar to that found in another model (Golomb and Rinzel, 1993); however, these clusters were not static and recurred randomly.

Synchronized rhythmicity could not be observed for proximally connected networks of RE neurons with $GABA_B$ synapses, despite exploring of a large range of parameters values, including different schemes of connectivity, different values for the conductances and the decay rates of IPSPs. For networks with a high degree of connectivity, there was no synchronization unless all neurons were connected. For example, synchrony was not observed in a $N = 400$ network in which each neuron received inputs from 181 neighbouring neurons, or 45% connectivity. However, Type 3 slow synchronized oscillations were observed in $N = 100$ networks with full connectivity, in which every neuron sent a connection to all other neurons in the network (not shown).

The mechanisms for generating rhythmic activity in the RE nucleus presented so far work well when the reversal potential of $GABA_A$ in RE cells is sufficiently hyperpolarized relative to the resting membrane potential, leading to hyperpolarizing IPSPs that may sustain rebound activity. However, there are circumstances when this may not occur. One possibility, that the $GABA_A$ reversal potential may be above the resting level, is treated in the next section. Another possibility, which may occur in slice preparations, is that the resting level may be too hyperpolarized to sustain rebound activity and is treated in a later section of this chapter.

6.2.6 Oscillations with depolarizing $GABA_A$ synapses

The reversal potential of $GABA_A$ in RE cells is in some preparations around -71 mV, which is depolarized compared to the resting potential of RE neurons (Ulrich and Huguenard, 1997a). Although hyperpolarizing fast IPSPs have been observed in RE cells (Fig. 6.9; Destexhe et al., 1994a; Bal et al., 1995b; Sanchez-Vives et al., 1997a, 1997b), there are circumstances when depolarizing $GABA_A$-mediated IPSPs can occur and affect rhythmicity in the RE nucleus. The model presented here (Bazhenov et al., 1999) explores the consequences of depolarizing IPSPs in the RE nucleus with simulation.

With depolarizing $GABA_A$ IPSPs, two-dimensional networks of RE cells displayed slow synchronized oscillations at a frequency of about 2.5 Hz (Fig. 6.20A). Bursts of action potentials were triggered by the depolarizing IPSPs and occurred synchronously in RE cells (compare single-cell and population traces in Fig. 6.20A). At more depolarized level (just below the $GABA_A$ reversal potential), the networks showed oscillations at a higher frequency around ~ 9 Hz (Fig. 6.20B). In this case, individual cells oscillated at ~ 4.5 Hz and the population activity showed oscillations at ~ 9 Hz (compare single-cell and population traces in Fig. 6.20B).

In this model of the RE nucleus, the occurrence of sustained oscillatory behaviour was paralleled by spatiotemporal patterns of activity consisting in travelling waves. The pattern of activity depended on the value of the maximal conductance of the T-current. For $\bar{g}_T = 2.2$ mS/cm^2,

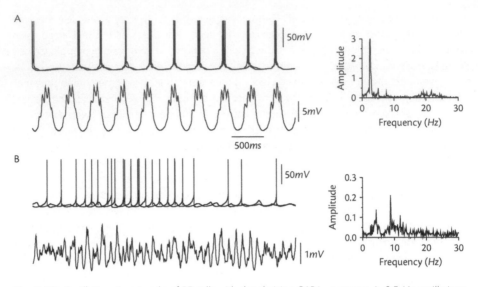

Fig. 6.20: Oscillations in networks of RE cells with depolarizing GABA$_A$ synapses. A. 2.5 Hz oscillations in a two-dimensional network of 33 × 33 RE cells. Cells had a hyperpolarized resting level ($E_L = -77$ mV) and depolarizing GABA$_A$ synapses ($E_{Cl} = -70$ mV). B. 9 Hz oscillations in the same network with a more depolarized resting level. For A and B, the different traces show individual cells in the network (top-left), the averaged population activity (bottom-left) and the power spectrum of population activity (right). Modified from Bazhenov et al., 1999.

the network supported travelling wave phenomena (Fig. 6.21A), but no self-sustained activity. For larger conductance values ($\bar{g}_T = 2.45 mS/cm^2$), self-sustained activity appeared as a rotating spiral-like wave (Fig. 6.21B). In this case, the periodicity of the spiral-like was related to the frequency of oscillation displayed by the network.

The similarity of spatiotemporal behaviour of this model with that in Section 6.2.2 is remarkable (compare Figs. 6.15 and 6.21B) given that the oscillations are based on quite different mechanisms (rebound bursts in Section 6.2.2 vs. depolarizing bursts here).

One of the advantages of the model with depolarizing GABA$_A$ IPSPs compared with a model based on depolarizing EPSPs is that the former network is inherently stable. If activity increase too rapidly, the cells depolarize and the IPSPs become hyperpolarizing, which tends to damp the oscillations and brings the network into a regime for which the models with hyperpolarizing GABA$_A$ IPSPs would apply. In contrast, a network with depolarizing EPSPs is unstable and tends to blow up without inhibitory feedback.

The models reviewed in this section thus show that the known intrinsic voltage- and calcium-dependent currents in RE neurons combined with their patterns of interconnectivity with GABA$_A$ synapses can account for the generation of sustained rhythmicity in the spindle frequency range and as well as the longer term waxing-and-waning. These models are therefore consistent with *in vivo* observations of rhythmicity in the deafferented RE nucleus (Steriade et al., 1987).

The next section will explore the alternative explanation that spindles are generated by an interaction between TC and RE cells and is supported by *in vitro* experiments (Figs. 6.5–6.6; von Krosigk et al., 1993). This exploration will also provide an explanation for the failure to observe oscillations in the isolated RE of the ferret slice preparation (von Krosigk et al., 1993).

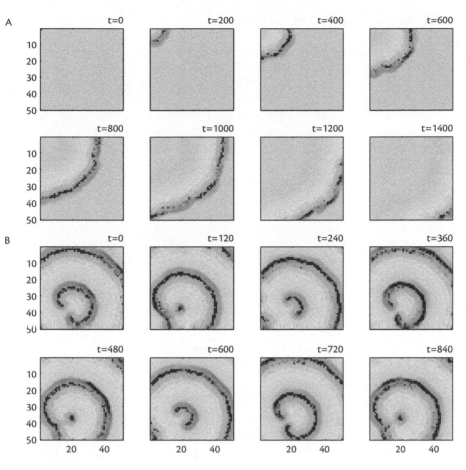

Fig. 6.21: Spatiotemporal patterns of activity in networks of RE cells with depolarizing $GABA_A$ synapses. A. travelling wave elicited by exciting the cell in top-left corner ($\bar{g}_T = 2.2\ mS/cm^2$). B. Rotating spiral-like wave patterns for stronger T-current conductances ($\bar{g}_T = 2.45\ mS/cm^2$). Modified from Bazhenov et al., 1999.

6.3 Models of rhythmicity arising from thalamic relay-reticular interactions

The observation of spindle waves in ferret thalamic slices (Figs. 6.5–6.6; von Krosigk et al., 1993; Bal et al., 1995a, 1995b) has made it possible to identify the synaptic interactions involved: AMPA (from TC→RE) and a mixture of $GABA_A$ and $GABA_B$ (from RE→TC). The known intrinsic properties of TC and RE cells, combined with the typical kinetics of the synaptic currents mediating their interactions, were sufficient to account for spindle oscillations with correct oscillation frequency in models. A simple TC-RE oscillator circuit was demonstrated by Destexhe et al. (1993b) that consisted of one TC and one RE cell interconnected with the synaptic receptors having the biophysical properties observed in thalamic slices. This model is reviewed below; a model of the thalamic slice will be considered in further sections.

6.3.1 Model of the TC-RE oscillator

Spindles are waxing-and-waning sequences of oscillations at 7–14 Hz separated by silent periods of 3 to 30 s. Can the mechanisms underlying the intrinsic waxing-and-waning slow oscillations (0.5–3.2 Hz) of TC cells (see Section 3.3) account for the waxing-and-waning of spindle oscillations (7–14 Hz) when the TC cell interacts with an RE cell?

The simplest TC-RE oscillator consists of a single cell of each type and is shown in Fig. 6.22A. The same set of intrinsic currents described in previous chapters were used in the model. TC cells contained I_T, I_h (as described in Chapter 3) and RE cells contained I_{Ts}, $I_{K[Ca]}$ and I_{CAN} (as in Chapter 4) and both cells contained I_{Na} and I_K responsible for action potentials (see Chapter 2). The equations used for describing the time evolution of the membrane potential were (Destexhe et al., 1993b):

$$C_m\dot{V}_T = -g_L\,(V_T - E_T) - I_T - I_h - I_{Na} - I_K - I_{GABA} \tag{6.6}$$

$$C_m\dot{V}_R = -g_L\,(V_R - E_R) - I_{Ts} - I_{K[Ca]} - I_{CAN} - I_{Na} - I_K - I_{GLU} \tag{6.7}$$

where V_T and V_R is the membrane potential of, respectively, the TC and the RE cell, $C_m = 1\,\mu F/cm^2$ is the specific capacity of the membrane, $g_L = 0.05\ mS/cm^2$ is the leakage

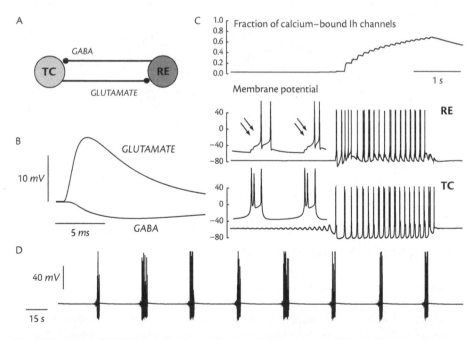

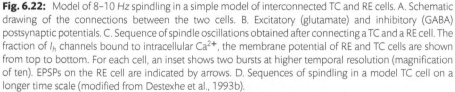

Fig. 6.22: Model of 8–10 Hz spindling in a simple model of interconnected TC and RE cells. A. Schematic drawing of the connections between the two cells. B. Excitatory (glutamate) and inhibitory (GABA) postsynaptic potentials. C. Sequence of spindle oscillations obtained after connecting a TC and a RE cell. The fraction of I_h channels bound to intracellular Ca^{2+}, the membrane potential of RE and TC cells are shown from top to bottom. For each cell, an inset shows two bursts at higher temporal resolution (magnification of ten). EPSPs on the RE cell are indicated by arrows. D. Sequences of spindling in a model TC cell on a longer time scale (modified from Destexhe et al., 1993b).

conductance, $E_T = -86\ mV$ and $E_R = -80\ mV$ are the leakage reversal potentials. The synaptic currents I_{GABA} and I_{GLU} represent respectively the GABA$_A$ synaptic current (inhibitory synapse from RE to TC), and the glutamate non-NMDA current (excitatory synapse from RE to TC), as identified in ferret thalamic slices (von Krosigk et al., 1993). These synaptic currents were modelled by pulse-based open/closed kinetic schemes (see Chapter 5) and the corresponding PSPs are shown in Fig. 6.22B.

6.3.2 Spindle oscillations in the simple TC-RE circuit

A TC cell model can generate slow oscillations by interactions between I_T and I_h (Chapter 3). When in the appropriate voltage range, these intrinsic oscillations wax and wane, due to the upregulation of I_h by intracellular calcium (Section 3.3; Destexhe et al., 1993a). Isolated RE cells can generate sequences of repetitive bursts at ~10 Hz terminated by a tonic tail of spikes, due to I_{Ts}, $I_{K[Ca]}$ and I_{CAN} (Chapter 4).

A simple TC-RE circuit was simulated consisting of a TC and an RE cell interconnected by glutamatergic and GABAergic receptors (Fig. 6.22A). This circuit displayed sequences of 8–10 Hz spindle oscillations (Fig. 6.22C). These 8–10 Hz spindles occurred when the TC cell displayed intrinsic waxing-and-waning oscillations. The spindle sequences were bursts of 8–10 Hz oscillations lasting a few seconds and separated by silent periods of 8–40s. The spindle oscillations began in the TC cell in a manner very similar to the waxing-and-waning slow oscillations of isolated TC cells (compare to Fig. 7.11B). As the oscillation began, the first burst of spikes in the TC cell elicited a series of excitatory PSPs which activated I_{Ts} in the RE cell. The RE cell started 8–10 Hz bursting and entrained the TC cell to this oscillation, but the excitatory feedback from the TC cell was necessary to maintain the 8–10 Hz rhythmicity. At each cycle of the oscillation, Ca^{2+} bound to the I_h channels in the TC cell and shifted the voltage activation curve for I_h. The oscillations in the circuit terminated by the same mechanism that caused the slow oscillations in isolated TC cells to wane (see Section 3.3 in Chapter 3; Destexhe et al., 1993a).

During the spindle oscillation, the two cell types were mirror images: spindle oscillations rode on a depolarizing envelope in the RE cell and a hyperpolarizing envelope in the TC cell (Fig. 6.22C). This is characteristic of intracellular recordings made in anesthetized cats in vivo (Steriade and Llinás, 1988) and in ferret slices in vitro (Fig. 6.5; von Krosigk et al., 1993).

The properties of spindle oscillations depended on the amplitude and kinetics of the synaptic conductances. If the glutamatergic or GABAergic conductances were too low, the coupling was insufficient to produce a spindle; if the GABAergic conductance was to high, the system produced a sustained 10 Hz oscillation because the strong hyperpolarization of the TC cells prevented the Ca^{2+}-bound I_h channels to terminate the oscillation; if the glutamatergic conductance was too high, the rapid depolarization of the RE cell led to a rapid burst and a higher frequency of oscillation. The oscillation frequency was also highly dependent on the kinetics of GABAergic currents, and slower oscillations at ~3 Hz were obtained (see Section 8.2.1) when the decay of the GABAergic current was within the range the decay time of GABA$_B$ currents. This is consistent with experiments showing that blockade of GABA$_A$ synapses by application of bicuculline transformed the spindle behaviour into 2–4 Hz oscillations (von Krosigk et al., 1993; see Fig. 8.10 and the model in Section 8.2.1 of Chapter 8).

The frequency of spindling depended not only on the decay of IPSPs, but also on the intrinsic properties of the RE cell. In the model of a single RE cell, blocking I_{CAN} resulted in a marked slowing down of the frequency of oscillation (see Chapter 4). When I_{CAN} was blocked in the coupled system during spindle oscillations, the frequency also decreased (Destexhe et al., 1993b).

The results of the models in Sections 6.2 and 6.3 demonstrate that the voltage-dependent currents generating the intrinsic properties of thalamic cells and the type of synaptic receptors mediating their synaptic interactions are consistent with spindle generation, both by TC-RE circuits and by the isolated RE nucleus. However, an explanation is still needed for why the isolated RE nucleus does not show spontaneous oscillations *in vitro* (Avanzini et al., 1989; von Krosigk et al., 1993; Huguenard and Prince, 1994b; Warren et al., 1994; Bal et al., 1995b). One possibility is that there are too few interconnected RE cells in the slice for the proposed mechanism to be effective (Steriade et al., 1993b). Another possible explanation for this discrepancy is based on the effect of neuromodulators on RE cells (Destexhe et al., 1994b), as reviewed in the next section.

6.4 Why does the RE nucleus oscillate *in vivo* but not *in vitro*?

One major difference in the thalamus between the slice preparation and the living brain is that there is a tonic level activity in neuromodulatory pathways *in vivo*, including acetylcholine (ACh), noradrenaline (NE) and serotonin (5HT). These pathways densely innervate thalamic nuclei and the release of neuromodulators alter channel conductances and the resting level of thalamic neurons (McCormick, 1992). We have explored the hypothesis (Destexhe et al., 1994b) that the presence of these neuromodulators, and their influence on RE cells, can explain why networks of RE cells spontaneously oscillate *in vivo* but not *in vitro*. In particular, we have considered the influence of noradrenaline and serotonin on RE cells, and how they could potentially influence RE oscillations at the network level.

6.4.1 Model of noradrenergic/serotonergic actions on RE cells

Electrophysiological experiments have shown that ACh affects the firing pattern of RE cells by activating a leak K^+ current (McCormick and Prince, 1986) whereas NE and 5HT depolarize thalamic cells by blocking a leak K^+ current (McCormick and Wang, 1991). Several other neuromodulators may also participate in the control of RE cells, including the glutamate metabotropic receptor (Ohishi et al., 1993). In the following, the generic term 'NE/5HT' will be used to refer to the neurotransmitter systems involved in the depolarization of RE cells via deactivation of a leak K^+ current.

To model this influence, a simple two-state kinetic model for the action of neuromodulators was introduced in Chapter 5 (see Section 5.2.7). The parameters of the model were estimated as follows. Application of noradrenergic and serotonergic agonists to RE neurons *in vitro* caused them to depolarize (McCormick and Wang, 1991); the response lasted for several minutes.

However, there is presently no *in vitro* data on the *synaptic* activation of these noradrenergic and serotonergic receptors. *In vivo* data indicate that brief stimulation of peribrachial cholinergic nuclei evoked a short lasting (about 2 s) muscarinic hyperpolarization in RE neurons (Hu et al., 1989); it is likely that these muscarinic receptors have the same G protein-based activation mechanism as noradrenergic and serotonergic receptors (Brown, 1990). Therefore, the parameters of the G protein kinetic model were chosen to obtain a slow depolarization of 2–3 s following a presynaptic spike (Eqs. 5.31–5.32 with $\alpha = 0.01 \ ms^{-1} \mu M^{-1}$, $\beta = 0.001 \ ms^{-1}$, pulse of [G] of 85 *ms* duration and $1 \mu M$ amplitude).

Single-compartment models were then used to evaluate the effect of NE/5HT synapses on the firing modes of RE cells. The model in Section 4.2.2 was used to investigate the effects of NE/5HT synapses on the firing modes of single RE cells at different levels of NE/5HT activity. First, the model neuron was hyperpolarized to −65 to −75 *mV*, similar to *in vitro* conditions when no NE/5HT synapses are activated (Fig. 6.23a). Next, the model was tested at a more depolarized resting level, from −60 to −70 *mV*, for which about 20% of the leak K^+ current was blocked, and which corresponds to weak NE/5HT activation (one out of five synapses was activated in Fig. 6.23b). In the third set of simulations, all NE/5HT synapses were activated and tonic spike activity resulted from the block of virtually all leak K+ current in the cell (Fig. 6.23c).

These three states also displayed different intrinsic firing properties: for both hyperpolarized and depolarized resting states, injection of hyperpolarizing current pulses resulted in a sequence of rebound bursts occurring rhythmically at a frequency of 6–10 Hz (Fig. 6.23a–b). This bursting activity resulted from the interaction between I_T and $I_{K[Ca]}$, as suggested by current clamp experiments on RE slices (Bal and McCormick, 1993) on which the model was based (see Section 4.2.2 in Chapter 4; see also Destexhe et al., 1994a). During tonic spike activity, injection of the same current pulse only slowed down the frequency of action potentials (Fig. 6.23c).

6.4.2 Neuromodulatory control of network oscillations in the RE nucleus

The effect of NE/5HT activity at the network level was investigated using networks of 100 RE neurons interconnected with $GABA_A$ synapses similar to Section 6.2. The oscillations arose from the interaction between the rebound burst property of RE cells with the GABAergic currents arising from intra-RE collaterals. The occurrence of these oscillations depended critically on the level of the membrane potential as shown in Fig. 6.24. During the first two seconds, all RE cells were brought to a depolarized resting state by activating 20% of NE/5HT synapses. When the NE/5HT activity was suppressed, all RE cells ceased to participate in oscillatory behaviour and relaxed to the hyperpolarized rest. Injection of hyperpolarizing or depolarizing current pulses failed to restore sustained oscillations.

The responses were highly robust to changes in parameters such as the reversal potential of GABAergic currents (E_{Cl}), the values of the synaptic conductances or the amount of leak K^+ current affected by NE/5HT synapses. When E_{Cl} and the resting level of the membrane potential were varied by 5 *mV* the standard values, the sustained oscillations arose only if there was sufficient 'driving force', of at least several millivolts (around 10–15 *mV*, depending on the maximal conductance of GABAergic synapses), between the resting membrane potential and E_{Cl}. If the membrane was hyperpolarized too close to E_{Cl}, the shunting inhibition between RE cells effectively prevents sustaining oscillations.

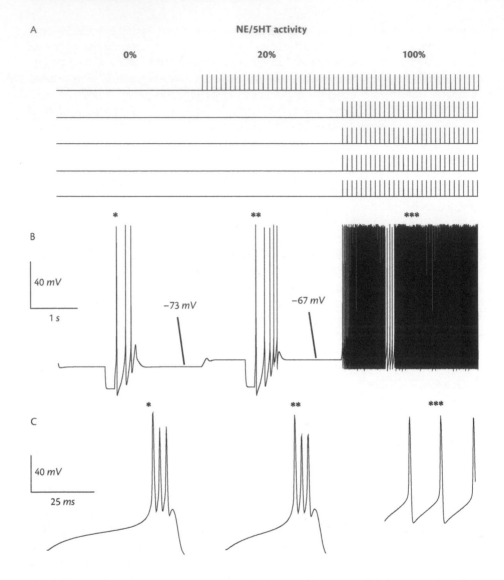

Fig. 6.23: Noradrenergic/serotonergic activation regulates the firing mode of thalamic reticular cells. A single model RE cell with five simulated noradrenergic synapses. A. Five top traces show the presynaptic activity of the NE/5HT synapses. B. Membrane potential of the RE cell. C. Bursts of action potentials shown at higher temporal resolution. Three successive 3-second periods of different level of NE/5HT activity (0/ and 100/injected in the RE cell. A hyperpolarizing current pulse revealed bursting responses in the first two periods, whereas it only slowed down the repetitive spiking in the third period.

The presence of a weak neuromodulatory drive may explain why the isolated RE nucleus oscillates *in vivo* but not *in vitro*. The model predicts that spontaneous sustained oscillations should be observed in slices of the RE nucleus if the resting level of RE cells could be brought to more depolarized values and the connections between RE cells in the slice were sufficiently

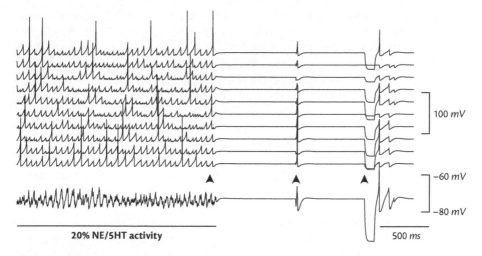

20% NE/5HT activity

100 *mV*

−60 *mV*

−80 *mV*

500 *ms*

Fig. 6.24: Dependence of RE oscillatory behaviour on the membrane potential. Simulation of a network with 100 RE cells locally interconnected through GABAergic synapses. The top ten traces represent the activity of ten neurons in the network and the bottom trace is the average membrane potential. 20% of NE/5HT synapses were initially activated (as in Fig. 6.23b). Under these conditions, the network showed self-sustained oscillations at a frequency of 10–16 *Hz* and the average membrane potential displayed waxing-and-waning amplitude fluctuations. After 2 seconds (first arrow), all NE/5HT synaptic activity was suppressed; the resulting hyperpolarization prevented the network from sustaining oscillations. Depolarizing (second arrow) or hyperpolarizing (third arrow) current pulses injected simultaneously in all neurons (with random amplitude) could not restore spontaneous oscillations. This simulation might correspond to the conditions of RE cells *in vitro* (modified from Destexhe et al., 1994b).

strong. This could be achieved by bath application of NE/5HT agonists at low concentrations to depolarize all RE neurons to the −60 to −70 *mV* range. Another prediction is that NE/5HT antagonists should suppress oscillatory behaviour in the isolated RE nucleus *in vivo*. The same results should also be found in other models of networks of inhibitory neurons displaying rebound bursts (Wang and Rinzel, 1993; Golomb et al., 1994).

6.5 Network model of spindle oscillations in ferret thalamic slices

We have shown above (Section 6.3) that a simple two-cell TC-RE model reproduced some characteristic features of spindle oscillations, such as the 8–10 *Hz* frequency, the waxing-and-waning envelope, the long interspindle silent periods of ~10 *s* and the typical mirror image between TC and RE cells. However, two features of the simple model were not consistent with experiments. The first feature is that the TC cell was a spontaneous oscillator, but both TC and RE cells stop oscillating if their interconnections are cut (Fig. 6.6; von Krosigk et al., 1993). Second, the TC cell in the model rebounded on every cycle of the spindle oscillation, but this is not typically observed; rather, 'subharmonic bursting' is found in intracellular recordings of TC cells during spindle oscillations (Fig. 6.5; see Andersen and Andersson, 1968; Steriade and Llinás, 1988; Muhlethaler and Serafin, 1990; von Krosigk et al., 1993; Bal et al., 1995a). Since RE cells receive barrages of EPSPs at ~10 Hz from several TC cells (Bal et al., 1995b), the population of TC cells may still generate a 10 Hz output although individual cells do not fire on every cycle. These types of interactions are examined in this section using a more detailed network model of TC and RE cells (Destexhe et al., 1996a).

6.5.1 Networks of TC and RE cells

Each cell in the network model had a single compartment, with the membrane potential governed by the following equations:

$$C_m \dot{V}_T = -g_L(V_T - E_L) - I_T - I_h - I_{KL} - I_{Na} - I_K - I_{GABA_A T} - I_{GABA_B} \tag{6.8}$$

$$C_m \dot{V}_R = -g_L(V_R - E_L) - I_{Ts} - I_{Na} - I_K - I_{AMPA} - I_{GABA_A R} \tag{6.9}$$

where V_T and V_R is the membrane potential of TC and RE cells respectively, and $C_m = 1\mu F/cm^2$ is the specific capacity of the membrane, g_L is the leakage conductance, E_L is the leakage reversal potential, I_T and I_{Ts} are the low-threshold calcium currents, I_h is the hyperpolarization-activated cation current, I_{KL} is a leak potassium current, I_{Na} and I_K are the sodium and potassium currents responsible for action potentials, and the synaptic currents are $I_{GABA_A T}$ and I_{GABA_B} associated respectively with GABA$_A$ and GABA$_B$ receptors in the synapses from RE to TC, I_{AMPA} for the AMPA/kainate receptors from TC to RE cells, and $I_{GABA_A R}$ for GABA$_A$ receptors between RE cells. These currents were described in previous chapters (Chapter 2 for I_{Na} and I_K, Chapter 3 for I_T and I_h, Chapter 4 for I_{Ts} and Chapter 5 for I_{GABA_A} and I_{GABA_B}). Details can also be found in Destexhe et al. (1996a).

6.5.2 Small circuits of thalamic reticular neurons

The reticular part of the model of spindle oscillations in thalamic slices reproduced all the features reviewed above for the isolated RE nucleus. The genesis of oscillations by the RE nucleus depended on the resting level of RE cells (see Section 6.4). For hyperpolarized resting levels similar to *in vitro* recordings, the reticular network did not generate oscillations, consistent with slice experiments (Fig. 6.6; von Krosigk et al., 1993). In this case, the burst discharges of RE cells were mostly determined by two conductances: the T-current conductance, as analysed in detail in Chapter 4, and GABA$_A$-mediated conductances, due to the presence of lateral inhibitory connections between RE cells. The role of GABA$_A$-mediated lateral inhibition was however different compared to the role evidenced above for oscillations: GABA$_A$ conductance acted as a 'brake' to limit the burst discharges generated by thalamic reticular cells (3–7 spikes in Fig. 6.25A2, top). These discharges were in agreement with intracellular recordings of PGN cells during spindle oscillations *in vitro* (Bal et al., 1995b). If GABA$_A$ interactions between RE cells were suppressed, mimicking the application of bicuculline, the same stimulus gave rise to strong burst discharges which may play an important role in some forms of epileptic seizures (see Chapter 8).

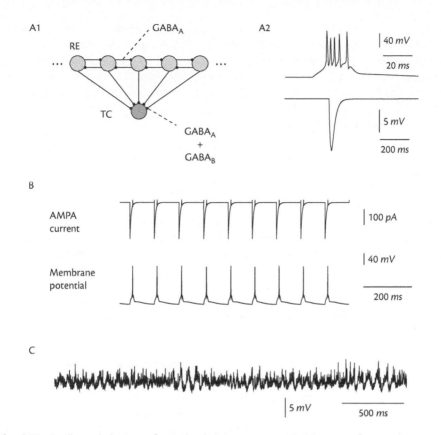

Fig. 6.25: Oscillatory behaviour of reticular thalamic neurons. A1. Schematic of a one-dimensional network of RE cells interconnected by GABA$_A$-mediated inhibitory synapses. Each RE cell contacted a TC cell through a mixture of GABA$_A$ and GABA$_B$ receptors. A2. Stimulation of each RE cell by a depolarizing current pulse under normal conditions produced a burst with only a few spikes, which evoked strong GABA$_A$ IPSPs in TC cells but only a weak GABA$_B$ component. The top trace indicates the burst in a RE cell, and the IPSP evoked in the target TC cell is shown at the bottom (top trace has ten times higher time resolution). B. Stimulation of a network of GABAergically-connected RE cells at 10 Hz through AMPA receptors can recruit RE cells to burst at the same frequency. C. If the membrane potential of RE cells was more depolarized, the network displayed synchronized oscillations at 12–16 Hz. The average membrane potential in a one-dimensional network of 100 RE neurons is shown; the 100 RE neurons were depolarized by partial block of leak K$^+$ currents, mimicking the effect of noradrenaline (see Section 6.4). Total conductances were of 0.2 μS for GABA$_A$ in RE cells, and 0.02 and 0.04 μS for GABA$_A$ and GABA$_B$ in TC cells, respectively (modified from Destexhe et al., 1996a).

A consequence of these limited burst discharges is that the IPSP generated onto TC cells was dominated by GABA$_A$ conductances (Fig. 6.25A2, bottom). The low number of spikes in RE cell bursts were insufficient to activate a significant GABA$_B$ component, consistent with the analysis presented in Chapter 5. The model therefore predicts that in conditions that prevail *in vitro*, the RE nucleus responds to excitatory stimulation by generating relatively moderate burst discharges, which in turn activate GABA$_A$-mediated IPSPs in TC cells.

This behaviour was also seen for repetitive synaptic stimulation of the model RE network. Fig. 6.25B shows a 10 Hz stimulation of RE cells through AMPA receptors in the same conditions. The cell shown in the figure responded by producing burst discharges at 10 Hz with a few spikes per burst (1–6 spikes). Because RE cells receive EPSPs from TC cells at roughly the same frequency during spindles, the model predicts that during spindle oscillations, the RE nucleus can be entrained to produce burst discharges at approximately 10 Hz, which activates repetitive GABA$_A$-mediated IPSPs at the same frequency in TC cells.

Whether the model of isolated RE nucleus sustained spontaneous oscillations depended on the membrane potential of RE cells (see Section 6.4). In the absence of external stimulation, or connections with TC cells, the network of RE cells would be quiescent, consistent with recordings of the isolated RE nucleus *in vitro* (Fig. 6.6; see Bal and McCormick, 1993; von Krosigk et al., 1993; Huguenard and Prince, 1994b; Warren et al., 1994; Bal et al., 1995a, 1995b). However, the model also produced sustained oscillations if the RE cells were more depolarized (Fig. 6.25C), as analysed in detail in Section 6.4 (Destexhe et al., 1994b). The same model is therefore consistent with the observation of oscillations in the deafferented RE nucleus *in vivo* (Steriade et al., 1987).

6.5.3 Subharmonic bursting of TC cells

When a TC cell was stimulated by GABAergic synaptic currents, the response depended on the type of receptor, the conductance and the frequency of stimulation. Fig. 6.26A–B shows repetitive stimulation of GABA$_A$ receptors at 10 Hz. Stimuli consisted of short bursts of presynaptic pulses occurring every 100 ms. These stimuli were inadequate to activate GABA$_B$ receptors, but GABA$_A$ IPSPs were activated at 10 Hz. For moderate to low GABA$_A$ conductances, the TC cell responded with rebound bursts once every other synaptic stimuli (Fig. 6.26A). This form of subharmonic bursting in model TC cells has been described previously (Kopell and LeMasson, 1994) and is due to the presence of I_T and I_h in TC cells (see also analysis by Wang, 1994).

This model also produced rebound bursts at 10 Hz for stronger GABA$_A$ conductances (Fig. 6.26B). This response pattern is similar to the simple TC-RE oscillator analysed previously (Section 6.3; Destexhe et al., 1993b). Although TC cells generally produce subharmonic burst responses once every 2–3 cycles of spindle oscillations, some types of TC cells typically burst on each cycle, such as thalamic intralaminar cells (Steriade et al., 1993a). These simulations therefore suggest that strong GABA$_A$ conductances may explain the patterns of spindling in thalamic intralaminar cells.

The analysis of TC cell's responses to GABA$_A$ stimulation therefore shows that TC cells can produce subharmonic burst responses typical of spindle oscillations for moderate GABA$_A$ conductances. We have shown above that GABA$_A$ conductances should be the output of the RE nucleus to excitatory stimulation. We combine these two features below to produce a TC-RE oscillator at the spindle frequency range.

6.5.4 Minimal circuit for spindle oscillations

Small circuits of interconnected TC and RE cells were investigated using GABA$_A$ receptor synapses between RE cells, AMPA receptor synapses from TC to RE cells, and a mixture of

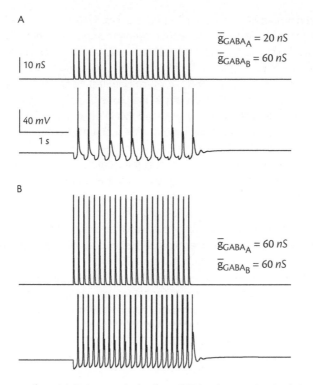

Fig. 6.26: Responses of model thalamocortical cells to GABAergic synaptic stimulation at 10 Hz. The stimulus was a 10 Hz train of 360 Hz presynaptic pulses mimicking the pattern of RE bursts (see Fig. 6.25A2). These presynaptic pulses affected both GABA$_A$ and GABA$_B$ receptors. A. Subharmonic response of the TC cell to 10 Hz stimulation for moderate GABAergic conductances. GABAergic receptors were stimulated using trains of three pulses at 360 Hz, repeated every 100 ms. The GABA$_A$ conductance is shown (top trace) along with the membrane potential (bottom). The TC cell produced rebound bursts at half of the stimulation frequency (5 Hz). B. Same simulation, but using three times stronger GABA$_A$ conductance. In this case, the TC cell produced bursting oscillations at 10 Hz (modified from Destexhe et al., 1996a).

GABA$_A$ and GABA$_B$ receptors at inhibitory synapses from RE to TC cells, as described above. In a previous model (Section 6.3; Destexhe et al., 1993b), an interacting pair of TC and RE cells showed 9–11 Hz waxing-and-waning oscillations similar to spindle oscillations. However, in order to elicit oscillations in a two-cell minimal model, the TC cell had to produce a rebound burst following each GABA$_A$ IPSP, which arrived at 10 Hz (as in Fig. 6.26B). The same type of oscillation was observed using strong GABA$_A$ conductances in the present model (not shown). Intracellular recordings of TC cells during spindling show subharmonic bursting with the TC cell producing rebound bursts every second or third GABAergic IPSP, a feature that cannot be reproduced in an isolated TC cell.

   In Section 6.3, blocking all GABAergic conductances did not eliminate all oscillatory activity in the two-cell model because the TC cell needed to have spontaneous oscillatory activity; this may be relaxed in models of larger networks where there is a range of intrinsic properties among the TC cells (Leresche et al., 1991). One possible source of heterogeneity in the intrinsic properties of TC cells is a difference in their resting membrane potential, as would occur if they had different values of I_h and leak K^+ conductances. Even if most TC cells are sufficiently depolarized to be in the 'relay' mode, with a resting membrane potential around -60 mV, some TC cells could still oscillate spontaneously due to a weaker I_h maximal conductance. This is consistent with current-clamp recordings from TC cells *in vitro* (McCormick and Pape, 1990a; Leresche et al., 1991; Soltesz et al., 1991), in which both resting and spontaneously oscillating TC cells were found. Such an inhomogeneity in the properties of TC cells may account for the initiation and propagation of spindle activity, as shown in the next section.

The minimal circuit for spindle oscillations with subharmonic bursting in TC cells must necessarily have several TC cells producing bursts in alternation, such that the population of TC cells bursts at $\sim$10 Hz. This situation is illustrated in a minimal circuit consisting of two cells of each type and their interconnection (Fig. 6.27, scheme). In this circuit, the weaker GABA$_A$ conductances recruited the TC cells to rebound once every two IPSPs, alternating with each other, such that the RE cells received EPSPs occurring at around 10 Hz although each TC cell burst at around 5 Hz. This alternating pattern of discharge is typical for TC cells during spindle oscillations. Dual intracellular recordings in ferret thalamic slices demonstrated such an alternating pattern of discharge (Fig. 6.28; Bal et al., 1995a).

In the model circuit, the two TC cells were different. There was an 'initiator' TC cell (TC1 in Fig. 6.27), which had a stronger I_h and was spontaneously oscillating, and a 'follower' TC cell (TC2 in Fig. 6.27), which was in a resting relay state and had weaker I_h. TC1 began oscillating and recruited the two RE cells, which in turn recruited TC2. The oscillation in the whole system was maintained at a frequency of 9–11 Hz during the spindle. After a few cycles, the Ca^{2+}-induced augmentation of I_h conductance depolarized the membrane in both TC cells and the oscillation stopped. After a silent period of 15–25 s, the initiator TC cell began to oscillate again and the cycle repeated.

The membrane potentials in TC and RE cells were mirror images of each other, which is typical of spindle oscillations (Fig. 6.27B). The bursts in the TC cell, rebound responses to IPSPs, rode on a hyperpolarizing envelope. In RE cells, EPSPs from TC cells activated I_T and elicited bursts, producing a depolarizing envelope during the spindle. This mirror image is characteristic of spindle oscillations recorded intracellularly in the two thalamic cells types (Fig. 6.5; Steriade and Llinás, 1988; von Krosigk et al., 1993; Bal et al., 1995a, 1995b).

The oscillation could also be started by extrinsic mechanisms, such as by stimulating any TC or RE cell in the system (not shown). The resting TC cell, TC2, acted here as a conditional oscillator; although it did not oscillate spontaneously, I_T and I_h were still present in the cell and it was able to participate in oscillatory behaviour. This system can be generalized to a large network where only one TC cell serves as initiator, while the majority of TC cells are 'followers' (see Section 6.5.4).

6.5.5 Oscillations in networks of TC and RE cells

A multielectrode investigation of spindle oscillations in ferret thalamic slices showed that spindle oscillations consistently propagate in the dorsal-ventral axis (Fig. 6.29; Kim et al., 1995). These experiments suggested that burst discharges in TC cells evoke barrages of AMPA-receptor mediated EPSPs in RE cells; that these EPSPs activate I_T and elicit burst discharges in RE cells, which in turn evoke GABAergic IPSPs in TC cells; and finally, that these IPSPs deinactivate I_T

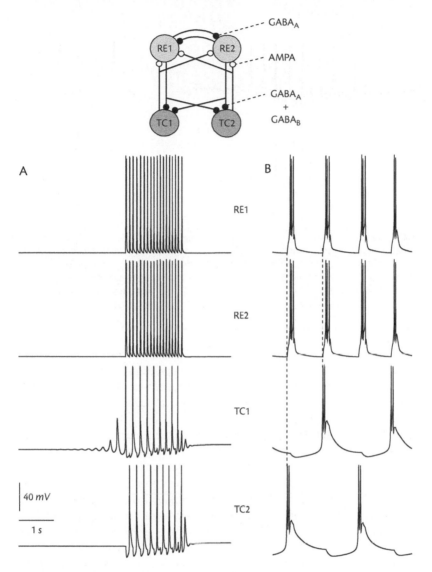

Fig. 6.27: Spindle oscillations in a four-neuron circuit of thalamocortical and thalamic reticular cells. Two TC and two RE cells were connected as shown in the diagrams above. One TC cell (TC1) spontaneously oscillated (initiator cell) and had a higher I_h conductance of $(\bar{g}_h = 0.025\ mS/cm^2; g_{KL} = 5\ nS)$. The second TC cell (TC2) had weaker I_h and g_{KL} $(\bar{g}_h = 0.015\ mS/cm^2; g_{KL} = 3\ nS)$, and was in a resting mode. The two RE cells were identical. Synaptic currents were mediated by AMPA/kainate receptors (from TC to RE; $\bar{g}_{AMPA} = 0.2\ \mu S$), a mixture of GABA$_A$ and GABA$_B$ receptors (from RE to TC; $\bar{g}_{GABA_A} = 0.02\ \mu S$ and $\bar{g}_{GABA_B} = 0.04\ \mu S$) and GABA$_A$-mediated lateral inhibition between RE cells $(\bar{g}_{GABA_A} = 0.2\ \mu S)$. A. Spindle oscillations began when the first TC cell (TC1) started to oscillate, recruiting the two RE cells, which in turn recruited the second TC cell. The oscillation was maintained for a few cycles and fell silent for 15–25 s before repeating. B. Initial bursts of the same cells shown at ten times higher time resolution. Each TC cell produced anti-phase bursts at around 5 Hz, but provided a common input at 9–11 Hz to the RE cells. A burst in the TC cell always preceded bursts in RE cells by a few milliseconds (reproduced from Destexhe et al., 1996a).

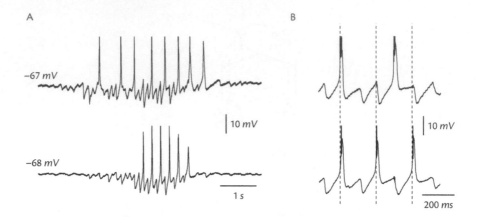

Fig. 6.28: Dual intracellular recordings of TC cells during spindle oscillations in ferret thalamic slices. Simultaneous recordings of two neighbouring LGN cells were performed in ferret thalamic slices (see details in Bal et al., 1995a). A. Dual recordings of spindle oscillations occurring spontaneously in the slice. The spindle started at different times between the two cells, consistent with the systematic propagation of these oscillations *in vitro* (Kim et al., 1995). B. Enlargement of A, showing that the two cells produced alternating burst discharges. Courtesy of T. Bal and D. A. McCormick (unpublished data).

and elicit burst discharges, and the same cycle starts again. A spindle could be started by the spontaneous discharge of either TC or RE cells and propagate through the recruitment of adjacent cells through localized axonal projections. Andersen and Andersson (1968) proposed a similar progressive recruiting mechanism based on local interactions between TC cells and inhibitory interneurons rather than RE cells.

The dynamics of interaction between TC and RE cells in the simple circuit of Fig. 6.27 was qualitatively similar in larger networks. In Fig. 6.30, a model with fifty TC and fifty RE cells is shown having the same types of cells and synaptic receptors as in Fig. 6.27. Based on anatomical data showing that axonal projections in the thalamic circuitry are local and topographic (Minderhoud, 1971; Sanderson, 1971; Jones, 1985; Fitzgibbon et al., 1995; Gonzalo-Ruiz and Lieberman, 1995), the structure of ferret thalamic slices was modelled by a two-layer one-dimensional network, one layer of TC cells and one layer of RE cells, in which connections were topographically local (Fig. 6.30A). Each cell type made axonal projections that contacted eleven neighbouring cells in the other layer (and also within the same layer for RE cells). All axonal projections were identical in extent and all synaptic conductances were equal. The total synaptic conductance on each neuron was the same for cells of the same type, independently of the size of the network, and was also the same as for cells in the four-cell circuit. Mirror boundary conditions were used to minimize boundary effects (see Section 6.2; see also Destexhe et al., 1994a).

The oscillatory behaviour of the population of TC cell was apparent in averages over local regions of ten TC cells, regularly spaced throughout the network, showed waxing-and-waning oscillations at around 10 *Hz*, indicating that the cells oscillated in synchrony although individual cells did not burst at every cycle (Fig. 6.30B1). As in the four-cell circuit, TC cells oscillated at a slower frequency around 5 *Hz* whereas RE cells tended to oscillate around 10 *Hz* but had more irregular behaviour (Fig. 6.30B2).

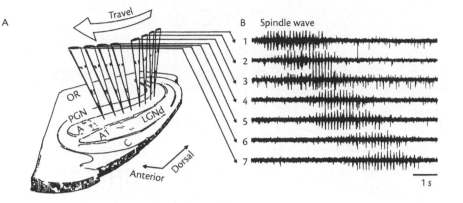

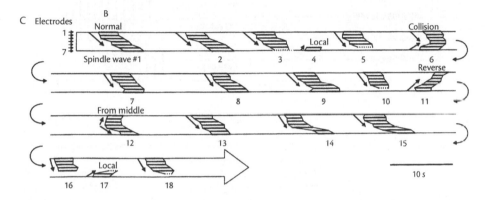

Fig. 6.29: Spindle waves are propagating oscillations in ferret thalamic slices. A. Drawing of a saggital ferret dorsal lateral geniculate nucleus (dLGN) slice with an array of eight multiunit electrodes arranged in lamina A. Electrodes were separated by 250–400 μm, extending over 2–3 *mm* in the dorso-ventral axis. B. Example of a propagating spindle wave. The spindle wave started in the dorsal end of the slice and propagated ventrally. Each spindle wave consisted in waxing-and-waning rhythmic action potential bursts with interburst frequency of 6–10 *Hz*. C. Schematic diagram of propagation for eighteen consecutive spindle waves. In most cases, spindles propagated in the dorsal→ventral direction, with propagation speed of 0.4–0.8 *mm/s* and interspindle periods of 13–15 *s*. Other spatiotemporal patterns were also seen, such as waves propagating in the reverse direction (11), locally propagating waves (4, 16, 17), spindle waves initiated from the middle of the slice (12) and collisions between two waves propagating in opposite directions (6). Figure reproduced from Kim et al., 1995.

The oscillation showed waxing-and-waning properties in individual cells (Fig. 6.30B2). The 'waxing' propagated as more and more cells were recruited. Similarly, the 'waning' also propagated, as TC cells progressively lost their ability to participate in rebound bursting activity. The resulting pattern had the appearance of 'waves' of oscillatory activity propagating through the network. Between waves, the network was quiescent for 15 to 25 *s*. These propagating oscillations were similar to the propagation of spindles found *in vitro* (Fig. 6.29; Kim et al., 1995).

The presence of propagating waves in the 100-cell network depended on the restricted topographic connectivity between the TC and RE layers. With local connectivity, the oscillation

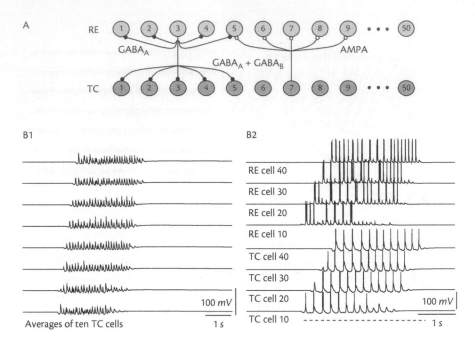

Fig. 6.30: Propagating spindle oscillations in a network of thalamocortical and thalamic reticular cells. A. Schematic diagram of connectivity of a one-dimensional network of fifty TC and fifty RE cells with localized axonal projections. Axons from TC cells ramified in the RE nucleus and contacted RE cells within a patch of 10% of the size of the network. Similarly, axons from RE cells contacted TC cells as well as had collaterals within the RE nucleus. All projections extended laterally to eleven neurons in the case shown here. TC1 was a spontaneously oscillating cell, whereas all other TC cells were in a resting mode; all RE cells were identical (same models as in Fig. 6.27). B1. Spindle oscillations at various sites of the network. Averaged membrane potentials were computed from ten neighbouring TC cells taken at eight equally spaced sites. B2. Membrane potential of four TC (four bottom traces) and four RE cells (upper traces) in the same simulation as in B1. Modified from Destexhe et al., 1996a.

started at one site, and propagated by recruiting more TC and RE cells on each cycle of the oscillation. The velocity of the propagation was directly proportional to the extent of the axonal projections (Golomb et al., 1996). Similar results were found with different sets of parameters as long as the connectivity was topographically organized (not shown). The average propagation delay between two neighbouring neurons was of about 19.4 *ms*. If the fan-out of the projections between the TC and RE cells in the model (eleven neurons) is assumed to be equivalent to 200 μm in the slice, then the velocity in the model is about 1.03 *mm/s* for spindles, which is within the range 0.28 to 1.21 *mm/s* observed experimentally.

Phase relations between TC and RE cells can be estimated from Figs. 6.27 and 6.30. The initiator TC cell (TC1) started oscillating first and recruited RE cells. However, the other TC cells always started spindling with an IPSP and followed the RE cells. If now these follower TC cells constituted the majority of the TC population, then comparing RE cells with this type of TC cell could lead to the incorrect conclusion that the RE cells initiated the spindle oscillations. The model therefore suggests that, even if the overwhelming majority of TC cells start spindling with IPSPs, the spindle sequence can still be initiated by a small minority of TC cells.

6.5.6 Spatiotemporal patterns of discharges

The spatiotemporal patterns of discharges during spindle waves in the model are shown in Fig. 6.31 and had the following properties: first, the 'waxing' and the 'waning' of the oscillation had a similar velocity. Second, during the oscillation, groups of cells segregated into localized clusters of activity discharging in alternation. This phenomenon is associated with the subharmonic bursting properties of TC cells, which does not allow the whole network to fire in unison at around 10 Hz. Third, individual clusters of activity also propagated with the same velocity as the leading edge. This phenomenon can be seen most clearly when the snapshots are played as a movie.[3] The simulated spindles appear to be composed of a sequence of successive clusters of activity that follow each other and propagate through the TC-RE structure. The cluster of activity at the front of the propagating oscillation can be seen clearly in the top series of frames in Fig. 6.31. These predicted patterns should be observable by imaging experiments.

All cells were identical in the TC cell layer, except one: the first TC cell, which had a higher I_h conductance and was spontaneously oscillating. All RE cells were identical. The presence of only one oscillating TC cell was sufficient to initiate a spindle oscillation in the entire network. The recruitment mechanism is illustrated in an inset in Fig. 6.31. The initiator TC cell (TC1) discharged first due to intrinsic oscillatory properties (arrow 1 in Fig. 6.31). AMPA-mediated EPSPs were evoked in neighbouring RE cells (arrow 2) following the discharge of TC1; these EPSPs were strong enough to activate I_t and evoke burst firing in RE cells (arrow 3). RE cell bursts then recruited neighbouring TC cells though GABAergic IPSPs (arrow 4). Following this inhibition, some TC cells produced rebound bursts (arrow 5) due to deinactivation of I_T, and the same sequence restarted. This back-and-forth series of interactions between cells in the TC and the RE layers recruited additional cells into the spindle oscillation on every cycle. After a few cycles, the Ca^{2+}-induced augmentation of I_h conductance in TC cells stopped the oscillation. Following a silent period of 15–25 s, the initiator TC cell began to oscillate again and the cycle repeated.

6.5.7 Refractoriness of the network

The calcium-mediated upregulation of I_h underlies the 'waning' of spindle oscillations. It builds up during the oscillation and progressively diminishes the tendency of TC cells to display rebound bursts, leading to cessation of oscillations after a few cycles. Upregulation of I_h also has other consequences at the network level, such as the presence of refractoriness in the network, the non-crossing of colliding waves, and sustained oscillations if I_h is altered. These conditions are examined below.

Two colliding spindle waves propagating in opposite directions do not cross each other, but rather merge into a single coherent oscillation, as shown experimentally in ferret thalamic slices (Fig. 6.32; Kim et al., 1995). This was also observed in the model because of the refractoriness of the cells (Fig. 6.33A). Two spindle waves, initiated from the opposite edges of the network with a 50 ms delay in order to produce oscillations that were not in-phase. The two oscillations propagated into the centre and collided. Upon collision, the two waves became synchronous

---

[3] Computer-generated movies can be viewed at: http://www.cnl.salk.edu/~alain or http://cns.iaf.cnrs-gif.fr.

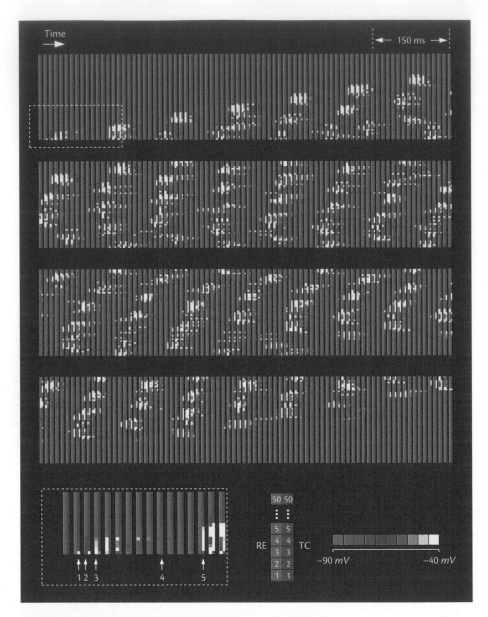

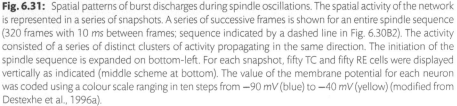

Fig. 6.31: Spatial patterns of burst discharges during spindle oscillations. The spatial activity of the network is represented in a series of snapshots. A series of successive frames is shown for an entire spindle sequence (320 frames with 10 ms between frames; sequence indicated by a dashed line in Fig. 6.30B2). The activity consisted of a series of distinct clusters of activity propagating in the same direction. The initiation of the spindle sequence is expanded on bottom-left. For each snapshot, fifty TC and fifty RE cells were displayed vertically as indicated (middle scheme at bottom). The value of the membrane potential for each neuron was coded using a colour scale ranging in ten steps from −90 mV (blue) to −40 mV (yellow) (modified from Destexhe et al., 1996a).

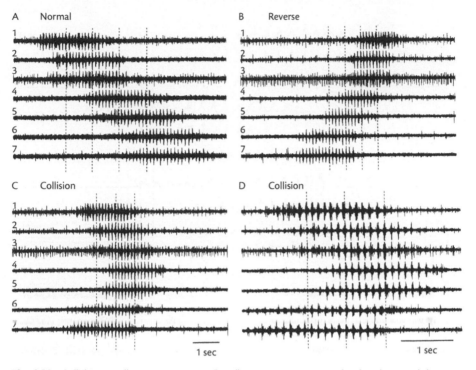

Fig. 6.32: Colliding spindle waves *in vitro*. A. Spindle wave propagating in the dorsal→ventral direction. B. 'Reverse' spindle wave propagating in the ventral→dorsal direction. C. Collision between two spindle waves propagating in opposite directions. D. Enlargement of C. In A–B, dashed lines indicate that bursts of action potentials were synchronous within a spindle wave. On the other hand, in C–D, the spindle waves were not synchronized initially (leftmost dashed line), but became synchronous during the collision (rightmost dashed line). Figure reproduced from Kim et al., 1995.

(compare dashed lines before and after collision in Fig. 6.33A). The waves did not cross due to upregulation of I_h in TC cells and the consequent refractoriness was imparted to the network.

Refractoriness in thalamic networks has been demonstrated in ferret slices experiments (Fig. 6.34; Kim et al., 1995), as well as in cats *in vivo* following cortical stimulation (see Fig. 7.18 in Chapter 7; Destexhe et al., 1998a). In these experiments, oscillations evoked by electrical stimulation depended sensitively on the timing of the stimulus relative to the previous spindle sequence. This dependency was also present in the model, as illustrated in Fig. 6.33B. Following a spindle sequence, there was a refractory period of around 8–10 s during which no propagating oscillations could be evoked. This phenomenon is therefore consistent with an activity-dependent upregulation of I_h. Following a spindle sequence, activation of I_h was increased to the point that TC cells were unable to sustain further oscillations, which lasted until I_h slowly recovered to its resting level.

Experiments in ferret thalamic slices (Kim et al., 1995) further showed that the refractory period was always shorter than the period for spontaneous oscillations. This was also true in the model: the network recovered full excitability after about 8–10 s although the spontaneous interspindle period was around 20 s. This property was due to the spontaneously oscillating initiator TC cells that imprinted their oscillation period to the whole network. At any time

A

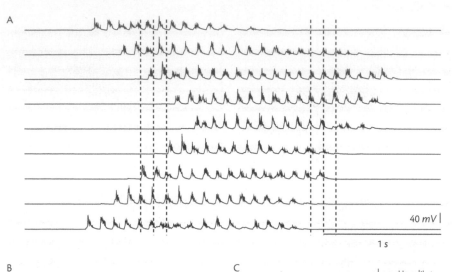

B   C

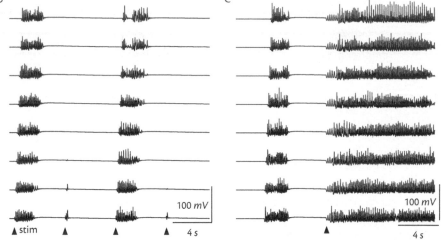

Fig. 6.33: Network properties of propagating spindle oscillations. A. Collision between two simulated spindle waves. Three TC cells at each edge of the network were stimulated by depolarizing current pulses (1 nA during 40 ms). One stimulus preceded the other by 50 ms in order to produce spindles that were initially not in-phase (first series of dashed lines). The two waves collided without crossing each other, and became synchronous (dashed lines on the left). The network was of 100 TC and 100 RE cells with identical connectivity and parameters as in Fig. 6.30A. B. Refractoriness of spindle oscillations. Repetitive stimulation was delivered to three TC cells at one edge of the network. Depolarizing current pulses of 1 nA during 40 ms were delivered every 5 s (arrows). Stimulation occurring 10 s after a spindle sequence successfully evoked propagating spindle oscillations, but those occurring after 5 s failed to evoke propagating oscillations. C. Transformation from spindle oscillations to sustained oscillations by reducing I_h conductances. In all TC cells, the maximal conductance was set to 0.004 mS/cm^2 (arrow), mimicking extracellular application of cesium. The spontaneous waxing-and-waning spindles (before the arrow) were transformed into sustained rhythmicity with similar frequency and cellular behaviour as spindles. A, B, C used identical parameters and averaging procedures as in Fig. 6.30B (from Destexhe et al., 1996a).

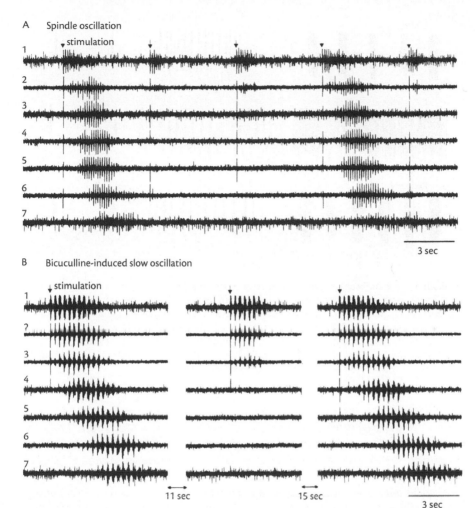

Fig. 6.34: Refractoriness of spindle waves *in vitro*. A. Repetitive stimulation of fibres in the optical radiation (OR in Fig. 6.29A) once every 5 s (three shocks at 100 *Hz*; 46 *μA*, 0.1 *ms* duration). Each stimulus elicited a spindle wave near the stimulation site (electrode 1), but propagation only occurred every third stimulus. The spontaneous interspindle period was about 15 s whereas the shortest evoked period of propagating waves was 12 s. B. Same protocol for oscillations after bath application of bicuculline. In this case, the refractory period for propagation was between 20 and 40 s. Figure reproduced from Kim et al., 1995.

between about 10 s following a spindle sequence and the next spindle sequence, a propagating oscillation could be initiated in the network by stimulating either TC or RE cells. This property may account for some differences between the spatiotemporal properties of oscillations *in vivo* and *in vitro* (see Chapter 7).

If spindles wax-and-wane due the upregulation of I_h, then suppressing this conductance should alter the waxing-and-waning. This is indeed what was observed in recent experiments in thalamic slices (Bal and McCormick, 1996; Luthi and McCormick, 1998). Application of extracellular cesium (which primarily affects I_h) transformed waxing-and-waning spindles into

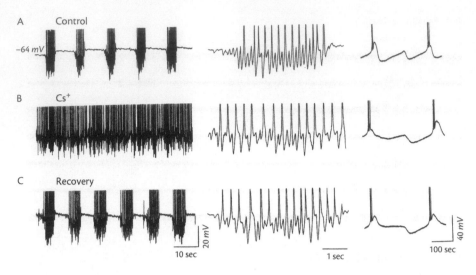

Fig. 6.35: Waxing-and-waning spindle waves can be transformed into sustained oscillations by blocking I_h. A. Control: spindle waves appear as waxing-and-waning oscillatory sequences recurring approximately once every 15 s. Expansion of intracellular recording in LGNd cell shows the rhythmic sequences of IPSP and rebound bursts. B. After extracellular application of Cs$^+$, the neuron hyperpolarized by about 5 mV and the silent periods progressively shortened until sustained oscillations occurred. Details show that the intracellular features of the oscillation inside the spindles were similar to the intact case. C. Recovery after washout of Cs$^+$. Figure reproduced from Bal and McCormick, 1996.

sustained rhythmicity at the same frequency (Fig. 6.35; Bal and McCormick, 1996). The same behaviour was also observed following application of specific blockers of I_h (Luthi and McCormick, 1998). This is also consistent with the present model as shown in Fig. 6.33C. Reducing I_h to a fraction of its initial conductance led to sustained oscillations of around 9 Hz. Although the remaining I_h was still upregulated by intracellular Ca^{2+}, it was not strong enough to suppress the rebound burst properties in TC neurons and consequently the oscillations remained sustained.

6.6 Intrathalamic augmenting responses

Repetitive stimulation of the thalamus at 7–14 Hz evokes responses of increasing amplitude in the thalamus and the areas of the neocortex to which the stimulated foci project, a phenomenon called the augmenting response (Morison and Dempsey, 1943). Decortication reduces but does not abolish the augmentation of repetitive responses recorded from the thalamus; however, removal of the thalamus abolished the repetitive and augmented responses in the cortex (Morison and Dempsey, 1943). The models developed here for thalamic spindling can be stimulated with 10 Hz trains to see if they too exhibit augmenting responses

(Bazhenov et al., 1998a). The underlying mechanisms can be explored in the model and later through experiments suggested by the model.

The IPSPs of thalamocortical (TC) cells following the first stimulus in a pulse-train at 10 Hz hyperpolarizes the cells and progressively deinactivates low-threshold Ca^{2+} currents (Jahnsen and Llinás, 1984a, b). TC cells located near the stimulating electrode receive sufficiently large excitatory postsynaptic potentials (EPSPs) that high-threshold currents are activated (Pedroarena and Llinás, 1997). This high-threshold type of augmenting response in the thalamus occurs only when the balance between synaptic excitation and inhibition is shifted towards excitation and occurs only in a limited region surrounding the stimulating electrode (Steriade and Timofeev, 1997); farther from the site of stimulation, there is a different, low-threshold type of augmenting response (Timofeev et al., 1996).

In *in vivo* intracellular recording from a TC cell during 10 Hz stimulation of the thalamus in unilaterally decorticated cats, the membrane potential becomes progressively hyperpolarized by IPSPs and the low-threshold currents become progressively deinactivated. This leads to an increase in the magnitude of the low-threshold spikes and an increase in the number of fast spikes in each burst (Fig. 6.36).

6.6.1 Model for the intrathalamic augmenting response

The TC and RE cell models used in these simulation were those used in Section 6.3.1 with one additional current in the TC cell model, a potassium A current I_A (Huguenard et al., 1991; Bazhenov et al., 1998a). The network consisted of two one-dimensional chains of twenty-seven RE and twenty-seven TC cells with each TC cell connected with nine nearest neighbours in the chain of RE cells and each RE cell made inhibitory synapses on the nine nearest neighbours from the layer of TC cells (Both $GABA_A$ and $GABA_B$) and also on the nine nearest neighbours within the layer of RE cells ($GABA_A$ only).

In the model, repetitive 10 Hz stimulation of both the RE and the TC cells elicited incrementing activity during the first three to four stimuli in a train of eleven shocks as seen in the progressively increasing number of spikes per burst in the TC cells and by the recruitment of more TC cells that fire action potentials (Fig. 6.37). TC cells remote from the stimulation site also participated in the augmenting response after a delay through TC-RE-TC interactions.

Fig. 6.38 shows expanded traces of three pairs of TC and RE cells located at different distances from the centre of stimulation. The first pair of cells (Fig. 6.38A) is located near the boundary of the chain and received low-intensity stimulation. This RE cell responded with a six-spike burst to the first shock and weaker responded (0–2 spikes) to the following shocks. The response decremented because the low-threshold Ca^{2+} current in RE cell partially deinactivated during the first spike burst. If the RE cell was depolarized by current injection, so that the I_T current was completely deinactivation before stimulation, then the RE cell displayed augmenting responses during the whole train of stimuli (not shown). The corresponding TC cell displayed a weak and delayed augmenting response with the number of spikes per burst increasing from 0 to 2. The second pair of RE-TC cells (Fig. 6.38B) is closer to the centre of the stimulation and received stronger stimulation. This RE cell showed a remarkable diminution of the response after the first stimulus followed by a slow augmentation of the burst. The paired TC cell displayed a strong but delayed augmentation of the responses during the train of stimuli. Finally, in the

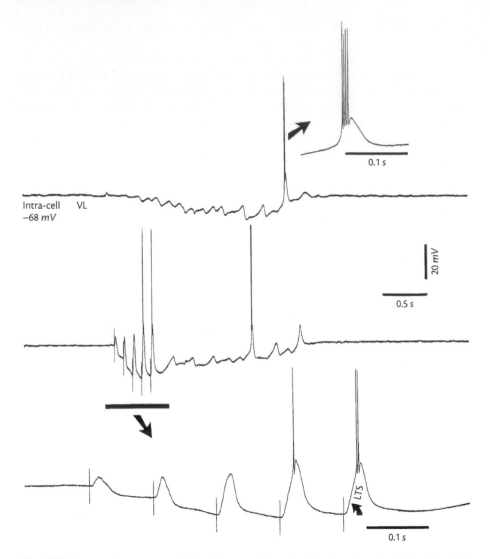

Fig. 6.36: Augmenting responses in VL thalamocortical cell arising from low-threshold spikes (LTS). Decorticated animal. Top trace, spontaneous spindle sequence (arrow points to expanded rebound burst). Below, five-shocks train at 10 *Hz* produced an augmenting responses stemming from hyperpolarized levels of membrane potential and followed by a spindle. Part of response indicated by horizontal bar is expanded below. Progressive hyperpolarization of TC cell led to progressive growing of low threshold responses. Oblique arrows on bottom trace indicate deflection between EPSP and LT spike. Note the similar shape of the low threshold spike during the spindle and that evoked by stimulation (from Bazhenov et al., 1998a).

third pair of RE-TC cells (Fig. 6.38C) located at the centre of stimulation, both of the cells (TC starting from the first shock and RE starting from the second shock) demonstrated strongly augmenting responses. The augmenting responses in the TC cell developed completely during first 3-4 stimuli.

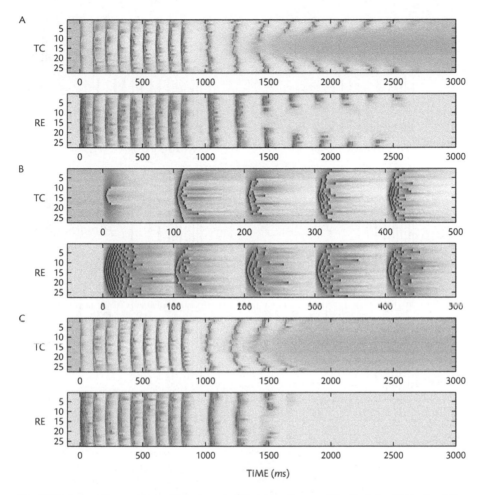

Fig. 6.37: Augmenting responses of twenty-seven TC and twenty-seven RE cells during 10 *Hz* stimulation in a chain of interacting RE and TC cells. Nine shocks were applied between $t = 0$ *ms* and $t = 800$ *ms*. Both RE and TC cells were stimulated simultaneously. The intensity of stimulation was maximal in the centre of the chain and decayed exponentially with distance from the centre. A, B. Diameter of connections for all projections was nine cells. Expanded traces of the panels (A) between $t = -50$ *ms* and $t = 500$ *ms* are given on (B). The first four shocks in the train of nine shocks evoked incremental responses in the TC cells. The RE cells demonstrated a diminished response to the second shock because of the partial deinactivation of low-threshold current in these cells and increasing of the responses to the following shocks. The train of stimuli was followed by slow 4–5 *Hz* post-stimulus oscillations. C. Diameter of connections for RE-TC, TC-RE projections was 17 cells. A smaller diameter for connections (nine cells) was used for RE–RE projections. Increasing the radius of RE-TC connections decreased the contribution of an individual cell to the summed postsynaptic potential. This resulted in the faster desynchronization of the network after a train of stimuli and termination of post-stimulus oscillations. The value of membrane potential for each neuron is coded in grey scale from -90 *mV* (white) to -30 *mV* (black). $g_{GABA_A(RE)} = 0.07$ μS, $g_{GABA_A(TC)} = 0.02$ μS, $g_{GABA_B} = 0.07$ μS, $g_{AMPA} = 0.07$ μS (from Bazhenov et al., 1998a).

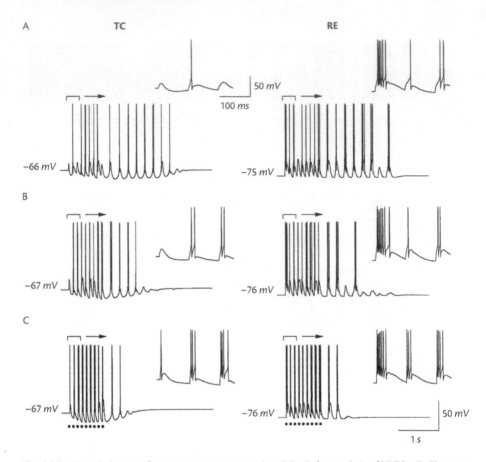

Fig. 6.38: Expanded traces of augmenting responses in three TC cells from a chain of RE-TC cells. The traces are expanded from Fig. 6.37. The insets show the responses to the first three stimuli at a different time scale. A. TC cell placed far from the centre of the chain (cell #3) obtained a low-intensity stimulation that resulted in a weak and delayed augmenting response followed by prolonged 3–5 Hz oscillations. Related RE cell showed a strong response to the first shocks and very weak responses to the next shocks. The post-stimulus oscillations were terminated as a result of desynchronization in the network evoked by variability in the parameters of the cells. B. TC cell placed closely to the centre of the stimulation (cell #7) showed stronger augmentation of the responses and faster termination of the post-stimulus oscillations. C. TC cells placed near the centre of stimulation (cell #14) showed a fast augmenting responses and just a few cycles of slow post-stimulus oscillations. RE cell displayed a diminished response to the second shock and augmentation of the responses to the following shocks (from Bazhenov et al., 1998a).

Following the simulated train of nine shocks to the linear chain of cells, a sequence of slow delta (3–4 Hz) oscillations was elicited by interactions between RE and TC cells. These oscillations terminated as a result of desynchronization in the network and depolarization of TC cells, which resulted from the Ca^{2+} regulation of I_h current (Destexhe et al. 1996a). The duration of the post-stimulus oscillations depended on the position of the cell in the chain. TC cells located far from the centre of stimulation and displaying weak burst discharges were involved in the most prolonged post-stimulus oscillations. In contrast, TC cells placed at the centre of the chain displayed powerful burst discharges. As a consequence, the intracellular

Ca^{2+} concentration increased rapidly during train of stimuli and poststimulus oscillations in these TC cells were terminated after 1–2 cycles.

In the RE–TC simulations, a strong $GABA_B$ component in the RE-evoked IPSPs is a necessary condition for deinactivation of I_T current in TC cells during repetitive stimulation. The kinetic scheme activated by the $GABA_B$ receptors involving G-proteins requires prolonged burst discharges in the RE cells to elicit powerful $GABA_B$ IPSPs, and the bursts depend on the low-threshold Ca^{2+} current. Strong depolarization of RE cell by positive DC current injection inactivated the I_T current and the RE cell responded only weakly during a train of stimuli (not shown). At smaller depolarizing DC levels, RE cells displayed augmenting responses. In this case, augmentation was obtained as a result of augmenting responses in TC cells. In the absence of DC current injection, the same RE cells displayed a powerful burst discharge for the first stimulus and decrementing responses to subsequent stimuli. The diminution of the burst discharges in the RE cell despite increasing EPSPs can be explained by the partial inactivation of the low-threshold current in this cell.

6.7 Further findings about thalamic networks

Several key experimental and modelling findings were obtained in the last years and are worth being mentioned.

6.7.1 Participation of inhibitory neurons in sleep spindles

The first recordings of inhibitory interneuron activity in cortex during natural sleep spindles were reported using extracellular recordings with tetrodes in rats during natural sleep (Peyrache et al., 2011; Fig. 9.8). The presumed inhibitory neurons were identified by their spike shape, auto-correlation and higher frequency discharge compared to presumed excitatory neurons. It was found that the presumed excitatory cells were less active than presumed interneurons during sleep spindles. This corroborates previously observations in intracellular recordings from cats (Contreras et al., 1997). Moreover, the inhibitory neurons fired phasically during sleep spindles, also confirming the previous intracellular evidence for strong inhibition during sleep spindles (Contreras et al., 1997). This confirms the prediction of the models that strong inhibition during sleep spindles reduces the spiking of cortical pyramidal neurons (see details in Chapter 9). A study in mice (Rovo et al., 2014) showed that the extrasynaptic $GABA_A$ receptors in TC cells are specifically activated by the burst firing of RE cells, and that this activation is sufficiently powerful to generate slow and spindle oscillations. This further emphasizes the need of RE cell bursting to robustly generate sleep spindles by thalamic circuits.

6.7.2 Electrically coupled thalamic neurons

Gap junctions were discovered between between RE cells in thalamic slices (Landisman et al., 2002). The presence of small, fast spikelets in RE neurons recorded intracellularly confirmed this discovery *in vivo* (Fig. 6.39; Fuentealba et al., 2004). In network models of the reticular nucleus incorporating gap junctions, the propensity of the network to oscillate in the spindle

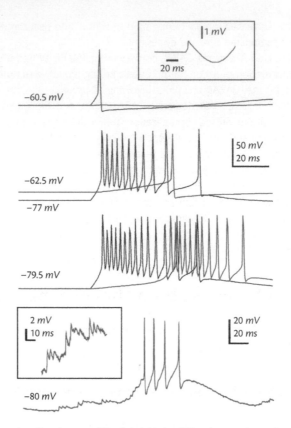

Fig. 6.39: Role of gap junctions between RE cells in initiating LTS and wave propagation. (A) Model of a pair of reciprocally connected RE cells. Spikelet (see inset) was induced in the postsynaptic cell (red line) by single spike in the presynaptic neuron (black line). Upon hyperpolarization of presynaptic neuron (second trace), a burst of spikes can trigger a single spike in the postsynaptic neuron. When both cells are hyperpolarized (third trace), a burst in the presynaptic cell can induce a delayed spike burst in the postsynaptic cell. Bottom trace represents intracellular recording of RE neuron *in vivo*. Adapted from Fuentealba et al., 2004.

frequency range was reinforced (Fuentealba et al., 2004). Thus, the model suggests that the intrinsic properties of thalamic RE neurons, combined with their electrical synaptic interactions, support the RE pacemaker hypothesis, thus confirming an earlier RE pacemaker model (Destexhe et al., 1994).

6.7.3 Simplified models of thalamic cells and spindle oscillations

Several models of thalamic cells and oscillations have been developed that are simpler than the Hodgkin-Huxley type models described in this chapter. An adaptive exponential integrate-and-fire model (Brette and Gerstner, 2005) generated rebound bursts similar to that of thalamic neurons (Destexhe, 2009). Spindle rhythmicity occurred in a network model of these interconnected simplified TC and RE cells. The simplified spindle model exhibited the typical mirror image between bursting in the TC and RE cells, with TC cells not producing rebound bursts on every cycle of the oscillation (Destexhe, 2009).

Models of TC-RE circuits were also studied in an 'open loop' configuration, an architecture for the facilitation of transmission of information to cortex (Willis et al., 2015). A thalamic model was also proposed for the generation of alpha oscillations, in thalamic circuits where high-threshold bursts occur (Vijayan and Kopell, 2012). Thalamic circuits were also studied with an integrate, fire and burst model using a population density approach (Huertas and Smith, 2006). This model reproduced the 7–14 *Hz* spindle-like oscillations, and various other patterns in the presence of external inputs.

6.8 Discussion

The focus of this chapter has been the oscillatory behaviour of thalamic systems. Computational models were used to explore the mechanisms underlying thalamic rhythmicity, starting with the initial model introduced by Andersen and Rutjord (1964), who emphasized the significance of inhibitory rebound bursts. The inhibitory neurons in the reticular nucleus of the thalamus have proved to be better suited for this role than the inhibitory interneurons that were initially proposed. Several models have been proposed for rhythmicity in the isolated reticular nucleus (Destexhe and Babloyantz, 1992; Wang and Rinzel, 1992, 1993; Destexhe et al., 1994a; Golomb et al., 1994; Bazhenov et al., 1999) and for spindle oscillations from reciprocal interaction between TC and RE cells (Destexhe et al., 1993b, 1996a; Golomb et al., 1996). Testable predictions were possible when the computational models were based on physiological and biophysical data. We summarize here the proposed mechanisms for thalamic rhythmicity, the predictions arising from these models, and directions for future studies.

6.8.1 Spindle oscillations in the intact thalamus

6.8.1.1 Mechanisms of rhythmicity

As we have reviewed earlier in this chapter, three hypothesis were proposed to explain the genesis of thalamic rhythmicity: the 'TC-interneuron' hypothesis of Andersen and Eccles (1962; see Fig. 3.2K); the 'TC-RE' hypothesis of Scheibel and Scheibel (1966a, 1966b, 1967) and the 'RE pacemaker' hypothesis of Steriade and collaborators (Steriade et al., 1987). The 'TC-interneuron' mechanism is probably not correct because it relies on intranuclear collaterals that were never seen experimentally. Nonetheless, it had the merit of pointing-out two essential features of thalamic oscillations: they are shaped by inhibitory mechanisms and the rebound property of thalamic neurons is critical. Thus the principles proposed by Andersen and Eccles (1962) for inhibitory feedback and rebound may be correct, but the implementation of these principles in thalamic circuits must require another cell type than interneurons.

The central role of the RE nucleus was firmly established by experiments realized on cats *in vivo* (Steriade et al., 1985, 1987; Fig. 6.3). The role of the RE nucleus was also demonstrated in more recent *in vitro* experiments (von Krosigk et al., 1993), which established that spindle waves depend on mutual interactions between TC and RE cells (Fig. 6.6), similar to the hypothesis of Scheibel and Scheibel.

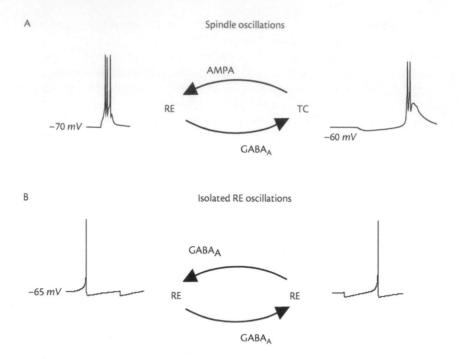

Fig. 6.40: Proposed mechanisms for different types of oscillations observed in the thalamic circuitry. A. ~10 *Hz* spindle oscillations can result from the interaction between TC and RE cells. In this case, there is a mirror image between the two types of cells. The TC cell is maintained depolarized around −60 *mV* and the LTS can be elicited by GABA$_A$-dominated IPSPs arising from the RE cell. The RE cell has a more hyperpolarized membrane potential, such that low threshold spikes can be triggered by AMPA-mediated EPSPs delivered from TC axons. B. Another type of oscillation at similar frequency can occur between RE cells if they are maintained at a more depolarized level. In this case, low threshold spikes can be produced in the RE cell following GABA$_A$-mediated IPSPs arising from neighbouring RE cells.

Different modelling studies investigated oscillatory behaviour in networks of TC and RE cells. These different models point to the conclusion that the intrinsic rebound burst properties of thalamic neurons, as well as the receptor types mediating their synaptic interactions constitute a robust excitable structure for oscillations (Fig. 6.40A). TC cells at a resting level around −60 to −70 *mV* receive strong IPSPs from RE cells. These IPSPs are ideally suited for triggering rebound bursts in these cells (Deschênes et al., 1984; Steriade and Deschênes, 1984; Huguenard and Prince, 1994b; Warren et al., 1994; Bal et al., 1995a). In RE cells, the resting level tends to be more hyperpolarized than TC cells and powerful EPSPs generate strong bursts (Huguenard and Prince, 1992; Contreras et al., 1993; Huguenard and Prince, 1994b; Warren et al., 1994; Bal et al., 1995a). These properties taken together make the interconnected TC-RE structure highly efficient in generating oscillations. The model also accounted for the mirror image between TC and RE bursts. This mirror image is a typical feature of spindle oscillations in the intact thalamus (Steriade and Llinás, 1988; Muhlethaler and Serafin, 1990; von Krosigk et al., 1993; Bal et al., 1995a, 1995b).

In addition to studying the mechanisms for rhythmicity, models were also used to investigate the mechanisms underlying the initiation, propagation/synchronization and termination of spindles, as well as augmenting responses.

6.8.1.2 Spindle initiation

The observation of spindles *in vitro* (Fig. 6.5; von Krosigk et al., 1993) led to several hypotheses concerning their initiation. First, a region of the thalamic RE nucleus might be more excitable and spontaneous firing could start an oscillation (Kim et al., 1995). This is supported by a report that stimulation of a single RE cell could initiate a spindle wave in the whole network (Kim et al., 1995). Second, there could be spontaneously oscillating TC cells in an otherwise homogeneous network, which would recruit the whole network into spindle rhythmicity (Destexhe et al., 1996a). These initiator TC cells can be as few as one in small networks of around a hundred cells. Blocking either excitatory or inhibitory transmission would result in a quiescent network, as observed experimentally (von Krosigk et al., 1993), except for a small minority of initiator cells, which should continue to display intrinsic oscillations. If several TC cells are needed to discharge one RE cell, then the network waits until the synchronous bursts occur 'by chance' in several TC cells that are sufficient to excite a target RE cell.

Other mechanisms have also been proposed for the initiation of spindle waves in the intact thalamocortical system *in vivo* (Destexhe et al., 1998a). In Chapter 7, we will show evidence for multiple thalamic initiation sites, as well as for the triggering of spindles by the cortex through corticothalamic feedback.

6.8.1.3 Spindle synchronization and propagation

The first proposition that spindles may propagate from an initiation site through local recruitment phenomena was advanced by Andersen and Andersson (1968) to explain the experiments of Verzeano and Negishi (1960). More recently, a multielectrode investigation of spindle waves in ferret thalamic slices (Kim et al., 1995) reported that these oscillations propagate through the dorso-ventral axis of the slice (Fig. 6.29). Models (Destexhe et al., 1996a; Golomb et al., 1996) successfully reproduced these experimental observations by postulating that TC and RE cells connections have some degree of divergence but are topographically organized between TC and RE layers. At each cycle of the oscillation, the spindle oscillation recruits, through the divergent axonal projections, more TC and RE cells. A systematic propagation occurs if these projections are topographically organized, as observed in morphological studies (Minderhoud, 1971; Sanderson, 1971; Gonzalo-Ruiz and Lieberman, 1995).

Thus models predict that spindle oscillations are synchronized within domains corresponding to the size of axon collaterals, and at each cycle of the oscillation, more cells are recruited into the oscillation, leading to a propagating pattern across the network. Within the assembly of cells that oscillate, the synchrony is maintained by local axon collaterals (TC-RE and RE-TC projections).

In the intact thalamocortical system *in vivo*, however, spindle oscillations display spatiotemporal patterns which are more simultaneous than in slices, although propagating patterns can be observed in some conditions (Contreras et al., 1997b; see Chapter 7). The mechanisms leading to the higher spatiotemporal coherence of spindles *in vivo* will be investigated in detail in the next chapter.

6.8.1.4 Spindle termination

Several possibilities have been proposed for the termination of spindles. It was proposed that the upregulation of I_h by intracellular Ca^{2+} might underlie spindle termination (Destexhe et al., 1993b; see Section 6.3.1). The termination might also result from the progressive hyperpolarization of RE cells during a spindle sequence (von Krosigk et al., 1993). Another possibility is that there might be an activity-dependent depression of GABAergic currents (von Krosigk and McCormick, 1992; Kim et al., 1995). *In vivo* data also indicate that the 'waning' of spindle waves is associated with a progressive desynchronization of the network (Contreras and Steriade, 1996), although this could also be a natural consequence of properties intrinsic to TC cells. Finally, spindles may terminate because of an enhancement of I_h due to its own activation kinetics[4].

We have investigated here the hypothesis that spindles terminate because of a Ca^{2+}-dependent upregulation of I_h. With this mechanism, the model can account for a number of experimental observations: (a) A few TC cells can display intrinsic waxing-and-waning oscillations (Fig. 3.12; Leresche et al., 1991), and those are sensitive to I_h (Fig. 3.13; Soltesz et al., 1991). These oscillations are characterized by an afterdepolarization (ADP), which lasts several seconds (Fig. 3.12; Leresche et al., 1991; see Fig. 3.16 for the model). (b) A cesium-sensitive ADP was also observed in TC cells following a spindle sequence, consistent with an augmentation of I_h conductance (Fig. 3.21; Bal and McCormick, 1996; see Fig. 3.22 for the model). Consistent with this, a progressive diminution of input resistance was measured in TC cells during the spindle oscillation *in vivo* (Nuñez et al., 1992). (c) Spindles triggered by electrical stimulation displayed a refractory period, as observed *in vitro* (Fig. 6.34; Kim et al., 1995; see model in Fig. 6.33B) and *in vivo* from cortically elicited spindles (see Chapter 7, Fig. 7.18A for experiments and Fig. 7.18B for the model). (d) Colliding spindle waves do not cross but rather merge into a unique oscillation (Fig. 6.32; Kim et al., 1995; see Fig. 6.33A for the model). (e) Spindle oscillations can be transformed into sustained oscillations following blocking of I_h (Fig. 6.35; Bal and McCormick, 1996; Luthi et al., 1998; see model in Fig. 6.33C). Taken together, these data are consistent with a progressive build-up of I_h during the spindle oscillation, until the TC cells can no longer be sufficiently hyperpolarized for the whole network to continue to oscillate.

The details of the biophysical mechanisms underlying the Ca^{2+}-dependent upregulation of I_h have been discussed in Chapter 3. They most likely involve the locking of I_h channels in the open state by intracellular Ca^{2+}, a mechanism with was initially proposed theoretically (Destexhe et al., 1993a), and later found as the best candidate to explain voltage-clamp experiments in thalamic neurons (Luthi and McCormick, 1999) as well as in other cell types (DiFrancesco, 1999).

6.8.1.5 Augmenting responses

The low-threshold mechanism for augmentation based on the deinactivation of I_T current explored here was highly robust to varying the parameters of synaptic and intrinsic currents. Except for a complete block of $GABA_B$ receptors, changing the synaptic conductances in the model had only a small qualitative effect on the augmenting responses in the TC cells (Bazhenov et al., 1998a). Augmentation is also observed over a wide range of the values of the intrinsic conductances. However, blocking the I_T currents in the TC and RE cells leads to

---

[4] As shown earlier (Section 3.3.4 in Chapter 3; Destexhe and Babloyantz, 1993), voltage-clamp data on I_h reveal a coexistence between fast and slow processes that can be modelled with two independent activation gates. In the presence of other currents, this dual activation kinetics can give rise to waxing-and-waning oscillations followed by an ADP without the need of calcium regulation (see Section 3.3.4 in Chapter 3).

non-augmenting responses during repetitive stimulation. Depolarizing TC cells had the same effect as blocking the I_T currents since depolarization inactivates the I_T current and prevents it from becoming deinactivation during RE-evoked.

The properties of augmenting responses in TC cells in the model depended on their position relative to the site of stimulation. TC cells located near the centre of stimulation displayed rapid augmentation followed by 1–2 cycles of slow oscillations. TC cells located far from the site of stimulation had weak augmentation and prolonged post-stimulus oscillations. The frequency of these self-sustained oscillations was 3–4 Hz, which is in the range of the delta rhythm (McCormick and Pape, 1990a; Leresche et al., 1991; Soltesz et al., 1991; Steriade et al., 1991). Delta oscillations can be generated in single TC cell that are hyperpolarized as a result of the interplay between low-threshold (I_T) and hyperpolarization-activated cation (I_h) currents. However, in the present model the generation of the slow post-stimulus oscillations depended on the inhibitory RE neurons. Synchronous burst discharges in RE cells induced a fast hyperpolarization in TC cells that activated the I_h current. This led to the depolarization of TC cells, followed by low threshold spikes and rebound bursts. Finally, burst discharges in TC cells evoked EPSPs and new bursts in RE cells. Progressive desynchronization of the network in the absence of external stimulation decreased the amplitude of the summed IPSPs in the TC cells and led to the termination of post-stimulus oscillations.

RE cells are critical in generating augmenting responses in TC cells. Powerful IPSPs delivered from RE cells deinactivated the low-threshold I_T current and set the conditions for augmentation to occur in the responses of the thalamic relay cells. The pattern of responses in RE cells during repetitive stimulation was not as stereotyped as in TC cells. The resting membrane potential of RE cells in the model was around -75 mV, based on in vitro recordings, slightly more negative than in vivo. The low-threshold Ca^{2+} current in RE cells was partially deinactivated at this level and the first stimulus in the train elicited powerful burst discharge (usually ~ 5–10 spikes). Depolarization of the RE cells inactivated I_T channels and the next stimulus evoked weaker responses. However, the buildup of TC-evoked EPSPs led to the slow augmentation of RE responses starting from the third stimulus. These features were especially prominent during low-intensity stimulation. In this case, the burst discharge evoked by the first EPSP in the train was followed by EPSPs without action potential during the rest of train. This result is in a good agreement with the in vivo data (Timofeev et al., 1996).

The augmentation of RE cell responses in the model during the entire train of high-intensity stimuli differed from that observed experimentally. One possible explanation is that the RE cells in the model were more hyperpolarized than those recorded in vivo. When positive DC current was injected, the RE cells depolarized and there was weaker but monotonically increasing responses that were more similar to in vivo responses (Bazhenov et al., 1998a). However, the depolarization in the RE cells produced weaker IPSPs in TC cells and delayed augmentation. These conflicting requirements may be resolved when multicompartment models of RE cells are used instead of the single compartment model used here. If the low-threshold Ca^{2+} current is concentrated in the distal dendrites of RE cells (Destexhe et al., 1996b), then hyperpolarization of the distal dendrites should lead to the deinactivation of the I_T current despite higher membrane potentials in the soma of RE cells.

6.8.1.6 Experimental consequences

A series of predictions were generated by the models that are verifiable by experiments.

1. With the resting membrane potential for TC cells at around -60 mV and -70 mV or lower for RE cells, then TC and RE cells interconnected with glutamatergic and GABAergic synapses form a highly excitable structure for generating sustained oscillations. This is consistent with intracellular recordings in both types of cells during spindle oscillations (Huguenard and Prince, 1992, 1994b; Bal et al., 1995a, 1995b). However, the differences observed in experiments are not as large as those predicted by the model, perhaps because the RE cells in the model had only a single compartment. As shown in Chapter 4, the evidence for a dendritic localization of I_T in RE cells is strong (Mulle et al., 1986; Destexhe et al., 1996) and in a multicompartment model of an RE cell burst discharges can be evoked by EPSPs even if the soma is more depolarized.

2. The models predicted an intrinsic mechanism in TC cells to provide the observed 'waning' of oscillations and that this mechanism may be an activity-dependent enhancement of I_h (Destexhe and Babloyantz, 1993; Destexhe et al., 1993a). This should result in an augmentation of I_h and an afterdepolarization (ADP) following the spindle sequence, leading to progressive hyperpolarization of the membrane during the silent period. This ADP and hyperpolarization have been observed in the intrinsic waxing-and-waning behaviour of TC cells (Fig. 3.12; Leresche et al., 1991). The fact that the same mechanism may underlie the waxing-and-waning of spindles (Destexhe et al., 1993b) predicted that AHP/hyperpolarizations should be observable also during spindles (see Fig. 6.22D). This prediction has been verified experimentally (Fig. 3.21; Bal and McCormick, 1996).

3. The prediction that the 'waning' of spindles is due to I_h predicted that alteration of this current should alter spindle waves (Destexhe et al., 1993b). This prediction has also been verified experimentally: in vitro experiments have demonstrated that waxing-and-waning spindles transform into sustained oscillations by blocking I_h (Fig. 6.35; Bal and McCormick, 1996; Luthi et al., 1998).

4. The biophysical mechanism underlying the enhancement of I_h was assumed in the model to be the direct binding of Ca^{2+} ions on the open state of the channel (Destexhe et al., 1993a). This prediction was shown to be incorrect, by whole-cell experiments showing that Ca^{2+} has no direct effect on I_h channels (Budde et al., 1997). However, an elevated level of intracellular Ca^{2+} does indeed enhance I_h (Fig. 3.23; Luthi and McCormick, 1998), probably acting indirectly via cAMP (Luthi and McCormick, 1999). A recent investigation provided evidence that cAMP regulates I_h channels by binding to and stabilizing the open state (DiFrancesco, 1999). Thus, the biophysical mechanism for I_h upregulation that has emerged is similar in spirit to the original model (Destexhe et al., 1993a) although based on more complicated molecular mechanisms.

5. The model shows that spindles can be initiated if only a few cells are intrinsic oscillators in the network (Destexhe et al., 1996a). If a subset of either TC or RE cells, or both, is spontaneous oscillator, then the model predicts initiation, propagation and termination of spindle waves compatible with experimental results. The model explored the possibility that as few as one TC cell is needed to produce intrinsic waxing-and-waning oscillations, as observed in a few cases of TC cells recorded from cats in vitro (Fig. 3.12; Leresche et al., 1991). The existence of such spontaneous waxing-and-waning TC cells has not yet been investigated in ferret thalamic slices, but spontaneous spindle oscillations typically started at the same foci (Fig. 6.29C; Kim et al., 1995), consistent with this hypothesis.

The model predicts that recording at one of these foci following the block of synaptic connections should reveal spontaneously oscillating TC cells.

6. To match the propagation velocity of spindle waves, models predicted that axonal projections between TC and RE nuclei must be topographically organized rather than interconnected to the entire thalamus (Destexhe et al., 1996a; Golomb et al., 1996). This prediction is consistent with morphological studies showing that in the thalamus, axonal projections arising from TC or RE cells extend at most over 100 to 500 μm (Harris et al., 1981; Jones, 1985; Cucchiaro et al., 1991; Uhlrich et al., 1991; Pinault et al., 1995), and with a topographic organization between reticular thalamus and other nuclei (Minderhoud, 1971; Sanderson, 1971; Ohara and Lieberman, 1985; Gonzalo-Ruiz and Lieberman, 1995; Liu et al., 1995).

7. In the model the spatiotemporal patterns of spindle oscillations were organized in localized clusters of discharging cells that propagated through the network with the same velocity as the wave front (Fig. 6.31). This property arose from the localized axonal projections combined with the subharmonic bursting properties of TC cells. In contrast, bicuculline-induced oscillations in the same network did not propagate in the same way, but showed more homogeneous patterns of discharge (Fig. 8.20; Destexhe et al., 1996a). At a time resolution of around 10 ms, it should be possible to directly observe these waves in optical recordings from ferret thalamic slices using voltage-sensitive dyes.

8. In the model of thalamic augmenting responses, activation of the GABA$_B$ synaptic interconnections from RE to TC cells was necessary for augmenting responses to occur in TC cells during repetitive stimulation. The model therefore predicts that blockade of GABA$_B$ receptors should transform the augmenting responses of TC cells into more stereotyped responses.

9. Simultaneous activation of RE and TC cells is necessary for augmenting responses in TC cells during repetitive stimulation. This predicts that during spontaneous sleep oscillations, intrathalamic stimulation activating both RE and TC neurons may result in immediate augmentation and self-sustained activity. Thus, stimulation of inputs that activate both RE and TC cells, such as corticothalamic volleys, may induce a transition from sleep rhythms to seizure activity.

10. The model predicts that intrathalamic mechanisms might make a major contribution to cortical augmenting responses during repetitive thalamic stimulation. In turn, cortical feedback could promote augmentation in the thalamus through mechanisms similar to those that are involved in the generation of spindles (Castro-Alamancos and Connors, 1996a, b). The contribution of cortical EPSPs to intrathalamic augmenting responses may be increased by intracortical short-term plasticity such as paired-pulse facilitation (Castro-Alamancos and Connors, 1996b).

6.8.1.7 Open questions and issues

Several parts of this framework remain to be investigated. The first is the question of the coexistence of spindle oscillations with other types of rhythmicity, such as delta waves or slower rhythms. The spindle oscillations appear in the model close to the resting 'relay' state of TC cells. A slight reduction of I_h transforms the relay state into waxing-and-waning oscillations;

with further reduction, TC cells show slower sustained oscillations. This is consistent with the fact that spindles always appear in the early phases of sleep, in association with a progressively reduced neuromodulatory drive, which is known to modify the amplitude and the voltage-dependence of I_h (McCormick and Pape, 1990b; Soltesz et al., 1991). This view is consistent with *in vivo* observations that spindle and delta oscillations cannot coexist because they are generated at different membrane potentials in thalamic neurons (Nuñez et al., 1992). Further exploration of the model could provide support for this view.

A second issue that should be investigated is the exact biophysical mechanism underlying upregulation of I_h in TC cells. As discussed above, it is now clear that: (a) I_h is enhanced during spindles and probably underlies its termination (Bal and McCormick, 1996; Luthi and McCormick, 1998); (b) rising intracellular Ca^{2+} also enhances I_h (Luthi and McCormick, 1998) but Ca^{2+} does not act directly on I_h channels (Budde et al., 1997); (c) cAMP induces a shift in the voltage-dependence of I_h (DiFrancesco and Totora, 1991) similar to the shift induced by intracellular Ca^{2+} (Luthi and McCormick, 1998); (d) cAMP seem to allosterically regulate I_h channels by stabilizing their open state (DiFrancesco, 1999), similar to the mechanism proposed for Ca^{2+} modulation of I_h (Destexhe et al., 1993a; see Section 3.3.2 in Chapter 3). As Ca^{2+} is actively involved in regulating adenylate cyclase activity (reviewed in MacNeil et al., 1985), a possibility is that thalamic neurons possess a Ca^{2+}-dependent adenylate cyclase, which would provide the missing link to fully explain the experiments. This possibility for a $Ca^{2+} \rightarrow$ cAMP $\rightarrow$ I_h pathway should be examined in future models.

The bidirectional excitatory and inhibitory interactions between TC and RE cells described above (Fig. 6.40A) also deserves further study. Other mechanisms for generating such an excitable structure are possible. For example, oscillations could also be sustained if both types of cells were hyperpolarized and interacted through excitation, or if both were depolarized and interacted through inhibition (such as in the isolated RE nucleus). What is the advantage of the proposed mechanism over the other possibilities? Is it more robust, more controllable? Is it related to information processing in the thalamus? These questions remain as challenges for future modelling studies.

6.8.2 Oscillations in the isolated reticular nucleus

6.8.2.1 Mechanisms of rhythmicity

The observation that the isolated RE nucleus oscillates *in vivo* (Steriade et al., 1987; see Fig. 6.4) has motivated models for rhythmogenesis in RE networks. A previous modelling study had shown that networks of inhibitory neurons exhibiting post-inhibitory rebound could generate rhythmic activity, explaining the oscillations in central pattern generators (Perkel and Mulloney, 1974). However, in this type of model, the mutual inhibitory-rebound interactions generate anti-phase oscillations, which does not explain the genesis of synchronized oscillations in networks of the RE nucleus.

Two hypothetical mechanisms were proposed independently to explain the genesis of synchronized oscillations by RE cells mutually connected with inhibitory synapses. First, a 'slow inhibition' mechanism was shown to produce synchronized oscillations when the decay of the synaptic current was sufficiently slow (Fig. 6.7; Wang and Rinzel, 1992, 1993). Another model investigated a 'fast inhibition' mechanism (Fig. 6.8; Destexhe and Babloyantz, 1992), according to which synchronized rhythms are observed with fast inhibition if the connectivity between neurons is not restricted to immediate neighbours, but rather to a more extended population of neighbouring cells. In both models the synaptic interactions graded transformations between the presynaptic membrane potential and the postsynaptic conductance of the synapse.

Subsequent models of RE oscillations were investigated based on more accurate biophysical representations of the intrinsic and synaptic currents (Destexhe et al., 1994a; Golomb and Rinzel, 1994; Bazhenov et al., 1999). The model presented in Section 6.2 (Destexhe et al., 1994a) was used to explore the effects of different patterns of connectivity and kinetics of inhibitory interactions on the oscillatory behaviour of RE networks. Several types of oscillations were identified corresponding to the 'slow inhibition' and 'fast inhibition' mechanisms proposed earlier. In large networks, oscillations with fast $GABA_A$-mediated inhibition showed waxing-and-waning patterns similar to the *in vivo* recordings (compare Fig. 6.16 with Fig. 6.4).

Oscillations in the isolated RE nucleus *in vivo* (Fig. 6.4; Steriade et al., 1987) have not yet been observed *in vitro* despite a large number of studies (Avanzini et al., 1989; Huguenard and Prince, 1992; von Krosigk et al., 1993; Huguenard and Prince, 1994b; Warren et al., 1994; Bal et al., 1995b). In models, networks of RE cells are unable to generate oscillations if the resting level of RE cells is too close to the $GABA_A$ reversal potential.[5] For oscillations to occur, the membrane potentials of RE cells may need to be depolarized (Destexhe et al., 1994b).

Neuromodulation can provide the necessary depolarization to generate oscillations in the RE nucleus (Section 6.4). In the presence of a depolarizing neuromodulatory drive, such as that provided by noradrenaline, serotonin, or glutamate acting at metabotropic receptors, the resting level of RE cells becomes more depolarized (McCormick, 1992). Under these conditions, the same IPSPs that were ineffective in the absence of neuromodulation can now trigger rebound bursts (Destexhe et al., 1994b). Interconnected RE cells therefore can oscillate via mutual inhibitory interactions if their resting level is depolarized (Fig. 6.40B). This is consistent with the observation of spindle rhythmicity in the isolated RE nucleus *in vivo* (Steriade et al., 1987).

Oscillations can also occur in the isolated RE nucleus if $GABA_A$ IPSPs are depolarizing (Fig. 6.20; Bazhenov et al., 1999). This can occur if the reversal potential for $GABA_A$ IPSPs is above the resting level for RE cells (Ulrich and Huguenard, 1997a).

6.8.2.2 Mechanisms for waxing-and-waning activity in the RE nucleus

In models of the isolated RE nucleus with appropriate resting level in RE cells, the network oscillates with waxing-and-waning fluctuations of amplitude in the average membrane potential, similar to field potential recordings in the isolated RE nucleus *in vivo* (compare Fig. 6.4 with Fig. 6.16). In the model, the waxing-and-waning activity was produced by the alternation of partially synchronized periods of oscillatory behaviour with desynchronized periods (Fig. 6.16). Similar waxing-and-waning oscillations were also found in a model based on depolarizing $GABA_A$ inhibitory interactions (Fig. 6.20B; Bazhenov et al., 1999). These models therefore suggest that waxing-and-waning oscillations can exist in the isolated RE nucleus, but are based on different mechanisms than spindles generated by TC-RE interactions.

---

[5] RE cells have been reported to have a resting level significantly more hyperpolarized than TC cells (Huguenard and Prince, 1992), which is also observed *in vivo* (Contreras et al., 1993).

6.8.2.3 Experimental consequences

Specific predictions about thalamic reticular cells allow the proposed mechanisms to be tested experimentally.

1. Models predicted that oscillations in the isolated RE nucleus can be generated by either a 'slow inhibition' mechanism (Wang and Rinzel, 1992) or by a 'fast inhibition' mechanism (Destexhe and Babloyantz, 1992). The latter was predicted to be $GABA_A$-receptor mediated (Destexhe et al., 1994a). *In vitro* experiments have indeed demonstrated that RE neurons interact together through $GABA_A$ receptors (Bal et al., 1995b; Ulrich and Huguenard, 1996; Sanchez-Vives et al., 1997a, 1997b; Zhang et al., 1997). However, weak postsynaptic $GABA_B$ conductances are also present (Ulrich and Huguenard, 1996) and in rats, $GABA_A$ currents have a slower decay time in reticular neurons compared to relay cells (Zhang et al., 1997). Experiments have therefore not decided which of the fast or the slow mechanism operates to generate thalamic reticular oscillations, except that the fast frequency of oscillations in the isolated RE nucleus (Steriade et al., 1997) would be more consistent with a mechanism based on $GABA_A$-receptor mediated interactions.

2. The model predicted that the waxing-and-waning of RE oscillations is based on mechanisms different from the waxing-and-waning of spindle waves in the TC-RE system. This should be observable in intracellular recordings made in the isolated RE nucleus *in vivo*. Such recordings should reveal an important difference from spindles recorded when the thalamus is intact: the isolated RE oscillations should show more sustained oscillatory patterns at the intracellular level (see Fig. 6.24). Although waxing-and-waning spindle oscillations have been recorded intracellularly in intact thalamocortical system (Andersen and Andersson, 1968; Steriade and Deschênes, 1984; Muhlethaler and Serafin, 1990; von Krosigk et al., 1993) intracellular recordings have not yet been made from neurons in the isolated RE nucleus *in vivo*.

3. Models predict that oscillatory behaviour in the isolated RE nucleus is highly sensitive to the resting level of RE cells. They must be sufficiently depolarized to produce rebound bursts in response to IPSPs (Destexhe et al., 1994c). The consequence is that oscillations should be restored *in vitro* if RE cells could be slightly depolarized. This could be achieved in slices by using bath application of noradrenergic or serotonergic agonists at concentrations sufficient to depolarize all RE neurons to the appropriate voltage range. Blocking neuromodulatory actions in the RE nucleus *in vivo* should suppress oscillatory behaviour.

6.8.2.4 Open questions and issues

Not only can neuromodulatory systems regulate the firing mode of RE cells, but they can also control the ability of interconnected cells to sustain oscillations. Neuromodulatory-induced oscillations in the RE nucleus might contribute to the ascending control of arousal by brain stem structures through extended projections to relay nuclei of the thalamus. The possibility that local regions of the RE nucleus, and associated relay nuclei, might selectively oscillate is worth exploring.

The models suggest that *in vivo* experiments should be performed to measure the resting level, GABA$_A$ reversal and role of neuromodulatory inputs in the isolated RE nucleus. The thalamus is capable of generating spindling in decorticate cats (see Fig. 6.1; Morison and Bassett, 1945; see also Fig. 7.3; Contreras et al., 1996). This preparation could be used to surgically isolate the RE nucleus with less technical difficulty than the previous experiments (see Fig. 6.4; Steriade et al., 1987). *In vitro* experiments should also be designed to investigate if the RE nucleus can oscillate following administration of low doses of neuromodulators, as predicted by the model.

6.9 Summary

The thalamus is an anatomical structure with only a few types of neurons, whose biophysical properties and synaptic interactions have been well characterized. This chapter examined the genesis of oscillatory behaviour by thalamic networks, in which individual neurons with complex intrinsic firing properties interact through synapses with multiple types of receptors. The oscillatory behaviour of thalamic networks varies in different preparations. We have shown here how computational models can be used to provide (a) a link between the biophysical properties of ion channels and the emergence of oscillatory behaviour at the level of networks of interconnected thalamic cells, (b) different hypotheses to explain why different preparations show different results; and (c) a unified framework in which all these experiments are compatible.

Models investigated several hypotheses that could explain the genesis of thalamic oscillatory behaviour in the spindle frequency range (7–14 *Hz*). First, oscillations can be generated by thalamic reticular neurons interacting through GABA$_A$ receptors (Section 6.2). The common core in the models that accounts for the genesis of oscillations by the isolated reticular nucleus was fast inhibition and a minimal amount of connectivity. Second, another type of oscillations can be generated by interconnected thalamic relay and reticular cells, interacting through AMPA and GABA$_A$ receptors (Sections 6.3 and 6.5.1). The 7–14 *Hz* frequency, phase relationships and the waxing-and-waning of spindle oscillations can be accounted for in models having accurate biophysical parameters for the kinetics of the intrinsic and synaptic currents.

Although these models can explain both TC-RE spindling activity and spindle rhythmicity in the isolated RE nucleus, they appear to be inconsistent with the observations that the isolated RE nucleus shows spontaneous oscillations *in vivo* but not *in vitro*. A possible explanation for these discrepancies is based on the effect of neuromodulators, which are present *in vivo* but not *in vitro* (Section 6.4). These neuromodulators may depolarize RE cells sufficiently to restore oscillatory behaviour that is absent when cells are too close to the equilibrium potential for the GABA$_A$ receptors. This model predicts that application of weak concentrations of noradrenaline or serotonin should restore oscillatory behaviour in slices of the RE nucleus. Similarly, blocking neuromodulation should suppress oscillations in the isolated RE nucleus *in vivo*. It is also possible in some preparations that the resting level of the RE cells is below the equilibrium potential for the GABA$_A$ receptors. In this case the GABA$_A$ currents will be depolarizing and can also support 7–14 *Hz* oscillations (see Section 6.2.6).

Finally, a network was investigated in which all the above properties were present together (Section 6.5.1). This model accounted for various network properties found in thalamic slices, such as the refractoriness of the network and the ratchet-like propagation of the oscillations. The model was also able to account for augmenting responses, which are observed when the thalamus is stimulated at about 10 *Hz* (Morison and Dempsey, 1943). The predictions made by the models can be further tested experimentally.

In the next chapter, the framework of interactions is expanded to include those between the thalamus and cerebral cortex.

Spindle oscillations in the thalamocortical system

Multisite recordings in the thalamus and the cerebral cortex have revealed the spatial and temporal structure of oscillatory activity in the thalamocortical system (Andersen and Andersson, 1968; Contreras et al., 1996, 1997a, 1997b). In this chapter, large-scale thalamocortical network models based on the mechanisms in previous chapters are matched to the spatiotemporal patterns of coherent activity observed in recordings of thalamocortical oscillations. In particular, a thalamocortical-loop framework is proposed to account for the initiation, synchronization and termination of spindle oscillations based on the known properties of thalamocortical interactions. This model of thalamocortical interactions can also be used to study pathological conditions, such as absence seizures (Chapter 8).

7.1 Experimental characterization of spindle oscillations in the thalamocortical system

7.1.1 Early studies

In their monograph, Andersen and Andersson (1968) summarized what was then known about how spindle oscillations are temporally organized and spatially distributed in the thalamo-cortical system (Fig. 7.1). They recorded spindle oscillations with multiple electrodes over the entire cortical mantle and found that spindles were characterized by a complex spatiotemporal organization, with highly variable sites of initiation (Fig. 7.1); however, the oscillations were synchronous over distances of several millimeters (see overlap of oscillatory sequences in Fig. 7.1).

To account for the large-scale synchrony and distribution of spindle oscillations, Andersen and Andersson (1968) proposed a model in which different oscillating sites in the thalamus were linked by 'distributor' TC cells, which would connect to other TC cells by intrathalamic excitatory

Thalamocortical Assemblies: Sleep Spindles, Slow Waves and Epileptic Discharges. Second Edition. Alain Destexhe and Terrence J. Sejnowski, Oxford University Press. © Oxford University Press 2023.
DOI: 10.1093/oso/9780198864998.003.0007

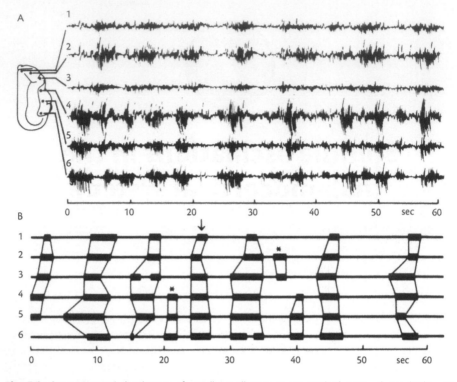

Fig. 7.1: Spatiotemporal distribution of spindle oscillations in cat cerebral cortex during barbiturate anesthesia. A. Spontaneous activity at electrode locations indicated. B. Duration of spindle sequences, recorded at same electrode locations as in A, plotted as horizontal bars. Thin lines signify the beginning and end of spindles at different electrodes. Note local (*) and nearly simultaneous (arrow) spindles. From Andersen and Andersson, 1968; after Andersen et al., 1967.

axon collaterals (Fig. 7.2). According to their view, spindling could start by the discharge of any neuron in a thalamic group and be transmitted to inhibitory interneurons through intra-nuclear recurrent collaterals; these in turn would project IPSPs back to the TC neurons, which would fire rebound spike bursts at the offset of the IPSPs. This mechanism explicitly predicted propagating patterns of spindle oscillations through the thalamus and consequently through the cortex (Andersen et al., 1967; Andersen and Andersson, 1968). It predicted the genesis of propagating waves through the progressive recruitment due to intrathalamic axonal projections, in line with early experimental observations (Verzeano and Negishi, 1960). Systematic propagation[1] was observed subsequently in thalamic slices (Kim et al., 1985) and analysed with computational models based on local intrathalamic axon collaterals (Section 6.5 in Chapter 6; Destexhe et al., 1996a; Golomb et al., 1996). However, the spatiotemporal patterns of spindles in the

---

[1] The term 'systematic propagation' is defined here as a systematic pattern of delay between two different sites, such that the initiation is progressively delayed for sites of increasing distances (see for example Fig. 6.5).

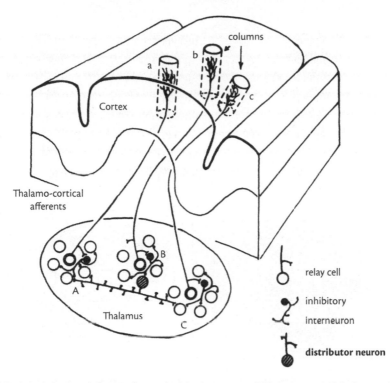

Fig. 7.2: Intrathalamic mechanism for synchronization proposed by Andersen and Andersson. Neurons of three thalamic nuclear groups (A, B, C) send their axons to the related parts of neocortex (columns a, b, c). Collaterals of these axons were postulated to excite inhibitory interneurons (black) that could then evoke synchronous IPSPs in a large number of thalamic neurons. The different thalamic groups may generate spindle oscillations autonomously and affect their cortical targets independently of each other. The postulated 'distributor thalamic neurons' (grey) would coordinate the oscillations across different thalamic and cortical sites. From Andersen and Andersson, 1968.

thalamocortical system observed recently *in vivo* (Contreras et al., 1997b; see below) are quite different from the systematic propagation observed in thalamic slices (Fig. 6.29; Kim et al., 1995).

In the model of Andersen and Andersson (1968), the synchrony of spindle oscillation was due exclusively to intrathalamic synchronizing mechanisms; synchronized oscillations would then be transferred to cortex. In this chapter, we review evidence that the observed synchrony is a consequence of mechanisms involving *both* the cortex and the thalamus. We begin by showing how corticothalamic feedback projections influence the synchronization of thalamocortical activity.

7.1.2 The influence of corticothalamic projections

The influence of the massive corticothalamic projection on the spatiotemporal coherence of spontaneous and global oscillations generated in the cat thalamus was investigated under barbiturate anesthesia. Recording of local field potentials from the thalamus, with eight tungsten

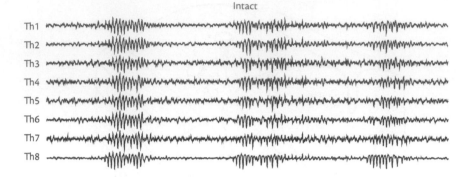

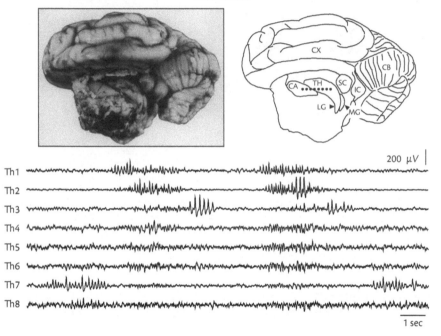

Fig. 7.3: Removal of the cerebral cortex affects the pattern of spindle oscillations in the thalamus. In an intact network under barbiturate anesthesia (upper panel), 3 spontaneous spindle sequences at 8–9 *Hz* and lasting for 1–3 *s* occurred at roughly the same time in the local field potentials recorded from eight tungsten electrodes (TH1–TH8). Tip resistances were 1 to 5 *MΩ* and interelectrode distances of 1 *mm*. Negativity downward. Cortex was removed by suction after careful cauterization with silver nitrate (Photo), exposing the head of the caudate nucleus (CA, in the drawing), most of the dorsal thalamus (TH), the lateral geniculate body (LG), the medial geniculate body (MG), the superior (SC) and inferior colliculli (IC). Also in the photograph, and represented in the drawing at right, are the intact contralateral cortex (CX) and the cerebellum (CB). The eight electrodes were held together and their tips lowered to the positions indicated by the black dots in the drawing. The two or three most anterior electrodes crossed through the head of the caudate nucleus to reach the thalamus. After decortication (lower panel), recordings from approximately the same thalamic location showed that spindling continued at each electrode site, but their coincidence in time was lost. The eight-electrode configuration was positioned at different depths within the thalamus (from −2 to −6) and different lateral planes (from 2 to 5); all positions gave the same result. Figure reproduced from Contreras et al., 1996.

electrodes (Fig. 7.3, TH1–TH8) and interelectrode distances of 1 *mm*, revealed that spindling was nearly simultaneous[2] in most of the thalamus (Fig. 7.3, Intact). In the example of Fig. 7.3, recordings with the cortex intact (upper panel) showed that spindling in the thalamus tended to start and finish within narrow time windows in all eight electrodes (TH1–TH8). The cortex was then removed by suction ($n = 8$) and the electrodes returned to approximately their same positions (Fig. 7.3, dots in the drawing). In the decorticated animal, spindle sequences were no longer nearly simultaneous in the different electrodes. The systematic propagation of spindles can be observed in the thalamus after decortication (see below) although spindle sequences were occasionally synchronous even among widely spaced thalamic territories (see Fig. 7.3, spindle sequence in the half-right part of the bottom panel; and Fig. 7.4, Seq. power spectra at around 27 *s* from start, in the decorticated panel).

The spatiotemporal characteristics of spindle oscillations in the thalamus with the cortex intact were compared with those after decortication. Spatiotemporal maps of activity (Fig. 7.4, upper panel) were constructed by plotting time (time runs from top to bottom in each column, arrow indicates 1 *s*), space (from left to right, the width of each column represents 8 *mm* distance in the antero-posterior axis of the thalamus), and local field potentials' voltage (from blue to yellow, colour represents the amplitude of the negative deflections of thalamic local field potentials). In the left panel, with the cortex intact, oscillatory activity was highly coherent over the entire thalamic area from which activity was recorded, as indicated by the formation of horizontal yellow (maximum local activity) and blue (local silence) stripes at 8–10 Hz crossing side to side the vertical columns. Spindle sequences not only initiated synchronously, but each oscillatory cycle formed continuous, uninterrupted yellow and blue stripes across the thalamus. These stripes were not perfectly horizontal, which indicates the existence of phase shifts among the thalamic sites. Removal of the cortex drastically diminished the spatiotemporal coherence, as shown by the disorganized pattern (Fig. 7.4, decorticated, right colour panel) with absence of stripes, indicating that oscillatory activity was no longer synchronized between thalamic sites located more than 2 or 3 *mm* away.

The coincidence of spindle oscillations among the eight electrodes was investigated by analysing the power spectra in sequential windows of 0.5 *s* and by plotting, against time, the total power between 7–14 *Hz*, normalized to the 100% of the highest peak obtained (Fig. 7.4, middle panel). The total power of spindling frequency increased and decreased coherently among the eight electrodes when the cortex was intact (Fig. 7.4, Seq. power spectra, left). After decortication, oscillatory activity was disrupted and no longer appeared as concerted activity among the electrodes.

The effect of cortex removal on thalamic synchronization was quantified by calculating the decay of the correlation with distance (Fig. 7.4, bottom panel), from 0 to 7 *mm* in steps of 1 *mm* as determined by the configuration of the recording electrodes. In the intact thalamus, correlations showed a limited decay with distance, with values around 0.7 for distances up to 7 *mm*. After decortication, the spatial correlation decreased stepwise to values of around 0.2 to 0.3 for distances greater than 1 *mm*.

Even when large-scale coherence was lost in the thalamus after decortication, the persistence of spindle oscillations in the local field potentials indicated that local synchrony was still

---

[2] The term 'nearly simultaneous' is defined here as a temporal overlap of oscillations within 100 ms in different electrodes during most of the oscillatory sequence (see for example Fig. 7.3, Intact).

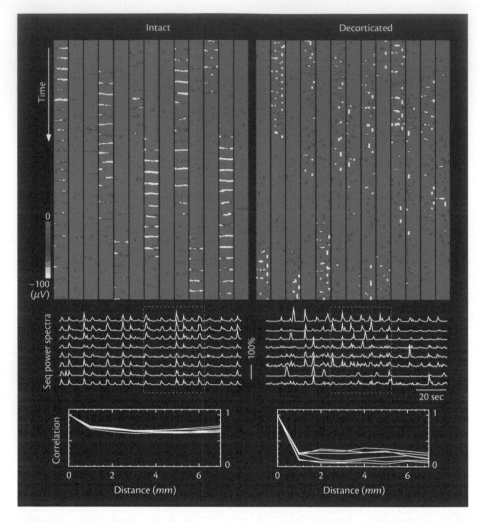

Fig. 7.4: Disruption of the spatiotemporal coherence of thalamic oscillations following removal of the cortex. Top panels: spatiotemporal maps of the distribution of electrical activity across the thalamus were constructed by assigning a colour to the value of the field potential at each electrode; the colour scale ranged in ten steps from the baseline (blue) to −100 μV (yellow). Time was divided in frames each representing a snapshot of 4 ms of thalamic activity and arranged in columns from top to bottom. A total of 40 s are represented (9880 frames, arrow is one second). Each frame consisted of 8 colour spots, each corresponding to the local field potential of one electrode from anterior to posterior (left to right). Middle panels: sequential power spectra evaluated at each site using a 0.5 s window. The total power of the 7–14 Hz frequency band was represented as a function of time (dashed box in red indicates the 40 seconds shown in top panels). Bottom panels: decay of correlation with distance. Cross-correlations were computed for all possible pairs of sites and the value at time zero from each correlation was represented as a function of the intersite distance for six different consecutive epochs of 20 s. Spatial correlation was calculated for thalamic recordings in the intact brain (left panels) and following removal of the cortex (right panels). Figure reproduced from Contreras et al., 1996.

maintained by intrathalamic connectivity. This can be achieved by the pattern of divergent connections between thalamic reticular (RE) and TC cells (Jones, 1985), as also demonstrated in a slice preparation (Fig. 6.5; von Krosigk et al., 1993). In order to determine local synchrony after decortication, dual intracellular recordings were performed (see Methods in Contreras et al., 1996) from pairs of TC cells in the decorticated thalamus at short (around 1 mm) and long (around 4 mm) interelectrode distances (Fig. 7.5). Intracellular recordings from TC cells revealed the typical pattern of intracellular events characteristic of spindling. Pairs of cells at 1 mm distance, recorded from the ventrolateral (VL) nucleus (Fig. 7.5A, scheme), showed spontaneous spindle sequences that were nearly synchronous between the two cells (Fig 7.5A, TC1 and TC2, upper panel). When the cells were recorded from the VL and the lateral posterior (LP) nucleus, at around 4 mm distance in the anteroposterior axis (Fig. 7.5B), spontaneous spindle sequences from both cells were no longer coincident in time. The initiation time of the IPSPs in one of the cells (TC1) was used to align the IPSPs of the other cell. Nearby cells showed almost simultaneous IPSPs, indicating that they received converging inputs from a pool of synchronized RE cells. Distant cells generally showed no consistent relationship between their intracellular activities related to spindling; also, the IPSP-triggered averaging in one cell gave rise to a flat line in the other cell, consistent with non-overlapping innervation from the RE cells (Fig. 7.5B).

The cortex may exert its global influence on the spatiotemporal organization of thalamic oscillations through projections of corticothalamic axons, which are more divergent than the reciprocal projections between the RE nucleus and the dorsal thalamus. The action of the cortex could act through direct excitation of TC cells, timing their output spike bursts by precipitating the offset of the cyclic IPSPs, through excitation of RE cells and synchronization of the onset of the IPSPs, or both.

Alternatively, the large-scale synchrony in the thalamus could be achieved within cortical circuits primarily through the widespread, long-range horizontal corticocortical projections within areas 5–7 (Avendano et al., 1988) and could be imposed on the thalamus secondarily. To investigate the role of intracortical connectivity on synchrony, multisite recordings were obtained from the suprasylvian cortex using the same array of electrodes as for thalamic recordings. In control conditions (Fig. 7.6A, Intact), spontaneous spindle oscillations occurred at 8–9 Hz almost simultaneously in the eight leads, reflecting the synchrony recorded in the thalamus with intact cortex (see Fig. 7.3). Following a deep coronal cut through the suprasylvian gyrus (Fig. 7.6A, Cut), leads Cx4 and Cx5 showed diminished activity due to local damage, but spontaneous oscillations still occurred simultaneously in the other leads. To quantify the effect of disruption of intracortical connections, the averaged cross-correlations was calculated between sites separated by increasing distances (Fig. 7.6B). As for thalamic recordings (Fig. 7.4), correlations showed a smooth decay with distance in the cortex. After the cut, a gap appeared in correlations between electrodes 1–4 and 1–5 due to tissue damage, but the same correlation patterns were seen between electrodes separated by 5 mm or more.

These results are consistent with the corticothalamic projections having a critical role in organizing the long-range synchrony and spatiotemporal patterns of oscillations generated in the thalamus. Although oscillations can occur simultaneously within cortical territories separated by ~7 mm (Fig. 7.3, Intact), the thalamocortical system in $vivo$ also shows signs of propagation, as shown in the next section.

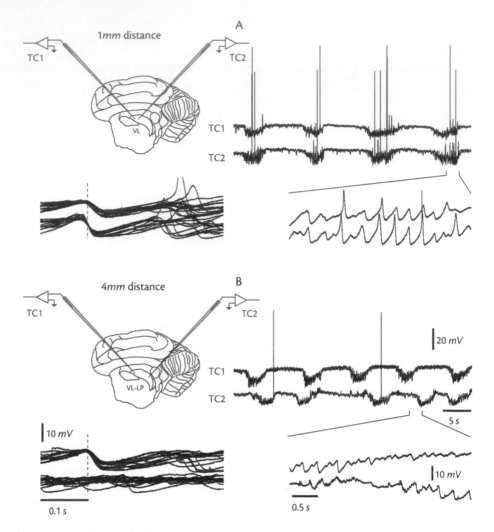

Fig. 7.5: Nearby TC cells displayed synchronized spindle sequences in the decorticated thalamus. (A) Intracellularly recordings from pairs of TC cells separated by no more than 1 *mm* (*n* = 5) located in the ventrolateral (VL) nucleus (TC1 and TC2 in the cartoon). Spindle sequences occurred simultaneously in both cells (spontaneous activity at right). The leftmost sequence expanded below shows the synchronous intracellular events characteristic of spindling. Spindling related IPSPs (*n* = 10) from TC1 were aligned using the time of their onset as a zero time reference (dotted line). The intracellular recording from TC2 was aligned to the same reference revealing the occurrence of IPSPs that were almost simultaneous with TC1. (B) In this pair of TC cells recorded from VL (TC1) and LP (TC2), separated by 4 *mm*, spindles occurring spontaneously in each cell showed no consistent temporal relationship. The spindling sequence expanded below shows that the termination of a spindle in TC1 coincides with the beginning of a spindle in TC2. Alignment of IPSPs (*n* = 10) from TC1 yielded a flat line in TC2. Figure reproduced from Contreras et al., 1996.

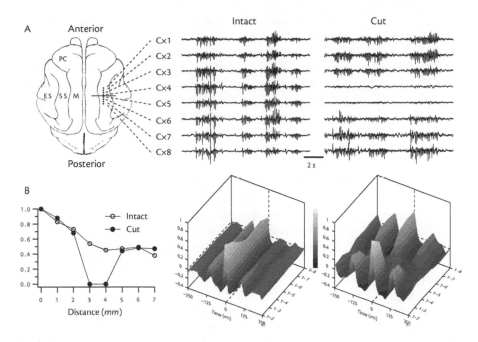

Fig. 7.6: Synchrony of spindle oscillations is not determined by intracortical connectivity. A. Multisite recordings from the depth (1 *mm*) of the suprasylvian (SS) gyrus using a similar electrode array (Cx1 to Cx8) as described in Fig. 7.3. Spontaneous spindle sequences occurred nearly simultaneously in control conditions (Intact). Following a 3 *mm*-deep coronal section (Cut) of the SS gyrus (horizontal line between electrodes Cx4 and Cx5 in the scheme), crossing laterally from the lateral aspect of the marginal gyrus (M) to the medial aspect of the ectosylvian gyrus (ES), did not disrupt simultaneity of oscillations. B. Synchronization was evaluated by calculating cross-correlograms between electrode Cx1 and the others. Correlograms from fifteen consecutive spindle sequences were averaged before and after the cut. The value of the averaged cross-correlation at time zero was represented as a function of distance with respect to the first electrode (left panel; • for intact cortex and ∘ after cut). Averaged cross-correlograms for each pair of electrodes were represented as surface plots for intact cortex (middle panel) and after cut (right panel). Correlation values were displayed using a grey scale ranging from −0.4 (black) to 1 (white; see grey scale bar). Secondary peaks around 120 *ms* indicate rhythmicity at 8–9 *Hz*. Figure reproduced from Contreras et al., 1996.

7.1.3 Propagating patterns of oscillations *in vivo*

Evidence for propagating patterns of spindle oscillations that had been obtained earlier in the thalamus *in vivo* (Verzeano and Negishi, 1960; Verzeano, 1972) were thought to reflect 'circulation of neuronal activity'. Those recordings were, however, from within restricted areas and the distance between electrodes was in the range 30–100 *μm* (Verzeano and Negishi, 1960). Propagating oscillations over several *mm* were later obtained in experiments with ferret thalamic slices (Fig. 6.29; Kim et al., 1995); they propagated through the progressive recruitment of neurons by intrathalamic axonal projections, through mechanisms similar to those proposed earlier by Andersen et al. (Andersen et al., 1967; Andersen and Andersson, 1968). The propagation could be quantitatively explained by the topographical structure of reciprocal connections between TC and RE cells and the upregulation of I_h (Section 6.5.5 in Chapter 6; Destexhe et al., 1996a).

Multielectrode recordings *in vivo* have shown that spindle oscillations occur nearly simultaneously if recorded in the thalamus using a linear array of eight equidistant electrodes spaced 1 *mm* apart (Fig. 7.7; Contreras et al., 1997b). This result is not incompatible with Verzeano's observations, which covered much smaller thalamic areas. Indeed, there is evidence for local propagation between Th1–Th6 in Fig. 7.7, although the spindle oscillations occurred approximately within the same time window at all sites.

In contrast to the synchronized spindles observed in the intact cortex, propagating spindles were observed routinely in the thalamus after decortication (Fig. 7.8; Contreras et al., 1997b). The propagation of spindles was visible in equally spaced extracellular recordings (Fig. 7.8, left panels) as well as from the corresponding rate metres (Fig. 7.8, right panels). These *in vivo* observations are consistent with the systematic propagation of spindle waves in slices (Fig. 6.29; Kim et al., 1995). However, there were no such signs of propagation in the presence of the cortex, suggesting that there is a tendency for activity to propagate in the isolated thalamus, but that this systematic propagation is not present in oscillations in the intact thalamocortical system.

Propagating patterns of oscillations were also observed in the intact thalamocortical system by using low-intensity electrical stimulation of the cortex (Fig. 7.9; Contreras et al., 1997b). Spindle sequences were consistently evoked by cortical stimulation. In Fig. 7.9, the EEG was recorded from the depth of the suprasylvian gyrus (~1 *mm*) with an array of eight electrodes (Cx1–Cx8, anterior to posterior, 1 *mm* interelectrode distance). Bipolar stimulating electrodes were positioned 3–4 *mm* anterior and posterior to the electrode array (see Contreras et al., 1997b). High intensity stimulation applied through the posterior stimulating electrode elicited an early response that decreased in amplitude with distance from Cx8 to Cx1 (top panel; early response is expanded at right, and stimulation is represented by dotted line) and was followed by a spindle sequence occurring synchronously in all eight electrodes (top panel, left). By decreasing the intensity to 15% of maximal stimulation, the evoked spindle sequence travelled away from the stimulating electrode (second panel, left), in which case the initial response barely reached more anterior electrodes (second panel, right). A mirror image sequence was obtained with anterior suprasylvian stimulation (bottom panel).

To test for the possibility that the propagation of spindles following cortical stimulation was mediated by intracortical horizontal connections, a deep coronal cut (similar to that depicted in Fig. 7.6) was performed between electrodes Cx4 and Cx5 and applied stimulation at both extremes of the suprasylvian gyrus (Fig. 7.10; $n = 6$). After the cut, the intensity of stimulation had to be increased, probably because of the tissue damage. The effect of the cut was visible in traces from electrodes 4 and 5, which became almost flat because of local depression; the early evoked potential was also strongly diminished, and in some cases disappeared, on the other side of electrodes 4 and 5 (see details at right). No effect, however, was visible in the propagation pattern of evoked spindles.

The hypothesis that the propagation of spindle sequences in the cortex reflects orderly propagation along the anteroposterior axis of the thalamus was tested by evoking spindles initiated by cortical stimulation similar to that used above to obtain cortical propagation while recording in the thalamus (Contreras et al., 1997b). In this case, low intensity stimuli to anterior suprasylvian cortex triggered a spindle sequence in the most anterior thalamic

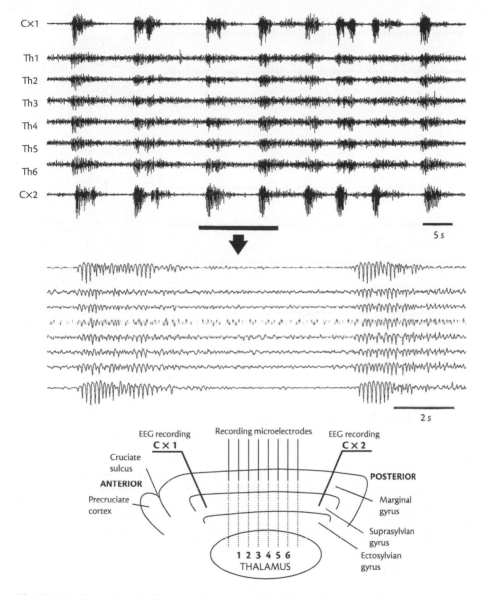

Fig. 7.7: Spindles synchronized between the cortex and the thalamus. The top panel shows spontaneous spindle sequences under barbiturate anesthesia, recorded by two bipolar electrodes located in the depth of the suprasylvian gyrus and separated by 15 *mm* (Cx1 and Cx2) and by six tungsten electrodes in the anteroposterior axis of the thalamus (Th1–Th6). The arrangement of recording electrodes is depicted in the scheme below; the thalamic electrodes penetrated through the marginal gyrus (dotted lines). A detail of two spontaneous spindle sequences, indicated by bar, is expanded below (arrow). Note that spindles occurred nearly simultaneously in all electrodes. Figure reproduced from Contreras et al., 1997b.

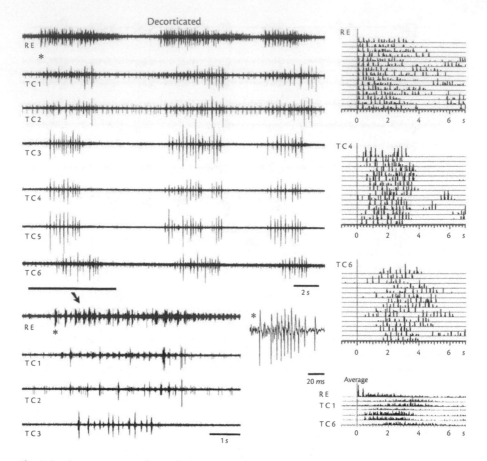

Fig. 7.8: Propagating spindle oscillations in decorticated thalamus *in vivo*. Multiunit recordings from the RE nucleus and six thalamic foci (TC1–TC6) were recorded simultaneously with 0.4 *mm* spacing between the electrodes. Spike bursts from three spontaneous spindles are shown in which firing of RE cells consistently preceded TC firing. The first spindle sequence (bar) is expanded below (arrow). Average rates were computed (bin size, 0.1 sec), and first burst from RE cells at each spindle sequence was taken as time t=0 to align the rate metres from the other cells. (Right column depicts RE, TC4 and TC6. Average includes rates from RE and all TC cells.) Fifteen consecutive spindle sequences show that spike bursts from RE cells preceded those from TC cells. A tendency for increased delay with distance suggested that thalamic propagation could occur in the absence of cortical influence. Figure reproduced from Contreras et al., 1997b.

electrode that travelled away from that electrode, and propagation in the opposite direction was obtained by low intensity stimulation to the posterior suprasylvian cortex. The synchronizing power of the corticothalamic projection became evident by increasing the intensity of stimuli applied to the anterior or the posterior suprasylvian sites: in each case that spindle sequence obtained was coherent over the entire extent of the thalamic recordings (see details in Contreras et al., 1997b).

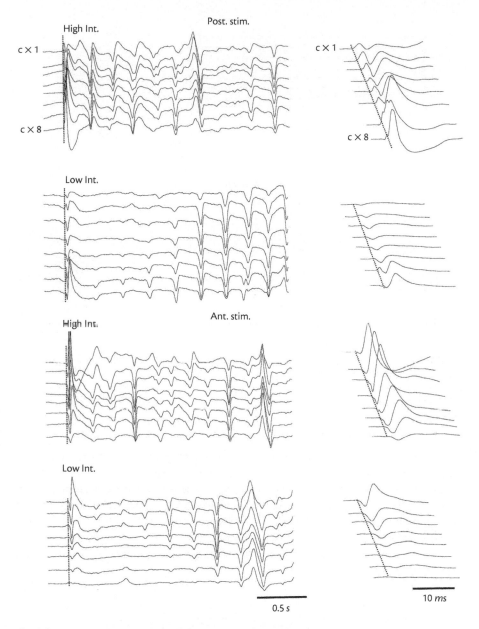

Fig. 7.9: Low-intensity cortical stimulation triggers spindle sequences that travel away from the site of stimulation. Local field potentials were recorded from eight tungsten electrodes (Cx1–Cx8) inserted in the depth (~1 mm) of the suprasylvian cortex. Stimuli were applied through bipolar electrodes situated 4 mm in front and 4 mm behind the electrode array (see similar location of recording cortical electrodes in the scheme of Fig. 7.6). Synchronized spindling was triggered by high intensity stimulation in the posterior part of the suprasylvian gyrus (top panel; initial response expanded at right, with sweeps displaced horizontally). Low intensity stimulation at the same posterior site triggered spindling that propagated to more anterior foci (second panel; initial response, expanded at right, barely reached electrode Cx1). Anterior stimulation also gave rise to full synchrony of evoked spindles on high intensity volleys (third panel), whereas propagation was from anterior to posterior sites by using low intensity stimuli (bottom panel). Figure reproduced from Contreras et al., 1997b.

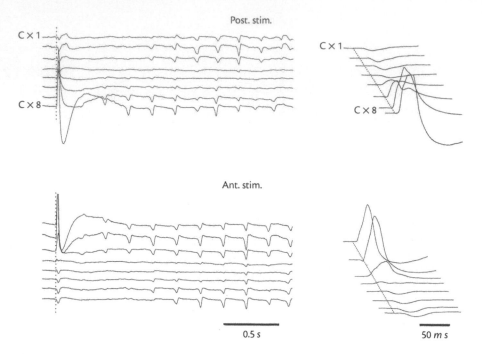

Fig. 7.10: Intracortical horizontal connections are not implicated in spindling propagation. Same experimental conditions as in Fig 7.9. A deep coronal cut was performed between electrodes 4 and 5, which caused a significant decrease in amplitude in those leads. Propagating spindle sequences were still observed in both directions (posterior-to-anterior and anterior-to-posterior). Figure reproduced from Contreras et al., 1997b.

These experiments show that the cortex has an important synchronizing impact on thalamic circuits, unlike the exclusively intrathalamic synchronizing mechanisms advanced by Andersen and Andersson (1968). This raises the question of how the strongly propagating activity observed in thalamic slices can be made compatible with the more synchronously occurring patterns of oscillation observed in the intact thalamocortical system. In the next section, we use computational models to suggest a possible explanation to reconcile these different experimental results, as well as a new view of how large-scale synchrony is generated in the intact thalamocortical system.

7.2 A thalamocortical network model of spindle oscillations

The large-scale synchrony observed in the intact thalamocortical system depends on the integrity of corticothalamic feedback. A computational model of the thalamocortical system is presented here to explore the determinant influence of the cortical feedback loop (Destexhe et al., 1998a). The mutual interaction between cortex and thalamus in this model is then analysed to determine possible cellular mechanisms responsible for the observed synchrony.

7.2.1 The thalamocortical network

Each cell in the model was described by an equation for the membrane current:

$$C_m \dot{V}_i = -g_L (V_i - E_L) - \sum_j I_{ji}^{int} - \sum_k I_{ki}^{syn} \tag{7.1}$$

where V_i is the membrane potential, $C_m = 1 \mu F/cm^2$ is the specific capacity of the membrane, g_L is the leakage conductance and E_L is the leakage reversal potential. Intrinsic and synaptic currents are respectively represented by I_{ji}^{int} and I_{ki}^{syn}.

Intrinsic currents were described within the Hodgkin-Huxley (1952) formalism:

$$I_{ji}^{int} = \bar{g}_j \, m_j^M h_j^N \, (V_i - E_j) \tag{7.2}$$

where the current is expressed as the product of, respectively, the maximal conductance, $\bar{g}_j$, activation (m_j) and inactivation variables (h_j), and the difference between the membrane potential V_i and the reversal potential E_j. Activation and inactivation gates follow the simple two-state kinetic scheme:

$$(closed) \, \underset{\beta(V)}{\overset{\alpha(V)}{\rightleftharpoons}} \, (open) \tag{7.3}$$

where α and β are voltage-dependent rate constants.

The set of intrinsic currents was different for each cell type and the parameter values for each current were obtained by matching the kinetic model to voltage-clamp data. The intrinsic currents were: I_T and I_h in TC cells (Chapter 3; Huguenard and McCormick, 1992; McCormick and Huguenard, 1992; Destexhe et al., 1996a); I_T in RE cells (Chapter 4; Destexhe et al., 1996b); I_M in cortical pyramidal (PY) cells (McCormick et al., 1993; see below) and, for all cells, $I_{Na} - I_K$ currents responsible for action potentials (Chapter 2; Traub and Miles, 1991). The Ca^{2+} dynamics in TC and RE cells and all other parameters were identical to those in a previously introduced model (Chapter 6; Destexhe et al., 1996a). The ionic mechanisms can be found in Chapters 2, 3 and 4, except the I_M current, which was described by (McCormick et al., 1993; Gutfreund et al., 1995):

$$I_M = \bar{g}_M \, m \, (V - E_K) \tag{7.4}$$

$$\frac{dm}{dt} = \alpha_m(V) \, (1 - m) - \beta_m(V) \, m \tag{7.5}$$

$$\alpha_m = \frac{0.0001 \, (V + 30)}{1 - \exp[-(V + 30)/9]} \tag{7.6}$$

$$\beta_m = \frac{-0.0001 \, (V + 30)}{1 - \exp[(V + 30)/9].} \tag{7.7}$$

Parameters for intrinsic currents and passive properties are summarized in Table 7.1.

The intrinsic properties of thalamic and cortical cells in the model are summarized in Fig. 7.11 (Intrinsic). Both TC and RE cells produced bursts of action potentials because of the presence of I_T. In TC cells, in addition to I_T, I_h conferred oscillatory properties and the upregulation of I_h by intracellular Ca^{2+} led to waxing-and-waning properties of these oscillations, as

Table 7.1 Parameters intrinsic to each cell type in the model. *extreme values used to randomize the properties of TC cells.

Parameter	Value
Cortical pyramidal cells (PY)	
membrane area	$29000\ \mu m^2$
$\bar{g}_L$	$0.1\ mS/cm^2$
E_L	$-70\ mV$
$\bar{g}_{Na}$	$50\ mS/cm^2$
$\bar{g}_K$	$5\ mS/cm^2$
$\bar{g}_M$	$0.07\ mS/cm^2$
Cortical interneurons (IN)	
membrane area	$14000\ \mu m^2$
$\bar{g}_L$	$0.15\ mS/cm^2$
E_L	$-70\ mV$
$\bar{g}_{Na}$	$50\ mS/cm^2$
$\bar{g}_K$	$10\ mS/cm^2$
Thalamic reticular cells (RE)	
membrane area	$14000\ \mu m^2$
$\bar{g}_L$	$0.05\ mS/cm^2$
E_L	$-90\ mV$
$\bar{g}_{Na}$	$200\ mS/cm^2$
$\bar{g}_K$	$20\ mS/cm^2$
$\bar{g}_{Ts}$	$3\ mS/cm^2$
Thalamocortical cells (TC)	
membrane area	$29000\ \mu m^2$
$\bar{g}_L$	$0.01\ mS/cm^2$
E_L	$-70\ mV$
$\bar{g}_{KL}$	$3\text{-}5^*\ nS$
$\bar{g}_{Na}$	$90\ mS/cm^2$
$\bar{g}_K$	$10\ mS/cm^2$
$\bar{g}_T$	$2\ mS/cm^2$
$\bar{g}_h$	$0.015\text{-}0.02^*\ mS/cm^2$

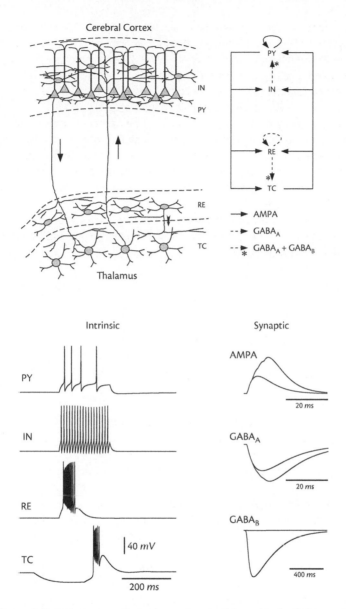

Fig. 7.11: Connectivity, intrinsic properties and synaptic potentials used in the thalamocortical network model. Schematic of four cell types and their connectivity are shown at the top: thalamocortical (TC) cells, thalamic reticular (RE) neuron, cortical pyramidal cells (PY) and interneurons (IN). Top-right: connectivity and receptor types used in the model. Bottom-left: intrinsic firing patterns of the four cell types: regular-spiking PY cell (depolarizing pulse of 0.75 nA during 200 ms; −70 mV rest), fast spiking IN (same pulse), RE cell burst (pulse of 0.3 nA during 10 ms) and rebound burst in a TC cell (pulse of −0.1 nA during 200 ms). Bottom-right: time courses of postsynaptic potentials for the three types of receptor used in the model. For each type of synaptic receptor type, the response to a single presynaptic spike is superimposed on the response to a high frequency burst of presynaptic spikes (300–400 Hz; four spikes for AMPA and GABA$_A$, ten spikes for GABA$_B$). Figure reproduced from Destexhe et al., 1998a.

in previous models (Chapter 3; Destexhe et al., 1993a, 1993b, 1996a). Models for cortical excitatory pyramidal (PY) and inhibitory (IN) cells were kept as simple as possible and were based on previous models (McCormick et al., 1993). Due to the presence of I_M, excitatory neurons generated adapting trains of action potentials, similar to 'regular spiking' pyramidal cells (Connors and Gutnick, 1990); the inhibitory interneurons only had the currents needed to generate action potentials and did not adapt ('fast spiking' cells).

All cells of the same type were identical and had homogeneous conductance parameters. The only exception was that the conductances for I_h and I_{KL} were randomly distributed over a physiologically realistic range in TC cells. This choice reflected the hypothesis that several initiation sites for oscillations occur in the thalamus because of the heterogeneity of the TC cells (see below). A subset of TC cells are spontaneous oscillators in thalamic slices (McCormick and Pape, 1990; Leresche et al., 1991) and *in vivo* (Curró Dossi et al., 1992). Because the TC cells had values of I_h conductance that covered a wide range, a few of these cells were intrinsic oscillators (see Table 7.1). These TC cells served as 'initiators' for spindle oscillations in the model (see Section 6.5; Destexhe et al., 1996a).

The synaptic currents obeyed:

$$I_{ki}^{syn} = \bar{g}_{ki}\, m_{ki}\, (V_i - E_{ki}) \tag{7.8}$$

where ki indicates the synaptic contact from neuron k to neuron i, $\bar{g}_{ki}$ is the maximal conductance and E_{ki} is the reversal potential of the postsynaptic current. The fraction of open receptors, m_{ki}, was modelled by a simple two-state kinetic scheme (Destexhe et al., 1994c; see Chapter 5):

$$(closed)\ +\ T(V_k) \underset{\beta}{\overset{\alpha}{\rightleftharpoons}} (open) \tag{7.9}$$

where $T(V_k)$ is the concentration of transmitter in the cleft. When a spike occurred in cell k, $T(V_k)$ was set to 0.5 mM for 0.3 ms, leading to the transient activation of the current. Receptor types such as AMPA and GABA$_A$ were described by two-state kinetic models for the activation variable m, whereas GABA$_B$ receptors had a more complex activation scheme based on the kinetics of G proteins (see details in Chapter 5; Destexhe and Sejnowski, 1995; Destexhe et al., 1996a).

All excitatory connections in this model were mediated by AMPA receptors. The results of simulations that included NMDA currents were not appreciably different from those shown here. A mixture of GABA$_A$ and GABA$_B$ receptors mediated all inhibitory synaptic interaction, except those within the RE nucleus, where GABA$_A$ receptors mediated the majority of synaptic interactions between RE cells (Bal et al., 1995b; Ulrich and Huguenard, 1996). When GABA$_B$ receptors were eliminated from the model, there was no effect on the synchrony and phase relationships between cells, but the precise time of onset of the spindle was changed, so GABA$_B$ receptors were therefore left in the model.

The time courses of the synaptic currents in the model were fit to whole-cell recordings of hippocampal pyramidal cells for GABA$_A$ (Otis and Mody, 1992), GABA$_B$ (Otis et al., 1993) and AMPA (Xiang et al., 1992) responses, as described previously (see Chapter 5; Destexhe et al., 1994c, 1998b). Since the present model was based primarily on experimental data obtained in barbiturate-anesthetized animals (pentobarbital), and since barbiturates prolong the decay time course of GABA$_A$ currents (Thompson, 1994), a larger decay time constant was used for GABA$_A$. Application of pentobarbital in hippocampal slices slowed down the normal rate of decay from around 5 ms to 7–25 ms, but the amplitude of the current remained unaffected (reviewed in Thompson, 1994). The rate of decay used here, 12.5 ms, was chosen from the middle of that range. A slower GABA$_A$ current provided IPSPs with time courses closer to that observed in intracellular recordings in animals anesthetized with pentobarbital (e.g. Steriade and Deschênes, 1984; Contreras and Steriade, 1996). Facilitation and depression of EPSPs and IPSPs occur in thalamus (Steriade and Deschênes, 1984; Deschênes and Hu, 1990; Thompson and West, 1991; Kim et al., 1997) but were not incorporated in this model. The time courses of these postsynaptic potentials are summarized in Fig. 7.11 (Synaptic).

The organization of the network was based on morphological studies suggesting monosynaptic feedback loops between thalamus and cortex. Ascending thalamocortical fibres have terminal arborizations in layers I, IV and VI of the cerebral cortex (White, 1986); given that layer VI pyramidal neurons constitute the major source of corticothalamic fibres, these projection cells therefore mediate a monosynaptic excitatory feedback loop (thalamus-cortex-thalamus; Hersch and White, 1981; White and Hersch, 1982). Stimulation of the thalamus produces both antidromic and monosynaptic responses in the same, deeply lying cortical cell (see Fig. 5 in Steriade et al., 1993b). Although all thalamic projecting layer VI pyramidal cells have axon collaterals in the RE nucleus, some lower layer V pyramids also project to thalamic nuclei, but do not leave collaterals in the RE nucleus (Bourassa and Deschênes, 1995). The latter were not included in the model, nor were inputs from thalamic intralaminar nuclei that project diffusely to the upper layers of the cerebral cortex, as well as receive projections from it (Jones, 1985). A simple, one-dimensional network of excitatory and inhibitory cortical cells was used to model layer VI. The other layers of the cortex were not included in this model since the focus was on the monosynaptic cortical feedback loop and the additional complexity was not needed.

In area 5 of cat cerebral cortex, the axon collaterals from a pyramidal cell form profuse and dense intracortical connectivity in a vertical column within a few hundred microns surrounding the apical dendrite (Avendaño et al., 1988). The connections of cells in the two layers of the cortical network model were organized such that each PY and IN cell projected densely to a localized area about 10% the size of the network (eleven cells centred on each presynaptic cell as in Fig. 7.15, scheme). The local projections used for intrathalamic connections had same range (see below).

The thalamic network consisted of two one-dimensional layers: one layer of TC cells and one layer of RE cells. Local interneurons were not included in the thalamic model based on evidence that thalamic interneurons are not involved in the generation of thalamic spindle oscillations (Steriade et al., 1985; von Krosigk et al., 1993). Axonal projections within the thalamus are local and topographic (Minderhoud, 1971; Jones, 1985; FitzGibbon et al., 1995); in the model, each thalamic cell type made topographic axonal connections with other cell types, contacting around 10% of the network (eleven cells; see Fig. 7.15, scheme). These intrathalamic connection patterns were identical to those used in a previous model (Section 6.5 in Chapter 6; Destexhe et al., 1996a).

Thalamocortical and corticothalamic projections are topographic (Robertson and Cunningham, 1981; Updyke, 1981; Avendaño et al., 1985; Jones, 1985) and are more divergent than either intrathalamic or intracortical connections (Landry and Deschênes, 1981; Freund et al., 1989; Bourassa and Deschênes, 1995; Rausell and Jones, 1995). As shown schematically in

Table 7.2 Synaptic conductances used for each type of connection. Each value represents the sum of all individual synaptic conductances of the same type converging on a given cell. The range of conductance values indicated gave rise to oscillations similar to those obtained for the optimal value. * range tested in Destexhe et al., 1996a.

Type of receptor	Location	Optimal conductance value	Range tested
AMPA	PY→PY	$0.6\ \mu S$	$0\text{–}0.9\ \mu S$
AMPA	PY→IN	$0.2\ \mu S$	$0.1\text{–}0.4\ \mu S$
$GABA_A$	IN→PY	$0.15\ \mu S$	$0.09\text{–}0.2\ \mu S$
$GABA_B$	IN→PY	$0.03\ \mu S$	$0\text{–}0.2\ \mu S$
AMPA	TC→RE	$0.2\ \mu S$	$0.1\text{–}1\ \mu S^*$
$GABA_A$	RE→RE	$0.2\ \mu S$	$0.05\text{–}0.4\ \mu S^*$
$GABA_A$	RE→TC	$0.02\ \mu S$	$0.01\text{–}0.04\ \mu S^*$
$GABA_B$	RE→TC	$0.04\ \mu S$	$0\text{–}0.15\ \mu S^*$
AMPA	TC→PY	$1.2\ \mu S$	$0.4\text{–}2.5\ \mu S$
AMPA	TC→IN	$0.4\ \mu S$	$0.1\text{–}0.6\ \mu S$
AMPA	PY→RE	$1.2\ \mu S$	$0.4\text{–}2\ \mu S$
AMPA	PY→TC	$0.01\ \mu S$	$0\text{–}0.07\ \mu S$

Fig. 7.15, each PY cell in the model projected topographically to twenty-one RE and twenty-one TC cells, twice the extent of intrathalamic and intracortical projections. Similarly, each TC cell projected topographically to twenty-one PY and twenty-one IN cells with equal synaptic strengths.

The pattern of axonal projections from each cell of a given type was identical and all the synaptic conductances were equal. The total synaptic conductance on each neuron was also the same for cells of the same type, regardless of the size of the network. Reflective boundary conditions were used to minimize boundary effects (see Chapter 6). Conductance values are given in Table 7.2.

In some simulations (Fig. 7.20), cortical pyramidal cells were subject to random synaptic bombardment in order to mimic the background activity that occurs *in vivo*. In this case, every PY cell had forty extra synapses (twenty AMPA and twenty $GABA_A$) with conductance values of $0.01\ \mu S$ and $0.0025\ \mu S$ for each AMPA and $GABA_A$ synapse, respectively. These extra synapses received random presynaptic action potentials computed using a Poisson spike generator (Press et al., 1986), with an average frequency of 15 *Hz*. This random synaptic bombardment of EPSPs and IPSPs produced a fluctuating voltage trace in PY cells with occasional spontaneous firing, similar to *in vivo* intracellular recordings during light barbiturate anesthesia.

The kinetics of intrinsic currents were estimated from voltage-clamp data in thalamic and cortical cells; in some cases, the relative conductance of each current could also be estimated from the published reports. The kinetics of synaptic currents were derived from fits to whole-cell recordings in thalamic, hippocampal and cortical cells. The simulated responses of single cells under current-clamp were compared to experimental data, an important check for the validity of the parameter values used. The amplitudes of the synaptic currents were larger than those in monosynaptic recordings since the model neurons typically had many fewer synapses than actual thalamocortical networks.

A rough estimate for synaptic conductance parameters were obtained from the sizes of EPSPs and IPSPs in various cells following extracellular stimulation. For example, compared with cortical stimulation, thalamic stimulation evoked powerful EPSPs in intracellularly recorded PY cells in vivo (Contreras et al., 1997c). In TC cells recorded intracellularly in vivo, cortical stimulation resulted in a small amplitude EPSP followed by a large IPSP (Fig 7.12A). These data constrained the relative strengths of the synapses, but the absolute amplitudes were adjusted so that the activity patterns in the network matched those observed in vivo.

Variability in the synaptic conductances was included for the cortical and thalamocortical connections. The parameters for the intrathalamic connections were the same as those used in Chapter 6, with the important exception that TC cell properties were randomly distributed in order to allow several initiation sites in the network. The total synaptic conductance per cell (the sum over all individual synaptic conductances of the same connection type) was used as a variable parameter. An extensive search of the synaptic conductance parameter space was done using the small thalamocortical circuit. The values obtained for the domain corresponding to spindle oscillations are given in the last column of Table 7.2. The behaviour of the small thalamocortical network (Fig. 7.14) is comparable to that of larger networks (as in Fig. 7.15). Many simulations of the large network ($N = 125$) were run for equivalent parameter values to check for robustness; the few differences that were observed could be attributed to the randomization of I_h conductance in TC cells. The range indicated in Table 7.2 is therefore also valid for the large network model, except perhaps for the most extreme values in the Table.

Spindling behaviour occurred over a wide range of most parameters, but some parameters were more restricted. For example, the intracortical $GABA_A$-mediated inhibition needed to be above some level to prevent avalanches of discharges in the system and epileptiform behaviour (see Chapter 8). Another critical parameter, which is the focus of the next section, was the conductance of AMPA-mediated cortical feedback in TC and RE cells.

7.2.2 Inhibitory dominance in thalamocortical cells

Cortical inputs influence the thalamus through projections onto both the TC and RE cells. In this section, experimental evidence is provided to support the hypothesis that the strongest inputs are on the RE cells, leading to strong feedforward inhibition onto TC cells, or 'inhibitory dominance' (Destexhe et al., 1998a). Later sections show how this property is essential in accounting for the spatial and temporal properties of spindle oscillations in the thalamocortical system.

According to inhibitory dominance, corticothalamic feedback to the thalamus by excites mainly RE cells, thereby recruiting TC cells through IPSPs rather than through direct cortical EPSPs. Although there are no quantitative data on the strength of cortical EPSPs on TC and RE cells, intracellular recordings of RE cells consistently show strong EPSPs of cortical origin that produce bursts of action potentials in response to electrical stimulation of the appropriate

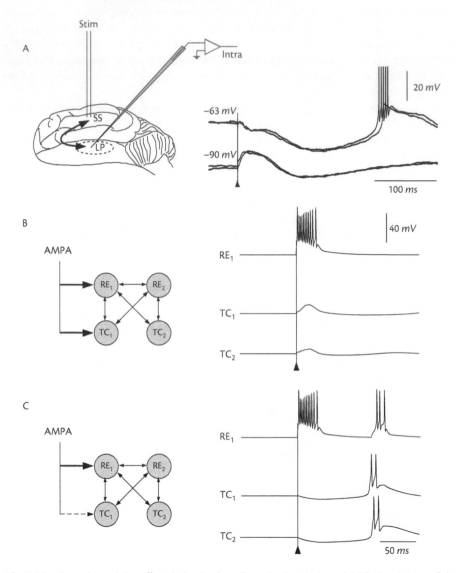

Fig. 7.12: Cortical stimulation affects thalamic relay cells predominantly through inhibition. A. Intracellular recording of a TC cell in the lateral posterior (LP) thalamic nucleus while stimulating the anatomically related part of the suprasylvian cortex in cats during barbiturate anesthesia. Cortical stimulation (arrow) evoked a small EPSP followed by a powerful biphasic IPSP. The IPSP gave rise to a rebound burst in the TC cell. This example represented the majority of recorded TC cells. B. Simulation of cortical EPSPs (AMPA-mediated) in a circuit of four interconnected thalamic cells. Cortical EPSPs were stimulated by delivering a presynaptic burst of four spikes at 200 Hz to AMPA receptors. The maximal conductance was similar in TC and RE cells (100 nS in this case) and no rebound occurred following the stimulation (arrow). C. Simulation of dominant IPSP in TC cell. In this example, the AMPA conductance of stimulated EPSPs in the TC cell was reduced to 5 nS. The stimulation of AMPA receptors evoked a weak EPSP followed by strong IPSP, then by a rebound burst in the TC cells, as observed experimentally. Figure reproduced from Destexhe et al., 1998a.

cortical area, even with low stimulus intensities (Mulle et al., 1986; Contreras et al., 1993). Stimulation of corticothalamic fibres has similar effects on RE cells in thalamic slices (Thomson and West, 1991). In contrast, intracellular recordings of TC cells in response to stimulating the appropriate cortical area show an EPSP-IPSP sequence dominated by the IPSP component (Fig. 7.12A). The majority of TC cells recorded in the lateral posterior nucleus (twenty-four out of twenty-six) had IPSP amplitudes of $11.1 \pm 1.2 \, mV$ (mean $\pm$ SE) at $-60 \, mV$ ($n = 26$). An EPSP was not apparent in a few cells ($n = 5$), but IPSPs always occurred. Cortical stimulation was able to fire the TC cell through EPSPs at the resting membrane potential ($-62.3 \pm 1.5 \, mV$) only occasionally ($n = 2$).

Inhibitory dominance in TC cells was investigated using simulations of thalamic networks. In a small circuit model with two TC interconnected with two RE cells (see scheme in Fig. 7.12B), the EPSPs on the RE and TC cells reproduced the EPSP/IPSP sequences observed experimentally provided that the cortical EPSPs on RE cells were stronger than the EPSPs on TC cells. In Fig. 7.12B, the conductance of AMPA-mediated cortical drive on TC and RE cells, as well as the GABA$_A$-mediated IPSP from RE cells were of the same order of magnitude. In this case, cortical EPSPs were shunted by reticular IPSPs and cortical stimulation did not evoke oscillations in the thalamic circuit. In contrast, when the EPSPs on TC cells had smaller conductances (5 nS compared to 100 nS), the EPSP-IPSP sequence was similar to intracellular recordings and cortical stimulation was effective in evoking oscillations (Fig. 7.12C).

These experimental observations and modelling results are compatible with a previous intracellular study *in vivo* in which the effect of corticothalamic feedback was compared before and after lesion of the thalamic reticular nucleus (Fig. 7.13; Deschênes and Hu, 1990). The

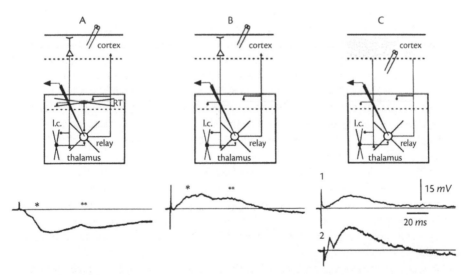

Fig. 7.13: Lesioning the RE nucleus reveals cortical EPSPs in TC cells A. In the intact thalamocortical system, stimulation of the cortex produced a dominant IPSP in TC cells (* and ** indicate different components of the IPSP). B. After lesion of the thalamic reticular nucleus, cortical stimulation induced a dominant EPSP consisting of several components. C. After acute capsular section combined with lesion of the RE nucleus, the corticothalamic pathway produced either a single-peaked EPSP (C1) or and EPSP interrupted by an IPSP (C2), depending on stimulation parameters. Reproduced from Deschênes and Hu, 1990.

'control' EPSP-IPSP sequence was transformed into a powerful EPSP after the RE nucleus was lesioned (Fig. 7.13; see also Steriade and Deschênes, 1988). These experiments show that the IPSP component of corticothalamic feedback is mostly originating in the RE nucleus, and this IPSP is powerful enough to mask the direct EPSPs from the cortex (Deschênes and Hu, 1990), consistent with the present view.

The concept of 'inhibitory dominance' is also supported by several *in vivo* observations. First, the dominance of inhibition has been observed previously in several thalamic relay neurons following stimulation of the corresponding cortical areas (Widen and Ajmone Marsan, 1960; Burke and Sefton, 1966; Steriade et al., 1972; Ahlsen et al., 1982; Lindström, 1982; Roy et al., 1984; Deschênes and Hu, 1990; Contreras and Steriade, 1996). Second, spindle oscillations could be robustly evoked by stimulating the cortex (even contralaterally to avoid backfiring of TC axons and collateral activation of RE cells, see Steriade et al., 1972; Roy et al., 1984; Contreras and Steriade, 1996). Third, during spontaneous oscillations, TC cells became entrained following an initial IPSP but rarely following an initial EPSP (Steriade and Deschênes, 1984). Fourth, dominant IPSPs have also been observed with other anesthetics, including ketamine-xylazine (Timofeev et al., 1996).

7.2.3 Inhibitory dominance is optimal for triggering thalamic oscillations

In vivo recordings consistently show that cortical stimulation is the most effective way of triggering spindle oscillations (Morison and Dempsey, 1943; Steriade et al., 1972; Roy et al., 1984; Contreras and Steriade, 1996). Here we show that this property is a natural consequence of inhibitory dominance. In a simplified model of a thalamocortical circuit, 9–11 *Hz* spindle oscillations could be initiated either by intrinsic mechanisms (Fig. 7.14A–B) or by electrical stimulation of cortical pyramidal cells (Fig. 7.14C–D). In the former case, the oscillation was initiated through one of the two TC cells that was intrinsically oscillatory (TC_1 in Fig. 7.14A–B). In the latter case, electrical stimulation of a PY cell could evoke spindle oscillations with all TC cells entrained to the oscillation by the initial IPSP (Fig. 7.14C–D). As previously described (Fig. 7.12B–C), cortically evoked rebound bursts in TC cells were observed only if cortical feedback activated TC cells through dominant inhibition.

The strength of cortical EPSPs on TC cells was critical in determining network activity. For weak EPSPs, TC cells were dominated by IPSPs from RE cells, and cortical discharges efficiently recruited spindle oscillations in the thalamus. When EPSPs and IPSPs were of comparable strength (as in Fig. 7.12B), it was not possible to evoke oscillations by stimulating the cortex due to the shunting effect of mixed EPSPs and IPSPs. As the cortically evoked EPSP was increased in strength, cortical discharges could evoke spikes in TC cells directly. In this case, sustained spiking of all cell types occurred in the network.

Whether oscillations were spontaneous or evoked by an initial cortical discharge, the network model exhibited spindles with similar characteristics: (a) The frequency range was 9–11 *Hz*; (b) All cell types discharged in phase after a single cycle, with the rebound bursts in TC cells starting the in-phase discharges, as observed experimentally for anatomically related cortical and thalamic territories (Contreras and Steriade, 1996); (c) The IPSPs were dominated by $GABA_A$ currents, since $GABA_B$-mediated currents were minimal following the moderate rate

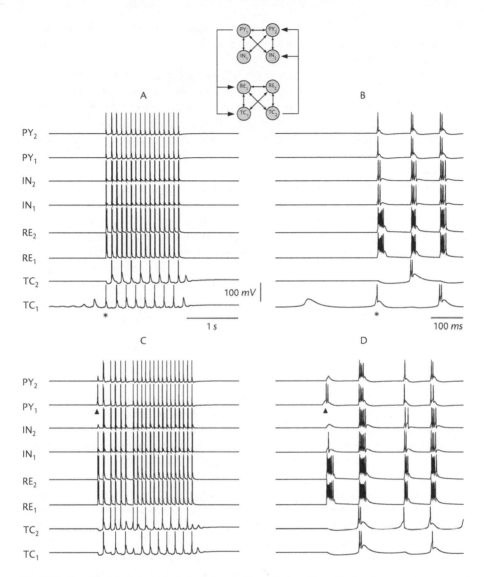

Fig. 7.14: Simplified thalamocortical circuit for spindle oscillations. Two cells of each type were connected according to the diagram (inset). Small arrows represent bidirectional coupling using the receptor types shown in Fig. 7.11. Large arrows indicate thalamocortical connectivity, with each PY cell connecting all thalamic cells and each TC cell connecting all cortical cells. A. Spontaneous spindle oscillation in the circuit. The oscillation began in one TC cell (TC$_1$—asterisk) and subsequently recruited the rest of the circuit. Oscillations were terminated due to upregulation of I_h in TC cells. B. Detail at ten times higher temporal resolution. C. Oscillations evoked by electrical stimulation of one PY cell (PY$_1$—arrow). D. Detail at ten times higher temporal resolution. Spindle oscillations occurred in the circuit only if reticular IPSPs dominated cortical feedback EPSPs in TC cells. Figure reproduced from Destexhe et al., 1998a.

of discharges of the cells; (d) The bursting pattern of thalamic cells was similar to that observed in thalamic circuits without cortical cells; in particular, the TC cells burst every other cycle (see Section 6.5.3 in Chapter 6; Destexhe et al., 1996a).

The range of parameter values over which spindle oscillations occurred in the model is given in Table 7.2 (last column). The spindling was quite robust to changes in conductance parameters within the ranges indicated and could be initiated either by intrinsic oscillatory behaviour in TC cells or by cortical stimulation. It was essential, however, to have strong corticothalamic feedback to RE cells and weak corticothalamic feedback to TC cells, consonant with Fig. 7.12C. The same parameter domain was also valid for larger networks with some reservations (see below).

Larger networks with 100 cells of each type were investigated in Fig. 7.15A. The intrinsic connectivity in the thalamus and the cortex had axonal projections to eleven neighbouring cells, similar to that in the model of thalamic networks (Section 6.5 in Chapter 6; Destexhe et al., 1996a). Projections between thalamus and cortex were more divergent than local interactions, consistent with anatomical studies (Jones, 1985). I_h conductance values were randomly distributed among TC cells throughout the network such that several TC cells oscillated spontaneously and served as initiation sites from which the oscillation spread to the entire network. The resulting network generated spindle oscillations with cellular discharge patterns similar to those in the simpler circuit, over a wide range of parameter values (Fig. 7.15B).

The model displayed oscillations that were nearly simultaneous in the sense that the network oscillated in unison during most of the spindle sequence. However, close scrutiny of single cells (Fig. 7.15B1–B2) revealed propagating patterns of discharges for some cells at neighbouring sites, separated by about 100 ms, similar to what has been observed in multielectrode studies of the cortex and thalamus in vivo (Verzeano and Negishi, 1960; Verzeano et al., 1965). However, these propagating patterns were local in both time and space and there was no evidence for systematic propagation. In addition to these local patterns of propagation, a prominent feature of these oscillations was that they became synchronized in less than 500 ms.

Local average membrane potentials were computed to compare the network activity with multisite field potentials recordings. In the model, each average trace represents the average over twenty-one adjacent PY cells. Different averages were taken at eight equally spaced cortical sites on the network. Local averaged potentials showed that the spindle oscillation began approximately simultaneously at several sites (see asterisks in Fig. 7.15C). The oscillations then propagated to the entire network and, after one or two cycles, merged into a unique synchronized oscillation. This scheme is similar to the 'collision' of spindle waves described in vitro (see Section 6.5.5; Kim et al., 1995). However, the synchronization of oscillations here was through corticothalamic interactions, which bypassed the intrathalamic synchronizing mechanism responsible for synchronization in the slice and set the network in a state of full synchrony within a few cycles.

The initiation of the oscillation occurred at different sites within time windows of about 0.2 s (see the average membrane potentials in Fig. 7.15C), similar to the patterns seen in field potential recordings during barbiturate anesthesia (compare Fig. 7.15C with Fig. 7.3). Another similarity with barbiturate anesthesia is that successive spindle sequences showed considerable variability in their initiation pattern, with 1–3 nearly simultaneous initiation sites. Possible explanations why the initiation sites should occur nearly simultaneously are given in the next section.

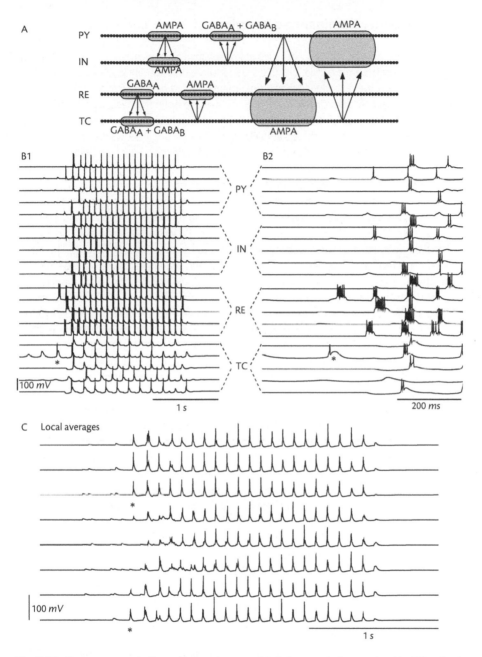

Fig. 7.15: Spontaneous spindle oscillations in a model thalamocortical network with 400 cells. A. Schematic connectivity. The network had four layers of PY, IN, RE and TC cells. Each cell is represented by a dot and the area to which it projects is depicted as a shaded area for a representative cell. Intrathalamic and intracortical connections were topographic with a divergence of eleven cells, whereas thalamocortical and corticothalamic projections were more extended, spanning over twenty-one cells. B1. Spontaneous spindle oscillation. Five cells of each type, equally spaced in the network, are shown (0.5 ms time resolution). The asterisks indicate an initiator TC cell. B2. Detail of spindle initiation. C. Locally averaged potentials. Twenty-one adjacent PY cells, taken at eight equally spaced sites on the network, were used to calculate each average. Asterisks indicate two nearly simultaneous initiation sites. Figure reproduced from Destexhe et al., 1998a.

7.2.4 Inhibitory dominance determines thalamic coherence

Thalamocortical network models were used to investigate how cortical feedback could organize the coherence of thalamic oscillations that were observed experimentally (Section 7.1.2; Contreras et al., 1996). We reproduced these experimental recording conditions in the model. The activity in individual thalamic cells as well as local average potentials were considerably more coherent in the presence of cortical feedback (Fig. 7.16): the *left panel* shows several spindle sequences using the same parameters as in Fig. 7.15. The *right panel* shows the same simulation with cortical cells removed. Without cortical feedback, different initiation sites for spindles were not coordinated. Some of them remained local, whilst others gave rise to systematic propagation of oscillations from one side of the network to the other (Fig. 7.16, bottom right panel). The randomly distributed I_h conductances among the TC cells was responsible for this diversity; as a consequence several TC cells were spontaneous oscillators and served as initiation sites.

The IPSP-dominated synaptic feedback is well-suited to recruiting I_h in TC cells. The mechanisms that modulate I_h, giving rise to the waxing and waning patterns, can be reset by corticothalamic feedback. During the initiation of spindle oscillations, such synchronized cortical feedback may synchronize the upregulation mechanisms for I_h in TC cells such that TC cells recover at roughly the same time, leading to several nearly simultaneous initiation sites for spindles. Thus, inhibitory dominance of cortical inputs could contribute to the spatial coherence of thalamic spindling *in vivo*. The refractoriness of individual TC cells in the model was varied to test this possibility. In these simulations, all TC cells had identical I_h that made them all 'initiators' (see Section 7.2.1). Only the decay rate of I_h modulation was randomized, such that initiator TC cells had silent periods distributed between 4 and 20 sec. Under these conditions, the thalamocortical network still displayed typical waxing and waning oscillations, but the patterns were more irregular: there was an increased occurrence of local oscillations, and initiation at multiple sites was rare (not shown). This shows that the waxing and waning of spindles does not depend on a perfect uniformity of refractoriness in TC cells, but that uniformity biases them more to restart together. Unlike conductance values that may vary from cell to cell because they depend on the density of channels in the membrane, the durations of the silent periods are not expected to be as diverse because they depend on kinetic rate constants, which are likely to be identical in all cells having the same biochemical reactions. This makes it at least plausible that different TC cells should have similar silent periods.

This mechanism is also consistent with a series of additional properties observed experimentally. First, dual intracellular recordings have shown that although there is diminished spatiotemporal coherence in decorticated animals, local synchrony is still present in nearby TC cells located less than about 1–2 *mm* apart (Fig. 7.5; Contreras et al., 1996; Timofeev and Steriade, 1996). This was also found in the model, where neighbouring TC cells were more synchronous than distant ones (Destexhe et al., 1996a; Golomb et al., 1996); the synchrony was a consequence of divergence in the intrathalamic reciprocal connections between TC and RE cells (Jones, 1985).

Second, spatiotemporal maps of activity from the model (Fig. 7.17) were strikingly similar to experimental data (compare with Fig. 7.4). In both models and experiments, the oscillations were initiated at different sites for each spindle sequence. In the presence of the cortex (Fig. 7.17, left colour panel), initiation occurred in a relatively narrow time window, with

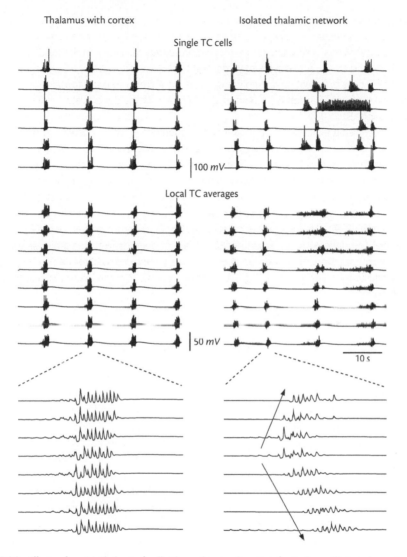

Fig. 7.16: Effects of corticothalamic feedback on the simultaneity of spindle oscillations in corticotha-lamic model. Spontaneous spindles are shown in the presence of the cortex (left panels) and in an isolated thalamic network (right panels) under the same conditions (same parameters as in Fig. 7.15). Single TC cells and local TC averages are shown for each case. Twenty-one adjacent TC cells, sampled from eight equally spaced sites on the network, were used to calculate each average. The bottom graphs represent averages of a representative spindle at ten times higher temporal resolution. The near-simultaneity of oscillations in the presence of the cortex is qualitatively different from the propagating patterns of activity in the isolated thalamic network (arrows). Figure reproduced from Destexhe et al., 1998a.

two sites sometimes starting at roughly the same time. Without the cortex (Fig. 7.17, right colour panel), the initiation sites led to local patterns of propagation and colliding waves. In some instances, when only a single site of initiation started oscillating in the isolated thalamic network, the model produced systematic propagation of spindle waves (last spindle sequence in Fig. 7.17). These propagating and colliding waves were similar to those found experimentally in thalamic slices (see Section 6.5.5; Kim et al., 1995) and in the decorticated thalamus *in vivo* (see Section 7.1.3).

The model displayed different spatiotemporal coherence in intact-cortex and decorticated conditions, mimicking the experimental results. These differences were apparent from the presence of horizontal yellow stripes in maps from the thalamic network with cortex, which were not present in maps of the isolated thalamic network (compare colour panels in Fig. 7.17). In a power spectrum analysis (not shown), power in the 7–15 Hz range increased together at distant sites in the presence of the cortex but not in decorticate animals (similar to Fig 7.4; Contreras et al., 1996). The differences in spatiotemporal coherence patterns were also studied with spatial correlations. As in experiments, spatial correlations from thalamic cells showed a more pronounced decay with distance when cortical feedback was removed compared to the intact thalamocortical network (Fig. 7.17, bottom graphs; compare with spatial correlations in Fig. 7.4).

7.2.5 Refractoriness of the corticothalamic network

Refractoriness is an essential property of the thalamic network and determines the spatiotemporal properties of thalamic oscillations (see Section 6.5.7). In thalamic slices, spindle oscillations were followed by a period of several seconds during which the thalamic network was refractory to further oscillations (Kim et al., 1995). Spindle waves *in vivo* were also followed by a prolonged period of refractoriness (Fig. 7.18A). The oscillations could be evoked by cortical stimulation only if a 2–8 s period of silence preceded the stimulation, depending on stimulation intensity (Contreras et al., 1997b).

Refractoriness was modelled as a Ca^{2+}-dependent modulation of I_h in TC cells (see Section 7.2.1), similar to the way that it was modelled in the isolated thalamic network model (Section 6.5.7 in Chapter 6; Destexhe et al., 1996a). Since the decay rate of I_h upregulation is quite slow, spindle oscillations are followed by a refractory period of several seconds. In Fig. 7.18B, stimuli delivered at moderate intensity every 4 seconds entrained the whole network model once every two stimuli. Thus, at this stimulus intensity the refractory period was greater than 8 seconds but less than 12 seconds, similar to that observed in ferret thalamic slices (Kim et al., 1995). The refractoriness and therefore the ability of the cortical input to trigger a spindle was confined to a short time window at the end of the interspindle lull.

7.2.6 Refractoriness influences propagation

Systematic propagation of spontaneous spindle waves has been observed in ferret thalamic slices (Fig. 6.29; Kim et al., 1995). The properties of the propagating activity could be explained by upregulation of I_h and the topographical structure of reciprocal connections between TC

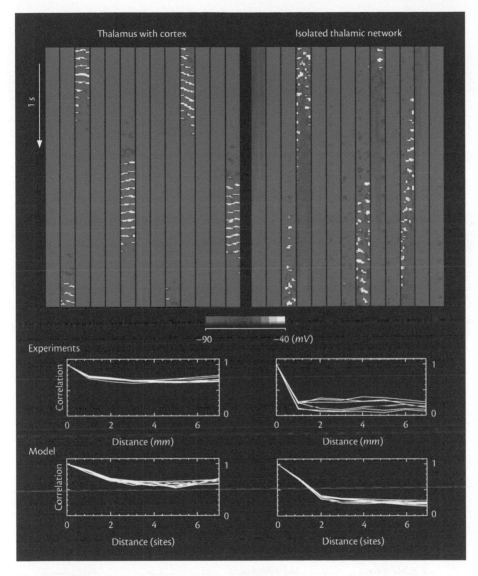

Fig. 7.17: Cortical feedback determines the spatiotemporal coherence of thalamic oscillations in a corticothalamic model. Top panels: spatiotemporal maps were constructed from local TC averages of spontaneous spindles in the presence of the cortex (left) and in an isolated thalamic network with same parameters (right). Each frame consisted of a horizontal stripe of eight colour spots representing the membrane potentials of TC averages shown in Fig. 7.16. Frames were arranged from top to bottom in thirteen columns (a total of 40 seconds of activity is shown). Colours ranged in ten steps from −90 or below (blue) to −40 mV or above (yellow; see colour scale). Same simulations and averaging procedures as in Fig. 7.16. Bottom panels: decay of correlation with distance. Cross-correlations were computed for all possible pairs of sites and the value at time zero from each correlation was represented as a function of the intersite distance. *Experiments*: decay of correlations of thalamic local field potentials in intact (left) and decorticate (right) cats under barbiturate anesthesia (modified from Contreras et al., 1996). *Model*: decay of correlation calculated for local averaged potentials in the presence of the cortex (left) and in the isolated thalamic network (right). Figure reproduced from Destexhe et al., 1998a.

A

Experiments

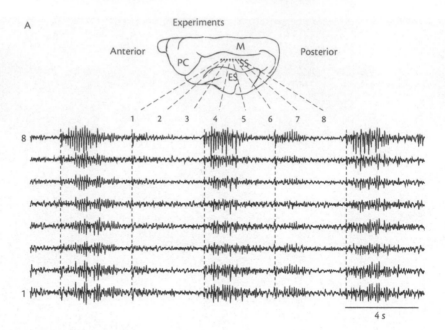

B

Model

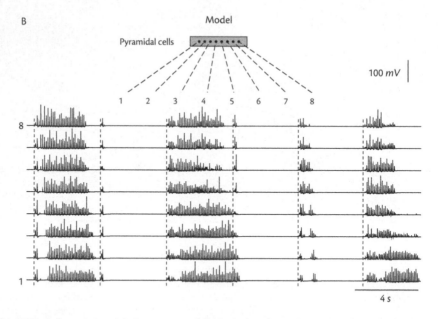

Fig. 7.18: Refractoriness of thalamic circuits following cortical stimulation. A. Refractoriness of spindle oscillations evoked by cortical stimulation during barbiturate anesthesia. Top schematic: electrode arrangement for multisite field potential recordings in cat suprasylvian cortex (SS suprasylvian gyrus, ES ectosylvian gyrus, M marginal gyrus, PC postcruciate gyrus). Bottom: field potential recordings in the cortex during repetitive cortical stimulation. Stimuli were repeated every 4 seconds through a stimulating electrode located adjacent to electrode 8. B. Refractoriness of spindle oscillations in the model. Top schematic: illustration of the eight equidistant sites chosen to calculate local average membrane potentials of PY cells. Bottom: local average potentials during cortical stimulation of PY cells repeated every 4 seconds. One out of five pyramidal cells were stimulated by injecting current pulses with random amplitude (0–1 nA) and random duration (0–100 ms). Dashed lines indicate the stimulus onset. Figure reproduced from Destexhe et al., 1998a.

and RE cells (Section 6.5 in Chapter 6; Destexhe et al., 1996a). Systematic propagation of spontaneous oscillations have, however, not been observed *in vivo*; the oscillations show variable initiation patterns with no systematic delay (Andersen and Andersson, 1968). However, spindles evoked by low-intensity cortical stimulation *in vivo* can show patterns of systematic propagation with an apparent velocity higher than that observed in vitro (Fig 7.9; Contreras et al., 1997b).

Although spontaneous oscillations occurred nearly simultaneously in the model (Fig. 7.15), systematic propagation could be evoked by reproducing experimental conditions of low-intensity cortical stimulation (Fig. 7.19). Cortical stimulation was mimicked by delivering depolarizing current pulses of random amplitude simultaneously to a localized population of PY cells (ten adjacent cells) (Fig. 7.19, Local). The initial discharge in a localized population of cortical cells recruited a local population of thalamic cells through excitation of the RE nucleus (same mechanism as in Fig. 7.14D). The corresponding TC cells then produced rebound bursts that re-excited a larger population of cortical cells and the same cycle restarted, leading to a progressive invasion of the network. Similar patterns of systematic propagation could also be evoked by stimulation of either TC or RE cells (not shown).

The model accounted for two additional features of cortically evoked oscillations. First, the oscillations evoked experimentally by high intensity did not show patterns of systematic

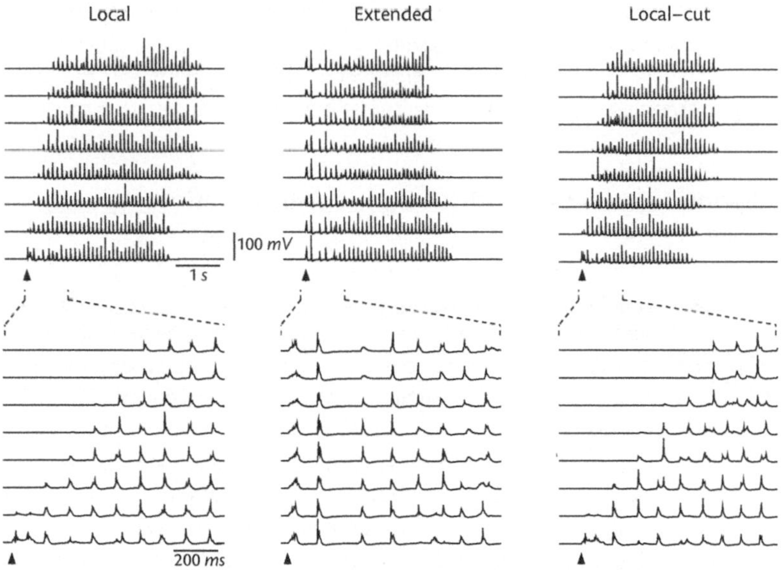

Fig. 7.19: Systematic propagation and nearly simultaneous spindle oscillations evoked by cortical stimulation in a corticothalamic model. The top traces show the local average potentials of PY cells (same procedures and parameters as in Fig. 7.15). The bottom traces show the beginning of the oscillatory sequence at five-times higher temporal resolution. Local: stimulation of ten adjacent pyramidal cells evoked a spindle oscillation that propagated away from the stimulus site. Extended: stimulation of ten widely distributed (one out of five) pyramidal cells evoked a nearly simultaneous oscillation. Local-cut: local stimulation with cortical cut produced patterns of systematic propagation similar to local stimulation in the intact network. The cut suppressed all connections crossing the centre of the network. Injected current pulses were of random amplitude (0–1 *nA*) and random duration (0–100 *ms*). Arrows indicate the stimulus onset. Figure reproduced from Destexhe et al., 1998a.

propagation but were nearly simultaneous (Fig. 7.9; Contreras et al., 1997b). In the model, when the stimulation was widely distributed throughout the network (ten cells, the same number of cells that was stimulated focally), there were nearly simultaneous spindle oscillation over the entire network (Fig. 7.19, Extended), consistent with the hypothesis that high-intensity cortical stimulation *in vivo* recruited a more extended population of cortical cells that project to a wider thalamic territory. Second, systematic propagation patterns were resistant to transection of horizontal intracortical connections (Fig. 7.10; Contreras et al., 1997b). The properties of evoked oscillations in the model also resisted interruption of cortico-cortical connections (Fig. 7.19, Local-cut), in agreement with the experimental results (compare with Fig 7.10; Contreras et al., 1997b).

It took around 600 *ms* for the wave of activity to propagate 100 cells in the network (Fig. 7.19A). In another simulation using a network with twice the number of cells (200 neurons of each type), and all axonal projections twice the extent described in Section 7.2.1, systematic propagation occurred took around 600 *ms* over 200 cells, around twice the absolute velocity, but the same velocity if expressed as a percentage of the size of the network. This shows that the propagation velocity does not depend on the absolute number of cells in the network but rather on the relative size of axonal projections compared to the total size of the network.

The following conclusions are suggested by these modelling results. First, because of thalamic refractoriness, patterns of systematic propagation can be generated at the end of the refractory period when a localized area of the thalamus is excited. Localized excitation is produced by localized low-intensity cortical stimulation. Second, the synchronization observed after high-intensity cortical stimulation is due to the activation of a more extended population of cortical cells at the same time. Third, the systematic propagation of evoked spindles in the cortex is due to the reciprocal cortico-thalamic connectivity and not to horizontal intracortical connections.

7.2.7 Spontaneous cortical discharges control spatiotemporal coherence

Another consequence of refractoriness *in vivo* is that it may lead to highly variable spatiotemporal patterns if the spontaneous activity is taken into account. The refractory period of spindle waves *in vivo* has approximately the same duration as the period of oscillations that occur spontaneously (4–10 s). In contrast, in thalamic slices, the refractory period (5–10 s) is shorter than the period of spontaneous oscillations (10–20 s; Kim et al., 1995). In the thalamic slice model (Section 6.5 in Chapter 6; Destexhe et al., 1996a), the refractory period was also shorter than spontaneous oscillation period. In effect, after recovery from refractoriness, the network 'waited' for a spontaneous event to occur to trigger the next spindle wave.

In vivo, the conditions could be quite different. In a model with the higher level of spontaneous activity observed *in vivo*, it is possible that spindle sequences initiate as soon as the network recovers from refractoriness. Here, initiation of thalamic oscillations by the cortex has an essential role. This possibility was simulated in the model by subjecting cortical PY cells to random synaptic bombardment such that they fired occasionally (Fig. 7.20). As a consequence, a sequence of oscillatory episodes were generated by the occasional spontaneous discharge of cortical PY cells (Fig. 7.20A, asterisk) through a process of recruitment involving the cortico-thalamic feedback, similar to Fig. 7.14D.

Spindle initiation in the model was consistent with several properties of spindles recorded *in vivo*: (a) The interspindle period was around 4–10 s; (b) the distribution of periods had high

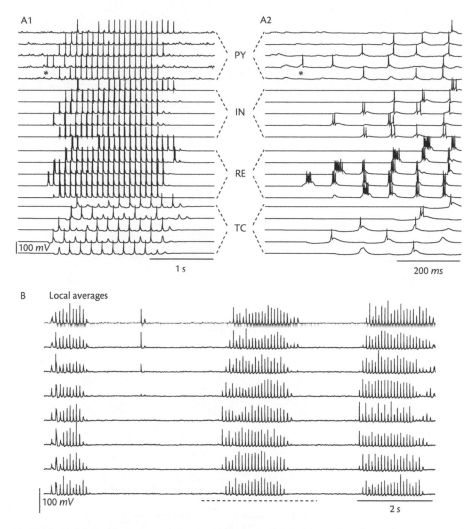

Fig. 7.20: Variability of spindle sequences in a thalamocortical model stimulated by spontaneous cortical activity. Same parameters and description as Fig. 7.15, except that every PY cell had forty extra synapses (twenty AMPA and twenty GABA$_A$) that received random presynaptic action potentials (see details in Section 7.2.1). A1. Spontaneous spindle oscillation in single cells illustrated by five cells of each type equally spaced in the network (same description as in Fig. 7.15B). The spindle oscillation was initiated by the spontaneous discharge of a PY cell (asterisk). A2. Detail of the spindle initiation. B. Local average membrane potentials of PY cells. Spontaneous spindle oscillations were highly variable with interspindle periods in the range of 4–10 s. Figure reproduced from Destexhe et al., 1998a.

variance; (c) local spindles occurred occasionally. Local spindles were triggered when spontaneous cortical discharges occurred before complete recovery from refractoriness, consistent with the shorter silent period that precedes local spindles *in vivo* (D. Contreras, A. Destexhe and M. Steriade, unpublished observations). Many of these properties had already been studied in an earlier model (Andersen and Andersson, 1968).

This model therefore provides a unifying framework to account for apparently inconsistent experimental observations: spontaneous spindle waves appear in different sites at approximately the same time *in vivo* (Fig. 7.3; Contreras et al., 1997b), but they typically show patterns of systematic propagation in thalamic slices (Kim et al., 1995). The model suggests that these differences are primarily due to presence of thalamocortical loops, which act through the RE nucleus (thalamocorticoreticular loops) and provide more global synchronization than can be achieved by intrathalamic loops alone (Fig. 7.21). Thalamocortical loops are highly efficient

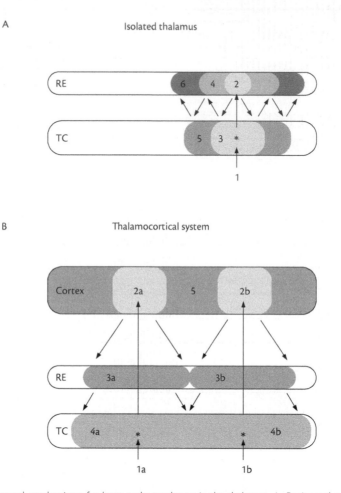

Fig. 7.21: Proposed mechanisms for large-scale synchrony in the thalamus. A. Reciprocal recruitment of thalamocortical (TC) relay cells and thalamic reticular (RE) cells synchronize cells in isolated thalamus. From an initial discharge of a TC cell (1; asterisk), a localized area of RE cells was recruited (2), which in turn recruited a larger neighbouring TC cells (3), etc. Progressively larger areas of the thalamus were recruited on each successive cycle (4, 5, 6, ...) through the topographic structure of the connectivity. An array or electrodes would record a 'systematic propagation' of the oscillation, as found in thalamic slices (Kim et al., 1995). B. Thalamic recruitment mechanism involving the cortex. Two approximately simultaneous initiation sites in the thalamus (1a, 1b) recruited localized cortical areas (2b, 2c), which in turn recruited the connected areas of the RE nucleus (3a, 3b), which in turn recruited larger areas of TC cells (4a, 4b), etc. At the next cycle, the entire cortical area was recruited (5). The corticothalamic connections overshadowed the local thalamic recruitment mechanisms shown in A and oscillations achieved large-scale synchrony within a few cycles, consistent with *in vivo* data (Contreras et al., 1996). Figure reproduced from Destexhe et al., 1998a.

at synchronizing large areas because the diverging cascade of axonal projection systems reaches a wide area of the cortex on a single cycle (cortex-to-RE, RE-to-TC, TC-to-cortex). In addition, spindles may be initiated at roughly the same time at different sites because of similar refractory periods in TC cells; as a consequence, oscillations may spread over large regions of the cortical mantle within a few cycles. Thalamic slices deprived of this powerful synchronizing mechanism display systematic propagation through the topographic structure of intrathalamic connections (Fig. 7.21A; see Destexhe et al., 1996a).

7.3 The large-scale synchrony of spindle oscillations during natural sleep

In the corticothalamic model, the large-scale synchrony of spindles arose from the mutual interactions between the cortex and the thalamus, with the corticothalamic feedback acting primarily through the RE cells, and intrathalamic synchronizing mechanisms having a limited role. This model accounted for properties of barbiturate spindles, such as their waxing-and-waning envelope, their refractory period and their relatively high variability. Several points regarding spindle oscillations, however, deserve further consideration. First, cortico-cortical connections have a negligible role in the large-scale synchrony of barbiturate spindles (Fig. 7.6; Contreras et al., 1996), but interhemispheric synchrony is reduced in cats following section of the corpus callosum (Bremer, 1956), suggesting that at least some intracortical connections are important. Second, during natural sleep, spindle oscillations are often characterized by strict simultaneity (see Fig. 7.22), in contrast to barbiturate anesthesia, which induces spindles with high variability (see Fig. 7.1; Andersen and Andersson, 1968). This suggests that the patterns of large-scale synchrony may be significantly different between these states.

In order to explore these issues, further experimental and modelling studies were undertaken to explore the large-scale synchrony of spindle oscillations during natural sleep, natural sleep following artificial depression of the cortex, and barbiturate anesthesia. These investigations led to the hypothesis that the modulation of coherence under these conditions occurs by intracortical mechanisms, through thalamocortical loops, but without the involvement of intrathalamic synchronizing mechanisms (Destexhe et al., 1999a).

7.3.1 Spatiotemporal analysis of spindle oscillations in the cortex

The highly coherent patterns of spindle oscillations that occur during natural sleep are shown in Fig. 7.22. The human EEG was analysed during sleep stage 2 ($n = 3$) from scalp locations indicated in the schematic in Fig. 7.22 (arrowheads). The waxing and waning sequences of 12–14 Hz human spindle waves recurred every 2–3 sec. Cross-correlations between the top trace (C3A2) and the other traces were calculated for fifteen consecutive spindle sequences. The averaged cross-correlations (CROSS, right side) were precisely superimposed, had a frequency near 14 Hz, and central peak values of 0.6–0.9. EEG recordings from the suprasylvian gyrus of chronically implanted naturally sleeping cats ($n = 4$) also revealed a high degree of synchronous spindling. In Fig. 7.22 (CAT), the EEG was recorded from the depth ($\sim$1 mm) of the suprasylvian

NATURAL SLEEP

HUMAN

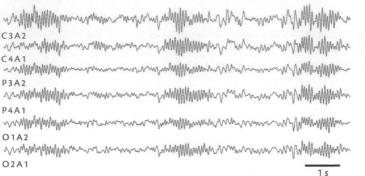

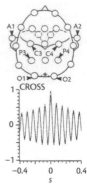

CAT

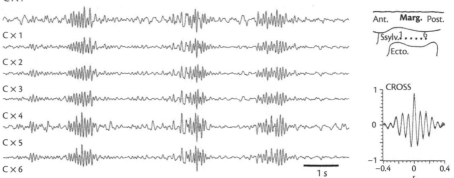

Fig. 7.22: Coherent cortical spindles during natural sleep. Top panel (HUMAN): spindles recorded from six standard EEG derivations (indicated in the schematic at right, arrowheads) in a normal subject during sleep stage 2. Cross-correlations of individual spindle sequences ($n = 15$) were calculated between C3A2 and each one of the other channels. Averaged correlations (CROSS) showed rhythmicity at 14 Hz and central peak values between 0.7 and 0.9. Bottom panel (CAT): EEG from a chronically implanted naturally sleeping animal. The EEG was recorded from six tungsten electrodes separated by 1 mm, inserted in the depth of the suprasylvian gyrus (Ssylv), represented by dots 1 to 6 in the scheme at right; the ectosylvian (Ecto.) and the marginal (Marg.) gyri are also indicated as well as the anterior (Ant.) and posterior (Post.) directions. The same procedure as for the human EEG was used to obtain the averaged cross-correlations depicted at right (CROSS), showing correlation at 14 Hz with central peaks between 0.75 and 0.9. Figure reproduced from Contreras et al., 1997b.

cortex by means of six tungsten electrodes separated by 1 mm (see the scheme at right). Spindles occurred every 2–3 sec, and the waves had frequencies of 12–14 Hz. Correlations between electrode 1 (Cx1) and each of the others were calculated from fifteen consecutive spindle sequences. The averaged cross-correlations superimposed and show central peak values of up to 0.85 for a frequency of 13 Hz (CROSS, right side). Similar results were obtained from the marginal gyrus (data not shown).

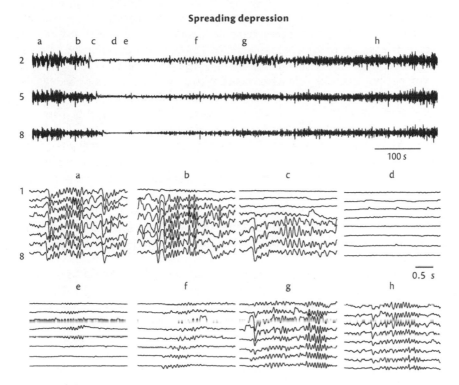

Fig. 7.23: Effect of spreading depression on spindling activity and synchrony. Upper three traces show spontaneous activity at electrodes 2, 5 and 8 during the spreading depression. Letters (a–h) mark expanded recordings at the points shown in top panel. Depression spread through the cortex until all electrodes had a flat trace (a to d). Spindling was less synchronized, occasionally displaying propagating patterns, during early recovery from cortical depression (e and f). As the EEG returned, waning spindles reappeared and became more synchronized (g and h). Figure reproduced from Contreras et al., 1997a.

To test the specific role of the cortex in generating these highly coherent patterns, cortical depression was induced in naturally sleeping cats by releasing a drop of highly concentrated potassium acetate (3 M) in the vicinity of electrode 1 ($n = 6$) (see Contreras et al., 1997a). The potassium solution induced a complete flattening of the local EEG, which spread slowly towards electrode 8 (Fig. 7.23, top panel, spreading depression), from the middle to the posterior suprasylvian gyrus (Fig. 7.23, a to d). After 5–8 sec of almost complete absence of cortical activity (Fig. 7.23, d), spindling activity reappeared, was first visible in the electrodes where the depression started, then progressed to the others, indicating that recovery started near the same electrode where the depression first started (Fig. 7.23, e to h). Spindling activity during the initial period of recovery after cortical EEG depression was much less synchronized than that occurring with normal cortical activity (compare e and f with g and h in Fig. 7.23) and propagation was occasionally observed. The return of normal cortical activity was marked by the appearance of depth-EEG positive waves, followed by sharp negative deflections (Fig. 7.23, g; electrodes 3 and 4). Once the cortex recovered, EEG patterns including the normal spindling of natural sleep reappeared and were synchronized on all electrodes (Fig. 7.23, h).

The first step in analysing these spindles patterns was to quantify the initiation of the oscillations, estimated as the first negativity-positivity complex that exceeded 25% of the total deflection of amplitude of the subsequent spindle. This estimate was performed for each site and the respective times of first negativity-positivity were plotted as a function of distance or electrode number. Several spindle waves were analysed using this procedure and the results were plotted together (Fig. 7.24B). This graph was also used to compute the distribution of initiation times (see Fig. 7.24C) from which an estimate of the 'time jitter' of initiation was quantified by computing the standard deviation (σ) of the distribution.

During natural sleep, the initiation of spindle oscillations was remarkably precise among the eight electrodes (Fig. 7.24A, Natural sleep). Spindle sequences were analysed from three naturally sleeping cats ($n = 24$) at the onset of slow-wave sleep. The relative initiation times of the oscillation at each electrode are illustrated in Fig. 7.24B (Natural sleep). Spindle sequences ($n = 24$) analysed from naturally sleeping cats showed initiation within time windows on the order of 0.05 ± 0.03 s (mean $\pm$ SD), corresponding to an average velocity of 140 mm/s. The distribution of initiation times (Fig. 7.24C, Natural sleep) was relatively narrow and had a standard deviation of $\sigma = 31$ ms.

The spindle oscillations that dominate the EEG activity under barbiturate anesthesia are superficially similar to those observed in the early phase of slow-wave sleep. These oscillations have slower frequencies than natural sleep spindles (6–8 Hz vs. 10–12 Hz), but they have similar cellular correlates (Steriade and Deschênes, 1984). They are also known to be highly coherent over the cortical hemisphere (Andersen and Andersson, 1968). However, upon closer analysis, barbiturate spindles were initiated with less precision than during natural sleep (Fig. 7.24A). The initiation times of these oscillations ($n = 35$; Fig. 7.24B, Barbiturate anesthesia) was about 0.20 ± 0.13 seconds (average velocity of 35 mm/s). Fig. 7.24B shows that there was also considerable variability in the number of initiation sites, their relative timing and their location from one sequence to the next. The distribution of initiation times (Fig. 7.24C, Barbiturate anesthesia) was broad (standard deviation of $\sigma = 139$ ms).

The highly coherent patterns of spindles that occur during natural sleep could be disrupted by the depression of the cortex induced by potassium acetate (Fig. 7.23). Spindles were the first sign of activity following the recovery from cortical depression. Spindle waves recorded in the period before the cortex had completely recovered were analysed as shown in Fig. 7.24A (Depressed cortex). The patterns of initiation of spindles following depression of the cortex during natural sleep were similar to those observed during barbiturate anesthesia (Fig. 7.24A).

The initiation times of spindles following the recovery period of spreading depression in two cats ($n = 27$) (Fig. 7.24B, Depressed cortex), were 0.18 ± 0.11 s (average velocity of 39 mm/s). During barbiturate anesthesia, there was considerable variability in the number, location and timing of spindle initiation sites (Fig. 7.24B). After complete recovery from cortical depression, the synchrony of spindles reverted back to values typical of natural sleep (Fig. 7.23; Contreras et al., 1997a) as did their patterns of initiation (not shown). The distribution of initiation times (Fig. 7.24C, Depressed cortex) was similar to that obtained during barbiturate anesthesia (standard deviation of $\sigma = 108$ ms).

Another way of representing the pattern of electrical activity in the cortex during spindles is through spatiotemporal maps similar to Fig. 7.4. Single frames of the spatial distribution of electrical activity across the cortex were generated by assigning a colour spot to the instantaneous

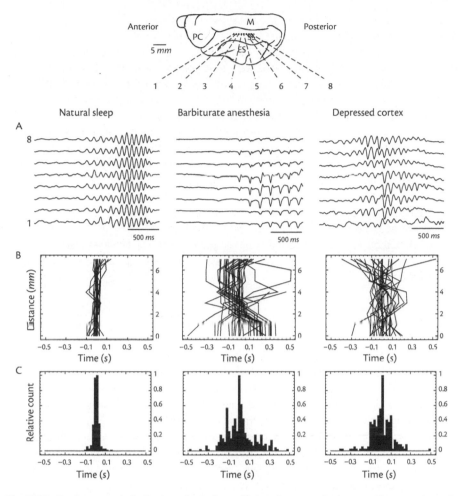

Fig. 7.24: Spatiotemporal distribution of spindle oscillations in cat cerebral cortex. The top schematic shows the localization of electrodes in area 5–7 of cat cortex (SS suprasylvian gyrus; PC postcruciate gyrus, ES ectosylvian gyrus, M marginal gyrus). The three panels from left to right indicate recordings in naturally sleeping animals (Natural sleep), in anesthetized animals (Barbiturate anesthesia), and in natural sleep following cortical depression (Depressed cortex). A. Multisite field potentials during spindle oscillations. The local field potentials recorded simultaneously at eight sites are shown during representative spindle sequences. B. Initiation patterns of spindle oscillations. The time of initiation of the oscillation was detected for each site. The relative initiation times were drawn as a line (vertical line = simultaneous), with one different line for each spindle sequence. C. Distribution of initiation times computed from the graphs in B. Figure reproduced from Destexhe et al., 1999a.

value of the local field potential at each electrode and arranging colour spots along a horizontal line (anterior-to-posterior axis). Successive snapshots were arranged vertically in columns, so that the local field potential is represented as a function of space and time. Synchronous events therefore appear as horizontal straight lines, while oblique lines signify propagating waves or systematic phase shifts.

Spatiotemporal maps of activity for cortical oscillations during natural sleep, barbiturate anesthesia and following depression in Fig. 7.24 are shown in Fig. 7.25 (colour panels). During natural sleep, oscillatory activity began almost simultaneously over the entire area recorded (Fig 7.25, Natural sleep). The oscillations were synchronized, as indicated by the formation of yellow (maximum local activity) and blue (local silence) horizontal stripes at 10–12 Hz. Barbiturate spindles occurred at lower frequency (6–8 Hz) and were less organized than natural sleep (Fig 7.25, Barbiturate). The oscillation began simultaneously at one or several sites and subsequently invaded the rest of the recorded area. The first spindle illustrated in the middle-top panel of Fig 7.25 shows two initiation sites, while the others show a single site for initiation. The last column reveals a spatially localized oscillation that did not invade the network. Blue-yellow stripes were not perfectly horizontal, which indicates the existence of phase shifts among the cortical sites. Cortical depression drastically reduced the spatiotemporal coherence of natural sleep spindles, as shown by the more disorganized pattern of activity (Fig. 7.25, Depressed cortex), similar to those seen under barbiturate anesthesia.

A third method to characterize the spatiotemporal coherence between electrodes is to compute the how the correlations decay with distance in the cortex. The spatial correlation was defined as:

$$C(x) = \frac{\sum_{i,j} v(r_i, t_j)\, v(r_i + x, t_j)}{\sum_{i,j} v(r_i, t_j)^2} \qquad (7.10)$$

where $v(r_i, t_j)$ is the normalized local field potential at site r_i and time t_j, and they were normalized by subtracting their average value and dividing by their standard deviation. Cross-correlations were then calculated between every possible pair of renormalized signals and the values at time zero were combined for all pairs with same intersite distance (x). The same procedure was repeated for several epochs of spindle activity and averaged together. The representation of the averaged correlation as a function of distance was used to measure the decay of correlation with distance in the anterior-posterior axis of the suprasylvian cortex.[3]

The decay of spatial correlations was always monotonically decreasing towards zero. A first-order decaying exponential term was used to fit the correlation curves:

$$C(x) = \exp(-x/\lambda) \qquad (7.11)$$

where the space constant, λ, is a measure of the spatial extent of the coherence of a spatially homogeneous phenomenon, similar to the coherence measure based on power spectra (Bullock and McClune, 1989). A closely related—but different—measure of spatial coherence is the falloff in the cross-correlation peaks as a function of distance. Both procedures yielded the same estimates because autocorrelations peaked at time zero.

These methods allowed the decrease in spatiotemporal coherence observed during anesthesia and cortical depression to be further quantified (Fig. 7.25, bottom panels). The decay of correlations was calculated for distances of 0 to 7 mm in steps of 1 mm as determined by the configuration of the recording electrodes. During natural sleep, correlations displayed only limited decay with distance, decreasing to only 0.75 for distances up to 7 mm. The

---

[3] The same procedure was also used in Fig. 7.4 (bottom graphs).

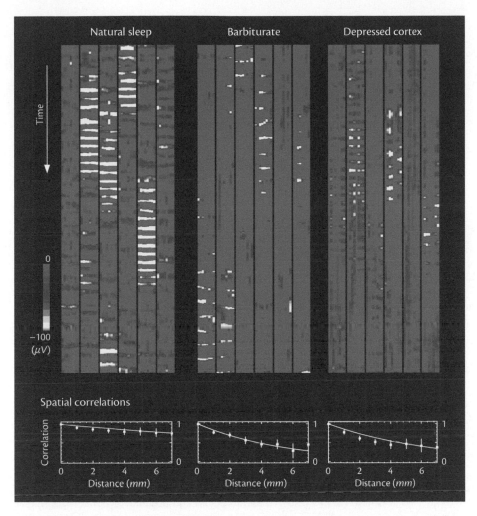

Fig. 7.25: Diminished spatiotemporal coherence of spindle oscillations during anesthesia and cortical depression. Top panels: spatiotemporal maps of the distribution of electrical activity across the cortex were constructed by assigning a colour to the value of the field potential at each electrode; the colour scale ranged in ten steps from the baseline (blue) to −100 μV (yellow). Time was divided in frames each representing a snapshot of 4 *ms* of cortical activity and arranged in columns from top to bottom. Each column is about 3 *s* of activity (arrow is one second). Each frame consisted of eight colour spots, corresponding to the local field potentials from eight electrode arrayed in a line from anterior to posterior (left to right). Bottom panels: decay of correlations with distance. Cross-correlations were computed for all possible pairs of sites and the value at time zero was represented as a function of the intersite distance in the cortex. Each point is an average over different combination of sites, and ten different epochs of 2 *s*; vertical bars indicate the standard deviation. Continuous lines indicate the best fit using a decaying exponential. Figure reproduced from Destexhe et al., 1999a.

space constant was $\lambda = 25$ mm (see Experimental Procedures). During barbiturate anesthesia or following cortical depression, spatial correlations decayed more rapidly with distance, decaying to values around 0.3–0.4 for distances up to 7 mm. The corresponding space constants ere $\lambda = 5.4$ mm and $\lambda = 6.5$ mm, respectively. When comparing natural sleep to barbiturate or depressed cortex, the space constant was reduced four to five times, similar to the reduction in the precision in initiation times (see above).

In conclusion, this analysis confirms that, during natural sleep, spindle oscillations appear nearly simultaneously over cortical distances of several millimeters. During anesthesia or following cortical depression, the oscillations are less coherent. A computational model of thalamocortical interactions that was used to help explain these results (Destexhe et al., 1999a) is reviewed below.

7.3.2 Computational models of cortical excitability

Andersen and Andersson (1968) suggested that intrathalamic mechanisms were responsible for the large-scale coherence of oscillations in the thalamocortical system (Fig. 7.2). This hypothesis seemed plausible, given that the oscillations are generated in the thalamus. According to this view, the cortex is passively driven by the thalamus. It assumes that the thalamic circuitry generates oscillations and controls their degree of coherence, and that this activity is transmitted to cerebral cortex. The experiments reviewed above (Section 7.1.2; Contreras et al., 1996), and in particular the decreased coherence following alteration of cortex alone (Section 7.3.1), demonstrate that a purely intrathalamic basis for the oscillations such as that postulated by Andersen and Andersson is untenable. In Section 7.2, we proposed an alternative thalamocortical-loop mechanism to generate large-scale coherent oscillations. This mechanism is further explored here, by considering first how the different states of the excitability of the cortex might influence the oscillations and second by investigating whether the same thalamocortical loops can also explain the experiments in which cortical excitability was altered.

The excitability of cortical pyramidal cells is regulated by a number of factors. First, the excitability of the cortical network is considerably higher during the awake state. Ascending activating systems from the brain stem, basal forebrain and posterior hypothalamus release neuromodulatory substances such as acetylcholine, noradrenaline, serotonin and histamine throughout the cortex and thalamus (Steriade and McCarley, 1990). Stimulation of these regions results in a marked depolarization of cortical pyramidal cells during intracellular recordings and a subsequent increase in excitability (Steriade et al., 1993a, 1996). Second, extracellular application of neuromodulatory substances in cortical slices induces a depolarization of cortical pyramidal neurons and increases their excitability (McCormick and Prince, 1986; McCormick and Williamson, 1989; McCormick et al., 1993). In particular, subthreshold stimuli can become suprathreshold in the presence of acetylcholine (McCormick and Prince, 1986). Third, neuromodulators such as acetylcholine and noradrenaline reduce leak K^+ currents and the conductances responsible for adaptation of spike firing rates in pyramidal cells (Krnjevic et al., 1971; McCormick, 1992).

Changes in cortical excitability were modelled by changes in pyramidal cells. First, with increasing levels of neuromodulators, the resting membrane potential of cortical PY cells was set to a more depolarized level (-60 mV instead of -70 mV), mimicking the effect of several neuromodulators in closing leak K^+ currents (McCormick, 1992). Second, the I_M conductance responsible for adaptation in cortical PY cells was reduced (from 0.07 mS/cm^2 to 0.02 mS/cm^2), to mimic the effect of acetylcholine, noradrenaline and metabotropic glutamate receptor agonists on spike rate adaptation (McCormick, 1992; Wang and McCormick, 1993). These two actions resulted in resting membrane potentials of PY cells closer to *in vivo* recordings during light barbiturate anesthesia (between −60 and −70 mV). All other parameters were kept constant to investigate specifically the effects of increasing the excitability of cortical pyramidal cells (the role of other cell types are investigated below in Section 7.3.5).

The effects of neuromodulators are shown in Fig. 7.26B. In control conditions, the resting membrane potentials of cortical pyramidal cells was around −70 mV and spike frequency adaptation was prominent (Fig 7.26B, Control). This behaviour resembled recordings from cortical pyramidal cells *in vitro* (Connors et al., 1982; Connors and Gutnick, 1990) and *in vivo* during barbiturate anesthesia (Steriade et al., 1993d). A state of enhanced excitability was simulated in pyramidal cells by reducing the I_M conductance, leading to less prominent spike frequency adaptation (Fig 7.26B, Enhanced excitability), as observed in intracellularly recorded neocortical neurons in awake cats (Baranyi et al., 1993). The resting membrane potential was also more depolarized, around −60 mV, as reported from intracellular studies in awake animals (Matsumara et al., 1978; Baranyi et al., 1993; Steriade et al., 1999; Steriade, 2000).

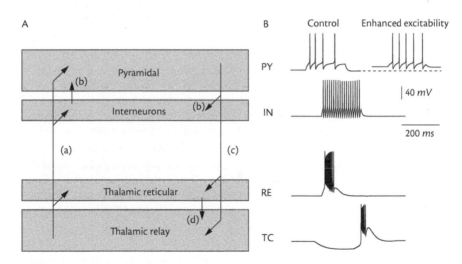

Fig. 7.26: Connectivity and firing patterns of thalamic and cortical cells in the model. A. Schematic diagram showing the connectivity between different cell types. Synaptic connections were simulated by kinetic models of postsynaptic receptors (AMPA receptors for excitatory connections and a mixture of GABA$_A$ and GABA$_B$ receptors for inhibitory connections). Thalamocortical (TC) relay cells do not contact each other but project to the other cell types in the network (a). Cortical pyramidal (PY) cells and interneurons (IN) form a network with local connectivity (b). PY cells contact all cell types in the network (c). Thalamic reticular (RE) cells form an inhibitory network and project exclusively to TC cells (d). B. Intrinsic firing patterns were simulated using Hodgkin-Huxley-like models. The PY cells were regular-spiking (depolarizing pulse of 0.75 nA during 200 ms; −70 mV rest), IN cells were fast spiking (same pulse), RE cells were bursting (pulse of 0.3 nA during 10 ms) and TC cells exhibited rebound bursts (pulse of −0.1 nA during 200 ms). The response of a cortical PY cell is shown in control conditions and with enhanced excitability (reduced I_M and more depolarized rest). Figure reproduced from Destexhe et al., 1999a.

7.3.3 Effect of enhancing the excitability of cortical pyramidal cells

The thalamocortical network model was identical to that introduced previously to study interactions between the cortex and the thalamus during spindle oscillations (Section 7.2; Destexhe et al., 1998a).[4] Each cell type was modelled by a single compartment containing the calcium- and voltage-dependent currents identified in these cells that account for their intrinsic firing properties (Fig. 7.26B). The thalamocortical network comprised four one-dimensional layers (TC, RE, IN and PY) of 100 cells each (Fig. 7.26A). The connectivity between cell types and the synaptic receptors involved in these connections were identical to the description given in Section 7.2 (Destexhe et al., 1998a).

Fig. 7.27 compares spindle oscillations with and without enhanced excitability in a tha-lamocortical network with 100 cells of each type. In control conditions, corresponding to barbiturate spindles, the oscillations started with the spontaneous firing of one or several TC cells (Fig. 7.27A). Several initiation sites occurred in the thalamus because of the heterogeneity of TC cells, as analysed previously (Section 7.2; Destexhe et al., 1998a). Cortical initiation was also possible (see Fig. 7.20). The 'initiator TC cells' of Fig. 7.27A (∗) recruited the rest of the network by stimulating the thalamus-cortex-thalamus recruitment loop, so that the oscillation generalized to the entire network within a few cycles. This model was shown to account for a number of different properties of barbiturate spindles, such as their patterns of initiation, synchrony and propagation following low-intensity cortical stimulation (see Section 7.2).

An enhanced excitability of cortical pyramidal cells was implemented through a decreased I_M conductance and depolarization to −60 mV. With enhanced excitability, the network still displayed spindle oscillations with cellular characteristics similar to those in control conditions (Fig. 7.27B). A marked difference, however, was that the oscillations spread throughout the network more rapidly, in about one cycle of the oscillation. This effect is clearly seen if local average membrane potentials of pyramidal cells are computed at equidistant sites (Fig. 7.28A and scheme). These average values reveal that in control conditions, the oscillation initiated almost simultaneously in two different sites (indicated by asterisks in Fig. 7.28A). With enhanced cortical excitability, the oscillation synchronized within a narrower time window.

In Fig. 7.27, the increased synchrony with enhanced cortical excitability could be attributed to the more depolarized resting level of PY cells. The contribution of the decreased I_M conductance was also significant, but less important in magnitude at the network level (not shown). A substantial proportion of corticocortical and thalamocortical EPSPs in PY cells were subthreshold at a resting membrane potential of −70 mV and became suprathreshold at −60 mV. The spontaneous firing of initiator TC cells recruited cortical cells, which in turn recruited the rest of thalamus into a synchronous oscillation. Consistent with this observation, suppressing PY→PY connections in the simulations shown in Fig. 7.27B led to similar initiation patterns as in Fig. 7.27A (not shown). Therefore, in the enhanced cortical excitability state, intracortical excitatory connections become more powerful, which results in a more compact network in which oscillations generalize more quickly.

---

[4] The only difference was that simulations of natural sleep and barbiturate anesthesia conditions differed in the kinetics of GABA$_A$ receptors. The rate of decay was 5 ms for natural sleep and 12.5 ms for barbiturate conditions, to account for the prolonged time course of GABA$_A$ currents by barbiturates (Thompson, 1994).

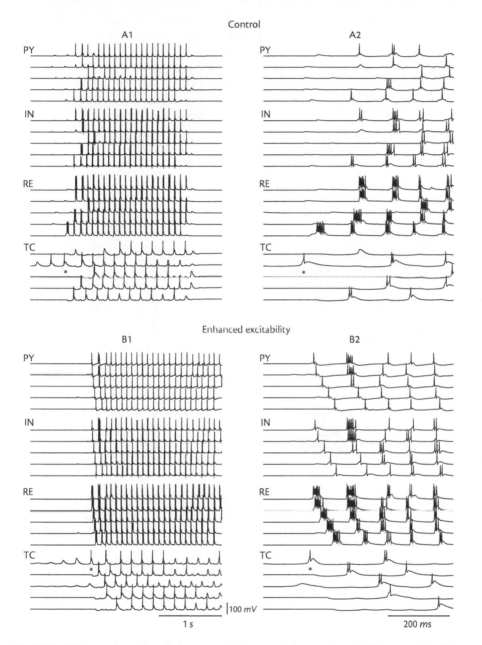

Fig. 7.27: Cellular patterns of oscillations in the thalamocortical network model. The network had four layers of PY, IN, RE and TC cells as organized in Fig. 7.26A. A1. Spontaneous spindle oscillation in control conditions. Five cells of each type, equally spaced in the network, are shown with 0.5 ms time step. A2. Detail of the initiation of the same spindle wave with five times higher temporal resolution. B1. Spontaneous spindle oscillation with enhanced cortical excitability. Same parameters and description as in A, except that PY cells had a more depolarized resting membrane potential of −60 mV and a reduced I_M current. B2. Same simulation at higher temporal resolution. In both cases, the oscillation initiated by the discharge of one or several TC cells (asterisks indicate an initiator TC cell). Synaptic conductances were 0.2 μS (TC-to-RE AMPA receptors), 0.2 μS (RE-to-RE GABA$_A$ receptors), 0.02 μS (RE-to-TC GABA$_A$ receptors), 0.04 μS (RE-to-TC GABA$_B$ receptors), 1.2 μS (TC-to-PY AMPA receptors), 0.4 μS (TC-to-IN AMPA receptors), 0.01 μS (PY-to-TC AMPA receptors), 1.2 μS (PY-to-RE AMPA receptors, 0.6 μS (PY-to-PY AMPA receptors), 0.2 μS (PY-to-IN AMPA receptors), 0.15 μS (IN-to-PY GABA$_A$ receptors) and 0.03 μS (IN-to-PY GABA$_B$ receptors). Figure reproduced from Destexhe et al., 1999a.

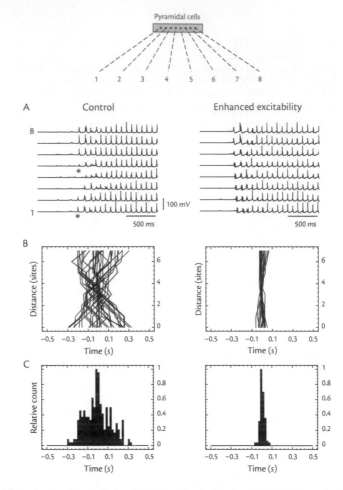

Fig. 7.28: Effect of enhancing cortical excitability on the spatiotemporal distribution of oscillations in the model. Simulated spindle oscillations are shown for control conditions and enhanced cortical excitability (left and right panels, respectively; simulations identical to Fig. 7.27). A. Local average membrane potentials. Twenty-one adjacent PY cells, taken at eight equally spaced sites on the network (see scheme on top), were used to calculate each average. Asterisks indicate two simultaneous initiation sites. B. Patterns of initiation. The onset of the oscillation was estimated from local averaged membrane potentials using the same procedure as in Fig. 7.24B. C. Distribution of initiation times computed from B, as in Fig. 7.24C. Figure reproduced from Destexhe et al., 1999a.

It was important that the corticothalamic feedback acted through dominant inhibition on TC cells. If the discharge of pyramidal cells evoked EPSPs on TC cells that were too strong, the resulting shunt between cortical EPSPs and reticular IPSPs could not recruit oscillations in the TC cell, and the oscillations would remain local. When corticothalamic feedback evoked strong IPSPs in TC cells, through discharging RE cells, then oscillations in the entire system were effectively evoked through the rebound burst property of TC cells, similar to the mechanism described in Section 7.2 (Destexhe et al., 1998a).

7.3.4 Spatiotemporal coherence of simulated oscillations

The local field potentials in the model shown in Fig. 7.28A were analysed with methods similar to those used in Section 7.3.1 for experimental data. Initiation times of simulated spindles were calculated from local averaged potentials using the same criteria as in Fig. 7.24B. In control conditions, spindle sequences ($n = 32$) were initiated within 0.2 ± 0.1 s (Fig. 7.28B, left panel). These values were comparable with those obtained during barbiturate anesthesia and following cortical depression. Spindle sequences ($n = 20$) with enhanced cortical excitability initiated within 0.06 ± 0.03 s (Fig. 7.28B, right panel), comparable to natural sleep spindles. The distributions of initiation times (Fig. 7.28C) also had standard deviations that were similar to those in experiments ($\sigma = 126$ ms and 24 ms for control and enhanced cortical excitability, respectively).

The similarity between the initiation times in the model and experiments (compare Fig. 7.24C with Fig. 7.28C) suggests that the intersite distance of eleven cells, used to calculate averages in the model, roughly corresponded to the interelectrode distance of 1 mm in the experiments. The propagation velocity of oscillations evoked by low-intensity cortical stimulation in experiments (Fig. 7.9; Contreras et al., 1997b) is comparable to that of models using the same correspondence between distance and number of model neurons (Section 7.2.6; Destexhe et al., 1998a), suggesting that the same model is consistent with both types of experiments, with one cell in the model representing about 315 neurons in the real cortex.

The increased coherence following enhancement of cortical excitability was also apparent in the spatiotemporal distribution of activity displayed in Fig. 7.29. In control conditions, oscillatory activity began in one or two sites and progressively invaded the network (Fig. 7.29, left colour panel). The synchrony and phase shifts of the oscillations were comparable to those during barbiturate anesthesia (compare with Fig. 7.25, Barbiturate). With enhanced cortical excitability, the simultaneity and synchrony of oscillations were enhanced, while phase shifts were reduced (Fig. 7.29C, right colour panel).

Finally, the influence of cortical excitability on spatiotemporal coherence also affected the decay of correlations with distance from local averaged potentials (Fig. 7.29, bottom panels). Spindles simulated in control conditions gave rise to correlation decay with a space constant of $\lambda = 4.36$ (in units of intersite distance). With enhanced cortical excitability, correlations had a space constant of $\lambda = 8.36$ and remained elevated for longer distances, although the correlations were not as high as in natural sleep. As in the experiments, however, enhancing cortical excitability increased spatiotemporal coherence and reduced the steepness of the correlation decay.

7.3.5 Effects of enhancing the excitability of different cell types

The excitabilities of other cell types were altered to compare their impact with the effects described above for changes in pyramidal cells (Fig. 7.30A). The excitability of cortical interneurons was changed by using a more depolarized resting level (-60 vs. -70 mV), which enhanced the discharge following depolarizing inputs (Fig. 7.30B, left). However, this change had no significant effect at the network level (Fig. 7.30B, right). On the other hand, decreasing the excitability of interneurons, by using more hyperpolarized resting level (-90 mV), led to

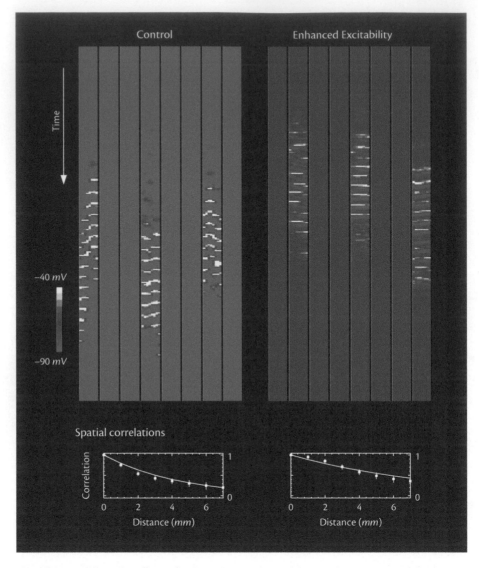

Fig. 7.29: Modelling the effects of enhanced cortical excitability on the spatiotemporal coherence of oscillations. Simulated spindle oscillations are shown for control conditions and enhanced cortical excitability (left and right panels respectively). Top panels: spatiotemporal maps of local averaged potentials. Maps were obtained from the simulations shown in Figs. 7.27), with time running from top to bottom (arrow = 1 sec) in steps of 10 ms. In each column, the membrane potential was represented with a colour scale ranging from −90 mV or below (blue) to −40 mV or above (yellow; see scale). Bottom panels: decay of correlations with distance, calculated from local averaged potentials. Spatiotemporal maps and correlations are displayed using as in Fig. 7.25. Figure reproduced from Destexhe et al., 1999a.

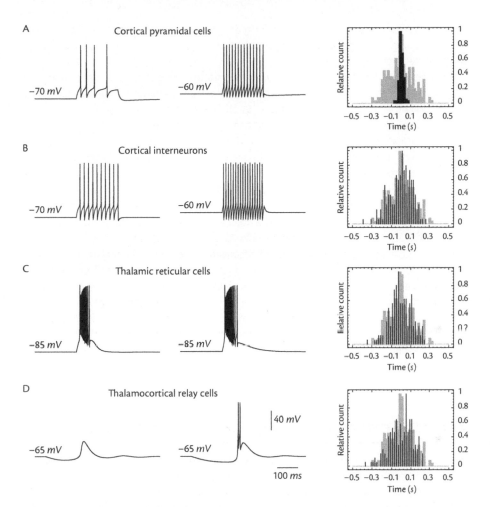

Fig. 7.30: Effects of enhancing the excitability of different cell types. One cell type was given enhanced excitability in simulation shown in Fig. 7.27A. The left side shows the effects of enhancing excitability on the discharge of the cell, and the right side shows the distribution of initiation times obtained for twenty simulated spindle sequences. A. Enhancing the excitability of PY cells (lower I_M conductance, from 0.07 to 0.02 mS/cm^2, and lower resting level, from -70 to -60 mV) had an effect both on cellular response (left; current pulse of 0.75 nA during 200 ms) and on network behaviour (right; same simulations as shown in Fig. 7.28). B. Enhancing the excitability of IN cells (lower resting level, from -70 to -60 mV) had an effect on cellular discharges (left; current pulse of 0.5 nA during 200 ms) but not on network behaviour (right). C. Enhancing the excitability of RE cells (lower leak conductance, from 0.05 mS/cm^2 to 0.02 mS/cm^2) affected burst responses (left; current pulse of 0.3 nA during 10 ms) but network behaviour was minimally affected (right). D. Enhancing the excitability of TC cells (lower leak K$^+$ conductance, from 3 nS to 1 nS) affected burst threshold (hyperpolarizing current pulse of -0.03 nA during 200 ms) but had little effect on network behaviour. All histograms shown in light grey correspond to the control simulation of Fig. 7.28C. Figure reproduced from Destexhe et al., 1999a.

epileptic-like discharges (not shown). Under these circumstances, the disinhibited pyramidal cells fired strong discharges and entrained the network in a ~3 Hz oscillation, similar to spike-and-wave seizures (see Chapter 8).

The effects of enhancing the excitability of thalamic cells were also investigated. In RE cells, the excitability was changed by using a smaller leak K^+ conductance, similar to the effect of noradrenaline on these cells (McCormick, 1992). This resulted in stronger burst discharges (Fig. 7.30C, left). In network simulations, enhancing the excitability of RE cells had a marked effect on oscillation frequency, as analysed in Chapter 8, but there was no prominent effect on the simultaneity and spatiotemporal coherence of oscillations (Fig. 7.30C, right). In TC cells, the excitability was enhanced by using smaller leak K^+ currents, mimicking the effect of acetylcholine and noradrenaline in these cells (McCormick, 1992). This resulted in lower burst threshold in response to hyperpolarizing current (Fig. 7.30D, left). At the network level, enhancing TC cell excitability led to significantly shorter spindle oscillations (not shown) but had no significant effect on spatiotemporal coherence (Fig. 7.30D, right).

In conclusion, this model shows that the circulation of activity in the thalamocortical loop can account for the highly coherent properties of natural sleep spindles. This mechanism emphasizes not only the 'pacemaker' properties of thalamic circuits, but also the powerful role of corticothalamic feedback in controlling these oscillations. As a consequence of thalamocortical loops, cortical excitability can affect large-scale coherence through recurrent excitatory connections. This framework supports Bremer's proposal that brain rhythms should not be described as the passive driving of cerebral cortex by impulses originating from pacemakers (Bremer, 1938a, 1958).

7.4 Thalamocortical augmenting responses

In Chapter 6, we showed that 10 Hz stimulation of the thalamic network model produced intrathalamic augmenting responses similar to those observed in the isolated thalamus (Bazhenov et al., 1998a). Augmenting responses are also observed when the intact thalamocortical system is stimulated at this frequency (see details in Bazhenov et al., 1998b).

7.4.1 Thalamocortical augmenting

When the thalamus is locally stimulated with a train of repetitive stimuli at 10 Hz, augmenting responses are consistently produced in both thalamic and cortical neurons (Bazhenov et al., 1998b). The top panel in Fig. 7.31 shows the augmenting responses in simultaneously recorded cortical and TC cells. The first thalamic stimulus evoked an EPSP followed by an IPSP in thalamic cells. Following the second stimulus, a spike burst appeared in the TC cell rising from $GABA_B$ IPSPs and a secondary EPSP appeared in the cortical cell. The TC cell reached its maximum hyperpolarization before the third and fourth stimuli, which evoked a stronger low threshold spike with five spikes in the burst. These spike bursts in the TC cells led to strong secondary excitation in the cortical cell. The depression of the IPSP and the activation of the I_h current slightly repolarized TC cell and the response to fifth stimulus was a burst of only four spikes which did not affect the shape of secondary depolarization in the cortical cell. The relationship

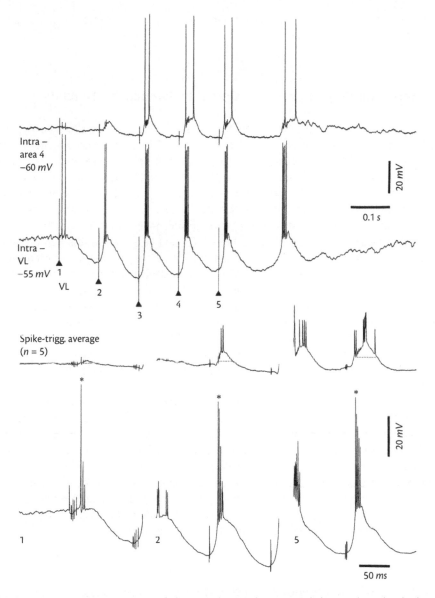

Fig. 7.31: Augmenting responses in thalamic and cortical neurons. Thalamic rebound spike bursts deinactivated by hyperpolarization during augmenting responses precede the depolarizing augmented responses in cortical neuron. Ketamine-xylazine anesthesia. Dual intracellular recording from thalamic VL and cortical area 4 neurons. VL stimulated at 10 Hz. Averages ($n = 5$) triggered by the first action potentials (asterisks) of the 1st, 2nd, and 5th responses of VL neuron show that they precede the late, augmented depolarization (marked by dotted line) in area 4 (from Bazhenov et al., 1998b).

of spike bursts in TC cell and secondary excitation in cortical cell is clearly seen in spike-triggered averages (Fig. 7.31, bottom panel).

7.4.2 Augmenting responses in simple thalamocortical circuits

The simplest thalamic network model that can generate a low-threshold augmenting response during repetitive stimulation is a reciprocal pair of RE-TC cells (Bazhenov et al., 1998a). A thalamocortical network model (similar to that used in preceding sections) displays the main features of augmenting responses observed in the cortex *in vivo*. Thalamic stimulation was modelled by AMPA EPSPs delivered to both RE and TC cells. Thalamic stimulation also produced monosynaptic excitation in cortical (CX and IN) neurons. This was modelled by cortical responses to stimuli that were only 10% of the intensity of RE-TC stimulation.

The external stimulus evoked EPSPs in RE, TC, CX and IN cells. The response of the TC cell in turn produced secondary EPSPs in CX and IN cells after a disynaptic latency (see Fig. 7.32). Feedback RE-evoked $GABA_A$–$GABA_B$ IPSP partially deinactivated the low-threshold Ca^{2+} current in the TC cell and the next stimulus evoked a low threshold spike leading to the augmented burst of spikes in the TC cell, which enhanced the secondary EPSP in the CX cell. Continued stimulation augmented the TC responses and, consequently, the secondary EPSPs in CX and IN cells. Thus, a simple network of four RE-TC-CX-IN cells could reproduce the main features of the augmenting responses—a two-component response with an augmenting second component—observed in cortical pyramidal cells during repetitive thalamic stimulation *in vivo* (Bazhenov et al., 1998a).

Cortical augmentation occurred in the model because of the growth of TC-evoked EPSPs in CX and IN cells. Strengthening the afferent TC→CX synaptic connections should result in even stronger cortical augmenting responses during repetitive thalamic stimulation. This was confirmed in simulations where the maximal conductance of the TC→CX connection was increased to almost twice its standard value (see Fig. 5 in Bazhenov et al., 1998b). In this case, stimulation elicited action potentials in the CX cell starting with the second stimulus in the train and the spike latency was much shorter compared with the experiment shown in the Fig. 7.32. Additional mechanisms affecting the strength of the cortical augmenting response were uncovered in more complex thalamocortical networks, as shown below.

7.4.3 Augmenting responses in thalamocortical networks

A network with four one-dimensional chains of RE, TC, CX, IN cells (see preceding sections) was analysed to determine the influence of geometry on the augmenting responses in the cortical CX and IN cells. Repetitive 10 *Hz* stimulation of RE-TC cells at 100% intensity and CX-IN cells at 10% intensity led to the augmentation of the responses in the TC, CX and IN layers during first three to four stimuli (Fig. 7.33). The number of spikes per burst and the number of cells firing action potentials increased. Nearly all of the cells fired after the second stimulus. In a larger network, the size of the active region increased gradually during a long train of stimuli, leading to more gradual growth of the secondary EPSPs in CX cells.

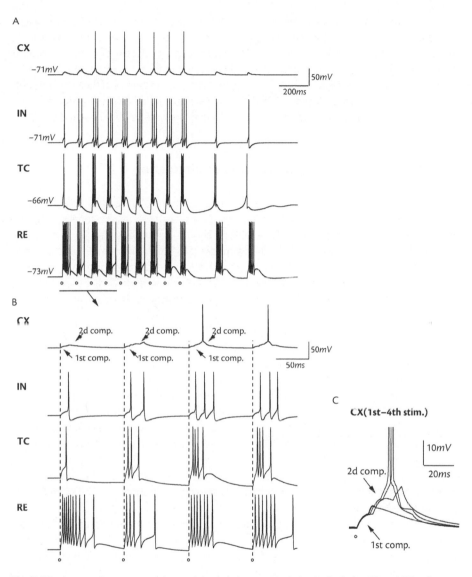

Fig. 7.32: Augmenting responses in a minimal thalamocortical circuit. Both the RE and TC cells were stimulated with 100% of maximal intensity and the CX and IN cells were stimulated with 10% of maximal intensity. (A) Monosynaptic stimulation of the CX cell elicited a non-augmenting component in the cortical EPSPs that occurred simultaneously with EPSPs in the RE and TC cells. Augmenting spike bursts in TC cell led to a growing secondary EPSPs in the CX cell. (B) Same response shown on a shorter time scale. (C) Superimposed traces of the first four EPSPs in a CX cell. Open circles indicate the time of thalamic stimulation. ($g_{AMPA} = 0.1$ μS from CX to IN, $g_{AMPA} = 0.1$ μS from CX to TC, $g_{AMPA} = 0.2$ μS from CX to RE, $g_{GABA_A} = 0.03$ μS from IN to CX, $g_{AMPA} = 0.035$ μS from TC to CX and $g_{AMPA} = 0.02$ μS from TC to IN; from Bazhenov et al., 1998b).

Note that three layers of the network (TC, CX, IN) displayed a similar form of augmentation during the train of stimuli, though with quite different patterns of spiking. The spikes in TC cells preceded the action potentials in CX-IN cells. In contrast, RE cells displayed a powerful response to the first stimulus that decremented in response to the second stimulus. The partial inactivation of the low-threshold Ca^{2+} current in RE cells reduced the size of their low threshold spikes (see details in Bazhenov et al., 1998a). The CX cells responded with a slowly augmenting response, starting with the second stimulus, to the increasing number of TC-evoked EPSPs.

7.4.4 Augmenting responses to cortical stimulation

Repetitive thalamic stimulation results in augmenting responses of CX and IN cells due to the enhancement of TC-evoked EPSPs in these cells. Based on these results, repetitive cortical stimulation in the presence of CX→TC→CX loop should lead to augmenting responses in CX cells. This suggestion was tested with the model (Fig. 7.34). The stimuli were delivered simultaneously to the CX and IN cells. The first four shocks in the train led to growing responses in the CX (from 1–3 to 1–4 spikes), IN (from 3–5 to 4–8 spikes) and TC cells (from 0 to 1–2 spikes). Fig. 7.34A, B show expanded traces of two TC-CX pairs in such a simulation. The same CX cells after removing of RE-TC network are shown in Fig. 7.34C.

The stimulation of cortical AMPA responses in CX and IN cells resulted first in 1–3 spike responses in the CX cells. Activation of CX→RE and CX→TC synapses evoked monosynaptic EPSPs followed by disynaptic IPSPs in TC cells. After the first shock, the CX-evoked EPSPs did not lead to action potentials in the TC network. However, RE-evoked hyperpolarization of TC cells deinactivated the low-threshold Ca^{2+} current in TC cells and, after a second stimulus, the CX-evoked EPSPs were followed by low threshold spikes and single spike responses in some of TC cells. These responses evoked secondary EPSPs in CX cells, which grew during the train of stimuli. The secondary EPSPs in the CX cells arrived when the CX cells still were depolarized following stimulus-evoked monosynaptic EPSPs (see Fig. 7.34A, B). This explains the robust effects of these relatively weak TC-evoked EPSPs. When the thalamic part of the network was 'lesioned', stimulation of the cortex produced only stereotyped responses in the same CX cells displayed (see Fig. 7.34C).

The strength of the cortical augmenting responses depended on the position of the cells relative to the centre of stimulation. The CX cells near the centre (Fig. 7.34B) displayed powerful (three spike) responses to the stimulus-evoked EPSPs and augmentation up to four spikes during the train of stimuli. Another CX cell located near the boundary (Fig. 7.34A) obtained weaker stimulation from both the external electrode and TC cells and displayed weak augmenting responses (from one to two spikes).

7.4.5 Mechanisms underlying augmenting responses following cortical stimuli

Repetitive 10 Hz stimulation of the cortical cells (CX, IN) induced augmenting responses in these cells through corticothalamic (CX→TC→CX) feedback. In the intact brain the intralaminar thalamic nuclei send widespread projections to cerebral cortex (Jones, 1985). This suggests that

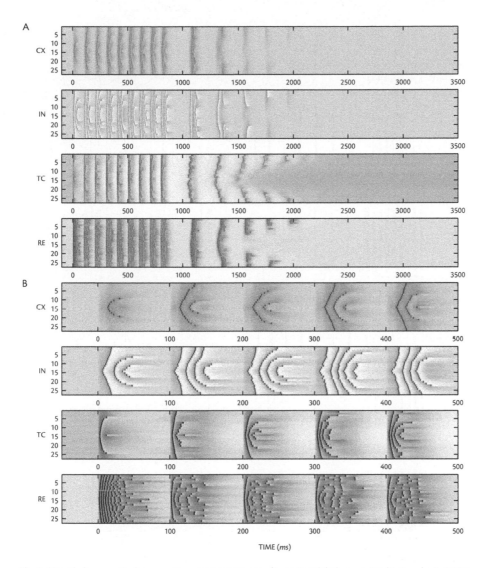

Fig. 7.33: Thalamocortical augmenting responses in one-dimensional thalamocortical networks. A. 10 *Hz* train of stimuli for 1 sec. Both RE and TC cells were stimulated at 100% maximal intensity and CX-IN cells were stimulated with 10% of maximal intensity. The intensity of stimulation was maximal at the centre of the network and decayed exponentially with distance from the centre. B. Expanded traces from (A) between $t = -50$ *ms* and $t = 500$ *ms*. The first four shocks in the train of nine shocks evoked a low-threshold augmenting response in the TC cells and an increasing number of spikes in CX cells (from 0–1 spikes to 1–3 spikes) and IN cells (from 1–3 spikes until 3–4 spikes). Slow (around 3 *Hz*) post-stimulus oscillations that occurred in the RE-TC network were echoed by the CX-IN cells. These oscillations terminated after 4–5 cycles as the spiking of the neurons in the network desynchronized. ($g_{AMPA} = 0.1$ μS between CX cells, $g_{AMPA} = 0.1$ μS from CX to IN, $g_{AMPA} = 0.1$ μS from CX to TC, $g_{AMPA} = 0.2$ μS from CX to RE, $g_{GABA_A} = 0.03$ μS from IN to CX, $g_{AMPA} = 0.08$ μS from TC to CX and $g_{AMPA} = 0.03$ μS from TC to IN; from Bazhenov et al., 1998b).

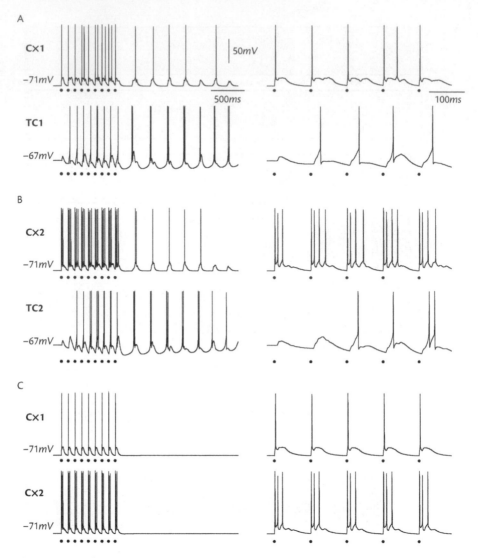

Fig. 7.34: Comparison of augmenting responses in the thalamocortical network and in an isolated cortical network following cortical stimulation. The responses to the first five stimuli on the right have an expanded time scale. A. Direct cortical stimulation evoked single spike responses in the CX cells and weak augmentation in a TC cell far from the centre of the network. B. Responses of a CX cell and a TC cell near the centre of the network. Augmentation of the TC responses produced a growing secondary EPSP in the cortical cell and additional fast spike. C. Stereotyped responses of the same CX cells after removing the thalamic (RE-TC) network. Filled circles indicate the time of cortical stimulation (from Bazhenov et al., 1998b).

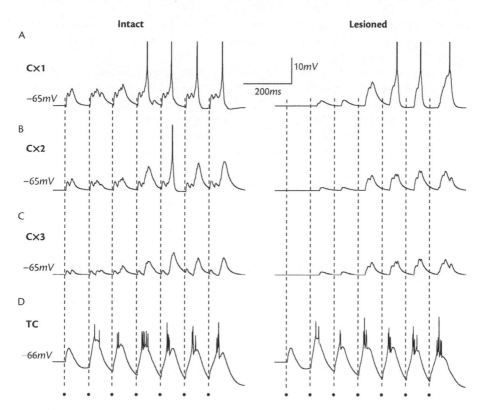

Fig. 7.35: The influence of the CX-TC-CX loop on the augmenting responses in cortical neurons distant from the site of stimulation. Repetitive 10 *Hz* stimulation of half of the cortical network (CX-IN cells from 1–13) produced augmenting responses of the CX cells from the second half of the cortical network: A. CX cell #15; B. CX cell #16; C. CX cell #17. Responses from an intact thalamocortical model shown on the left. Secondary EPSPs in CX cells was from activation of the lateral (CX→CX) and cortico-thalamo-cortical (CX→TC→CX) connections. Lesion of the lateral connections between the two halves of the cortical network eliminated the initial EPSPs in CX cells as shown on the right. D. The averaged responses of the TC cells contributing to the EPSPs in the CX cells shown in (A-C) (TC cells #7–#25). Filled circles indicate the time of cortical stimulation. ($g_{AMPA} = 0.1$ μS between CX cells, $g_{AMPA} = 0.1$ μS from CX to IN, $g_{AMPA} = 0.1$ μS from CX to TC, $g_{AMPA} = 0.2$ μS from CX to RE, $g_{GABA_A} = 0.03$ μS from IN to CX, $g_{AMPA} = 0.1$ μS from TC to CX and $g_{AMPA} = 0.03$ μS from TC to IN; from Bazhenov et al., 1998b).

the cortico-thalamo-cortical loop could also induce augmenting responses in cortical areas remote from the site of stimulation. To test this hypothesis, we stimulated only half of a cortical network. Fig. 7.35A–C shows the responses of the three cortical cells from the non-stimulated half of a cortical network. Fig. 7.35D shows the average response of the TC cells contributing to the secondary EPSPs in these CX cells. Two models were examined: one with intact lateral connections between the two halves of the cortical network and the same model with the lateral connections lesioned (Fig. 7.35). In the former case, the CX cells displayed strong monosynaptic EPSPs starting with the first stimulus in the train. The fast response arose from lateral AMPA connections from the CX cells in the directly stimulated half of the cortical network. These EPSPs disappeared after the lateral interconnections were cut separating CX cells in the two halves of the network.

In both networks, the CX cells displayed secondary EPSPs starting from the second stimulus in the train. These EPSPs arose from the CX→TC→CX loop that was activated when some of TC cells responded with action potentials following CX-evoked stimulation (see Fig. 7.35D). The growth of the secondary EPSPs in the non-stimulated half of the cortical network occurred because there were more spikes per burst in the TC cells and more TC cells were recruited to fire actions potentials. The secondary effect was more prominent because the weak CX-evoked EPSPs in TC cells limited the number of spikes in TC responses to only 1–2.

These results show how the thalamic network, through the activation of TC cells, could produce augmenting responses in spatially distant cortical areas.

7.5 Further findings on thalamocortical networks

The last years were rich in further experimental and theoretical results, that are worth to be mentioned.

7.5.1 Participation of cortical interneurons during cortical oscillations

Single-unit recordings have recently been obtained from the human cortex in epilepsy patients during sleep spindle oscillations (Andrillon et al., 2011), delta waves, Up/Down states, and fast oscillations during wakefulness and REM sleep (Peyrache et al., 2012; Dehghani et al., 2016; Le Van Quyen et al., 2016). In the latter studies, it was possible to distinguish between excitatory and inhibitory neurons, confirming the participation of cortical inhibitory cells in all types of oscillatory activity in neocortex. Thus, the strong involvement of inhibition during spindles is not an exception, but rather represents a general rule.

7.5.2 Two spindle systems in human cortex

Research on animal models of sleep spindles has focused on the projections from the specific or core thalamic nuclei to layer 4 of the neocortex. These include the lateral geniculate nucleus that projects to the primary visual cortex and the ventral posterior nucleus that projects to layer 4 of the primary somatosensory cortex. Recently, using laminar recording electrodes, a second thalamocortical spindle system has been identified in humans that projects from the intralaminar nuclei to layer 1 in the cortex (Hagler, 2018). These projections are more diffuse and are called the nonspecific or matrix system. The spindles in these two systems can occur independently but can also occur simultaneously, though not always in the same phase. A large-scale model of these two thalamocortical spindle pathways could account for these recordings, including increased synchrony of the spindle activity due to the more diffuse thalamic projections to the superficial cortical layers in the model (Bonjean et al., 2012; Krishnan et al., 2018). In agreement with cortical recordings, focal spindles originating in the core, spread to the matrix through cortical coupling between these two systems. This led to much more spatially coherent matrix spindles, as observed in EEG recordings. The significance of these two thalamocortical spindle systems is not known.

7.5.3 Spindles are circular travelling waves

Intracranial electrocorticogram recordings from humans have made it possible to examine the large-scale spatial organization of sleep spindles across the cortex. Earlier recordings from a few sites had concluded that sleep spindles were synchronous across the cortex. Analysis of sleep spindle recordings from an array of electrodes on the lateral surface of the cortex of epilepsy patients have revealed phase differences across the cortex corresponding to circular travelling waves (Fig. 7.36; Muller et al., 2016). Single cycle analysis is necessary because of cycle-to-cycle differences, and averaging across spindle cycles washes out the patterns. However, some circular patterns repeat hundreds of times during a night.

Travelling waves have been observed at many other frequencies (Muller et al., 2012, 2018), including the theta (4–8 *Hz*) (Lubenov and Siapas, 2009; Patel et al., 2012), alpha (~10 *Hz*) (Muller et al., 2014) and gamma (30–80 *Hz*) bands (Lê van Quyen et al., 2016). The speed of these travelling waves is consistent with the time delays of the long-range association fibres that connect distant parts of the cerebral cortex. The time delay allows the presynaptic input to arrive before the postsynaptic spindle burst, which could trigger spike-time dependent plasticity after pairing at 10–14 *Hz* (Markram and Sakmann, 1997).

7.5.4 Cortical deafferentation during sleep spindles

In a recent study with multi-electrode recordings, the cortex of naturally sleeping rats was unresponsive to external inputs during sleep spindles (Fig. 7.37; Peyrache et al., 2011). Similar findings were reported in an fMRI study in humans (Dang-Vu et al., 2011). This is consistent with the high-level of inhibitory activity during sleep spindles, which effectively 'deafferents' cerebral cortex during the sleep spindle, and is consistent with the computational models of sleep spindles reviewed in this chapter. The cortical cell model, together with experiments, predicted that sleep spindles are characterized by high levels of synaptic excitation and inhibition. External input should have less impact on the cortical circuit during such intense intrinsic inputs, as reported experimentally. Models of thalamocortical interactions have explored cortical feedback to the thalamus during sensory stimulation (Yousif and Denham, 2007).

7.5.5 Initiation and termination of spindle oscillations

Both initiation and termination of spindle oscillations were thought to originate in the thalamus. Based on a computer modelling and *in vivo* multisite recordings from the cortex and the

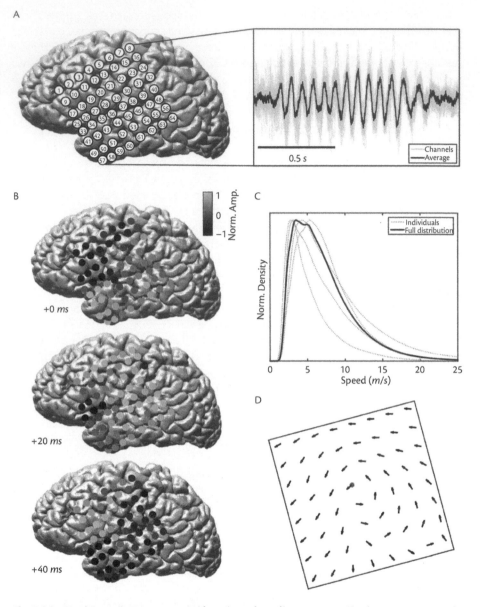

Fig. 7.36: Circular travelling wave recorded from the surface of human cortex. The three time points on the left cover around 50% of a single spindle cycle, colour coded for phase. The arrows in the phase diagram on the lower right point in the direction of maximum phase increase. The measured speeds of rotation in the upper right are compatible with measured time delays on long-range association fibres between cortical areas (adapted from Muller et al., 2016).

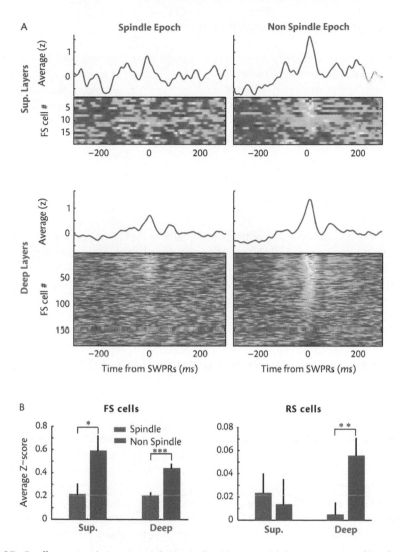

Fig. 7.37: Deafferentation during natural sleep spindles. Hippocampal sharp-wave events (SPWs) evoked a strong response in prefrontal cortex (right), but this response was nearly abolished during cortical sleep spindles (left). A. Average (top panels) and individual (bottom panels) z-scored cross-correlograms of superficial FS cells relative to SPWs occurrence time during spindle oscillations episodes (left panels) and non spindle state (right panels). Cells are ordered relative to the strength of activation during non spindle states. B: Summary statistics of FS cell (left panel) and RS cell (right panel) modulation by SPWs. Modified from Peyrache et al., 2011.

thalamus in cats, the initiation and termination of spindle sequences also involves corticothalamic influences (Bonjean et al., 2011). Although the propagation of spindles depended on synaptic interaction within the thalamus, the new mechanism for spindle termination involved the desynchronization of firing between thalamic and cortical neurons.

7.5.6 Large-scale thalamocortical models

Large-scale two-dimensional thalamocortical models have been explored to examine the effects of neuromodulators on different sleep states (Krishnan et al., 2016), and spontaneous activity, sensitivity to changes in individual neurons, and the emergence of waves and rhythms (Izhikevich and Edelman, 2008). Other thalamocortical models have been used to investigate synchrony during anaesthesia (Sheeba et al., 2008; Dover et al., 2010) and other sleep oscillations, including the transition to wakefulness (Hill and Tononi, 2005).

7.6 Discussion

The experimental data on the spatiotemporal patterns of spindle oscillations indicate that the cerebral cortex has a significant influence on thalamic oscillations in the thalamocortical system (Contreras et al., 1996, 1997a, 1997b, 1997c; Destexhe et al., 1998a, 1999a). The computational models of thalamocortical oscillations explored in this chapter show how known cortical mechanism can account for the data and make further predictions that can be used to test the validity of the model.

7.6.1 The coherence of spindle oscillations in the thalamocortical system

7.6.1.1 Experimental data

Experiments that have been performed *in vivo* (summarized in Section 7.1) have revealed a number of fundamental properties of thalamocortical spindles that need to be explained (Contreras et al., 1996): (a) Spindle oscillations appear coherently across large regions of the thalamus and cerebral cortex (Figs. 7.1, 7.3, 7.22); (b) removal of the cortex abolishes the large-scale coherence of oscillations in the thalamus (Figs. 7.3 and 7.4); (c) local synchrony (<2 *mm*) is, however, still preserved in the thalamus following decortication (Fig. 7.5); (d) propagating patterns of spindle oscillations occur in the thalamus in the absence of the cortex (Fig. 7.8), similar to the systematic propagation found in thalamic slices (Fig. 6.29; Kim et al., 1995); (e) the large-scale synchrony of oscillations observed under barbiturate anesthesia does not depend on intact intracortical connections (Fig. 7.6).

   Although spontaneous spindles appear coherently over large distances, systematic propagation could be elicited by low-intensity electrical stimulation of the cortex (Fig. 7.9; Contreras et al., 1997b). Like the coherent spindles, the propagating spindle waves were insensitive to cortical transection (Fig. 7.10; Contreras et al., 1997b), suggesting the hypothesis that propagation either is generated in the thalamus and projected to the cortex, or it arises through reverberations in the thalamocortical loops.

   The role of the cortex in shaping the coherence of thalamic oscillations was further investigated by studying natural sleep and barbiturate anesthesia (Contreras et al., 1997a; Destexhe et al., 1999a). During natural sleep, oscillations were remarkably coherent (Fig. 7.22), but this coherence was markedly lower following application of extracellular potassium (Fig. 7.23) or under barbiturate anesthesia (Fig. 7.24), both of which depressed cortical activity. The key observation is that altering the cortex disrupted large-scale coherence in the thalamus even though the oscillations were generated in the thalamus. Taken together, these experiments suggest that the properties of spindles, such as the large-scale coherence and propagation, depend on thalamocortical loops and cannot be supported by local intrathalamic interactions alone.

7.6.1.2 Inhibitory dominance of corticothalamic feedback

The mechanisms underlying the thalamocortical oscillations were investigated with a thalamocortical network model based on accurate biophysical properties of intrinsic cellular properties and synaptic interactions (Destexhe et al., 1998a). The thalamic model was identical to the model in Chapter 6 (Section 6.5) developed to explain *in vitro* data showing systematic propagation of spindles (Fig. 6.29). The cortex was represented by a network of layer VI cortical pyramidal cells that project to the thalamus and also receive connections from ascending thalamocortical axons, forming a monosynaptic feedback loop between cortex and thalamus (White and Hersch, 1982). Cortical inhibitory neurons were also included.

The model was used first to investigate corticothalamic interactions. In order to reproduce the experimental data in the model, the effect of corticothalamic feedback on TC cells had to be strongly inhibitory (Fig. 7.12). 'Inhibitory dominance' occurred when the cortical EPSPs were more effective in recruiting RE cells compared to their direct excitatory effects on TC cells. The bursting of RE cells and subsequent feedforward inhibition produced large IPSPs in the TC cells (see scheme in Fig. 7.12). This type of interaction allowed cortical EPSPs to trigger oscillations in the thalamus, which is also consistent with experimental observations (Steriade et al., 1972; Roy et al., 1984; Contreras and Steriade, 1996).

Inhibitory dominance is supported by several experimental observations. Cortical stimulation evokes responses in TC cells that are dominated by inhibition *in vivo* (see Fig. 7.12A; Widen and Ajmone Marsan, 1960; Burke and Sefton, 1966; Steriade et al., 1972; Ahlsen et al., 1982; Lindström, 1982; Roy et al., 1984; Deschênes and Hu, 1990; Contreras and Steriade, 1996). Could the dominance of IPSPs over EPSPs be an artefact of barbiturate anesthetics? This is unlikely since barbiturates prolong the decay of GABAergic currents but do not affect their peak or the rise time (Thompson, 1994), so the early EPSP should not be affected. Furthermore, inhibitory dominance has also been observed under ketamine-xylazine anesthesia (Timofeev et al., 1996). *In vitro* experiments are also consistent with inhibitory dominance: stimulation of the internal capsule in thalamic slices (Thomson and West, 1991) or cortical stimulation in thalamocortical slices (Kao and Coulter, 1997) produced EPSP-IPSP sequences dominated by inhibition in a significant proportion of the recorded TC cells.

Additional lines of evidence support inhibitory dominance. First, cortical synapses on TC cells end on the most distal parts of the dendritic tree but inhibitory synapses from the RE nucleus are more proximal (Liu et al., 1995; Erisir et al., 1997b). Consequently, the total number of cortical synapses on TC cells cortical EPSPs are likely to be shunted by reticular IPSPs. Second, cortical synapses on TC cells are less numerous than previously thought because many of the excitatory synaptic terminals on TC cells actually arise from brainstem structures (Erisir et al., 1997a, 1997b). Third, the oscillatory entrainment of TC cells occurs through an initial IPSP (Steriade and Deschênes, 1984). Fourth, during cortical seizures, 60% of TC cells are hyperpolarized while RE cells are strongly excited (Steriade and Contreras, 1995). Finally, the powerful dendritic T-current in RE cells make them particularly responsive to cortical EPSPs (Destexhe, 2001; see also Section 4.3 in Chapter 4; Destexhe et al., 1996b). Consistently, IPSPs are observed in TC cells even with low-intensity cortical stimuli (D. Contreras and M. Steriade, personal communication). The need to be highly sensitive to cortical EPSPs may be one of the reasons why of RE cells have a high concentration of T-current on their distal dendrites.

7.6.1.3 A cellular mechanism for large-scale synchrony

Because spindle oscillations are generated in the thalamus, it was initially thought that thalamocortical synchrony depended exclusively on intrathalamic mechanisms (Fig. 7.2; Andersen and Andersson, 1968). However, the experimental data reviewed here demonstrate that the large-scale synchrony depends on the integrity of corticothalamic and thalamocortical connections. In computational models, the cortex recruited the thalamus primarily through the RE nucleus, demonstrating how 'inhibitory dominance' may orchestrate synchrony based on thalamocortical loops. The Cortex→RE→TC→Cortex loops (Fig. 7.21) account for the following experimental observations: (a) Spindles are 'nearly simultaneous' in the thalamocortical system (Fig. 7.15); (b) removing the cortex disrupts the large-scale synchrony of oscillations in the thalamus (Figs. 7.16 and 7.17); (c) low-intensity cortical stimulation evokes propagating oscillations (Fig. 7.19); (d) spontaneous activity in the cortex induces variability consistent with *in vivo* data (Fig. 7.20); (e) there is a refractory period following cortical stimulation (Fig. 7.18).

The thalamocortical model offers a plausible explanation for apparently inconsistent observations, namely that spindle waves are nearly simultaneous *in vivo* (Fig. 7.22; Contreras et al., 1997b) although they systematically propagate in thalamic slices (Fig. 6.29; Kim et al., 1995). These differences are due to the combined action of several factors.

First, thalamocortical loops recruit larger thalamic areas than intrathalamic loops (see Fig. 7.21), because the divergence of different axonal projection systems are additive (cortex-to-RE, RE-to-TC, TC-to-cortex). As illustrated in Fig. 7.38, intrathalamic loops can generate spindle oscillations (Fig. 7.38A) but in the intact thalamocortical system (Fig. 7.38B), the oscillations are enhanced by additional loops involving the cortex.

A second factor that could enhance synchrony is the presence of several simultaneously active initiation sites in the thalamus (* in Fig. 7.15). Initiation can be synchronized if all TC cells have the same refractory period in the model, such that initiator TC cells tend to start oscillating at roughly the same time. The length of the refractory period depends on biochemical rate constants that are likely to be similar in all TC cells that possess this mechanism. Therefore, the presence of similar refractory periods throughout the thalamocortical system would allow the network to 'learn' large-scale synchrony by setting these mechanisms in phase.

A third factor promoting large-scale synchrony is an increase in cortical excitability, as analysed in detail in Section 7.3. Modulating the state of the cortex had a major influence on the coherence of thalamic activity (Depressed cortex in Fig. 7.24). A cortical network with enhanced excitability provided more powerful and coherent feedback to the thalamus, resulting in a more rapid spread of the discharges (Fig. 7.29). In contrast, depressing cortical responsiveness, either by using extracellular potassium or with barbiturates, increased the threshold for excitatory inputs needed to fire pyramidal cells and therefore decreased coherence because fewer cells were recruited on each cycle of the oscillation.

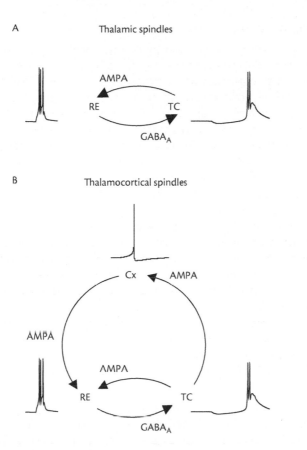

Fig. 7.38: Proposed mechanisms for thalamic and thalamocortical spindles. A. ~10 *Hz* spindle oscillations resulting from the interaction between TC and RE cells (see Chapter 6). B. Spindle oscillations in the intact thalamocortical system. In addition to intrathalamic loops, TC cells also excite RE cells via the monosynaptic feedback loop mediated by layer VI pyramidal cells. If the corticothalamic feedback is mostly effective on RE cells (Cx→RE), then this thalamocortical loop can explain the large-scale synchrony of oscillations in the thalamocortical system.

7.6.1.4 Propagating oscillations

The thalamic model was originally constructed to account for the propagating spindle oscillations in thalamic slices (see Chapter 6). Systematic propagation of spindles also occurred in the cortex *in vivo* when they were elicited by low-intensity cortical stimuli (Fig. 7.9; Contreras et al., 1997b). This propagation was insensitive to cortical cuts (Fig. 7.10) suggesting that propagation occurred either in the thalamus or through thalamocortical loops. The model provided support for the second of these hypotheses (Fig. 7.19). When a localized cortical area was stimulated before oscillations in the thalamus had restarted, the oscillation propagated through the topographic structure of the network, 'bypassing' the intrinsic initiation mechanisms of the thalamus (Fig. 7.18).

7.6.1.5 Predictions provide a way to test the models

1. *Propagation velocity.* The propagation velocity in the model was proportional to the spread of the TC-to-cortex, cortex-to-RE, RE-to-TC axonal projections. The propagation velocity in the model matched the experiments (compare Fig. 7.19 with Fig. 7.9) when the intersite distance of eleven cells, used to calculate the averages in Fig. 7.19, corresponded roughly to the interelectrode distance of 1 *mm* in the experiments. With this assumption, the size of the thalamocortical projections in the cortical network (twenty-one cells in Fig. 7.15A) would correspond to 1.9 *mm* in the cortex. Although this is large for thalamocortical arborizations in cortical area 5–7, it is consistent with current estimates that in somatosensory cortex the terminal arborizations of an ascending TC axon have a diameter of 600 μm, and neighbouring TC cells can diverge as much as 1500 μm (Landry and Deschênes, 1981; Rausell and Jones, 1995).

   Thus, one cortical cell in the model corresponds to a population of cells in cortex in a 90 μm diameter (1mm = eleven cells), or about 315 cortical neurons (assuming a density of 100,000 neurons per mm^3). In a 200 μm diameter piece of thalamus there are around eleven neurons (see Section 6.5.5 in Chapter 6). Therefore, a thalamic cell in the model represents a volume 20 μm in diameter in the thalamus. The ensemble of axons from TC cells in this volume project to the cortex, contacting neurons within an area having a diameter of approximately 1900 μm (90 × 21). Similarly, a group of cells within 90 μm diameter in cortex sends axons to a region of the thalamus with a diameter of about 380 μm. These projections are consistent with anatomical measurements (Landry and Deschênes, 1981; Freund et al., 1989; Bourassa and Deschênes, 1995; Rausell and Jones, 1995).

2. *Inhibitory dominance.* If inhibitory dominance determines spatiotemporal coherence, then local injection of a GABA$_A$ blocker in a thalamic relay nucleus should lead to dramatic changes in the large-scale coherence of oscillations. This could also be achieved by preventing RE cells from firing bursts. This mechanism could also be tested by locally applying synaptic antagonists in slices: the ability of corticothalamic axons to evoke thalamic oscillations should be unchanged by blocking excitatory synapses on TC cells; but the oscillations should be profoundly altered by blocking GABA$_A$-mediated inhibition on TC cells.

3. *Excitability of cortical cells.* In neocortical slices, local stimulation of the white matter produces limited horizontal propagation (Chagnac-Amitai and Connors, 1989; Albowitz and Kuhnt, 1993; Contreras et al., 1997d), but the application of GABA$_A$ antagonists generates epileptic discharges that propagate across the cortex (Chagnac-Amitai and Connors, 1989; Albowitz and Kuhnt, 1993). The model suggests that, in addition to GABA$_A$ antagonists, horizontal propagation should be markedly enhanced in cortical slices by using extracellular application of neuromodulators, such as acetylcholine or noradrenaline, through an increase of excitability of cortical pyramidal cells. The velocity of these rapidly propagating discharges (100–200 *mm/s*) depends on the level of excitability (see Fig. 4 in Destexhe et al., 1999a) and therefore should be dependent on the concentration of neuromodulator. Rapidly propagating cortical discharges should also be detectable by high-resolution optical recording methods *in vivo*.

A recent cortical slice preparation has been proposed in which the cortical excitability was larger than conventional slices (Sanchez-Vives and McCormick, 2000). If neurons can be maintained at their depolarized (most excitable) level during a sufficiently long time, then this preparation could be used to investigate the presence of rapidly propagating spike discharges.

4. *Resting membrane potentials.* The model predicts that natural sleep and barbiturate anesthesia correspond to different levels of resting membrane potential in cortical pyramidal cells. The model predicts that intracellular recordings performed in neocortical pyramidal cells during natural sleep should have relatively depolarized resting level, close to −60 mV, compared to the hyperpolarized resting level of −70 mV to −80 mV typical of barbiturate anesthesia. Recent results (Steriade et al., 1999; Steriade, 2000) suggest that this prediction might be correct.

5. *Propagating oscillatory waves.* The model predicts that systematic propagation should not occur during natural sleep (Fig. 7.9). In contrast, low-intensity electrical stimulation of the cortex can induce propagating oscillatory waves during barbiturate anesthesia (Fig. 7.9; Contreras et al., 1997b). These propagating oscillations persisted after cortical cuts (Fig. 7.10; Contreras et al., 1997b), suggesting that horizontal intracortical connections were not responsible. However, intracortical connections should have an important role in supporting the simultaneity of natural sleep spindles. This directly implies that, during natural sleep, propagating oscillations evoked by cortical stimulation should not occur. A corollary to this prediction is that cortical cuts should affect the spatiotemporal patterns and synchrony of natural sleep spindle oscillations. This is supported by the observation of diminished interhemispheric synchrony of spindles following callosal transection (Bremer et al., 1956).

7.6.1.6 Future studies

The present mechanisms are compatible with the measurement of large-scale synchrony in area 5–7 of cat cerebral cortex during barbiturate anesthesia (see Section 7.1; Andersen and Andersson, 1968; Contreras et al., 1997b). However, spindles occurring during natural sleep have a higher degree of coherence (Section 7.3.1; Destexhe et al., 1999a). To account for synchrony over distant cortical areas, interareal connections must be taken into account as well as corticothalamic axons that project to several thalamic nuclei (Bourassa and Deschênes, 1995). Another possibility is based on the diffuse connectivity in the rostral pole of the RE nucleus (Steriade et al., 1984), which may account for the occasional occurrence of nearly simultaneous oscillations in the decorticated thalamus (see Fig. 8, panel 2 in Contreras et al., 1997b). More detailed modelling studies are needed to explore this issue.

The thalamocortical model presented here can account for large-scale synchrony in thalamocortical systems over distances of approximately 7 mm in cerebral cortex. Can these mechanisms also explain synchrony observed over larger cortical regions? One possibility is that EPSPs from the rostral intralaminar thalamic neurons become suprathreshold when pyramidal cells become sufficiently depolarized, leading to enhanced large-scale synchrony. Thalamic intralaminar neurons project over widespread neocortical areas and tend to contact cortical neurons distally in layer I (Jones, 1985). It has been proposed that intralaminar neurons play an important role in coordinating oscillations over the entire cortical mantle (Ribary et al., 1991; Llinás et al., 1994). The observation that these neurons discharge robust spike bursts within each cycle of spindle oscillations (Steriade et al., 1993b) suggests that they may be responsible for coordinating the global coherence of spindles. In the model, cortical pyramidal cells are sensitive to intralaminar EPSPs only when they are in an excitable state, with resting level around -60 mV. More detailed knowledge about the electrophysiological properties and connectivity of intralaminar cells are needed to further investigate this hypothesis.

Finally, the models make a general prediction about the function of corticothalamic interactions. Corticothalamic projections, which have only modulatory influence in the relay mode, could be responsible for the long-range synchrony in the entire thalamocortical system during rhythmic activity. The model predicts that that the recruitment of the thalamic circuitry occurs through the RE nucleus. This mechanism is consistent with the long-range synchrony observed in the cortex for slow oscillations during deep sleep but is incompatible with the local synchrony that occurs during fast oscillations (20–60 Hz) in activated periods associated with REM and alertness (Steriade and Amzica, 1996; Steriade et al., 1996; Destexhe et al., 1999b). To allow local synchrony, corticothalamic feedback should instead recruit TC cells with dominant EPSPs and weak IPSPs (Destexhe, 2001). This switch may be facilitated by the cholinergic inputs to the thalamus, which have a dual effect on thalamic cells (depolarizing TC cells but hyperpolarizing RE cells; see McCormick, 1992). In the bursting mode, the corticothalamic feedback would recruit bursts in RE cells, thereby favouring dominant IPSPs in TC cells, leading to long-range synchrony. In activated states, cholinergic modulation would promote a mode that prevents bursts in RE cells (Hu et al., 1989), thereby favouring EPSPs on TC cells and the relay of information to the cerebral cortex. The local synchrony may be responsible for enhancing the flow of information through the thalamus by increasing the signal to noise ratio. Models to explore the relay mode of the corticothalamic system need to be developed and explored.

7.7 Summary

The focus of this chapter has been on models of spindle oscillations in thalamocortical assemblies. The experimental data have demonstrated that corticothalamic feedback may be important in organizing the large-scale synchrony of oscillations (Section 7.1.2; Contreras et al., 1996, 1997b). Analysis of multisite field potentials (Section 7.3.1) further characterized the remarkable degree of large-scale synchrony of oscillations during natural sleep and how this can be disrupted by cortical depression or by anesthesia (Contreras et al., 1997a; Destexhe et al., 1999a). Computational models of thalamocortical networks proposed a possible mechanism for large-scale thalamocortical synchrony (Section 7.2; Destexhe et al., 1998a). The models emphasized the importance of refractoriness in thalamic neurons, as observed in thalamic slices (Kim et al., 1995; Bal and McCormick, 1996; see Chapter 6), and suggested a mechanism by which thalamocortical loops interact with thalamic refractoriness to produce typical patterns of spindle oscillations observed *in vivo*. The same model therefore accounts for the known features of spindle oscillations *in vivo* and *in vitro*.

The model also made predictions about the influential role of the cortex in triggering and synchronizing oscillations generated in the thalamus through corticothalamic feedback projections (Section 7.3; Destexhe et al., 1999a). Intracortical mechanisms may be responsible for synchronizing oscillations over cortical distances of several millimeters through cortex-thalamus-cortex loops, even though the generators of the oscillations are in the thalamus. According to this view, the neocortex shapes and controls the spatial pattern of thalamic oscillations.

This view of the thalamocortical system in which the corticothalamic feedback has a determinant influence may also be used to undertand pathological situations. In the next chapter (Chapter 8), we consider similar corticothalamic interactions to investigate the genesis of some types of epileptic discharges.

Thalamocortical mechanisms for spike-and-wave epileptic seizures

Paroxysmal oscillations occur in thalamic and thalamocortical circuits during petit-mal or absence epilepsy. A large body of experimental data suggest that these seizures are generated by mechanisms that are similar to those that generate spindle oscillations. However, the results of *in vivo* and *in vitro* experiments do not appear to be compatible. These experimental results are reviewed in the context of the thalamocortical-loop framework introduced in Chapter 7. Computational models provide an integrated view for the genesis of these pathological oscillations and resolve the apparent contradictions in the interpretation of the experimental data.

8.1 Experimental characterization of paroxysmal oscillations

8.1.1 Experimental models of absence seizures

Absence seizures are particularly common in children. During a seizure, which is accompanied by a brief loss of consciousness that typically lasts a few seconds, high-amplitude spikes in the electroencephalogram (EEG) alternate with slow positive waves at a frequency of about 3 *Hz* (Fig. 8.1). These generalized 'spike-and-wave' seizures appear suddenly and invade the entire cerebral cortex nearly simultaneously. Similar spike-and-wave EEG patterns are observed in other neurological disorders as well as in experimental models of absence seizures in cats, rats, mice and monkeys.

The sudden appearance and global nature of spike-and-wave oscillations suggests that they are generated in a central structure projecting widely to the cerebral cortex. The involvement of the thalamus in spike-and-wave seizures was initially suggested by Jasper and Kershman (1941)

Thalamocortical Assemblies: Sleep Spindles, Slow Waves and Epileptic Discharges. Second Edition. Alain Destexhe and Terrence J. Sejnowski, Oxford University Press. © Oxford University Press 2023.
DOI: 10.1093/oso/9780198864998.003.0008

A

B

O1

FP2

O2

FP1

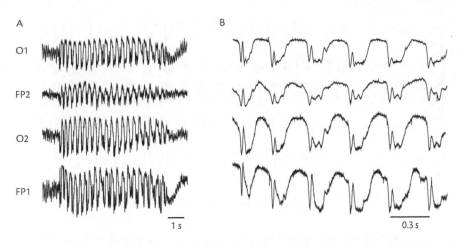

1 s

0.3 s

Fig. 8.1: Human absence seizure consisting of spike-and-wave oscillatory patterns at ~3 *Hz*. A. Electroencephalogram (EEG) recording of a human absence seizure, which consisted of approximately five seconds of 3 *Hz* oscillations. The oscillation appeared nearly simultaneously in all EEG leads. B. Higher temporal resolution reveals that each cycle of the oscillation consists of interleaved 'spikes' and 'waves'. Channels FP1 and FP2 measure the potential differences between frontal and parietal regions of the scalp whereas channels O1 and O2 correspond to the measures between occipital regions. Modified from Destexhe, 1992.

and is now supported by several findings. First, simultaneous thalamic and cortical recordings in humans during absence attacks directly demonstrated thalamic participation during the seizures (Fig. 8.2; Williams, 1953). Not only did the thalamus participate in the seizures, it typically started oscillating before spike-and-wave paroxysms appear in the EEG, suggesting that the seizure started in the thalamus and spread to the cortex.[1] Second, thalamic activation in human absence seizures was demonstrated with positron emission tomography (Prevett et al., 1995). Third, electrophysiological recordings in experimental models of spike-and-wave seizures support a strong thalamic involvement: cortical and thalamic cells fire prolonged discharges in phase with the 'spike' component, while the 'wave' is characterized by a silence in all cell types (Fig. 8.3; Pollen, 1964; Steriade, 1974; Avoli et al., 1983; McLachlan et al., 1984; Buzsaki et al., 1988; Inoue et al., 1993; Seidenbecher et al., 1998). Electrophysiological recordings also indicate that spindle oscillations, which are generated by thalamic circuits (Steriade et al., 1990, 1993; see Chapter 6), can be gradually transformed into spike-and-wave activity. Furthermore, all interventions that promote or antagonize spindles have the same effect on spike-and-wave activity (Fig. 8.4; Kostopoulos et al., 1981a, 1981b; McLachlan et al., 1984). Finally, spike-and-wave patterns disappear following thalamic lesions or after the thalamus is inactivated (Fig. 8.5; Pellegrini et al., 1979; Avoli and Gloor, 1981; Vergnes and Marescaux, 1992).

Although these results may suggest a thalamic origin for spike-and-wave seizures, the cortex is centrally involved. In cats, thalamic injections of high doses of GABA$_A$ antagonists, such as penicillin (Ralston and Ajmone-Marsan, 1956; Gloor et al., 1977) or bicuculline (Steriade

---

[1] In contrast, cortical mechanisms are implicated in experimental models of absence seizures induced in cats treated with penicillin (see Gloor and Fariello, 1988).

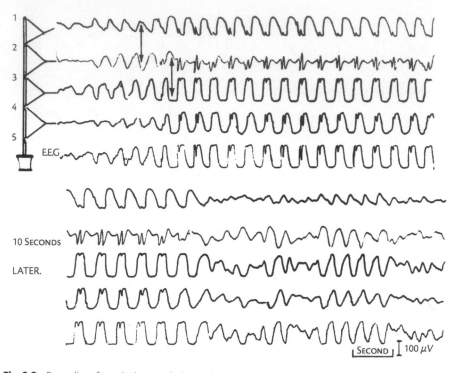

Fig. 8.2: Recordings from the human thalamus during absence seizures. A five-electrode assembly (shown on the left) was inserted stereotaxically such that the tip was in the thalamus (contacts 1–2) and the hilt near the scalp. These ~3 Hz spike-and-wave discharges were recorded from children during absence attacks. Here, the seizure began simultaneously in thalamus and neocortex, but in many cases the oscillations started in the thalamus, 1–2 seconds before any sign of spike-and-wave discharge at the cortical level. Figure reproduced from Williams, 1953.

and Contreras, 1998), led to highly synchronized 3–4 Hz oscillations, but no spike-and-wave discharge[2] (Fig. 8.6). On the other hand, injection of the same drugs into the cortex, without altering the thalamus, resulted in full-blown spike-and-wave seizures (Gloor et al., 1977; Steriade and Contreras, 1998). The threshold for epileptogenesis was extremely low in the cortex compared to the thalamus (Steriade and Contreras, 1998). Finally, it was shown that a diffuse application of a diluted solution of penicillin to the cortex results in spike-and-wave seizures although the thalamus was intact (Fig. 8.7; Gloor et al., 1977).

Although spike-and-wave activity typically requires the presence of the thalamus, in some experiments, a purely cortical spike-and-wave activity was observed. A slow type of spike-and-wave, with a less prominent 'spike' component, was recorded in the isolated cortex or athalamic preparations in cats (Marcus and Watson, 1966; Pellegrini et al., 1979; Steriade and Contreras, 1998). However, intracortical spike-and-wave appears not to occur in rats

---

[2] Similar observations have been made in rats, where thalamic injections of bicuculline led to hypersynchronous ~3 Hz oscillations (Castro-Alamancos, 1999).

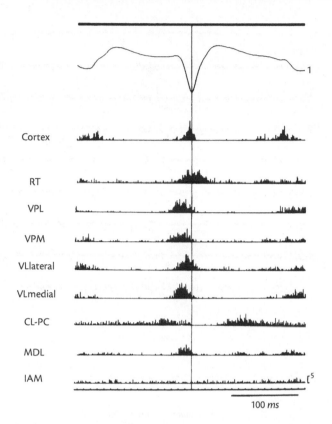

Fig. 8.3: Discharge patterns of thalamic and cortical neurons during spike-and-wave seizures. Cumulative unit activities from twenty spike-and-wave discharges from various locations from seven rats. The excitation from most of the recorded cells was coincident with the 'spike' component while the 'wave' was correlated with absence of firing. Thalamic relay cells (VPL, VPM, VL) tended to discharge before cortical (Cortex) and thalamic reticular (RT) cells. Other thalamic nuclei such as centrolateral and paracentral (CL-PC), mediodorsal lateral (MDL) and interanteromedial (IAM) were less correlated. Figure reproduced from Inoue et al., 1993.

(Vergnes and Marescaux, 1992) and has never been reported in neocortical slices. Further experiments should characterize the differences between intracortical and thalamocortical spike-and-wave activity. These experiments further confirm the importance of the cortex in generating seizure activity, even though the typical spike-and-wave EEG patterns of generalized seizures require both cortex and thalamus.

8.1.2 Involvement of GABA$_B$ receptors

A series of pharmacological results suggest that γ-aminobutyric acid$_B$ (GABA$_B$) receptors are critical in the genesis of spike-and-wave discharges in rats. GABA$_B$ agonists exacerbate seizures, while GABA$_B$ antagonists suppress them (Hosford et al., 1992; Snead, 1992; Puigcerver et al., 1996; Smith and Fisher, 1996). Antagonizing GABA$_B$ receptors in thalamic nuclei leads to the

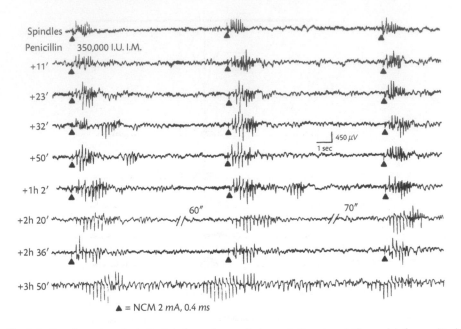

Fig. 8.4: Transformation of spindles in spike-and-wave discharges in the cat penicillin model of generalized epilepsy. Recordings from the middle suprasylvian gyrus of cats before and after intramuscular injection of penicillin (post-injection time indicated on the left). Arrows indicate stimuli to nucleus centralis medialis (NCM). Samples 2h20 and 3h50 are spontaneous activity. Figure reproduced from Kostopoulos et al., 1981.

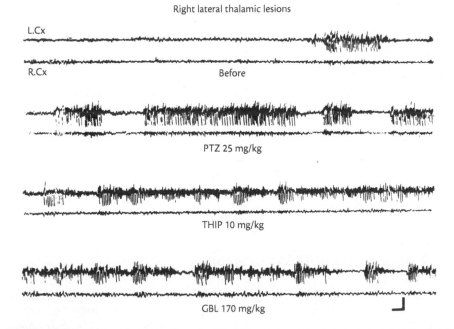

Fig. 8.5: Spike-and-wave discharges disappear following massive thalamic lesions. Left (L Cx) and right (R Cx) cortical EEG of a rat following callosal transection and a large lesion in the right thalamus. No spike-and-wave discharges appeared in the right cortex. Injection of various convulsants (A, before injection; B, PTZ; C, THIP; D, GBL) evoked spike-and-wave discharges from the intact hemisphere but never from the lesioned side. Figure reproduced from Vergnes and Marescaux, 1992.

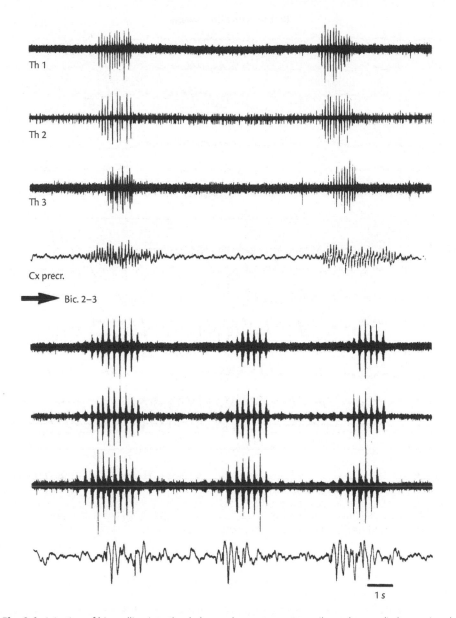

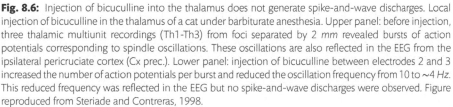

Fig. 8.6: Injection of bicuculline into the thalamus does not generate spike-and-wave discharges. Local injection of bicuculline in the thalamus of a cat under barbiturate anesthesia. Upper panel: before injection, three thalamic multiunit recordings (Th1-Th3) from foci separated by 2 *mm* revealed bursts of action potentials corresponding to spindle oscillations. These oscillations are also reflected in the EEG from the ipsilateral pericruciate cortex (Cx prec.). Lower panel: injection of bicuculline between electrodes 2 and 3 increased the number of action potentials per burst and reduced the oscillation frequency from 10 to ~4 Hz. This reduced frequency was reflected in the EEG but no spike-and-wave discharges were observed. Figure reproduced from Steriade and Contreras, 1998.

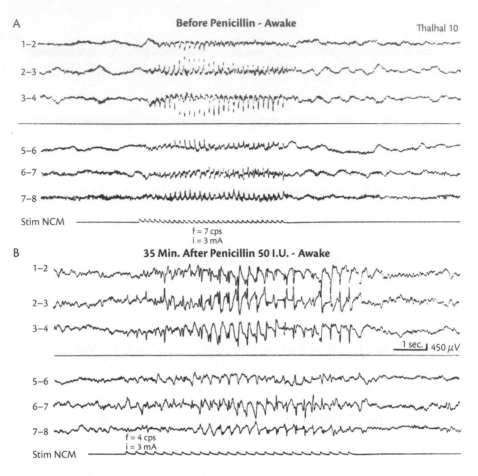

Fig. 8.7: Spike-and-wave discharges following diffuse application of penicillin to the cortex. A. Control: stimulation of nucleus centralis medialis (NCM; 7 *Hz*) induced a recruiting response in the cortex. B. After diffuse application of diluted solution of penicillin to the cortex (50 IU/hemisphere) in the same animal: 4 *Hz* stimulation of NCM elicited bilaterally synchronous spike-and-wave activity. Similar spike-and-wave discharges also occurred spontaneously. Figure reproduced from Gloor et al., 1977.

suppression of spike-and-wave seizures (Liu et al., 1992), further suggesting that the thalamus is critically involved.

A role for thalamic $GABA_B$ receptors was also shown using the anti-absence drug clonazepam in thalamic slices. This drug diminishes $GABA_B$-mediated inhibitory postsynaptic potentials (IPSPs) in thalamocortical (TC) cells (Fig. 8.8; Huguenard and Prince, 1994a), reducing their tendency to burst in synchrony. Clonazepam reinforces $GABA_A$ receptors in the RE nucleus (Fig. 8.9; Huguenard and Prince, 1994a; Gibbs et al., 1996; Hosford et al., 1997). Indeed, there is a diminished frequency of seizures following reinforcement of $GABA_A$ receptors in the RE nucleus in rats (Liu et al., 1991).

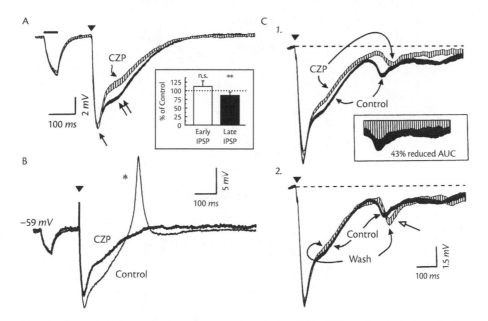

Fig. 8.8: The anti-absence drug clonazepam acts by reducing GABA$_B$ IPSPs in TC cells. Bath application of clonazepam reduces IPSPs in ventrobasal (VB) thalamic relay neurons of rats. A. Biphasic IPSP evoked by stimulation of the RE nucleus (triangle) consists in an initial GABA$_A$ component (arrow) and a late GABA$_B$ component (double arrow). Only the GABA$_B$-mediated component was reduced by bath application of clonazepam (see inset). B. Similar protocol in another cell, the biphasic IPSP evoked a rebound burst (control). Application of clonazepam resulted in depression of both GABA$_A$ and GABA$_B$-mediated components of the IPSP and abolished the rebound burst response. C. Secondary 'feedback' IPSPs (open arrow in C2) occur at a latency of $\sim$350 ms and are reduced by clonazepam (C1, 43% reduction—see inset) in a reversible fashion (C2). The reduction of secondary IPSPs was greater than the GABA$_A$ and GABA$_B$ IPSPs (43% compared to 7% and 15%, respectively). Figure reproduced from Huguenard and Prince, 1994a.

Perhaps the strongest evidence for the involvement of the thalamus was that, in ferret thalamic slices, spindle oscillations can be transformed into slower and more synchronized oscillations at $\sim$3 Hz following block of GABA$_A$ receptors (Fig. 8.10; von Krosigk et al., 1993). This behaviour is reminiscent of the gradual transformation of spindles into spike-and-wave discharges in cats found by Kostopoulos et al. (1981a, 1981b). Moreover, like spike-and-waves, the $\sim$3 Hz paroxysmal oscillations in thalamic slices are suppressed by GABA$_B$ receptor antagonists (Fig. 8.10; von Krosigk et al., 1993).

These experiments show that both GABA$_A$ and GABA$_B$ receptors are actively involved in generating spike-and-wave discharges, but the exact mechanisms are unclear. In the next sections, we review models for thalamic $\sim$3 Hz paroxysmal oscillations and for thalamocortical oscillations with spike-and-wave field potentials.

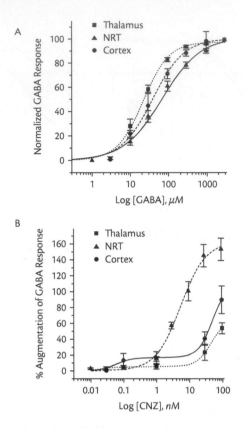

Fig. 8.9: The thalamic reticular nucleus is highly sensitive to clonazepam. Comparison of dose-response curves for extracellularly applied GABA and clonazepam in adult rat cortical and thalamic neurons *in vitro*. A. Comparison of dose-response curves for the currents evoked by application of GABA to thalamic relay, thalamic reticular (NRT) and cortical neurons (somatosensory cortex). The potency of GABA in these regions was thalamic>cortical>NRT. B. Dose-response curves for clonazepam in the same brain regions. The data indicate the % augmentation of GABA currents following application of clonazepam. The RE nucleus clearly showed the highest sensitivity to clonazepam. Figure reproduced from Gibbs et al., 1996.

8.2 Modelling the genesis of paroxysmal discharges in the thalamus

8.2.1 Early models

Spindle waves *in vitro* can be transformed into ~3 *Hz* oscillations by blocking $GABA_A$ receptors (Fig. 8.10) (von Krosigk et al., 1993). This oscillation is sensitive to blockade of $GABA_B$ receptors by saclofen (Fig. 8.10) and can also be suppressed by AMPA-receptor antagonists (von Krosigk et al., 1993). These *in vitro* experiments thus suggested that paroxysmal thalamic oscillations at ~3 *Hz* are mediated by $GABA_B$ IPSPs (RE→TC) and AMPA EPSPs (TC→RE). A model showing

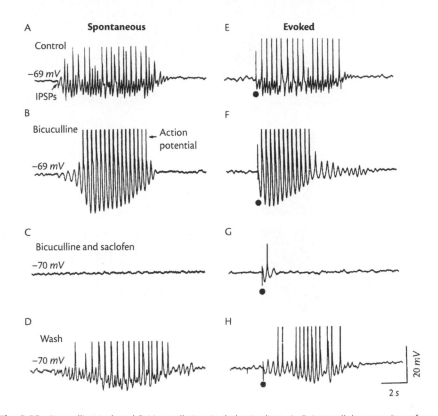

Fig. 8.10: Bicuculline-induced 3 *Hz* oscillation in thalamic slices. A, E. Intracellular recording of a relay neuron in the dorsal lateral geniculate nucleus in ferret thalamic slices, during spindle waves that occurred either spontaneously (A) or were evoked by electrical stimulation of cortical fibres (B). B, F. Blocking GABA$_A$ receptors by bath application of bicuculline slowed the oscillation to 2–4 *Hz* and markedly increased rebound burst activity. C, G. Further application of the GABA$_B$ receptor antagonist saclofen abolished the slowed oscillation. D, H. Recovery after washout of antagonists, showing that these effects are reversible. Figure reproduced from von Krosigk et al., 1993.

a similar transformation of spindle oscillations into ~3 *Hz* oscillations had been proposed (Destexhe et al., 1993b) and is reviewed here.

The starting point for modelling the *in vitro* experiments was the computational model of the simple TC-RE circuit investigated in Chapter 6 (Section 6.3; Destexhe et al., 1993b). The circuit (scheme in Fig. 8.11) displayed a transformation between 8–10 *Hz* spindle oscillations to ~3 *Hz* oscillations as the decay time of the GABAergic current was varied (Fig. 8.11). The results summarized in Table 8.1 show how the decay time of inhibition (β in Eq. 5.12) affects the frequency of the spindle oscillations, with slow decay corresponding to low frequencies. When the decay was adjusted to match experimental recordings of GABA$_B$-mediated currents (obtained from Otis et al., 1993), the circuit oscillated at around 3 *Hz* (Fig. 8.11B; Destexhe et al., 1993b).

Several mechanisms have been proposed to account for the effects of blocking of GABA$_A$ receptors in thalamic circuits (Wallenstein, 1994b; Wang et al., 1995; Destexhe et al., 1996;

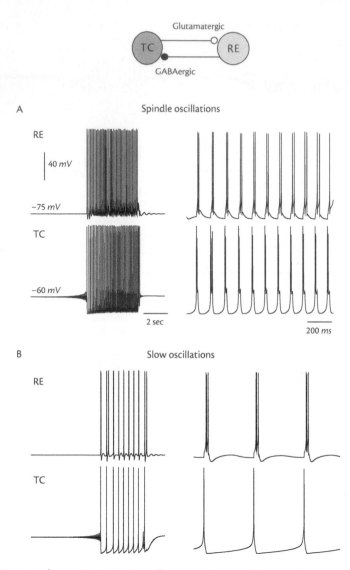

Fig. 8.11: Transition from 8–10 *Hz* spindle oscillations to ~3 *Hz* oscillations by slowing down the kinetics of GABAergic currents. A simple circuit was simulated consisting of an interconnected pair of TC and RE cells, in which connections were mediated by glutamatergic and GABAergic receptors (scheme on top). A. 8–10 *Hz* spindle oscillations in the circuit with fast GABAergic synapses. The left panel shows a detail of a few cycles within the oscillation at ten times higher resolution. The time course of synaptic currents was similar to AMPA receptors for TC→RE and to GABA$_A$ receptors for RE→TC (decay rate constant $\beta = 0.1\ ms^{-1}$). B. Slower oscillations for slow GABAergic synapses. The decay rate constant of the GABAergic synapse was $\beta = 0.003\ ms^{-1}$. Modified from Destexhe et al., 1993b.

Table 8.1 Frequency of thalamic oscillations for different values of the decay of inhibitory synaptic currents.

GABA decay β (ms^{-1})	Oscillation frequency (Hz)
0.1	10.0
0.03	9.0
0.01	5.5
0.005	3.6
0.003	2.5

The decay time of IPSPs was varied from fast decay, similar to $GABA_A$ IPSPs ($\beta = 0.1$ ms^{-1}) to slow decay, similar to $GABA_B$ IPSPs ($\beta = 0.003$ ms^{-1}). All other parameters were identical to those used in Fig. 6.22 in Chapter 6 (modified from Destexhe et al., 1993b).

Golomb et al., 1996). Wallenstein (1994b) explored the suggestion of Soltesz and Crunelli (1992) that disinhibition of interneurons projecting to TC cells with $GABA_B$ receptors result in stronger discharges when $GABA_A$ receptors are antagonized. A model including TC, RE and interneurons (Wallenstein, 1994b) reproduced the stronger discharges in TC cells following application of bicuculline. Although it is possible that this mechanism has some role in thalamically generated paroxysmal discharges, it does not account for experiments showing the decisive influence of the RE nucleus in preparations devoid of interneurons (Huguenard and Prince, 1994a, 1994b).

Increased synchrony and stronger discharges were also found in the model of Wang et al. (1995). In this case, the synchronous state coexisted with a desynchronized state of the network, which has never been observed experimentally. In other models, the cooperative activation proposed for $GABA_B$ receptors (Destexhe and Sejnowski, 1995; see Sections 5.2.6 and 5.3.4 in Chapter 5) produced robust synchronized oscillations and traveling waves at the network level (Destexhe et al., 1996a; Golomb et al., 1996; see Section 6.5 in Chapter 6), similar to those observed in thalamic slices (Fig. 6.29; Kim et al., 1995). The transformation of spindles to ~3 Hz paroxysmal oscillations was observed in the model following block of $GABA_A$ receptors (Section 8.2.4; Destexhe et al., 1996a). These modelling studies concluded that the transition from spindle to paroxysmal patterns could be achieved by cooperativity in $GABA_B$ responses. This mechanism is analysed in more detail below.

8.2.2 Models of the activation properties of $GABA_B$ responses

Paroxysmal discharges in the thalamus depend critically on $GABA_B$ responses (Section 8.1.2). The first step in exploring these mechanisms was to design appropriate biophysical models. In the biophysical model investigated in Section 5.3.4, the $GABA_B$ responses depended on the presynaptic pattern of activity; in particular, $GABA_B$ IPSPs were seen only for longer presynaptic bursts (Fig. 5.12B). This accounted for the specific stimulus dependence of $GABA_B$ responses, both in hippocampus and thalamus (Section 5.3 in Chapter 5; Destexhe and Sejnowski, 1995).

This property has consequences at the network level and is critical to explain the genesis of paroxysmal discharges in thalamic slices.

In Fig. 8.12A, the properties of GABAergic responses in thalamic slices were simulated using models of RE cells that included a low-threshold calcium current (see Chapter 4) and reciprocal GABA$_A$-mediated synaptic interactions within the RE nucleus (see Chapter 6). Under normal conditions, stimulation in the RE nucleus evokes biphasic IPSPs in TC cells, with a rather small GABA$_B$ component (Fig. 8.12B). An increase of intensity was mimicked in the model by increasing the number of RE cells discharging. The ratio between GABA$_A$ and GABA$_B$ IPSPs was independent of the intensity of stimulation in the model (Fig. 8.12D), as observed experimentally (Huguenard and Prince, 1994). However, this ratio could be changed by blocking GABA$_A$ receptors locally in the RE nucleus, leading to enhanced burst discharge in RE cells and a more prominent GABA$_B$ component in TC cells (Fig. 8.12C). This is consistent with the effect of clonazepam described in Figs. 8.8–8.9.

These simulations suggest that, because of the characteristic properties of GABA$_B$ receptors, the output of the RE nucleus onto TC cells is determined by the efficiency of GABA$_A$ interactions between RE cells. The presence of these GABA$_A$ synapses restricts the bursts of RE cells to few spikes, and leads to IPSPs dominated by GABA$_A$ in TC cells. However, when RE lateral inhibition is suppressed, model RE cells produced prolonged bursts and evoked IPSPs dominated by GABA$_B$ in TC cells. How this mechanism could generate 3 *Hz* oscillations in thalamic circuits is discussed below.

8.2.3   Genesis of ∼3 *Hz* oscillations in thalamic circuits

The bursting dynamics of reticularis neurons is a fundamental component of the mechanism for the genesis of ∼3 *Hz* paroxysmal oscillations in thalamic circuits. The lateral GABA$_A$-mediated inhibition within the RE nucleus controls the burst discharge of RE neurons and may explain the effects of clonazepam in thalamic slices (Fig. 8.12). This has consequences for the participation of the RE nucleus in oscillatory dynamics at low frequencies, as illustrated in Fig. 8.13. The output of the RE nucleus in a simple one-dimensional network of RE cells (Fig. 8.13A) was monitored in response to a synchronous AMPA-mediated excitation, mimicking a synchronized input from TC neurons. In control conditions, this AMPA-mediated EPSP occurred simultaneously in all RE cells and evoked relatively mild burst discharges of 3–7 spikes (Fig. 8.13A2, top). In contrast, the bursts became prolonged (15–25 spikes) after suppression of GABA$_A$ receptors between RE cells (Fig. 8.13A3, top). Because of the highly nonlinear properties of GABA$_B$ responses, these two types of burst discharges gave IPSP sequences with quite different GABA$_A$/GABA$_B$ ratio in TC cells (bottom traces in Fig. 8.13A2, A3).

In response to repetitive stimuli (simultaneous AMPA-mediated EPSPs in all RE cells), the RE network reliably entrained to the 10 *Hz* stimulation (Fig. 8.13B1). When GABA$_A$ receptors were suppressed, however, the RE neurons generated prolonged burst discharges to the same AMPA-mediated EPSP and it was therefore difficult for the network to follow the 10 *Hz* stimulus. However, the network was entrained to stimulation at 3.3 *Hz* (Fig. 8.13B2).

A second component in the mechanism responsible for generating paroxysmal oscillations in the thalamic network is the nature of RE-induced IPSPs on TC cells. The rebound response of TC cells to GABAergic synaptic stimulation was analysed in Chapter 6 (see Fig. 8.14). A model TC cell was stimulated with presynaptic bursts of action potentials acting on GABA$_A$ and GABA$_B$

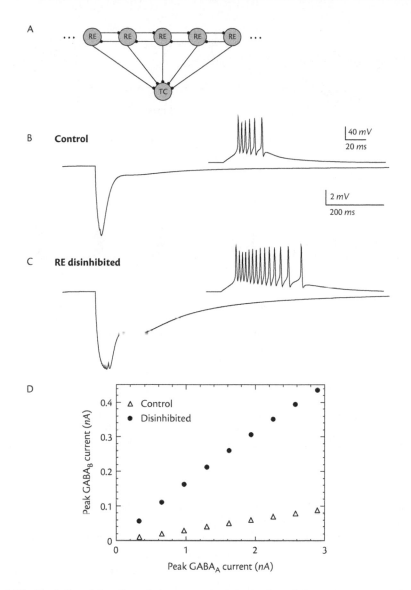

Fig. 8.12: Simulation of the effect of clonazepam in thalamic slices. GABA$_B$ response were enhanced in thalamocortical cells through disinhibition in the thalamic reticular nucleus. A. Connectivity: a simple network of RE cells was simulated with GABA$_A$ receptor-mediated synaptic interactions. All RE cells project to a single TC cell with synapses containing both GABA$_A$ and GABA$_B$ receptors. Model of the RE cells taken from Destexhe et al. (1994a). B. In control conditions, the bursts generated in RE cells by stimulation have 2–8 spikes (inset) and evoke in TC cells a GABA$_A$–dominated IPSP with a small GABA$_B$ component. C. When GABA$_A$ receptors are suppressed in RE, the bursts become much larger (inset) and evoke in TC cells a stronger GABA$_B$ component. D. Peak GABA$_A$ versus peak GABA$_B$ current for increasing numbers of RE cells stimulated. Reproduced from Destexhe and Sejnowski, 1995.

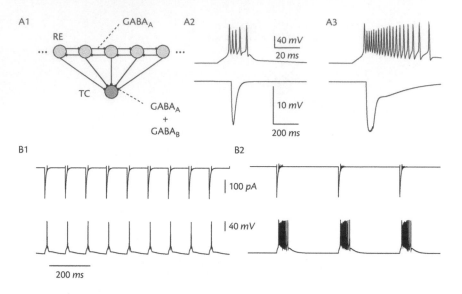

Fig. 8.13: Bursting dynamics of reticular thalamic neurons. A1. Schematic of a one-dimensional network of RE cells interconnected by $GABA_A$-mediated inhibitory synapses. Each RE cell contacted a TC cell through a mixture of $GABA_A$ and $GABA_B$ receptors. A2. Stimulation of each RE cell by a depolarizing current pulse under normal conditions produced a burst with only a few spikes, which evoked strong $GABA_A$ IPSPs in TC cells but only a weak $GABA_B$ component. The top trace: burst in a RE cell; bottom trace: IPSP evoked in the target TC cell (top trace at ten times higher time resolution than the bottom trace). A3. Same simulation without intra-RE inhibition. The stimulation evoked prolonged burst discharges in RE cells, which, in turn, evoked IPSPs in TC cells that contained a strong $GABA_B$ component (see Destexhe and Sejnowski, 1995). B1. Stimulation of a network of GABAergically connected RE cells at 10 Hz through AMPA receptors can recruit RE cells to burst at the same frequency. B2. Same simulation in the absence of intra-RE inhibition. The 3.3 Hz stimulation recruited RE cells in repetitive prolonged burst discharges (modified from Destexhe et al., 1996a).

receptors (Fig. 8.14) in response to the output of the thalamic reticular network (Fig. 8.12B–C and Fig. 8.13B). For brief bursts (three spikes at 360 Hz), mimicking the output of the RE nucleus in control conditions (Fig. 8.12B), the TC cell produced subharmonic bursting similar to spindle oscillations (Fig. 8.14A). When intra-RE $GABA_A$ conductances were suppressed, the RE cells produced prolonged discharges. When these prolonged discharges were used as the presynaptic signal (seventeen spikes at 360 Hz), mimicking the output of the disinhibited RE nucleus (Fig. 8.12C), strong $GABA_B$ IPSPs were activated. The TC cell could follow a stimulation at 3.3 Hz (Fig. 8.14B) and the TC-cell bursts were larger because I_T was more completely deinactivated by the $GABA_B$ IPSPs.

The block of $GABA_A$ receptors by application of bicuculline transforms spindle behaviour into a slower (3–4 Hz) oscillation of high synchrony and this slow oscillation is sensitive to $GABA_B$ receptors (Fig. 8.10; von Krosigk et al., 1993; Bal et al., 1995a; Kim et al., 1995). Combining the mechanisms discussed above (Figs. 8.12–8.14), the model suggests a cellular mechanism for generating paroxysmal slow oscillations in thalamic circuits. In a simple model (Fig. 8.15), the slow oscillations were a consequence of the highly nonlinear properties of $GABA_B$

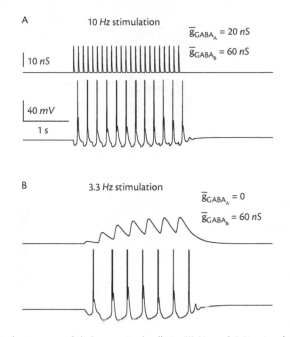

Fig. 8.14: Simulated responses of thalamocortical cells to 10 *Hz* and 3 *Hz* stimulation of GABA$_A$ and GABA$_B$ receptors. A. 10 *Hz* trains of three pulses at 360 *Hz*, occurring every 100 *ms*. The GABA$_A$ conductance is shown (top trace) along with the membrane potential (bottom). B. 3.3 *Hz* stimulation of only the GABA$_B$ receptors (the GABA$_B$ conductance is on top). In this case, seven successive bursts were simulated with an interburst period of 300 *ms*; each burst was a train of eighteen pulses at 360 *Hz*. In contrast to the 10 *Hz* stimulus that evoked a weak GABA$_B$ component in the IPSP (see Fig. 8.12), the 3 *Hz* stimulus evoked strong GABA$_B$-mediated currents and the TC cell was recruited in secure rebound bursts responses. These TC bursts were larger due to the more complete deinactivation of I_T provided by GABA$_B$ IPSPs. Modified from Destexhe et al., 1996a.

responses as described above. Following removal of GABA$_A$-mediated inhibition, the RE cells produced prolonged bursts that evoked strong GABA$_B$ currents in TC cells (as described in Fig. 8.13A3). These prolonged IPSPs evoked robust rebound bursts in TC cells (as in Fig. 8.14B), and TC bursts in turn elicited bursting in RE cells through EPSPs. This mutual TC-RE interactions entrained the system into a 3–4 *Hz* oscillation (Fig. 8.15B–C), with characteristics similar to those of bicuculline-induced paroxysmal oscillations in ferret thalamic slices (see Fig. 8.10). The oscillation was caused by the same 'TC-RE' mechanism described in Chapter 6 for spindle oscillations, but oscillations were slower and more synchronized in larger networks (see below).

In this model, the silent period of the oscillation was significantly longer for paroxysmal oscillations (26 ± 5 *s*) compared to spindle waves (20 ± 6 *s*). The length of the silent period was sensitive to the maximal conductance of I_h and the unbinding rate of calcium-binding protein from I_h channels (k_4 in Eq. 3.20). For a given set of parameters, the period of paroxysmal oscillations was always longer than the period of spindles. This was consistently observed experimentally (Kim et al., 1995) but the difference was less pronounced in the model. The relative duration of the silent period could be matched more closely with other

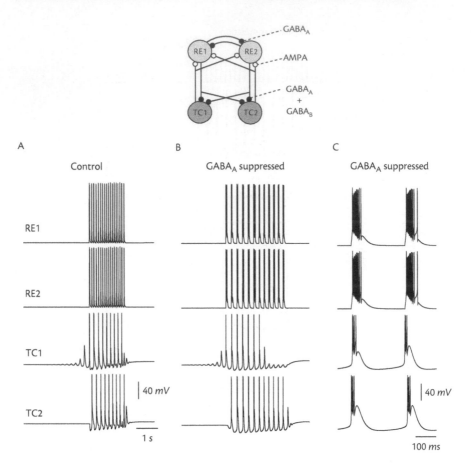

Fig. 8.15: Transformation from spindle to slow oscillations by blocking GABA$_A$ receptors in a four-neuron circuit of thalamocortical and thalamic reticular cells. The circuit consisted of two TC and two RE neurons interconnected using AMPA, GABA$_A$ and GABA$_B$ receptors (scheme on top). A. In control conditions, the circuit generated 8–10 Hz spindle oscillations (same simulation as described in Fig. 6.27). The first TC cell (TC1) started to oscillate, recruiting the two RE cells, which in turn recruited the second TC cell. B. Slow 3–4 Hz oscillation obtained after suppressing GABA$_A$ receptors, mimicking the effect of bicuculline. The mechanism for recruitment was the same as that for spindle oscillations in A, but the oscillations were more synchronized, with a lower frequency and had a 15% longer silent period. The burst discharges were prolonged due to the loss of lateral inhibition in the RE. C. First bursts of the same cells shown at 10 times higher time resolution. TC cells were in phase and phase-led RE cells by a few milliseconds. Modified from Destexhe et al., 1996a.

parameter values, but the results were highly variable and not robust to small changes in the parameters. This sensitivity may be reflected in the variability that has been reported between experiments (Kim et al., 1995). The longer interspindle periods was a consequence of the stronger upregulation of I_h, as also reflected in the higher amplitude of the ADP during paroxysmal oscillations (see below).

8.2.4 ~3 *Hz* paroxysmal oscillations in thalamic networks

Network models with fifty TC and fifty RE cells were investigated that were similar to the models described in Section 6.5.5 of Chapter 6. As in the four-cell circuit, TC and RE cells oscillated synchronously at a frequency around 3 *Hz* (Fig. 8.16). In individual cells, the oscillation showed waxing-and-waning properties similar to those described in Section 6.5.5 for spindle oscillations. The paroxysmal oscillations supported propagating-wave phenomena, similar to spindle waves, when the connectivity between the TC and RE layers was topographically organized (Fig. 8.16, scheme). In this case, the synchrony was significantly higher inside the oscillatory sequence, as shown by the amplitude of averaged activities (Fig. 8.16A; compare to spindles in Fig. 6.30) and by comparing individual membrane potential traces across the network (Fig. 8.16B).

This increase of synchrony was confirmed by cross-correlation analysis. The values of cross-correlations were significantly larger for paroxysmal oscillations compared to spindles (Fig. 8.17). In addition, there was a progressive phase shift that increased with the distance between sites, reflected in the time shift of the maximum correlation in Fig. 8.17A–B. The higher synchrony for paroxysmal oscillations and the progressive phase shifts have been observed in ferret thalamic

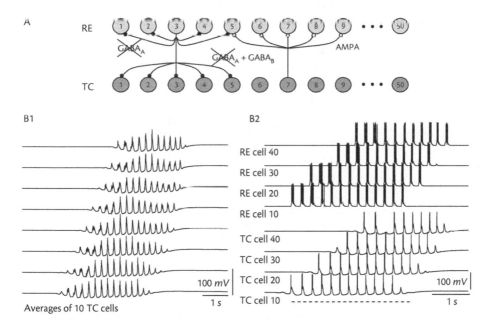

Fig. 8.16: Propagation of paroxysmal oscillations in a network of thalamocortical and thalamic reticular cells. A network with fifty TC and fifty RE cells with localized axonal projections was used, as described in Fig. 6.30A. A. Scheme of the connectivity in the network. Each axonal projection ramified and contacted all neurons within a diameter of about 10% of the size of the network (eleven cells were contacted). B1. Average membrane potentials of slow paroxysmal oscillations in the absence of GABA$_A$ receptors. Averaged membrane potentials were computed from ten neighbouring TC cells taken at eight equally spaced sites. B. Membrane potential of four TC (four bottom traces) and four RE cells (upper traces) in the same simulation. Modified from Destexhe et al., 1996a.

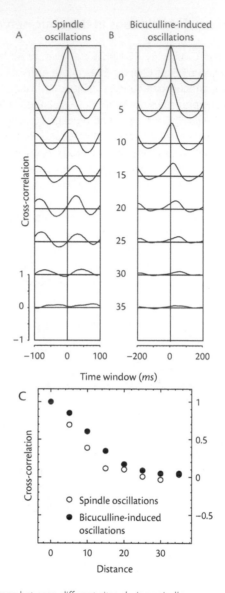

Fig. 8.17: Cross-correlations between different sites during spindle waves and paroxysmal oscillations. Correlations were evaluated between one reference site (TC cell 10) and sites taken at progressively larger distances (units of intercellular distances in the network; intersite distance shown for each graph). A. Cross-correlations during spindle oscillations. B. Cross-correlations during paroxysmal oscillations. In both cases, the correlation was calculated between averages of TC cells (same simulation and averaging procedures as in Fig 6.30). C. Space correlation calculated as the cross-correlation at time zero, plotted as a function of the distance between sites. For both oscillations, the correlations decayed with distance, but paroxysmal oscillations always had higher correlation than spindles and were therefore more synchronized. Reproduced from Destexhe et al., 1996a.

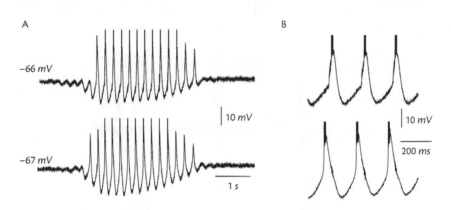

Fig. 8.18: Dual intracellular recordings of TC cells reveal phase shifts during bicuculline-induced oscillations. Simultaneous recordings from two neighbouring lateral geniculate nucleus cells in ferret thalamic slices. A. Dual recordings of the slow oscillation obtained after addition of bicuculline. Both cell types generated prolonged discharge patterns. B. Enlargement of A, showing that the two cells did not burst perfectly in phase, but one cell tended to lead the other by 10–20 ms. Courtesy of T. Bal and D. A. McCormick (unpublished data).

slices (Kim et al., 1995). An example of phase shift during the slow oscillations is shown in Fig. 8.18. Correlations showed a progressive decay with distance, with a rate of decay that was similar for both types of oscillations (Fig. 8.17C), as expected for propagating oscillations.

For all parameter values and patterns of connectivity tested, the propagation velocity for paroxysmal oscillations was slower than for spindle oscillations, as consistently observed in ferret thalamic slices (Kim et al., 1995). The average propagation delay between two neighbouring neurons was approximately 19.4 ms for spindles and 55 ms for paroxysmal oscillations. Assuming that the fan-out of the projections between the TC and RE cells in the model (eleven neurons) is equivalent to 200 μm in the slice, the velocity in the model is about 1.03 mm/s for spindles, which is within the range 0.28 to 1.21 mm/s observed experimentally, and the velocity for paroxysmal oscillations was 0.36 mm/s, compared to the range 0.22 to 0.95 mm/s observed experimentally. Although these values approximately match, the model systematically produced a larger difference between the spindle velocity and the paroxysmal oscillation velocity than observed experimentally, presumably due to the small size of the network and the homogeneous patterns of connectivity (see analysis by Golomb et al., 1996).

The stronger bursts during paroxysmal oscillations allow more calcium entry and led to a more pronounced upregulation of I_h. This has been observed in intracellular recordings from TC cells in ferret thalamic slices, which revealed a prominent ADP occurring in the inter-spindle silent period (Fig. 8.19A–top trace). This ADP was more prominent during paroxysmal oscillations (Fig. 8.19A–bottom trace). The same phenomenon occurred in the model (Fig. 8.19B) and was due to stronger activation of I_T underlying prolonged burst discharges, leading to more calcium entry and a more pronounced upregulation of I_h.

8.2.5 Spatiotemporal properties of thalamic paroxysmal oscillations

The spatiotemporal patterns shown for spindle oscillations in Chapter 6 (Fig. 6.31) dramatically change when GABA$_A$-mediated currents are suppressed (Fig. 8.20). Under this condition, the

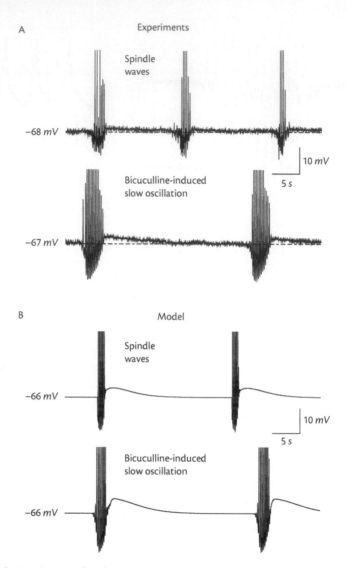

Fig. 8.19: Afterdepolarization (ADP) in TC cells during spindle and paroxysmal oscillations. A. Comparison of spindle and paroxysmal oscillations in an intracellularly recorded TC cell in ferret thalamic slices. Before (top trace) and after (bottom trace) application of bicuculline, the silent period between oscillations showed an ADP, which was more prominent during paroxysmal oscillations. B. Same oscillations simulated using a network of fifty TC and fifty RE cells (from Figs. 6.30 and 8.16), in normal conditions (top trace) and after block of $GABA_A$ receptors (bottom trace). The ADP occurred in all TC cells of the network due to upregulation of I_h. A more pronounced enhancement of I_h was responsible for the larger ADP in paroxysmal oscillations. Sodium spikes were truncated in all traces. Modified from Destexhe et al., 1996a.

waxing and the waning of paroxysmal oscillations both propagated with a lower velocity compared to spindles. The bursts were not organized in distinct clusters but instead involved a much larger population of cells with more prolonged discharges (compare Fig. 6.31 with Fig. 8.20). In this sense, the oscillation were more synchronized. However, bursting discharges did not appear simultaneously in the whole network but rather started at a focus and invaded the network progressively, leading to curved patterns in the space-time diagram (Fig. 8.20). This phase shift was also observed in other simulations (not shown) and had no preferential direction.

8.3 Model of spike-and-wave oscillations in the thalamocortical system

The experiments reviewed in Section 8.1 show that the thalamus is essential for generating 3 *Hz* spike-and-wave seizures, and indeed thalamic slices display paroxysmal oscillations at ~3 *Hz* following application of convulsants (GABA$_A$ antagonists), which mechanism was analysed in detail in Section 8.2. However, other experiments suggest that this thalamic 3 *Hz* oscillation is a phenomenon distinct from spike-and-wave seizures. Injections of GABA$_A$ antagonists into the thalamus with an intact cortex failed to generate spike-and-wave discharges (Fig. 8.6; Ralston and Ajmone-Marsan, 1956; Gloor et al., 1977; Steriade and Contreras, 1998). In these experiments, suppressing thalamic GABA$_A$ receptors led to 'slow spindles' around 4 *Hz*, quite different from spike-and-wave oscillations. On the other hand, spike-and-wave discharges were obtained experimentally by diffuse application of GABA$_A$ antagonists to the cortex (Fig. 8.7; Gloor et al., 1977). Thus, *in vivo* experiments indicate that spindles can be transformed into spike-and-wave activity by altering only cortical inhibition, without any change in the thalamus. Possible mechanisms to explain these observations and relate them to the 3 *Hz* thalamic oscillation were investigated using a thalamocortical model (Destexhe, 1998).

8.3.1 The thalamocortical model

The genesis of spike-and-wave oscillations was investigated in a thalamocortical network model. The network model was similar to that introduced in Chapter 7 and included layers of thalamic TC and RE cells, and a simplified version of the deep layers of the cortex, including pyramidal (PY) cells and interneurons (IN).

Postsynaptic currents mediated by glutamate AMPA and NMDA receptors, as well as GABAergic GABA$_A$ and GABA$_B$ receptors, were simulated using kinetic models of postsynaptic receptors (see Chapter 5). The models of slow GABA$_B$ receptor-mediated inhibition required a more complex scheme to capture the nonlinear properties of this type of interaction (Chapter 5; Destexhe and Sejnowski, 1995). The model of GABA$_B$ was the same used for the thalamic networks (see Chapter 6; Destexhe et al., 1996a). Intrinsic calcium- and voltage-dependent currents were modelled using kinetic models of the Hodgkin-Huxley (1952) type, as described in the thalamocortical model for spindles (Chapter 7; Destexhe et al., 1998a).

Extracellular field potentials were calculated from postsynaptic currents in single-compartment models, according to the model of Nunez (1981):

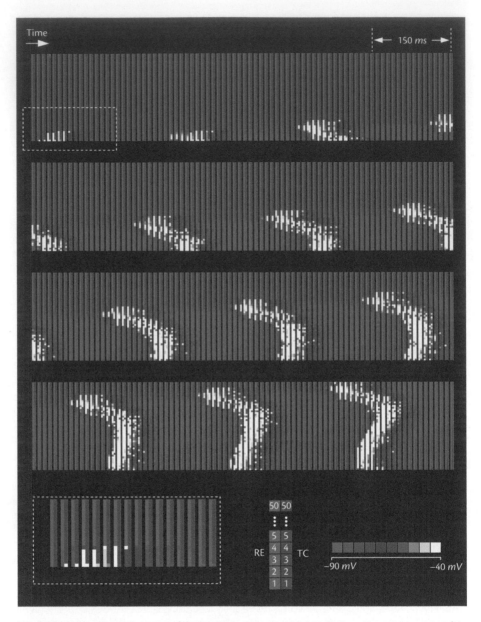

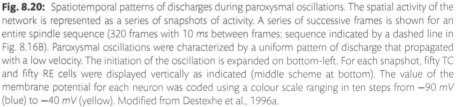

Fig. 8.20: Spatiotemporal patterns of discharges during paroxysmal oscillations. The spatial activity of the network is represented as a series of snapshots of activity. A series of successive frames is shown for an entire spindle sequence (320 frames with 10 *ms* between frames; sequence indicated by a dashed line in Fig. 8.16B). Paroxysmal oscillations were characterized by a uniform pattern of discharge that propagated with a low velocity. The initiation of the oscillation is expanded on bottom-left. For each snapshot, fifty TC and fifty RE cells were displayed vertically as indicated (middle scheme at bottom). The value of the membrane potential for each neuron was coded using a colour scale ranging in ten steps from −90 *mV* (blue) to −40 *mV* (yellow). Modified from Destexhe et al., 1996a.

$$V_{ext} = \frac{R_e}{4\pi} \sum_j \frac{I_j}{r_j}$$ (8.1)

where V_{ext} is the electrical potential at a given extracellular site, $R_e = 230\ \Omega$cm is the extracellular resistivity, I_j are the postsynaptic currents and r_j is the distance between the site of generation of I_j and the extracellular site.

Field potentials were calculated from a single cell receiving 200 simulated synapses (100 excitatory synapses had AMPA and NMDA receptor types, and 100 inhibitory synapses had GABA$_A$ and GABA$_B$ receptors; see scheme in Fig. 8.21B). In this case, trains of presynaptic action potentials were generated individually for each synapse. To avoid possible artefacts due to the coincident timing of action potentials at different synapses, a random jitter of $\pm 1\ ms$ was added to the timing of each presynaptic action potential.

Field potentials were calculated from network activity including only the contributions of cortical pyramidal cells. The field potential at a given extracellular site, calculated from the post-synaptic currents from neurons equally spaced by 20 μm along one dimension, was given by:

$$V_{ext} = \frac{R_e}{4\pi} \sum_{i,k} \frac{I_{syn}^{ki}}{r_i}$$ (8.2)

where r_i is the distance between each PY cell and the extracellular site.

In some cases, the contribution of the voltage-dependent current I_M to field potentials was evaluated according to the relation:

$$V_{ext} = \frac{R_e}{4\pi} \sum_i \frac{(I_M^i + \sum_k I_{syn}^{ki})}{r_i}$$ (8.3)

where I_M^i is the voltage-dependent K$^+$ current responsible for adaptation of repetitive firing in the ith PY cells.

The mechanism underlying spike-and-wave oscillations will be explained in three stages: (a) The nonlinear activation properties of GABA$_B$ responses can lead to the generation of spike-and-wave field potentials; (b) intact thalamic circuits can be forced into a ~3 Hz oscillation by corticothalamic feedback; (c) the combination of these two factors can generate ~3 Hz oscillations with spike-and-wave field potentials in thalamocortical networks. These points are considered in turn.

8.3.2 Possible role of GABA$_B$ receptors in generating spike-and-wave field potentials

We show first that the nonlinear stimulus dependence of GABA$_B$ currents is fundamental in explaining the genesis of spike-and-wave field potential patterns, as illustrated in Fig. 8.21A. An isolated presynaptic spike does not evoke detectable GABA$_B$ current (Fig. 8.21A1), consistent with the absence of GABA$_B$-mediated miniature events (Otis and Mody, 1992; Thompson and Gahwiler, 1992; Thompson, 1994). However, a burst of 5–10 high-frequency spikes evokes a powerful GABA$_B$ responses (Fig. 8.21A2). This is consistent with the observation that GABA$_B$

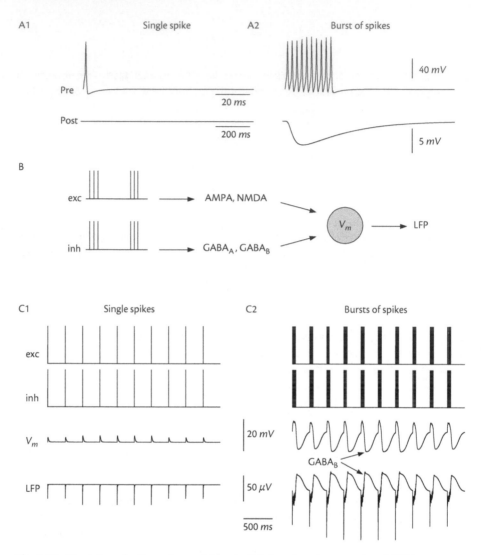

Fig. 8.21: Simulation of spike-and-wave field potentials based on the properties of GABA$_B$ receptors. A. Nonlinear activation properties of GABA$_B$ receptors. Because the binding of four G-proteins is needed to activate the K$^+$ channels associated to GABA$_B$ receptors, the GABA$_B$-mediated inhibitory responses depend nonlinearly on the number of presynaptic action potentials. A1. With a single presynaptic spike, no GABA$_B$ response was detectable. A2. A burst of presynaptic spikes led to sufficient accumulation of G-proteins to generate the slow IPSP. B. Scheme for the model of local field potentials. Excitatory and inhibitory presynaptic trains of action potentials stimulated AMPA, NMDA, GABA$_A$ and GABA$_B$ postsynaptic receptors at 100 synapses of each type on a single compartment model and used to calculate the extracellular field potential at a distance of 5 μm from the simulated neuron. C. Field potentials generated by single spikes and bursts of spikes. With single spikes (C1), the mixed EPSP/IPSP sequence led to negative deflections in the field potentials. With bursts of spikes (C2), fast spiky components alternate with slow positive deflections, similar to spike-and-wave patterns. These slow positive waves were due to the activation of GABA$_B$-mediated currents (arrows). Conductance values were 4 nS, 1 nS, 1.5 nS and 4 nS for individual AMPA, NMDA, GABA$_A$ and GABA$_B$ synapses, respectively. Reproduced from Destexhe, 1998.

responses require a high-intensity stimulus (Dutar and Nicoll, 1988; Davies et al., 1990) and that bursts of high-frequency action potentials evoke $GABA_B$ currents (Huguenard and Prince, 1994b; Kim et al., 1997). In the model, this occurred because a sufficiently high level of G-protein must accumulate before a significant K^+ current is generated (see analysis in Chapter 5, Section 5.3.4).

A simple model for generating extracellular field potentials was used to investigate how the activation properties of $GABA_B$ currents could affect the genesis of spike-and-wave discharges. A passive single-compartment model was simulated with postsynaptic currents generated by 100 excitatory synapses (AMPA and NMDA receptors) and 100 inhibitory synapses ($GABA_A$ and $GABA_B$ receptors; see scheme in Fig. 8.21B). The field potentials generated by this model for different stimulus conditions are shown in Fig. 8.21C. With presynaptic trains consisting of single spikes, the voltage showed mixed EPSP/IPSP sequences and the field potential was dominated by negative deflections (Fig. 8.21C1). In contrast, bursts of high-frequency presynaptic spikes produced mixed EPSP/IPSPs followed by large $GABA_B$-mediated IPSPs in the cell (Fig. 8.21C2). In the latter case, the fast EPSP/IPSPs generated spiky field potentials, followed by a slow positive wave due to $GABA_B$ currents. This simple model shows that synchronous high-frequency discharges of excitatory and inhibitory cells in the presence of $GABA_B$ receptors is sufficient to generate field potentials waveforms resembling interleaved 'spikes' and 'waves'.

Fig. 8.22A shows how the morphology of simulated spike-and-wave complexes depended on the detailed properties of the synapses. When excitatory synapses discharged earlier than inhibitory synapses (2 ms and 5 ms latency), the 'spike' component was enhanced. The 'spike' and 'wave' components were also influenced by synaptic conductances: AMPA and NMDA conductances affected primarily the negative peak of the 'spike' component (Fig. 8.22B, top trace), but the positive peak was influenced mostly by $GABA_A$ conductances (Fig. 8.22B, middle trace). The $GABA_B$ conductances component had only minor effects on the 'spike', but significantly affected the 'wave' component (Fig. 8.22B, bottom trace). Other slow voltage-dependent K^+ currents had the same effect (see below).

Thus, highly synchronized prolonged discharge patterns that activate strong K^+ currents in cortical neurons, as invariably found experimentally during seizures, can by itself generate spike-and-wave field potentials, but only if the mechanism that recruits the K^+ currents is nonlinearly sensitive to the strengths of the discharges, as found for the activation of $GABA_B$ receptors: the 'spike' corresponds to the mixed fast EPSPs/IPSPs during the discharge, and the 'wave' is generated by the slow K^+ currents.

8.3.3 Intact thalamic circuits can be forced into ~3 Hz oscillations due to $GABA_B$ receptors

To investigate how spike-and-wave field potentials can be generated in the thalamocortical system, the first step is to examine the behaviour of thalamic circuits and how they are controlled by the cortex. Isolated thalamic networks have a propensity to generate 7–14 Hz spindle oscillations (Steriade et al., 1993; von Krosigk et al., 1993; see Fig. 6.5). Although these oscillations originate in the thalamus, they can be triggered by the neocortex (Steriade et al., 1972; Roy et al., 1984; Contreras and Steriade, 1996) through the corticothalamic feedback, which also organizes the spatial and temporal patterns of thalamic oscillations (Contreras et al., 1996) (Chapter 7).

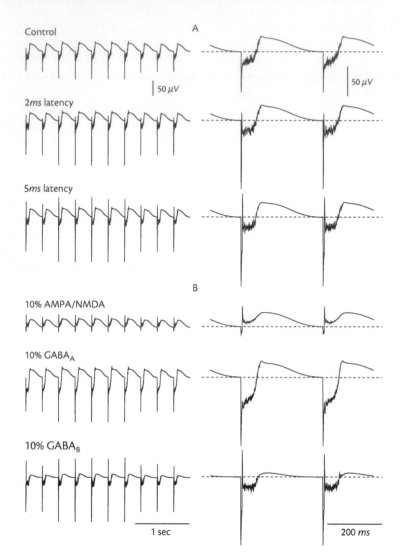

Fig. 8.22: Factors determining the morphology of simulated spike-and-wave complexes. A. Effect of a latency between the firing of excitatory and inhibitory synapses. Control: simulation identical to Fig. 8.21C2. If excitatory synapses discharged earlier than inhibitory synapses (2 ms and 5 ms latency), the 'spike' component was enhanced. B. Effect of synaptic receptor types. The '5 ms latency' simulation from A was repeated here with different values for synaptic conductances. A 90% reduction of either AMPA/NMDA conductances (top trace), GABA$_A$ conductances (middle trace) or GABA$_B$ conductances (bottom trace) changed the morphology of the spike-and-wave patterns. The right column shows the third set of complexes at higher resolution. Reproduced from Destexhe, 1998.

The corticothalamic EPSPs on RE cells had to be more powerful than those on TC cells in order to reproduce the cortical control of thalamic spindles in computational models (Chapter 7; Destexhe et al., 1998a). Under these conditions, excitation of corticothalamic cells led to mixed EPSPs and IPSPs in TC cells, in which the IPSP was dominant, consistent with experimental observations (Burke and Sefton, 1966; Deschênes and Hu, 1990). If cortical EPSPs and IPSPs from RE cells were of comparable conductance, cortical feedback could not evoke oscillations in the thalamic circuit due to shunting effects between EPSPs and IPSPs (Fig. 7.12; Destexhe et al., 1998a). The most likely reason for 'IPSP dominance' in TC cells is that RE cells are extremely sensitive to cortical EPSPs (Contreras et al., 1993), probably due to a powerful T-current in dendrites (Section 4.3 in Chapter 4; Destexhe et al., 1996b). In addition, cortical synapses contact only the distal dendrites of TC cells (Liu et al., 1995) and are probably attenuated. Taken together, these experimental observations and modelling results suggest that corticothalamic feedback operates primarily by eliciting bursts in RE cells, which in turn evoke powerful IPSPs on TC cells that largely overwhelm the direct cortical EPSPs.

The effects of corticothalamic feedback on the thalamic circuit are shown in Fig. 8.23A. Simulated cortical EPSPs evoked bursts in RE cells (Fig. 8.23B, arrow), which recruited TC cells through IPSPs, and triggered a $\sim$10 Hz oscillation in the circuit. During the oscillation, TC cells rebounded once every two cycles following GABA$_A$-mediated IPSPs and RE cells only discharged a few spikes, evoking GABA$_A$-mediated IPSPs in TC cells with no significant GABA$_B$ currents (Fig. 8.23B). These features are typical of spindle oscillations (Steriade et al., 1993; von Krosigk et al., 1993; see Fig. 6.5).

With high intensity 3 Hz stimulation (fourteen spikes every 333 ms; Fig. 8.23C), however, all cell types discharged in synchrony at $\sim$3 Hz. In contrast, repetitive stimulation at 3 Hz with low intensity produced spindle oscillations (Fig. 8.23D) similar to Fig. 8.23A. Strong-intensity stimulation at 10 Hz led to quiescence in TC cells (Fig. 8.23E), due to sustained GABA$_B$ currents, similar to a previous analysis (see Fig. 12 in Lytton et al., 1997).

In conclusion, these modelling results suggest that strong corticothalamic feedback at 3 Hz can produce a different type of oscillation in intact thalamic circuits (Destexhe, 1998). Cortical EPSPs can recruit large bursts in RE cells (Fig. 8.23C, arrows), making it possible to activate GABA$_B$ responses (see Fig. 8.21A). As a consequence, the TC cells were 'clamped' at hyperpolarized levels by GABA$_B$ IPSPs during $\sim$300 ms before they could rebound. The nonlinear properties of GABA$_B$ responses could therefore account for the coexistence between two types of oscillations in the same circuit: weak 3 Hz corticothalamic feedback produced $\sim$10 Hz spindle oscillations, while strong 3 Hz feedback entrained the intact circuit to the same frequency due to the nonlinear activation properties of intrathalamic GABA$_B$ responses.

Recent experiments performed in ferret thalamic slices have verified this prediction of the model (Fig. 8.24; Bal et al., 2000). In this experiment, the activity of thalamic relay cells was used to trigger the electrical stimulation of corticothalamic fibres (Fig. 8.24A). With this feedback, the activity in the slice depended on the stimulus strength. For mild feedback, the slice generated normal spindle oscillations (Fig. 8.24B). However, for strong stimulation of corticothalamic fibres, the activity switched to slow synchronized oscillations at $\sim$3 Hz (Fig. 8.24C). This behaviour was dependent on GABA$_B$ receptors, as shown by its sensitivity to GABA$_B$ antagonists (Bal et al., 2000). These results suggest that strong corticothalamic feedback can force physiologically intact thalamic circuits to oscillate synchronously at 3 Hz. The same results were also obtained in another study (Blumenfeld and McCormick, 2000).

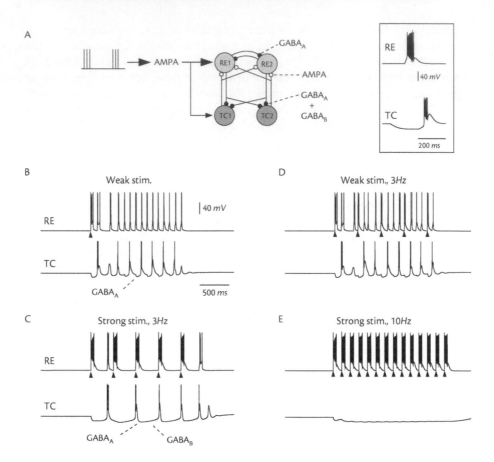

Fig. 8.23: Corticothalamic feedback can force thalamic circuits into ~3 Hz oscillations due to the properties of GABA_B receptors. A. Scheme of connectivity and receptor types in a circuit of thalamocortical (TC) and thalamic reticular (RE) neurons. Corticothalamic feedback was simulated through AMPA-mediated synaptic inputs (shown on the left of the connectivity diagram; total conductance of 1.2 μS to RE cells and 0.01 μS to TC cells). The inset shows simulated burst responses of TC and RE cells following current injection (pulse of 0.3 nA during 10 ms for RE and −0.1 nA during 200 ms for TC). B. A single stimulation of corticothalamic feedback (arrow) entrained the circuit into a 10 Hz mode similar to spindle oscillations. C. Following 3 Hz stimulation with strong intensity (arrows; fourteen spikes/stimulus), RE cells were recruited into large bursts, which evoked IPSPs onto TC cells dominated by GABA_B-mediated inhibition. In this case, the circuit could be entrained into a different oscillatory mode, with all cells firing in synchrony. D. Weak stimulation at 3 Hz (arrows) entrained the circuit into spindle oscillations (identical intensity as in B). E. Strong stimulation at 10 Hz (arrows) led to quiescent TC cells due to sustained GABA_B current (identical intensity as in C). Reproduced from Destexhe, 1998.

8.3.4 Suppression of intrathalamic GABA_A inhibition does not generate spike-and-wave

A thalamocortical network was simulated to explore the contribution of intrathalamic GABA_A inhibition to 3 Hz oscillations. The network included layers of thalamic TC and RE cells, and a simplified representation of the deep layers of the cortex with pyramidal cells and

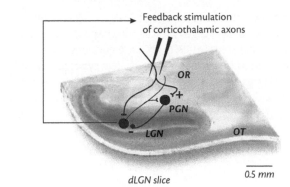

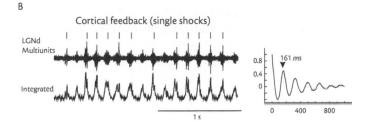

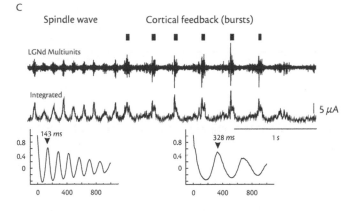

Fig. 8.24: Control of thalamic oscillations by corticothalamic feedback in ferret thalamic slices. A. Schematic thalamic slice. Corticothalamic axons run in the optic radiation (OR) and connect thalamocortical cell in the LGN layers and GABAergic interneurons in the perigeniculate nucleus (PGN). Bipolar stimulating electrodes were placed in the OR (OT: optic tract). B. Weak (single shock) stimulation at a latency of 20 *ms* after the detection of multiunit bursts activity (upper trace). Lower trace: smooth integration of the multiunit signal. C. A 7 *Hz* control spindle is robustly slowed to 3 *Hz* oscillation by the feedback stimulation (five shocks; 100 *Hz*; 20 *ms* delay). Modified from Bal et al., 2000.

interneurons (Fig. 8.25A). In control conditions (Fig. 8.25B), the network generated synchronized spindle oscillations and cellular discharges with all cell types oscillating in phase, as observed experimentally (Contreras and Steriade, 1996) and in simulations (Destexhe et al., 1998a; see Chapter 7). TC cells discharged on average once every two cycles following $GABA_A$-mediated IPSPs, while all other cell types discharged on roughly every cycle at ~10 Hz, consistent with the typical features of spindle oscillations observed intracellularly (Steriade et al., 1990; von Krosigk et al., 1993; see Fig. 6.5). The simulated field potentials displayed successive negative deflections at ~10 Hz (Fig. 8.25B), in agreement with the pattern of field potentials recorded during spindle oscillations (Steriade et al., 1990). Consistent with the analysis of Fig. 8.21C1, this pattern of field potentials was generated by the limited discharge in PY cells, which fired roughly one spike per oscillation cycle.

In thalamic slices, when $GABA_A$ receptors are blocked spindles are transformed into paroxysmal oscillations at ~3 Hz (Fig. 8.10; von Krosigk et al., 1993; see model in Sections 8.2.3–8.2.4). In simulations of the thalamocortical network where $GABA_A$ receptors were suppressed in thalamic cells (leaving cortical inhibition intact), spindle oscillations were also transformed into slower 3–5 Hz oscillations at (Fig. 8.25C). There was an increase in synchrony, and the TC cells fired on every cycle of the oscillation. RE cells generated prolonged burst discharges, leading to $GABA_B$-mediated IPSPs in TC cells, and consequently a slow oscillation frequency. However, this oscillation did not generate spike-and-wave field potentials, but rather consisted in successive negative deflections (Fig. 8.25C, bottom), similar to spindles. This pattern in the field potentials was generated by PY cells that discharged single spikes on almost each cycle of the oscillation (similar to Fig. 8.21C1). This simulation therefore suggests that removing intrathalamic $GABA_A$-mediated inhibition affects the oscillation frequency, but does not by itself generate paroxysmal spike-and-wave patterns, because pyramidal cells are still under the strict control of cortical fast inhibition. This is consistent the injection of $GABA_A$ antagonists into the thalamus *in vivo*, which elicits slow oscillations with increased thalamic synchrony, but no spike-and-wave patterns in the field potentials (Fig. 8.6; Ralston and Ajmone-Marsan, 1956; Gloor et al., 1977; Steriade and Contreras, 1998).

8.3.5 Suppression of intracortical $GABA_A$ inhibition leads to ~3 Hz spike-and-wave

A diffuse application of penicillin, a $GABA_A$ antagonist into the cortex, with no change in thalamus, leads to spike-and-wave oscillations (Fig. 8.7; Gloor et al., 1977). In the model, decreasing $GABA_A$ conductances in cortical cells, leaving the thalamus intact, transformed the spindle oscillations into 2–3 Hz oscillations (Fig. 8.26). The field potentials generated by these oscillations consisted of interleaved 'spikes' and 'waves' (Fig. 8.26, bottom).

When intracortical fast inhibition was decreased by 50%, the occurrences of prolonged high-frequency discharges increased during spindle oscillations (Fig. 8.26A). In field potentials, these events tended to generate large amplitude negative deflections, followed by small-amplitude positive waves (Fig. 8.26A, bottom). When the $GABA_A$-mediated inhibition in the cortex was totally suppressed, the network generated a slow oscillation at 2–3 Hz with field potentials similar to spike-and-wave (Fig. 8.26B). Field potentials displayed one or several negative/positive sharp deflections, followed by a slowly developing positive wave (Fig. 8.26B, bottom). During

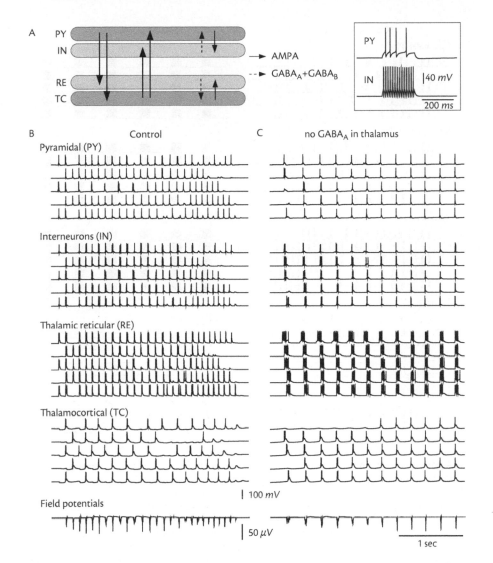

Fig. 8.25: Transformation of spindle oscillations into ~4 Hz oscillations by blocking thalamic inhibition in thalamocortical networks. A. Schematic of the connectivity between different cell types: 100 cells of each type were simulated, including TC and RE cells, cortical pyramidal cells (PY) and cortical interneurons (IN). The connectivity is represented by continuous arrows, for AMPA-mediated excitation, and dashed arrows for mixed GABA$_A$ and GABA$_B$ inhibition. In addition, PY cells were interconnected using AMPA receptors and RE cells were interconnected using GABA$_A$ receptors. The inset shows the repetitive firing properties of PY and IN cells following depolarizing current injection (0.75 nA during 200 ms; −70 mV rest). B. Spindle oscillations in the thalamocortical network in control conditions. Five cells of each type, equally spaced in the network, are shown (0.5 ms time resolution). The field potentials, consisting of successive negative deflections at ~10 Hz, is shown at the bottom. C. Oscillations following the suppression of GABA$_A$-mediated inhibition in thalamic cells with cortical inhibition intact (all GABA$_A$ conductances postsynaptic to RE cells were suppressed). The network generated synchronized oscillations at ~4 Hz, with thalamic cells displaying prolonged discharges. The pattern of discharges in PY cells were resembled spindles but at a lower frequency, as reflected in the field potentials (bottom). Reproduced from Destexhe, 1998.

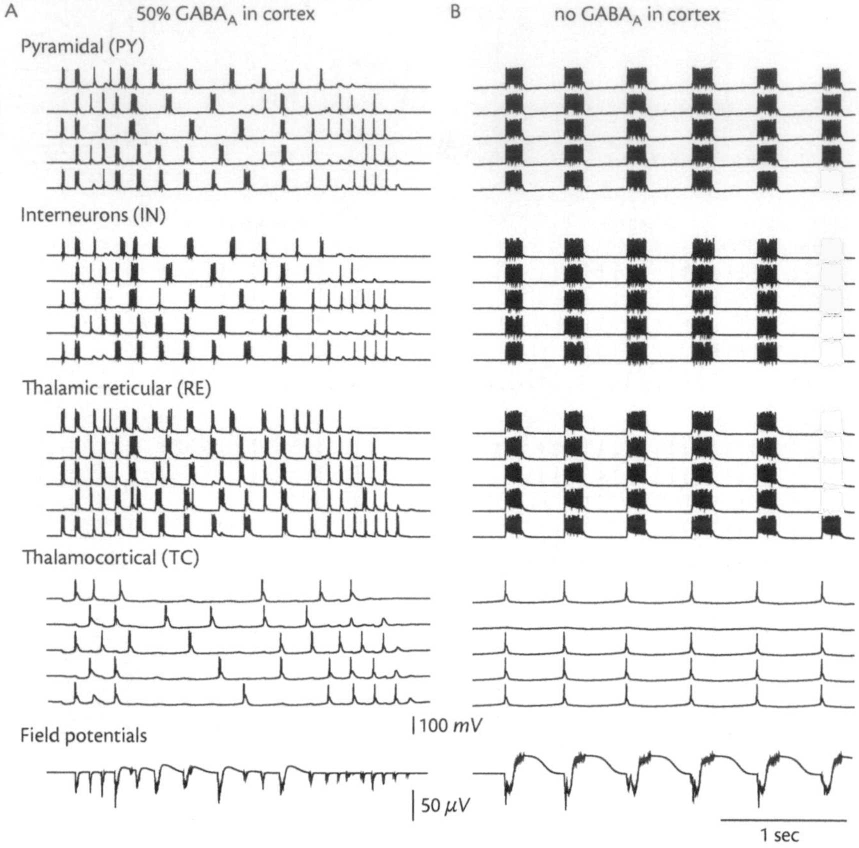

Fig. 8.26: Transformation of spindle oscillations into ~3 Hz oscillations with spike-and-wave field potentials by reducing cortical inhibition. Similar arrangement of traces as in Fig. 8.25B–C. A. Oscillations with 50% decrease of GABA$_A$-mediated inhibition in cortical cells (0.075 μS, IN→PY). Stronger burst discharges appeared within spindle oscillations, leading to large amplitude negative spikes followed by small positive waves in the field potentials (bottom). B. Oscillations following suppression of GABA$_A$-mediated inhibition in cortical cells. All cells had prolonged, in phase discharges, separated by long periods of silence, at a frequency of ~2 Hz. GABA$_B$ currents were maximally activated in TC and PY cells during the periods of silence. Field potentials (bottom) displayed spike-and-wave complexes. Thalamic inhibition was intact in A and B. Reproduced from Destexhe, 1998.

the 'spike', all cells fired prolonged high-frequency discharges in synchrony; in contrast, all cell types were silent during the 'wave'. This activity profile is typical of cortical and thalamic cells recorded during spike-and-wave patterns (Fig. 8.3; Pollen, 1964; Steriade, 1974; Avoli et al., 1983; McLachlan et al., 1984; Buzsaki et al., 1988; Inoue et al., 1993; Seidenbecher et al., 1998). Some TC cells stayed hyperpolarized during the entire oscillation (second TC cell in Fig. 8.26B), which has also been observed experimentally (Steriade and Contreras, 1995). A similar oscillation occurred when GABA$_A$ receptors were suppressed in the entire network (not shown).

Fig. 8.27 shows the progressive transformation from spindles to spike-and-wave oscillations in the model. With intact cortical inhibition, the discharges of cells in the network were limited to few spikes. Consequently, IPSPs in PY cells were almost exclusively GABA$_A$-mediated, leading to field potentials consisting solely of negative deflections (Fig. 8.27, 100%). When intracortical inhibition was partially reduced, prolonged discharges increased along with an increased contribution of GABA$_B$ IPSPs in PY cells, leading to small positive waves in field potentials (Fig. 8.27, 50%). With further reduction of intracortical GABA$_A$-mediated inhibition, the system

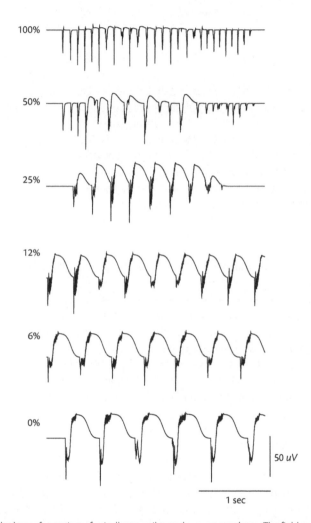

Fig. 8.27: Gradual transformation of spindles to spike-and-wave complexes. The field potentials obtained from simulations similar to those in Fig. 8.26 are shown from top to bottom. The conditions were identical except that intracortical GABA$_A$-mediated inhibition (IN→PY) was reduced, with total conductance values of 0.15 μS (100%), 0.075 μS (50%), 0.0375 μS (25%), 0.018 μS (12%) and 0.009 μS (6%). At 100% GABA$_A$ intracortical inhibition there was a spindle sequence (as in Fig. 8.25B) and at 0% there were fully developed spike-and-wave complexes (as in Fig. 8.26B); intrathalamic inhibition was intact in all cases. Reproduced from Destexhe, 1998.

displayed fully developed spike-and-wave complexes in field potentials with 2–3 Hz oscillation frequencies (Fig. 8.27, 25% to 0%). The frequency of spike-and-wave oscillations was roughly proportional to the amount of fast inhibition still present in the cortex. The occurrence of a positive 'spike' was also correlated with intracortical fast inhibition (Fig. 8.27), consonant with the effects of GABA$_A$ conductances in Fig. 8.22B.

These simulations suggest that spindles can be transformed into spike-and-wave oscillations by altering cortical inhibition alone, without any change in the thalamus. This is in agreement with spike-and-wave discharges obtained experimentally by diffuse application of dilute penicillin to the cortex (Fig. 8.7; Gloor et al., 1977). The generation of ~3 Hz oscillation in this model depends on the thalamocortical loop since both cortex and thalamus are necessary, but neither generates the 3 Hz rhythmicity alone (see below).

Waxing-and-waning in the model for spike-and-wave oscillations (Fig. 8.27, 25%) occurs through the same mechanisms that cause waxing-and-waning of spindle oscillations, the entry of calcium during low threshold calcium activation and calcium-dependent upregulation of I_h in TC cells (Destexhe et al., 1993a, 1996a), as presented in Chapter 3. This mechanism underlies the waxing-and-waning of spindles, both in thalamic networks (Destexhe et al., 1993b, 1996a; Bal and McCormick, 1996; Luthi and McCormick, 1998b; Chapter 6) and in thalamocortical networks (Destexhe et al., 1998a; Chapter 7). The calcium-dependent upregulation of I_h in TC cells is responsible for the temporal modulation of spike-and-wave oscillations in which bursts of several cycles of spike-and-wave oscillations are interleaved with long periods of silence (~20 sec), as observed experimentally in sleep spindles and spike-and-wave seizures. This is another example of the similarity between these two types of oscillation.

8.3.6 A thalamocortical loop mechanism for ~3 Hz spike-and-wave oscillations

Fig. 8.28 shows the phase relationships between the different types of cell in the model of spike-and-wave oscillations. The pattern displayed by the network is similar to the pattern in Fig. 8.21C2: High-frequency discharges generated the 'spike' components in the field potentials, whereas the 'wave' components were generated by GABA$_B$ IPSPs in PY cells due to the prolonged firing of cortical interneurons. The hyperpolarization of PY cells during the 'wave' also contained a significant contribution from the voltage-dependent K$^+$ current, I_M, which was maximally activated due to the prolonged discharge of PY cells during the 'spike'. The 'wave' component was therefore generated by two types of K$^+$ currents, intrinsic and GABA$_B$-mediated. The relative contribution of each current to the 'wave' depends on their respective conductance values.

The 'spike' component was generated by the concerted discharge of all cell types. However, the prolonged discharges were not perfectly in-phase, as indicated in Fig. 8.28B. There was a significant phase advance of TC cells, as observed experimentally (Inoue et al., 1993; Seidenbecher et al., 1998). This phase advance was responsible for the initial negative spike in the field potentials, which coincided with the first spike in the TC cells (Fig. 8.28B, dashed line). This precedence of EPSPs over IPSPs in the PY cell generates spike-and-wave complexes, as shown above (Fig. 8.22A). The simulations therefore suggest that the initial spike of spike-and-wave complexes are due to thalamic EPSPs that precede other synaptic events in PY cells. Thalamic EPSPs may also trigger an initial avalanche of discharges from pyramidal cell firing

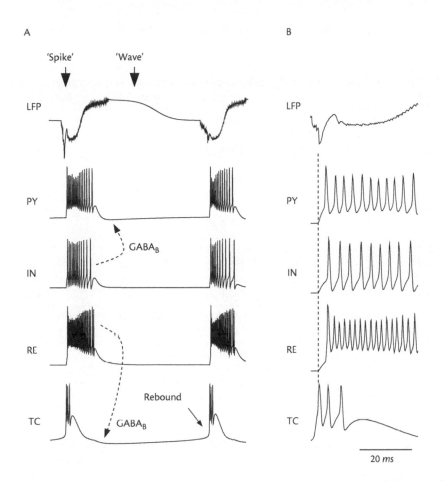

Fig. 8.28: Phase relationships during simulated 3 *Hz* spike-and-wave discharges. A. Local field potentials (LFP) and representative cells of each type during spike-and-wave oscillations. *Spike*: all cells displayed prolonged synchronous discharges, leading to spiky field potentials. *Wave*: the prolonged discharge of RE and IN neurons evoked maximal GABA$_B$-mediated IPSPs in TC and PY cells respectively (dashed arrows), terminating the firing of all neurons over the 300 to 500 *ms* interval, and generating a slow positive wave in the field potentials. The next cycle restarted due to the rebound of TC cells following GABA$_B$ IPSPs (arrow). B. Phase relationships in the thalamocortical model. TC cells discharged first, followed by PY, RE and IN cells. The initial negative peak in the field potentials coincided with the first spike in TC cells before PY cells started firing, and was generated by thalamic EPSPs in PY cells. Reproduced from Destexhe, 1998.

before IPSPs arise, which would also produce a pronounced negative spike component in field potentials.

8.3.7 Determinants of 3 *Hz* spike-and-wave oscillations

The parameter space of conductances in the thalamocortical model was searched to determine the regions that produced spike-and-wave oscillations. Table 8.2 gives the range of synaptic conductances giving rise to spike-and-wave for each type of connection in the absence of

Table 8.2 Range of values for synaptic conductances and their effects on spike-and-wave oscillations.

Receptor type	Location	Optimal value	spike-and-wave range	spike-and-wave freq. (50%)	spike-and-wave freq. (200%)
AMPA	PY→PY	0.6 μS	0.3–0.9 μS	3.0 Hz	–
AMPA	PY→IN	0.2 μS	0.06–2 μS	1.3 Hz	1.9 Hz
GABA$_B$	IN→PY	0.03 μS	0.02–5 μS*	1.3 Hz	2.0 Hz
AMPA	TC→RE	0.2 μS	0–5 μS*	1.7 Hz	1.7 Hz
GABA$_A$	RE→RE	0.2 μS	0–1.2 μS	1.4 Hz	2.5 Hz
GABA$_A$	RE→TC	0.02 μS	0–1 μS	1.7 Hz	1.8 Hz
GABA$_B$	RE→TC	0.04 μS	0.01–5 μS*	2.1 Hz	1.3 Hz
AMPA	TC→PY	1.2 μS	0.15–5 μS*	1.7 Hz	1.7 Hz
AMPA	TC→IN	0.4 μS	0–5 μS*	1.7 Hz	1.7 Hz
AMPA	PY→RE	1.2 μS	0.4–5 μS*	2.2 Hz	1.5 Hz
AMPA	PY→TC	0.01 μS	0–0.1 μS	1.7 Hz	1.7 Hz

Conductance values represent the sum of all individual synaptic conductances of the same type converging on a given cell. The range of conductance values giving rise to spike-and-wave oscillations are indicated in the fourth column (∗: no value larger than 5 μS was tested because sodium spike inactivation may occur for too high conductance values). The minimal oscillation frequency is indicated in the last two columns when each of the respective conductances was set to 50% and 200% of its optimal value: if the range was the same as the control (1.7 Hz) then this parameter had no detectable influence on spike-and-wave. All simulations were run using the thalamocortical network model with suppressed GABA$_A$-mediated inhibition in cortical cells (as in Fig. 8.26B).

intracortical GABA$_A$-mediated inhibition. It must be borne in mind that this model used highly simplified single-compartment models of thalamic and cortical neurons, with minimal sets of intrinsic currents, no dendrites (and therefore no dendritic currents and no dendritic synapses) and simplified models for intrinsic and synaptic currents. The conductance values do not allow quantitative comparisons with physiological values and must be interpreted qualitatively.

The two last columns of Table 8.2 indicate the minimal frequency of spike-and-wave bursts when each parameter was varied within 50%–200% around its optimal value. This provides an estimation for the sensitivity of a given parameter for affecting spike-and-wave oscillations. The synaptic conductances that were most influential on spike-and-wave were: PY→PY, PY→IN, IN→PY, RE→RE, RE→TC (GABA$_B$), PY→RE and a weak effect for RE→TC (GABA$_A$). TC→RE, TC→PY, TC→IN and PY→TC had minimal effect. As expected, the recurrent excitation between pyramidal cells (PY→PY), the excitation of interneurons (PY→IN), as well as the inhibitory feedback on PY cells (IN→PY), were effective in influencing spike-and-wave activity because these conductances determined the excitability of the cortical network. Less expected was the role of cortical excitatory feedback on RE cells (PY→RE), intra-RE inhibition (RE→RE) and the GABA$_B$ inhibition from RE onto TC cells (RE→TC). These factors are examined in more detail below.

1. *Intra-RE GABAergic connections.* Fig. 8.29A shows the transition that occurs from spike-and-wave oscillations to spindle waves, as a function of intracortical GABA$_A$ inhibition, similar

to Fig. 8.27. Reinforcing intra-RE GABA$_A$ inhibition significantly reduced spike-and-wave oscillations in favour of spindles (Fig. 8.29A, compare triangles to filled circles), whereas decreasing this inhibition had the opposite effect (squares). In the model, reinforcing intra-RE GABA$_A$-mediated inhibition diminished the tendency of RE cells to produce bursts of action potentials, which reduced GABA$_B$-mediated IPSPs in TC cells and damped the tendency for spike-and-wave oscillations. This behaviour is consistent with the presumed role of the anti-absence drug clonazepam, which may reduce the tendency of the network to produce spike-and-wave by specifically acting on GABA$_A$ receptors in the thalamic RE nucleus (Fig. 8.9; Huguenard and Prince, 1994; Gibbs et al., 1996; Hosford et al., 1997).

2. *Corticothalamic feedback on RE cells*. Reducing the AMPA conductance of cortical EPSPs on RE cells significantly shifted the balance from spike-and-wave oscillations in favour of spindles (Fig. 8.29B, compare triangles to filled circles) and increasing this conductance had the opposite effect. The model therefore predicts that diminishing the impact of corticothalamic EPSPs on RE cells would be an effective way to reduce the threshold for spike-and-wave in the network. The need for larger conductances from cortical EPSPs onto RE cells vs. TC cells ('inhibitory dominance') was also indicated in the context of spindle oscillations (Section 7.2 in Chapter 7; Destexhe et al., 1998a). To have spike-and-wave oscillations coexisting with spindles required at least four times larger AMPA conductances on RE cells (0.4 μS for PY→RE and 0.1 μS of PY→TC in Table 8.2). This is consistent with the anatomical observation that cortical synapses are the predominant type in thalamic reticular neurons (Liu and Jones, 1999).

3. *T-current conductance in RE cells*. Reducing the T-current of RE cells significantly reduced spike-and-wave activity and increased spindle activity (Fig. 8.29C, compare triangles to filled circles); reinforcing this current had the opposite effect (squares). The effect of reducing T-current amplitude was similar to that of reinforcing GABAergic inhibition in the RE nucleus, consistent with the experimental finding that the T-current is selectively increased in RE cells in a rat model of absence epilepsy (Tsakiridou et al., 1995).

4. *GABA$_B$ conductance in TC cells*. The frequency of spike-and-wave discharges could be effectively controlled by GABA$_B$-mediated IPSPs on TC cells (Fig. 8.29D, filled circles). Changing the decay of intrathalamic GABA$_B$ currents (parameter K_4 in Eq. 5.30) affected only the frequency, with minimal changes in the bursting patterns of the different cell types (not shown). This effect was because, in this model, the duration of the 'wave' was mainly determined by GABA$_B$ IPSPs in TC cells, longer IPSPs leading to slower spike-and-wave by further delaying the rebound of TC cells. The frequency varied from 1–5 *Hz* for decay values of 50% to 250% of the control value, suggesting that the different frequencies of the spike-and-wave bursts in different experimental models may be a consequence of the differences in the kinetics of GABAergic inhibition in TC cells. This point is analysed in detail in Section 8.3.8.

5. *T-current conductance in TC cells*. The strength of this current also affected the spike-and-wave frequency (Fig. 8.29D, squares). Stronger T-current conductances led to earlier rebound and faster frequencies. In contrast, the T-current amplitude in RE cells had minimal effect on spike-and-wave frequency (Fig. 8.29C, triangles). Reducing the T-current in TC cells had only a weak effect on the spike-and-wave threshold (not shown), but T-current reduction greater than 40% in TC cells led to the suppression of oscillatory behaviour, consistent with the effect of the anti-absence drug ethosuximide in reducing the total T-current conductance in TC cells (Coulter et al., 1989).

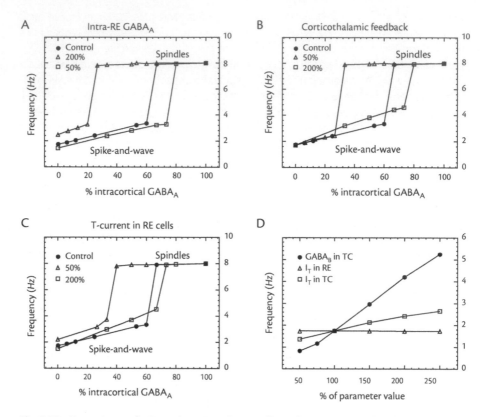

Fig. 8.29: Determinants of spike-and-wave oscillations. Effects of GABA$_A$-mediated inhibition between RE cells. Lower bound on the frequency of spike-and-wave complexes as a function of the amount of GABA$_A$ inhibition in cortex (simulations similar to Fig. 8.27). In control conditions (filled circles), the frequency of spike-and-wave oscillations increased steadily up to 60% of cortical GABA$_A$, then a transition occurred to spindle oscillations (lowest frequency of ~8 Hz). When the intra-RE GABA$_A$ conductances were reduced to half the control size (squares), this transition occurred at ~75% cortical GABA$_A$. When intra-RE GABA$_A$ conductances were doubled, the domain of spike-and-wave was significantly smaller, with a transition occurring at about 20% of cortical GABA$_A$ (triangles). B. Effect of corticothalamic feedback on RE cells. With reinforced AMPA conductance in PY→RE synapses (200% of control value), the domain of spike-and-wave discharges was significantly reduced (squares), whereas reducing the cortical EPSPs had the opposite effect (triangles). Filled circles: same control as in A. C. Effect of the T-current conductance in RE cells. With reinforced T-current (200% of control value), the transition occurred at about 75% of cortical GABA$_A$(squares), whereas with diminished T-current (50% of control value), the domain of spike-and-wave was significantly reduced (triangles). Filled circles: same control as in A. D. Dependence of spike-and-wave frequency on the decay of intrathalamic GABA$_B$ currents (filled circles), the T-current conductance in TC (squares) and the T-current in RE cells (triangles) (simulations from Fig. 8.26B). Parameters were represented as % of their control value (100% = control). Reproduced from Destexhe, 1998.

8.3.8 'Fast' 5–10 *Hz* spike-and-wave oscillations

Recent intracellular recordings from the thalamus in the 'generalized absence epilepsy rat from Strasbourg' (GAERS) reported that, during spike-and-wave discharges, TC cells are paced by $GABA_A$ IPSPs (Pinault et al., 1998). In contrast, in the above model TC cells are paced by $GABA_B$ IPSPs. However, in this experimental model, absence seizures consist of spike-and-wave discharges at a fast frequency (5–10 *Hz*) compared to the slow (~3 *Hz*) spike-and-wave analysed above.

The computational model was used to determine whether a different balance of GABAergic conductances in TC cells might explain both the fast (5–10 *Hz*) and slow (2–3 *Hz*) type of spike-and-wave oscillations based on similar thalamocortical mechanisms (Destexhe, 1999a). The above thalamocortical model of spike-and-wave was used, with three differences: (a) TC cells had a depolarized resting membrane potential of −56 *mV*, as observed experimentally in GAERS (Pinault et al., 1998); (b) the $GABA_B$ conductance from RE→TC was smaller than in the previous model (0.015 μS vs. 0.04 μS); the $GABA_A$ conductance from RE→TC was larger than in the previous model (0.03 μS vs. 0.02 μS).

In 'control' conditions, the network generated 8–12 *Hz* spindle oscillations, in which all cell types produced moderate rates of discharge approximately in phase, while the field potentials displayed successive negative deflections (Fig. 8.30A). These features are in agreement with experimental observations in thalamic and cortical neurons during sleep spindles (Steriade et al., 1990; see Chapter 7). In the model, these oscillations were not critically dependent on the strengths of $GABA_A$ and $GABA_B$ conductances in TC cells, as shown in Fig. 8.30A.

The excitability of the cortical network was increased by decreasing the effectiveness of $GABA_A$-mediated intracortical inhibition, as in the previous model, but the network generated a different type of oscillation (Fig. 8.30B) in which cortical (PY, IN) and thalamic RE cells fired prolonged discharge patterns in synchrony, interleaved with periods of silence that occurred simultaneously in all cell types. This cellular pattern generated spike-and-wave field potentials: the 'spike' component was generated by fast EPSPs followed immediately by $GABA_A$-mediated IPSPs in PY cells, while the positive 'wave' was due to activation of slow K^+ currents ($GABA_B$-mediated and voltage-dependent I_M) in PY cells.

The oscillatory pattern of discharge depended on the positive feedback in the corticothalamic loop, which was also essential in the 3 *Hz* spike-and-wave model. The 5–10 *Hz* oscillation shown in Fig. 8.30B differed, however, from the 2–4 *Hz* frequency in Fig. 8.26B. The fast oscillation frequency of the discharge of TC cells was shaped by $GABA_A$-mediated IPSPs (arrows in Fig. 8.30B2). $GABA_B$ receptors also contributed to the oscillation but produced a sustained hyperpolarization in TC and PY cells, a feature that has also been observed experimentally (Pinault et al., 1998).

Although the oscillation was paced by $GABA_A$ conductances, $GABA_B$ conductances were also involved. The time course of GABAergic conductances during spike-and-wave activity (Fig. 8.31A) shows that $GABA_B$ conductances were tonically activated owing to their slow kinetics, generating a sustained hyperpolarization in TC cells. This sustained hyperpolarization contributed to maintaining the oscillations, since smaller $GABA_B$ conductances led to a markedly reduced tendency to oscillate (Fig. 8.31B, top trace), and larger $GABA_B$ conductances led to a slower, 2–3 *Hz* spike-and-wave oscillation (Fig. 8.31B, bottom trace). This behaviour occurred for all parameter values that supported the faster spike-and-wave. The effect of

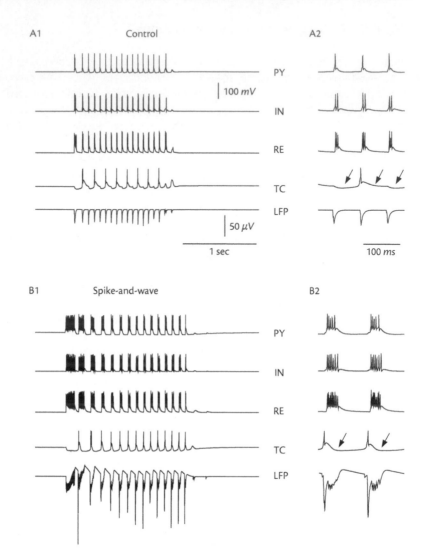

Fig. 8.30: Fast spike-and-wave oscillations with stronger GABA$_A$ conductances in TC cells. A1. Control spindle oscillations elicited by injection of depolarizing current into PY cells (1 nA during 20 ms). All cell types displayed moderate discharges at 10–12 Hz accompanied by negative deflections in the local field potential (LFP). B1. Spike-and-wave oscillations following increase of cortical excitability (same simulation as in A with intracortical GABA$_A$ conductances decreased from 0.15 μS to 0.04 μS). All cell types displayed synchronized discharges at 5–10 Hz and the field potentials consisted of spike-and-wave patterns. The right panels (A2, B2) show two oscillation cycles at higher temporal resolution. Arrows indicate GABA$_A$ IPSPs in TC cells (modified from Destexhe, 1999a).

completely suppressing the GABA$_B$ conductances depended on the strength of GABA$_A$ IPSPs: for weak GABA$_A$ conductances (<0.028 μS), suppressing GABA$_B$ blocked oscillations (Fig. 8.31C, top trace) whereas for stronger GABA$_A$ conductances (>0.028 μS), suppressing GABA$_B$ reduced the number of cycles and increased the frequency (Fig. 8.31C, bottom trace).

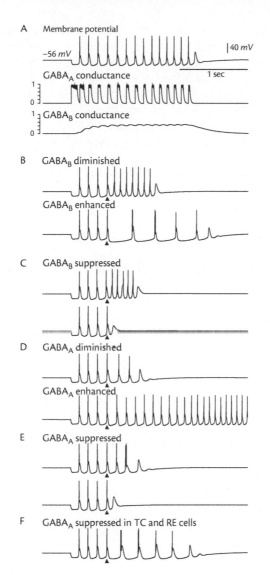

Fig. 8.31: Properties of the faster spike-and-wave model. Manipulations of TC cells were based on the simulations to Fig. 8.30B. A. Time course of GABAergic conductances in TC cells: membrane potential of a representative TC cell (top trace); fraction of GABA$_A$ conductance activated (middle trace); fraction of GABA$_B$ conductance activated (bottom trace). Same simulation as in Fig. 8.30B (maximal conductances were g_{GABAA} = 0.03 μS and g_{GABAB} = 0.015 μS). B. GABA$_B$ receptors prolong oscillatory activity. Reducing GABA$_B$ conductances (arrow) in all TC cells (top trace; g_{GABAB} = 0.005 μS) shortened oscillations but increasing GABA$_B$ conductances had the opposite effect (bottom trace; g_{GABAB} = 0.04 μS). C. Blocking GABA$_B$ conductances (arrow) resulted in either shortening (top trace; g_{GABAA} = 0.04 μS) or suppressing the oscillatory behaviour (bottom trace; g_{GABAA} = 0.02 μS). D. GABA$_A$ conductances enhance oscillatory behaviour. Diminishing GABA$_A$ conductances (arrow) in TC cells (middle trace; g_{GABAA} = 0.01 μS) shortened oscillations but augmenting them led to sustained oscillations (bottom trace; g_{GABAA} = 0.06 μS). E. Blocking GABA$_A$ conductances (arrow) in TC cells either shortened the oscillatory sequence (top trace; g_{GABAB} = 0.015 μS) or suppressed oscillatory behaviour (bottom trace; g_{GABAB} = 0.005 μS). F. Suppressing GABA$_A$ conductances in both TC and RE cells (arrow) resulted in spike-and-wave oscillations at 3–4 Hz (modified from Destexhe, 1999a).

GABA$_A$ conductances in TC cells are of central importance for generating fast spike-and-wave oscillations in this model. Reducing the conductance of the GABA$_A$ IPSPs markedly reduced oscillatory sequences (Fig. 8.31D, top trace) and increasing them led to prolonged oscillations (Fig. 8.31D, bottom trace). This GABA$_A$ effect was observed for all parameters tested. Suppressing GABA$_A$ conductances in TC cells had two possible effects depending on the value of GABA$_B$ conductances: for strong GABA$_B$ conductances (>0.01 μS), blocking GABA$_A$ shortened the oscillatory sequence (Fig. 8.31E, top trace), but if GABA$_B$ conductances were too small (<0.01 μS), blocking GABA$_A$ suppressed oscillatory behaviour (Fig. 8.31E, bottom trace). Finally, when GABA$_A$ conductances were suppressed in both TC and RE cells, the behaviour of the system reverted to approximately 3–4 Hz spike-and-wave oscillations (Fig. 8.31F), but only when GABA$_B$ conductances were sufficiently strong (>0.01 μS); this leads to experimentally testable predictions, as discussed below.

8.4 Further findings on absence seizures

Calcium channels have been long known to be involved in absence seizures. Further evidence came from blocking Cav3.1 channels, which abolishes thalamocortical absence-like seizures (Kim et al., 2001). These results strengthen the involvement of cortico-thalamic interactions in absence seizures (Destexhe, 1998), as we reviewed in this chapter.

An intriguing discovery, potentially important for understanding the mechanisms underlying absence seizures, was made in the Wag-Rij rat genetic model of absence epilepsy (Meeren et al., 2002). In this experimental model of absence seizure activity, large-scale recording from cerebral cortex by multisite electrodes revealed that the seizure started in the somatosensory area, evidence for a cortical focus. In GAERS rats, another experimental model of absence seizure, deep (layers 5-6) cortical neurons in somatosensory cortex were found to be hyperexcitable and began discharging at the beginning of ictal activity (Polack et al., 2007).

These findings are consistent with our model of absence seizure, in which the spike-and-wave activity occurs within a loop formed by an over-excitable cortex and a physiologically intact thalamus. It thus explains very well why an increase of excitability in the thalamocortical system (here the somatosensory cortex) leads to hypersynchronized spike-wave discharges in that system. Our model can thus explain how the seizure starts and develops in the somatosensory cortex, based on the excessive corticothalamic feedback provided by the hyperexcitable layer 6 cells. This is consistent with the model developed in this chapter. What remains to be explained is how this pathological activity invades the 'normal' cortex, leading to generalized seizures, and remains a challenge for future experiments and models (for a recent study, see Depannemaecker et al. 2022).

The thalamus and cortex may not be the only structures involved in absence seizures: experiments also implicate the basal ganglia (Paz et al., 2007). A model proposed to reproduce these findings was unfortunately incomplete and did not include the thalamocortical loop or the synaptic receptors present in thalamocortical circuits (Arakaki et al., 2016). There have been no tests of the predictions made by this model. It would be interesting in the future to investigate models of absence including circuits from the thalamus, basal ganglia and cortex, in order to examine their respective role.

Another mechanism that has not yet been incorporated in absence models is the possible influence of GABA receptors on presynaptic terminals (Waldmeier and Baumann, 1990;

Sperk et al., 2004). Inhibition of synaptic release could mediate phenomena such as disinhibition at GABAergic terminals, which could potentially have convulsant effects. The possible influence of disinhibition on epileptic discharges should also be included in future models.

8.5 Discussion

The focus of this chapter was the genesis of paroxysmal discharges in the thalamus and how they lead to spike-and-wave seizures in the thalamocortical system. A large body of experimental data obtained *in vitro* as well as during induced seizures *in vivo* were accurately mimicked in detailed computational models that incorporated the key ionic mechanisms found in thalamic and cortical neurons. We summarize here the proposed mechanisms, how they account for experiments and possible ways to test them.

8.5.1 The importance of the nonlinear activation of GABA$_B$ responses

A major component in the model for spike-and-wave discharges is the nonlinear stimulus-response relationship of GABA$_B$ responses, which was needed to reproduce both 3 *Hz* oscillations in thalamic circuits and spike-and-wave oscillations in the thalamocortical system. One or a few presynaptic spikes do not activate GABA$_B$ currents, whereas high-frequency bursts of more than five action potentials activate full GABA$_B$ responses as a consequence of a cooperative intracellular signalling mechanism based on G-proteins (see response curve in Fig. 5.13B in Chapter 5). As a consequence, the GABA$_B$ IPSPs appear only under conditions of intense synaptic release, such as during paroxysmal discharges.

The fact that GABA$_B$ responses should depend nonlinearly on the number of presynaptic spikes is a property that was first predicted in modelling studies (Destexhe and Sejnowski, 1995; see Chapter 5) and was later confirmed experimentally (Kim et al., 1997; Thomson and Destexhe, 1999). Two independent modelling studies (Destexhe et al., 1996a; Golomb et al., 1996) showed that this kind of cooperativity in GABA$_B$ responses could reproduce the precise properties of oscillations in thalamic circuits. In particular, the models predicted that the nonlinear dependence of GABA$_B$ responses on the number of presynaptic spikes is a necessary property for the genesis of paroxysmal discharges. This conclusion holds regardless of the exact biophysical mechanisms underlying this nonlinearity.

A variety of explanations have been proposed for why GABA$_B$ receptors are only stimulated by a strong stimuli: (a) GABA spills over from neighbouring synapses; (b) Several transmitters co-release in response to different patterns of stimulation, and (c) GABA$_A$ and GABA$_B$ receptors arise from different populations of inhibitory neurons (see details in Chapter 5; see also reviews by Mody et al., 1994; Thompson, 1994; Benardo, 1994). The alternative explanation that we proposed is that GABA$_B$ responses are *intrinsically* sensitive to the number of presynaptic spikes. We suggested a kinetic model of GABA$_B$ receptors that postulated the presence of multiple binding sites of G-proteins to activate K$^+$ channels (see Section 5.2.6 in Chapter 5). The accumulation of a threshold level of G-proteins is required in the model to activate detectable K$^+$ currents. The activation of GABA$_B$ responses by high-frequency stimulation in the model has important consequences at the network level, in both thalamic circuits and the thalamocortical system, as discussed below.

8.5.2 Paroxysmal discharges in the thalamus

At the level of thalamic circuits, the the nonlinear $GABA_B$ response can explain the enhancement of $GABA_B$ IPSPs following application of $GABA_A$ antagonists in thalamic slices[3] (Soltesz and Crunelli, 1992; Huguenard and Prince, 1994a, 1994b; Bal and McCormick, 1995b; Kim et al., 1997). Several explanations have been proposed for this observation, such as the hypothesis that different populations of interneurons may mediate $GABA_A$ and $GABA_B$ responses (Soltesz and Crunelli, 1992) or that there may be an enhanced action potential discharge due to increase of synchrony and disinhibition of inhibitory neurons (Huguenard and Prince, 1994a, 1994b; Bal and McCormick, 1995b). The present model fully supports the later suggestion. If inhibitory neurons contact each other via $GABA_A$ receptors, then their discharges would be stronger following the blockade of this inhibition. These enhanced discharges would then provide the strong stimulus needed to fully evoke $GABA_B$ currents, as illustrated in Fig. 8.12.

The activation characteristics of $GABA_B$ receptor-mediated currents are critical in explaining the genesis of 3 *Hz* oscillations by thalamic circuits in the presence of $GABA_A$ antagonists, as observed *in vitro* (Fig. 8.10; von Krosigk et al., 1993) and *in vivo* (Fig. 8.6; Steriade and Contreras, 1998; Castro-Alamancos, 1999). The model suggests that the suppression of $GABA_A$-mediated inhibition promotes the genesis of large bursts in RE cells and entrains the network in a different type of oscillation that is slower and more synchronized. The prolonged bursts of RE cells activate large $GABA_B$-mediated IPSPs in TC cells; TC cells rebound following these IPSPs, re-exciting RE cells, and the system is entrained to a slow oscillation generated by the TC-RE loop. This reciprocal recruitment mechanism is similar to that underlying spindle generation, as illustrated in Fig. 8.32 (see also Chapter 6). Another modelling study of slow thalamic oscillations independently reached similar conclusions (Golomb et al., 1996).

The models confirm and extend experimental observations that thalamic circuits are capable of generating two types of oscillations, spindle waves and slow highly-synchronous oscillations. The model suggests that the mechanisms for initiation, propagation and termination of slow oscillations are similar to those for spindle oscillations, as discussed in Chapter 6, but with some differences. First, in contrast to spindle oscillations where TC cells do not fire on every cycle, during slow oscillations all TC cells burst in synchrony: this is because the strong $GABA_B$ IPSPs deinactivate I_T more completely and produce a more secure rebound burst. Second, it follows that the two cell types would produce stronger burst discharges. Third, the amplitude of the afterdepolarization produced in TC cells is higher during slow oscillations than during spindles because of the stronger activation of I_T and, as a consequence, a more pronounced upregulation of I_h. Fourth, the interspindle silent period is longer than during spindles owing to longer recovery following a more pronounced upregulation of I_h. Fifth, the propagation velocity is slower, due to the slower oscillation frequency. All these properties have been observed in

---

[3] The nonlinear activation can also explain a similar phenomenon observed in hippocampal slices (see Newberry and Nicoll, 1985 for experiments and Destexhe and Sejnowski, 1995 for the model; see also Section 5.3 in Chapter 5).

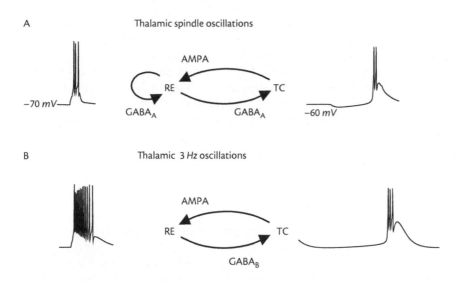

A   Thalamic spindle oscillations

B   Thalamic 3 *Hz* oscillations

Fig. 8.32: Proposed mechanisms for two types of oscillations in thalamic circuits. A. ~10 *Hz* spindle oscillations can result from the interaction between TC and RE cells. The presence of GABAergic interactions between RE cells limit their discharges, resulting in only a few spikes per burst and a $GABA_A$-dominated IPSP in the target TC cell. B. ~3 *Hz* paroxysmal oscillations after removal of $GABA_A$ receptors. In this case, the suppression of GABAergic interactions between RE cells promote their bursting activity, resulting in bursts with a large number of spikes (20–30 spikes/burst). This stimulus optimally activates $GABA_B$ IPSPs in TC cells, which in turn excite RE cells, thus forming a TC-RE loop. The low frequency is a consequence of the slow kinetics of $GABA_B$ IPSPs in TC cells.

thalamic slices (Kim et al., 1995), and were reproduced in the model because of the activity-dependent upregulation of I_h, as discussed in Chapter 6 for spindle generation.

Models also corroborate experiments suggesting that reinforcing lateral inhibition in the RE nucleus should diminish the occurrence of synchronized burst discharges. This mechanism may explain the action of some anti-absence drugs, such as clonazepam (Figs. 8.8–8.9; Huguenard and Prince, 1994a; Gibbs et al., 1996). In the presence of lateral inhibition, RE cells *in vitro* discharge only a small number of spikes per burst (Huguenard and Prince, 1994b; Bal et al., 1995b); this is insufficient to evoke strong $GABA_B$ currents and the IPSPs on TC cells are thus dominated by the $GABA_A$ component (Fig 8.13A2). In the absence of lateral inhibition, RE cells produce prolonged burst discharges that evoke a prominent $GABA_B$ component in the target TC cells (Fig 8.13A3), as observed in thalamic slices (Huguenard and Prince, 1994a; Bal and McCormick, 1995; Kim et al., 1997; Sánchez-Vives et al., 1997a, 1997b). Finally, enhancing $GABA_A$ receptors in the RE nucleus diminishes the frequency of seizures *in vivo* (Liu et al., 1991; Hosford et al., 1997). The role of $GABA_A$ receptors in mediating the lateral inhibition between RE cells was explored with models, which could account for all of the experimental results (Fig. 8.12; Destexhe and Sejnowski, 1995). This is an important confirming step since pharmacological manipulation in a highly complex system may affect more than one mechanism and make the interpretation of the results uncertain.

The same model accounted for many *in vitro* experiments with only a few simple mechanisms. The main postulate was that the $GABA_B$ responses in the TC cells needed to be nonlinearly dependent on the number of presynaptic spikes, a property that also has consequences for the genesis of spike-and-wave oscillations in the thalamocortical system (see below). The *in vitro* finding that the thalamic spindles can be transformed into highly synchronous oscillations at ~3 *Hz* following application of GABA$_A$ antagonists (Fig. 8.10; von Krosigk et al., 1993) appeared to corroborate previous *in vivo* studies reporting that sleep spindle oscillations can be transformed into 3 *Hz* spike-and-wave oscillations (reviewed in Gloor and Fariello, 1988; Avoli et al., 1990). Although highly synchronous 3 *Hz* oscillations can be found in thalamic slices, these are different from spike-and-wave seizures because thalamic injections of GABA$_A$ antagonists fail to produce seizures *in vivo* (Fig. 8.6; Ralston and Ajmone-Marsan, 1956; Gloor et al., 1977; Steriade and Contreras, 1998). The model suggests a possible way to explain both of these experiments, as discussed below.

8.5.3 3 *Hz* spike-and-wave in thalamocortical networks

The model reviewed in Section 8.3 (Destexhe, 1998) proposed a thalamocortical-loop mechanism for the genesis of highly synchronous ~3 *Hz* oscillations with spike-and-wave field potentials. This model was based mostly on experimental observations in the penicillin model of absence seizures in cats (reviewed in Gloor and Fariello, 1988), which reveals that spike-and-wave seizures can be initiated by reducing inhibition in the cortex if—and only if—the thalamus is intact. In the model: (a) The cortical excitability was increased up to the point of generating a strong corticothalamic activity; (b) The abnormally strong corticothalamic feedback was capable of 'forcing' physiologically intact thalamic circuits to generate a highly synchronized 3 *Hz* oscillations (Fig. 8.23); (c) The resulting ~3 *Hz* oscillations invaded the entire thalamocortical network through thalamocortical loops (Fig. 8.26); (d) The generalized ~3 *Hz* oscillation generated spike-and-wave field potentials (Fig. 8.26). During the 'spike', thalamic and cortical cells produced prolonged discharges in synchrony, while the 'wave' is generated by a mixture of voltage-dependent and GABA$_B$-mediated K$^+$ currents; (e) The ~3 *Hz* frequency of the network was determined by intrathalamic GABA$_B$-mediated inhibition (Fig. 8.29D).

Fig. 8.33 summarizes the thalamocortical mechanisms for spindle and spike-and-wave oscillations in this model. During spindles, the oscillation is generated by intrathalamic interactions (TC-RE loop in Fig. 8.33A). Oscillations can also be generated by a thalamocortical loop (TC-Cx-RE loop in Fig. 8.33A) (Destexhe et al., 1998a; see Section 7.2 in Chapter 7). The combined actions of intrathalamic interactions and thalamocortical loops provide a moderate excitation of RE cells, which evokes GABA$_A$-mediated IPSPs in TC cells and sets the frequency to ~10 *Hz*. During spike-and-wave oscillations (Fig. 8.33B), an increased cortical excitability provides strong corticothalamic feedback that is able to force prolonged burst discharges in RE cells, which in turn evoke IPSPs in TC cells dominated by the GABA$_B$ component. The ~3 *Hz* frequency is determined by the duration of the prolonged inhibition. The thalamus is physiologically intact during the 3 *Hz* oscillation generated by the thalamocortical loop (TC-Cx-RE loop in Fig. 8.33B). Therefore, if the cortex is inactivated during spike-and-wave, the model predicts that the thalamus should resume generating 10 *Hz* spindle oscillations, as was observed experimentally in cats treated with penicillin (Gloor et al., 1979).

The thalamocortical loop model is consistent with a number of experimental results on spike-and-wave seizures: (a) Thalamic and cortical neurons discharge in synchrony during the 'spike' while the 'wave' is characterized by neuronal silence (Pollen, 1964; Steriade 1974; Avoli et al.,

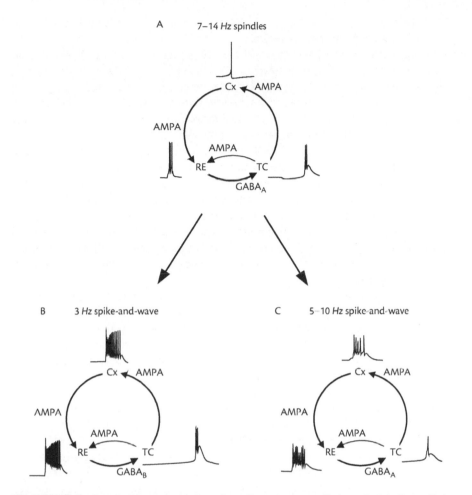

Fig. 8.33: Thalamocortical loop mechanisms for spike-and-wave oscillations. A. Spindle oscillations resulting from a mutual recruitment of thalamic TC and RE cells (thick lines), in which TC cells rebound following fast $GABA_A$-mediated IPSPs, at a frequency of approximately ~10 Hz. Here, the oscillation is generated in the thalamus and is reinforced by the thalamocortical loop (thin lines). B. Proposed mechanism for ~3 Hz spike-and-wave oscillations. In this case, the corticothalamic feedback is much stronger due to increased cortical excitability, forcing thalamic cells to display prolonged burst discharges, which evoke $GABA_B$-mediated IPSPs in TC cells. This prolonged inhibition prevents cells from firing during ~300 ms and sets the frequency to ~3 Hz. C. Proposed mechanism for fast (5–10 Hz) spike-and-wave oscillations. If $GABA_A$ conductances are stronger in TC cells, they will mediate IPSP-rebound sequences. The same thalamocortical loop mechanism as in B can operate in this case, except that $GABA_A$-mediated IPSPs pace TC cells, which results in a faster oscillation frequency but similar discharge patterns and phase relationships between cells.

1983; McLachlan et al., 1984; Buzsaki et al., 1990; Inoue et al., 1993), similar to Figs. 8.25E and 8.28A; (b) TC cell firing precedes the firing of cortical and RE cells (Inoue et al., 1993), similar to the phase relationships in the present model (Fig. 8.28B); (c) Spike-and-wave patterns disappear following either removal of the cortex (Avoli and Gloor, 1982) or the thalamus (Pellegrini et al., 1979; Vergnes and Marescaux, 1992), as predicted by the present mechanism; (d) Antagonizing thalamic $GABA_B$ receptors suppresses spike-and-wave seizures (Liu et al., 1992), consistent with the model; (e) Spindle oscillations can be gradually transformed into spike-and-wave discharges (Kostopoulos et al., 1981a, 1981b), as described in Fig. 8.27.

In a study of spike-and-wave seizures in cats treated with penicillin, it was reported that cortical inhibition was minimally affected (Kostopoulos et al., 1983), which is apparently inconsistent with the present model. However, this study did not distinguish between $GABA_A$ and $GABA_B$-mediated inhibition. In the present model, even when $GABA_A$ is blocked, IPSPs remain approximately the same size because cortical interneurons fire more strongly (Fig. 8.25D–E) which induces $GABA_B$ currents. The compensatory effect of $GABA_B$-mediated IPSPs (not shown) may give the appearance that inhibition is preserved. The model predicts that spike-and-wave discharges should also occur if there is excess intracortical excitation, rather than a deficit of inhibition as explored here.

With regard to thalamic oscillations, the RE nucleus has a central role in this model of spike-and-wave. Reinforcing $GABA_A$-mediated inhibition in the RE nucleus reduces the large burst discharges evoked in RE cells by corticothalamic EPSPs, blocking the genesis of $GABA_B$-mediated IPSPs in TC cells, and thereby antagonizing the development of spike-and-wave discharges (Fig. 8.29A). This scenario is consistent with the reduction of seizures observed *in vivo* following reinforcement of $GABA_A$ receptors in the RE nucleus (Liu et al., 1991; Hosford et al., 1997). It is also consistent with the action of the anti-absence drug clonazepam in slices, which acts preferentially by enhancing $GABA_A$ responses in the RE nucleus (Fig. 8.9; Gibbs et al., 1996; Hosford et al., 1997), leading to diminished $GABA_B$-mediated IPSPs in TC cells (Fig. 8.8; Huguenard and Prince, 1994a; Gibbs et al., 1996).

Injection of $GABA_A$ antagonists in the thalamus with an intact cortex fails to generate seizure activity *in vivo* (Fig. 8.6; Ralston and Ajmone-Marsan, 1956; Gloor et al., 1977; Steriade and Contreras, 1998). In the model, suppressing thalamic $GABA_A$ receptors led to 'slow spindles' around 4 Hz, which were highly synchronized, but different from spike-and-wave oscillations (Fig. 8.25C). In this case, the discharge of PY cells was extremely brief because cortical $GABA_A$-mediated inhibition was preserved and no $GABA_B$ IPSPs could be evoked. This result is consistent with the powerful control exerted on pyramidal cells by intracortical $GABA_A$-mediated inhibition, as shown by intracellular recordings and modelling studies (Contreras et al., 1997c).

The model suggests how the highly synchronized ~3 Hz oscillation induced by bicuculline in thalamic slices and the ~3 Hz spike-and-wave oscillations in the thalamocortical system can be integrated. The explanation is that a similar 3 Hz oscillation can be evoked in the physiologically intact thalamus by an abnormally strong corticothalamic feedback. If there is a deficit in inhibition (or excess of excitation) in the cortex, the thalamocortical system will sustain 3 Hz oscillations. This mechanism explains spike-and-wave seizures in a way that is consistent with both *in vivo* and *in vitro* experiments.

8.5.4 Faster (5–10 *Hz*) spike-and-wave in thalamocortical networks

The spike-and-wave oscillations of absence seizures are significantly faster in rodents[4] (5–10 *Hz*) than in cats (2–4 *Hz*). Fast spike-and-wave oscillations can be maintained in the model when the balance between GABA$_A$ and GABA$_B$ receptors in TC cells is changed (Fig. 8.30). In this case, the corticothalamic feedback can force the thalamus into a different oscillatory mode, not yet identified in slices, consisting of highly synchronized 5–10 *Hz* oscillations paced by GABA$_A$ IPSPs in TC cells (see Fig. 8.33C). This type of oscillation is, however, different from spindle oscillations, because all cell types generate prolonged discharges, and TC cells do not display the typical out-of-phase patterns characteristic of spindles (compare A and B in Fig. 8.30).

The mechanism for faster spike-and-wave discharges displays striking similarities with experimental models of generalized absence epilepsy in rodents. Intracellular recordings from neurons in GAERS rats (Charpier et al., 1998; Pinault et al., 1998) agree with the properties of thalamic neurons predicted by the model (Fig. 8.30B): (a) The TC cells had a relatively depolarized resting level of approximately −56 *mV*; (b) The TC cells had a relatively moderate discharge during the 'spike' component of spike-and-wave; (c) There were GABA$_A$ IPSPs in TC cells during the 'wave' component; (d) TC and PY cells had a sustained hyperpolarization, which in the model was due to tonic activation of GABA$_B$ receptors (see Fig. 8.31A); (e) There was an afterdepolarization in TC cells following the seizure, caused by the upregulation of I$_h$.

In addition, the model is consistent with several general properties of rodent spike-and-wave seizures: (a) All cell types discharged during the 'spike' (with TC cells firing slightly earlier), while the 'wave' coincided with neuronal silence (Buzsaki et al., 1988; Inoue et al., 1993; Seidenbecher et al., 1998; see Fig. 8.30B2); (b) Antagonizing GABA$_B$ receptors in TC cells antagonized spike-and-wave (Liu et al., 1992; see Fig. 8.31B-C); (c) Enhancing GABA$_A$ receptors exacerbates spike-and-wave activity (Vergnes et al., 1984; Hosford et al., 1997; see Fig. 8.31D); (d) The oscillation is generated by thalamocortical loops, which agrees with the observation that the integrity of both the cortex and thalamus are necessary for generating seizures in GAERS (Vergnes and Marescaux, 1992).

The model suggests a plausible explanation for apparently incompatible experimental observations about the firing of TC cells during absence seizures. Some experiments have reported burst firing of TC cells (McCormick and Hashemiyoon, 1998; Seidenbecher et al., 1998) but others have not (Steriade and Contreras, 1995; Pinault et al., 1998). In the model, TC cells fired low-threshold spike rebound bursts in response to GABA$_A$ IPSPs. However, these bursts were weak and often consisted of single spikes (Fig. 8.30B2). This was due to the relatively depolarized level of TC cells (−56 *mV*) so that GABA$_A$ IPSPs could only partially deinactivate the T-current. Such 'weak' rebound bursts would therefore be difficult to identify, which may explain the conflicting experimental observations. Rebound bursts are central to the oscillatory mechanism because they trigger the spike-wave discharge in the entire network (Fig. 8.30B2),

---

[4] Although the frequency of oscillations in rodent seizures is generally between 5–10 *Hz*, infusion of GABA$_A$ receptor antagonists in rats has been reported to evoke oscillations at either 3 *Hz* or ~12 *Hz*; the former is sensitive to GABA$_B$ receptor antagonists but not the latter (Castro-Alamancos, 1999).

in agreement with the experimental observation that TC cells discharge in advance of other cell types (Inoue et al., 1993; Pinault et al., 1998; Seidenbecher et al., 1998).

The similarity of this mechanism with that proposed for 3 *Hz* spike-and-wave (Destexhe, 1998) suggests that fast and slow types of spike-and-wave oscillations may be generated by similar thalamocortical recruitment mechanisms; namely, the IPSP-rebound sequences in TC cells and differences in the frequency dependence of the $GABA_A$ and $GABA_B$ receptors that mediate these IPSPs.

8.5.5 Predictions

The same model accounts for a large body of data from *in vivo* and *in vitro* experiments, but, more importantly, it generates a number of predictions that can be tested experimentally.

1. The mechanisms for spike-and-wave oscillations in the model, at both 3 *Hz* or 5–10 *Hz*, involve inhibitory-rebound sequences in TC cells. This predicts that blocking the T-current in TC cells should suppress seizures. This is consistent with presumed effect of the anti-absence drug ethosuximide on reducing the effectiveness of the T-current in thalamic neurons (Coulter et al., 1989).

2. In the model, the 'wave' component of spike-and-wave is generated by massive K^+ currents, arising from $GABA_B$-mediated currents and intrinsic voltage-dependent K^+ currents. This predicts that applying $GABA_B$ antagonists to the cortex should lead to significant alteration of the 'wave' component in field potentials and that intracellular injection of cesium should abolish the hyperpolarization. Applying substances that block voltage- and Ca^{2+}-dependent K^+ currents should have similar effects. One must, however, bear in mind that other mechanisms, not taken into account here, may also participate in the 'wave' component, such as the activation of Ca^{2+} spikes and Ca^{2+}-dependent K^+ currents in dendrites (Traub and Miles, 1990).

3. The genesis of spike-and-wave discharges in the model depends on an abnormally strong corticothalamic feedback. This therefore predicts that during seizures, there should be an increased output of cortical layer VI neurons projecting to the thalamus. This increased output could result from either an increase in the discharge of individual neurons, or an increase in the synchrony of the population of neurons in layer VI that project to the thalamus. This prediction could be tested in cortical slices.

4. The model predicts that physiologically intact thalamic circuits can be entrained into ~3 *Hz* oscillations by strong stimulation of corticothalamic feedback. This experiment could be performed in slices or in decorticated animals by stimulating the corticothalamic fibres. The intensity should be high enough to force large bursts in RE cells, evoking $GABA_B$ IPSPs in TC cells and delaying their rebound by ~300 *ms*. Another way to perform this experiment would be to artificially create a feedback loop in which the discharge of TC cells directly triggered the stimulation of cortico-thalamic fibres. The model predicts a switch from ~10 *Hz* oscillations to ~3 *Hz* when stimulation intensity is increased. These predictions have recently been confirmed (Bal et al., 2000; Blumenfeld and McCormick, 2000).

5. In the model for the fast (5–10 Hz) type of spike-and-wave oscillation in rodents, the frequency is higher than in cats because of a different balance between $GABA_A$ and $GABA_B$ receptors in TC cells. The mechanism is the same as for slow the (~3 Hz) spike-and-wave. Therefore the same feedback paradigm outlined above should lead in rodent thalamic slices to highly synchronized 5–10 Hz oscillations that are different than spindles.

6. Inhibition between the RE and TC cells in the thalamus is critical in generating oscillations. The model predicts that fast (5–10 Hz) and slow (~3 Hz) spike-and-wave oscillations should be transformable into each other by manipulating GABAergic conductances in TC cells: (a) Enhancing $GABA_B$ conductances in TC cells should slow down the frequency of spike-and-wave to ~3 Hz (Fig. 8.31B); (b) blocking $GABA_B$ receptors in TC cells should reduce or suppress seizures (Fig. 8.31C); (c) suppressing thalamic $GABA_A$ conductances should either complete suppress the seizures or should slow down the faster spike-and-wave discharges (Fig. 8.31E–F).

7. In the model, the waxing-and-waning of spike-and-wave oscillations is due to upregulation of I_h in TC cells. This predicts that local injection of specific I_h blockers, such as ZD7288, into the thalamus should transform waxing-and-waning spike-and-wave bursts into continuous spike-and-wave activity.

8. Finally, the model depends on the nonlinear relationship between the number of presynaptic spikes and the magnitude of the $GABA_B$ IPSP. In the model the independent binding of four G-proteins were required to activate the K^+ channels underlying $GABA_B$ responses (Section 5.3 in Chapter 5; Destexhe and Sejnowski, 1995). This was implemented by assuming that: (a) The K^+ channels associated with $GABA_B$ receptors are tetramers of four identical subunits; (b) Each subunit has a G-protein binding site; (c) All four subunits must be bound for the channel to open. There are other possible ways to implement this nonlinearity. The main prediction of the model is that paroxysmal discharges can be explained if the nonlinearity arises from mechanisms *intrinsic* to the molecular machinery underlying $GABA_B$ responses.

8.5.6 Future directions

Despite the success of the model in accounting for a wide range of experimental data on the genesis of spike-and-wave oscillations, it is still possible that other mechanisms are responsible, perhaps in addition to the thalamocortical loop scenario. Although most experimental data favour a mechanism involving both the thalamus and the cortex (see Section 8.1), some evidence points to an intracortically generated form of spike-and-wave activity. There is a form of spike-and-wave discharge in isolated cortex or athalamic preparations in cats (Marcus and Watson, 1979; Pellegrini et al., 1979; Steriade and Contreras, 1998). This type of paroxysmal oscillation has a lower 1–2.5 Hz frequency and a morphology that is different from that of the typical 'thalamocortical' spike-and-wave oscillation (Pellegrini et al., 1979; Destexhe et al., 2001). Intracortical spike-and-wave discharges have not been observed in athalamic rats (Vergnes and Marescaux, 1992) and have never been reported in neocortical slices.

Since there are no intracellular recordings that have been made during paroxysms in cat isolated cortex, it is not clear if this oscillation represents the same spike-and-wave paroxysm observed when the thalamocortical system is intact. Nevertheless, the isolated cortex is known to display intrinsic oscillations generated by bursting neurons (Silva et al., 1991) and pyramidal cells displaying low-threshold spike (LTS) activity have been observed in some cortical areas (de la Pena and Geijo-Barrientos, 1996). It may be that these properties are sufficient to sustain a form of purely cortical spike-and-wave, through a sequence of GABA$_B$ IPSPs and rebounds, similar to the mechanism analysed here. In a recent model of intracortical spike-and-wave, a small number of LTS pyramidal cells was sufficient to generate paroxysmal oscillations with spike-and-wave field potentials in the disinhibited isolated cortex (Destexhe et al., 2001). The spike-and-wave oscillations in this model, as in experiments, had a lower frequency (1.8–2.5 Hz) and a different morphology from that in the thalamocortical model. As more experimental data become available, particularly from intracellular recordings, models for purely cortical spike-and-wave discharges should be pursued.

Another interesting direction to explore is the effectiveness of cortical EPSPs on RE cells. The present model, if correct, predicts that spike-and-wave paroxysms arise because abnormally strong cortical EPSPs are capable of overcoming the lateral GABA$_A$ inhibition in the RE nucleus and forcing RE cells to generate prolonged burst discharges. Because RE cells are characterized by dendritic T-currents (see Chapter 4), they may have a particularly high sensitivity to cortical EPSPs because of local interactions taking place in their dendrites (Destexhe, 2001). Altering these dendritic interactions is a promising new avenue for suppressing seizure activity. Drugs that selectively diminish the effectiveness of cortical EPSPs on RE dendrites could protect against spike-and-wave discharges (see Fig. 8.29B).

8.6 Summary

In this chapter, the models developed for investigating normal thalamocortical interactions in Chapter 7 were used to understand the conditions that led to spike-and-wave discharges. The models exhibited qualitative characteristics consistent with the patterns of activity observed in several experimental models of spike-and-wave epilepsy, as well in thalamic slices; both fast (5–10 Hz) and slow (~3 Hz) types of spike-and-wave activity were generated by similar thalamocortical recruitment mechanisms. The frequency of the spike-and-wave oscillations was determined mainly by intrathalamic mechanisms, based on IPSP-rebound sequences in TC cells, with the frequency depending on whether GABA$_A$ or GABA$_B$ receptors mediated these IPSPs. The threshold for initiating spike-and-wave discharges was determined by intrinsic biophysical mechanisms and the divergence of the anatomical projections in the thalamo-cortico-thalamic loop.

The computational model shows how cortico-thalamic feedback can trigger powerful ~3 Hz thalamic bursting by activating intrathalamic GABA$_B$-mediated inhibition of TC cells. The nonlinear activation properties of the GABA$_B$ receptors combined with the complex intrinsic firing properties of thalamic cells were primarily responsible for generating cortico-thalamic spike-and-wave discharges. Because thalamic RE cells generate bursts through dendritic T-currents (see Chapter 4), these dendrites may be finely tuned to cortical EPSPs. A new class of anti-absence drugs could be developed that selectively interfere with these dendritic interactions and thereby suppress seizure activity.

The large-scale synchrony of thalamocortical oscillations (Chapter 7) and the emergence of pathological oscillations in the models were dependent on the cortex, which dominated the activity in the thalamus. The models provide working hypotheses for further investigation and new approaches to suppressing seizure activity. In the next chapter, we build on these foundations to explore possible physiological functions for sleep oscillations.

A physiological role for sleep oscillations

The focus of the preceding chapters has been on the ionic and network mechanisms responsible for spindles in thalamic and thalamocortical networks and on the pathological conditions underlying absence seizures. The models, based closely on experimental observations, have led to specific interpretations of experimental observations and clear predictions that would test these interpretations. In this final chapter, we broaden the focus to other types of thalamocortical oscillations in slow-wave sleep (SWS) and conclude with some hypotheses on possible functions for these oscillations.

In the first part of this chapter, a model is used to explore the effects of the thalamic input on neocortical pyramidal neurons based on intracellular recordings *in vivo*. The model suggests that spindle oscillations are highly effective at inducing Ca^{2+} entry in the dendrites of pyramidal neurons, which might prime the neuron for biochemical events that later could lead to permanent changes in the neuron. In the second part, we analyse the spatiotemporal distribution of activity associated with different events of wake and sleep states. We then propose a model for the genesis of slow waves in cortical networks that provides insight into how synapses are selected for permanent changes. The overall view of sleep that emerges is one of widespread biochemical activity within neurons and structural reorganization of connections between neurons, as reviewed in the last section on mechanisms for memory consolidation.

9.1 Impact of thalamic inputs on neocortical neurons

As shown in previous chapters, spindle oscillations appear in the early phases of SWS, they are highly synchronized and they are generated by intrathalamic and thalamocortical loops, in which the rebound bursts of thalamocortical (TC) cells are central. EPSP/IPSP sequences are

Thalamocortical Assemblies: Sleep Spindles, Slow Waves and Epileptic Discharges. Second Edition. Alain Destexhe and Terrence J. Sejnowski, Oxford University Press. © Oxford University Press 2023.
DOI: 10.1093/oso/9780198864998.003.0009

typically evoked by synchronized thalamic inputs in neocortical pyramidal neurons during sleep spindles. The dendrites of neocortical pyramidal cells therefore receive highly synchronized and powerful excitatory inputs from the thalamus. However, despite the potentially powerful nature of these synchronized thalamic inputs, pyramidal neurons have a relatively low rate of discharge during spindle oscillations (Evarts, 1964; Steriade et al., 1974). The possibility raised here is that although strong EPSPs do indeed occur in pyramidal cell dendrites, they are invisible in somatic recordings because of strong feedforward inhibition. This possibility was tested by combining computational models and intracellular recordings (Contreras et al., 1997c), as reviewed below.

9.1.1 Experimental characterization of the thalamic input in neocortical pyramidal neurons

Intracellular recordings were performed in the suprasylvian gyrus of cats under barbiturate anesthesia (see Methods in Contreras et al., 1997c). Fig. 9.1A shows the moderate discharge of pyramidal neurons during typical barbiturate spindles. Hyperpolarizing events are clearly visible (Fig. 9.1A, right) and reversed around -70 mV (Fig. 9.1B). These presumed $GABA_A$ IPSPs had amplitudes of 7.4 ± 1.4 mV (mean $\pm$ SE; measured at -54 mV), duration of 81.2 ± 8.1 ms, and reversal of $-69.4 + 3.2$ mV.

To demonstrate that spindle-related IPSPs were indeed triggered by thalamic inputs, spontaneous IPSPs were compared with IPSPs evoked by thalamic (VP) stimulation (Fig. 9.1B, left). Thalamic stimulation triggered an EPSP in an SI neuron, followed by an IPSP under DC depolarization (-57 mV); the IPSP was completely reversed by DC hyperpolarization (-80 mV). The same reversal to current injection occurred at around -70 mV for the IPSPs during spontaneous spindle sequences (Fig. 9.1B, right). At rest (middle traces), spontaneous PSPs were barely visible, but thalamic stimulation revealed a clear EPSP-IPSP sequence. During spontaneous spindle waves, excitation in cortical neurons was less powerful than that elicited by thalamic stimulation, but the IPSPs looked almost identical.

To reveal the amount of Cl^--mediated GABAergic inhibition during each cycle of a spindle oscillation, cortical cells from SI and anterior suprasylvian area 5 were recorded with KCl filled pipettes ($n = 23$). The injection of Cl^- ions into the cells should shift the reversal potential for Cl^--mediated $GABA_A$ responses to more positive values. Shortly after impalement, cells that previously fired scattered single action potentials during spindling, started to fire 1–3 action potentials per spindle wave on the top of growing depolarizing potentials, in phase with the depth-EEG recorded from the vicinity (Fig. 9.2, 1' after impalement). After a few minutes (Fig. 9.2, 10' after impalement), the cell was firing powerful bursts of 4–7 action potentials at $\sim$100 Hz with spike inactivation, in phase with spindle waves.

To determine whether the increased amplitudes and duration of PSPs, produced by the leakage of Cl^- inside the cell, was confined to a certain phase of the spindle oscillatory cycle, the changing PSPs were compared with the EEG activity recording from the cortical depth. The top trace in the right column of Fig. 9.2 illustrates superimposed oscillatory cycles, taken from spontaneous spindle sequences occurring at different times after the impalement. The negative peak of the depth-EEG was taken as reference time for the alignment of the intracellular traces. An increased duration and amplitude appeared uniformly at both sides of the centreline, showing that although IPSPs usually dominated the second half of spindle-related synaptic events,

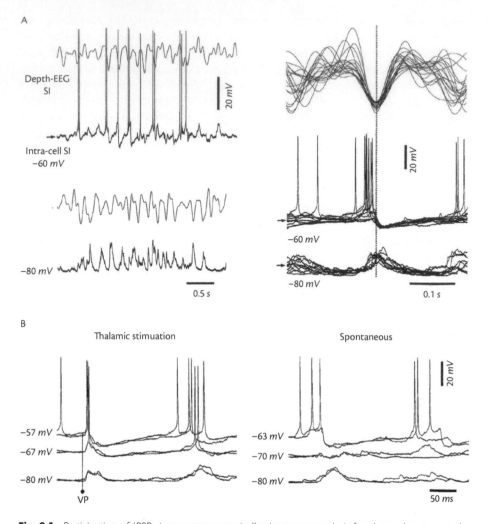

Fig. 9.1: Participation of IPSPs in spontaneous spindles in neocortex. A: Left column shows examples of spindle sequences under barbiturate anesthesia in a neuron from primary somatosensory cortex (SI), recorded together with the depth-EEG from its vicinity (Depth-EEG SI). Upper traces were under depolarizing DC (−60 mV) and lower traces under DC hyperpolarization (−80 mV). Right column: superimposed traces (ten cycles for each Vm) are from successive cycles of the same cell, at −60 mV and −80 mV; traces were aligned to the negative peaks of the depth-EEG (top traces, n = 20; time zero indicated by dotted line; ten sweeps correspond to intracellular recording at −60 mV, while the remaining ten sweeps to −80 mV; note stable EEG spindles in both cases). B: A different SI cell responded to VP stimulation (Thalamic stimulation, dot) while the Vm was changed by means of DC (resting Vm = −70 mV; two traces superimposed for each stimulus). VP-evoked responses were compared to spontaneous PSPs during spindling, aligned by the depth-EEG peak negativities (Spontaneous; EEG not depicted; see panel A) and under similar DC levels as for VP-evoked responses. Reproduced from Contreras et al., 1997c.

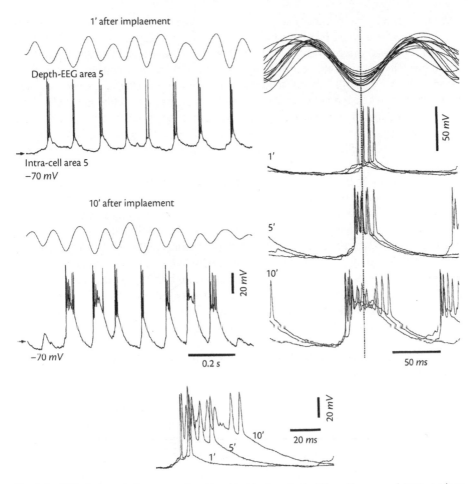

Fig. 9.2: IPSPs during spindles are mediated by chloride. A cortical cell from the suprasylvian gyrus (area 5) was recorded with a KCl (3M) loaded pipette at rest (−70 mV), simultaneously with the EEG from its vicinity (Depth-EEG area 5). Left column shows two spontaneous spindle sequences at 1′ (above) and 10′ (below) after impalement. Right column shows superimposed individual spindle waves recorded at 1′, 5′ and 10′ after impalement; alignment to the negative peak of the depth-EEG (dotted line). Bottom panel shows three individual spindle waves from the superposition at right, aligned by their initiation times. Reproduced from Contreras et al., 1997c.

they were present from the beginning of each oscillatory cycle. Spindle-related depolarizing potentials progressively changed from 7–8 mV amplitude and 30 ms duration (Fig. 9.2, right column, 1′ after impalement) to 25 mV and 50 ms (at 5′), finally reaching values of up to 45 mV in amplitude and 100 ms duration (10′). These alterations were observed without changing the resting Vm (−70 mV). A superposition of individual spindle-related synaptic potentials at 1, 5 and 10 minutes after impalement is illustrated at the bottom of Fig. 9.2. The traces were artificially aligned by the times of their departures from baseline.

Although thalamocortical excitation is synchronized over large cortical territories during spindling, the inhibitory inputs that control spindle-related excitatory events are presumably generated locally. This possibility was tested by performing dual intracellular recordings. Fig. 9.3A shows an example of two pyramidal cells from SI, recorded and stained simultaneously. The cell at left was located in upper layer V, with a pyramidal-shaped soma and a prominent apical dendrite that bifurcated early, giving rise to three trunks that ramified in layer I (see reconstruction in Fig. 9.4). Basal dendrites from this layer V cell ramified extensively into layers V and VI. The cell on the right was located in layer VI: it had an ovoidal soma with a thin apical dendrite that did not reach layer I, and two prominent basal dendritic trunks, one of them coursing far into the white matter. Both cells (Fig. 9.3B, Cell 1 = layer VI, Cell 2 = layer V) showed a mixture of PSPs leading to occasional firing during synchronized spindle oscillations in the EEG from the vicinity of Cell 1.

Single, high-amplitude IPSPs, occurring in relation to spindle waves, were confined to one of the two cells at a time (Fig. 9.3C). In the left panel of Fig. 9.3C, the EEG from the vicinity of Cell 1 shows two cycles of a spindle oscillation in which both cells displayed synchronized PSPs leading to spike firing, but only Cell 2 displayed a clear IPSP following the excitatory potential that was synchronized with Cell 1. In the right panel of Fig. 9.3C, Cell 1 displayed a high-amplitude IPSP, corresponding to the depth-EEG spindle wave. This result indicates that, although synchronized excitatory inputs from thalamic bursts trigger concomitant inhibition in virtually all cortical targets, the machinery responsible for the inhibition is local and able to generate slight differences among various sites in the circuit.

9.1.2 Computational models of thalamic inputs in neocortical pyramidal neurons

Computational models were used to explore the possibility that thalamic inputs evoke in pyramidal neurons a mixture of direct excitatory inputs and feedforward inhibition. The first step was to account for bursts of action potentials in the recordings shown above following chloride injection (Contreras et al., 1997c). The model was then used to infer possible physiological consequences of these EPSP/IPSP sequences on pyramidal neurons.

9.1.2.1 Morphology

Neurons were impaled for intracellular recording, stained with Neurobiotin (Fig. 9.3A) and reconstructed using a computerized tracing system (Eutectic), as shown in Fig. 9.4. The layer V cell had nine primary branches with a total dendritic length of 22,173 $\mu m2$ and 91,620 $\mu m2$ total area. In the layer VI cell, these parameters were respectively, 7576 $\mu m2$ and 31,225 $\mu m2$. The geometry of these cells was incorporated in the NEURON simulation environment, leading to 529 compartments for the layer V cell and 198 compartments for the layer VI cell.

Layer V neocortical pyramidal cells have a high density of dendritic spines, approximately 8,000–14,000 spines per cell (Larkman, 1991) representing 25–45% of the total cell surface (DeFelipe and Fariñas, 1992; Mungai, 1967). In the model, the dendritic surface was corrected by assuming a uniform spine density that increased the dendritic surface by 45% (~30% of the cell surface). Surface correction was made by rescaling the size of each compartment or by rescaling the values of the specific membrane capacitance (C_m) and conductances in dendrites by a factor of 1.45. The two methods gave similar results.

An axon was modelled on the basis of serial electron microscopic measurements of neocortical pyramidal cells (Fariñas and DeFelipe, 1991b), as modelled previously (Mainen et al., 1995).

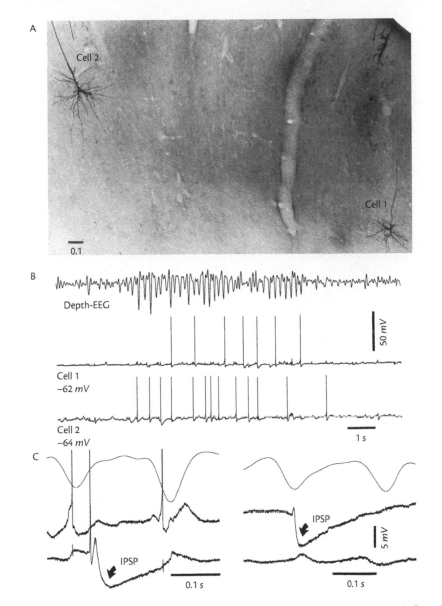

Fig. 9.3: IPSPs during spindles are generated locally. A: Two cells were recorded simultaneously from the postcruciate SI cortex with pipettes containing 2% Neurobiotin. Both were found on the same section: Cell 1 (at right) was in layer VI and Cell 2 was in layer V. B: Example of a spontaneous spindle sequence in both cells. Depth-EEG recorded from the vicinity of cell 1. C: Two individual spindle waves from a different sequence as shown in B, showing IPSP visible only in Cell 2 (left column, arrow) and in Cell 1 (right column). Depth-EEG recorded from the vicinity of Cell 1. Reproduced from Contreras et al., 1997c.

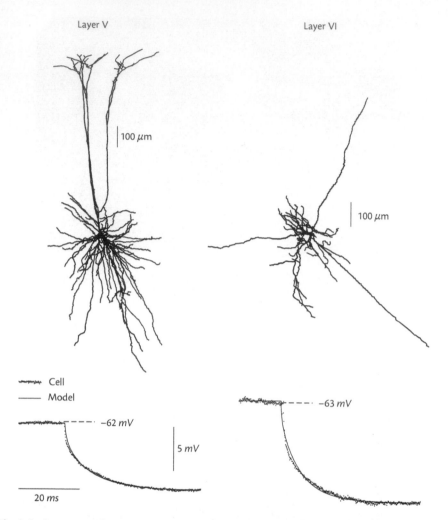

Layer V

Layer VI

100 μm

100 μm

⌇⌇⌇⌇ Cell

—— Model

$-62\,mV$

$-63\,mV$

5 mV

20 ms

Fig. 9.4: Geometry and passive properties of simulated cortical pyramidal cells. Two neocortical neurons stained during dual intracellular recordings (Fig. 9.3A) were reconstructed using a tracing system and incorporated into NEURON for simulations. The membrane area was corrected for spines assuming a 45% increase in dendritic surface due to spines (Mungai, 1967; DeFelipe and Fariñas, 1992). The passive parameters of the model were estimated by fitting simulations to the recordings of passive responses to hyperpolarizing current pulses in the specific cells that were stained (lower panels). Reproduced from Contreras et al., 1997c.

The axon consisted of a 10 μm hillock, a 15 μm unmyelinated initial segment and a 504 μm of axon proper. The axon hillock had 10 compartments arranged in a conical form, tapering from 4.3 μm at the soma level to 1 μm at the level of the axon initial segment. The initial segment consisted of ten compartments (1 μm diameter). The axon proper was divided into five myelinated compartments of 100 μm, interleaved with nodes of Ranvier (length of 1 μm) having a higher density of Na+ channels (see details in Mainen et al., 1995). In an alternative model, the axon was simulated using a single equivalent cylinder for the hillock and a single compartment for the initial segment; only minimal differences were found with results from the detailed model in the context of the present simulations.

9.1.2.2 Passive properties

In order to obtain the correct time constant and input resistance, passive parameters were estimated by fitting the model to the voltage trace obtained in the same cell that was reconstructed. The fit was performed using a simplex fitting algorithm (Press et al., 1986). If the geometry of the cell is known, this procedure should in principle lead to a unique set of passive parameters (leak conductance g_{leak}, leak reversal potential E_{leak}, specific capacitance C_m, specific intracellular resistivity R_i) assuming they are distributed uniformly in the cell membrane (Rall et al., 1992). We found that multiple solutions were possible for a given cellular geometry with uniform passive properties. The fitting was therefore further constrained by forcing the capacitance to $C_m = 1 \mu F/cm^2$, a typical value for neuronal membranes. The fitting procedure then converged to a unique set of passive parameters from different initial conditions, and led to passive responses closely matching that of the recorded cells (Fig. 9.4, lower panels). Thus the input resistance and time constant of the model closely matched that of the cell recorded experimentally. The optimal parameters were: $g_{leak} = 0.097 \ mS/cm^2$, $E_{leak} = -61.2 \ mV$, $C_m = 1 \ \mu F/cm^2$, $R_i = 384 \ \Omega cm$ for the layer V cell, and $g_{leak} = 0.13 \ mS/cm^2$, $E_{leak} = -63.25 \ mV$, $C_m = 1 \ \mu F/cm^2$, $R_i = 184 \ \Omega cm$ for the layer VI cell.

9.1.2.3 Active properties

Active currents were inserted into the soma, dendrites and axon with different densities in accordance with available biophysical evidence. In light of patch-clamp data showing that the soma and dendrites of neocortical and hippocampal pyramidal cells have similar Na^+ channel densities (Huguenard et al., 1989; Stuart and Sakmann, 1994; Magee and Johnston, 1995a), the Na^+ conductance was set to 70 $pS/\mu m^2$ (20–100 $pS/\mu m^2$ tested) in dendritic and somatic compartments. This value corresponds to the estimated value for Na^+ current densities in adult pyramidal neurons of the hippocampus. Axonal densities used were in agreement with binding and immunohistochemical studies (Black et al., 1990) indicating that there is a much higher Na^+ channel density in the initial segment and nodes of Ranvier compared to the internodal, somatic and dendritic membranes. A density of Na^+ channels of 30,000 $pS/\mu m^2$ was chosen for the axon initial segment and nodes of Ranvier. This high density was required to reproduce the overshooting somatic action potentials observed in most *in vivo* intracellular recordings, while being consistent with the lower density in soma and dendrites (see also Mainen et al., 1995; Rapp et al., 1996). The Na^+ channel density of the internodal segments was set to the same value as in the soma and dendrites. This type of distribution falls within the range of estimated Na^+ channel density in the axon (Black et al., 1990) and is also in agreement with a recent model of the axon (Mainen et al., 1995).

High-threshold Ca^{2+} currents (I_{CaL}) were inserted in the soma (30 $pS/\mu m^2$) and dendrites (30 $pS/\mu m^2$ in the proximal regions up to 50 μm from soma, and 15 $pS/\mu m^2$ elsewhere) according to the densities estimated in hippocampal pyramidal cells by patch-clamp recordings (Magee and Johnston, 1995). The delayed-rectifier K^+ conductance (I_{Kd}) was inserted in dendrites (100 $pS/\mu m^2$) in the soma (200 $pS/\mu m^2$, tested range, 20–300 $pS/\mu m^2$) and the axon initial segment and nodes of Ranvier (2,000 $pS/\mu m^2$). A non-inactivating K^+ conductance (I_M)

was also included in the soma and dendrites with a density of 5 $pS/\mu m^2$. This value was required to reproduce the repetitive firing behaviour of neocortical pyramidal neuron observed *in vivo*.

9.1.2.4 Kinetics of intrinsic and synaptic currents

The kinetics of the voltage-dependent currents (I_{Na}, I_{CaL}, I_{Kd} and I_M) were described by Hodgkin-Huxley type equations (Hodgkin and Huxley, 1952) and were simulated using NEURON (Hines, 1993). Kinetic parameters for I_{Na} and I_{Kd} were taken from models of hippocampal pyramidal cells (Traub and Miles, 1991; see Chapter 2), with reversal potentials of +50 mV and −90 mV, respectively. Previous models of neocortical pyramidal cells were used for the kinetics of I_{CaL} (Reuveni et al., 1993) and I_M (Gutfreund et al., 1995). Intracellular Ca^{2+} dynamics was modelled in a 0.1 μm shell beneath the membrane (see Destexhe et al., 1993): influx of Ca^{2+} occurred through Ca^{2+} channels and Ca^{2+} efflux was a fast decaying signal (time constant of 5 ms) to an equilibrium intracellular Ca^{2+} concentration of 240 nM. The extracellular Ca^{2+} concentration was kept at 2 mM and the Ca^{2+} reversal potential was calculated using the Nernst relation (equilibrium reversal potential of about +120 mV). All simulations corresponded to a temperature of 36 °C (temperature factor for time constants was $Q_{10}=3$).

AMPA, NMDA and GABA$_A$ receptors were modelled using two-state kinetic models (Chapter 5; Destexhe et al., 1994d). Other types of synaptic responses were not included because they were not required to account for the behaviour of the cells and also because GABAergic and glutamatergic synapses account for the vast majority of synapses in the cerebral cortex (White, 1989; DeFelipe and Fariñas, 1992). Based on whole-cell recordings of hippocampal pyramidal and dentate gyrus cells (Otis and Mody, 1992; Xiang et al., 1992; Hessler et al., 1993), kinetic models of GABA$_A$, NMDA and AMPA receptors were fit to experimental data using a simplex procedure (see details in Chapter 5).

9.1.2.5 Density of synapses in different regions of the cell

The relative density of glutamatergic and GABAergic synapses in each region of the cell was constrained by morphological data (reviewed in DeFelipe and Fariñas, 1992). Electron microscopic observations of neocortical pyramidal cells have established that the soma, axon hillock and initial segment exclusively form symmetric synaptic contacts (Jones and Powell, 1970; Peters et al., 1981) that are most likely GABAergic (White, 1989; DeFelipe and Fariñas, 1992). Quantitative studies have revealed that the density of symmetric synapses on the soma of deep pyramidal cells is 10.6±3.7 per 100 μm^2 while there are 20–24 symmetric synapses on the initial segment (Fariñas and Defelipe, 1991a, 1991b). In the model the surface area of the initial segment was 47.1 μm^2, which yields a GABAergic synaptic density of ~45 per 100 μm^2. A similar density was assumed for the axon hillock.

About 16% of the total number of synapses found on neocortical pyramidal cells are symmetric: 7% of these contact the soma and 93% are on dendrites (DeFelipe and Fariñas, 1992). In the layer V neuron illustrated in Fig. 9.4, the soma surface was 3200 μm^2 and the dendritic surface was 91,620 μm^2, yielding a ratio of about 2.3:1 in favour of the soma. In this case, the relative densities of GABAergic synapses were 0.26, 0.6, 2.5 and 2.5, for the dendrites, soma, axon hillock and initial segment, respectively.

In neocortical pyramidal neurons, the vast majority of asymmetric synapses are found on dendritic spines (White, 1989). Thus, in models, excitatory synapses were exclusively located in the dendrites, with a uniform density, except for the most proximal dendritic segments (up to 40 μm from soma), which are devoid of spines (Jones and Powell, 1968; Peters and Kaiserman-Abramof, 1970) and mostly form symmetric synapses (Jones and Powell, 1970). Optical microscope inspection of the two cells modelled here confirmed the absence of spines in proximal dendrites. Thus, in the model, excitatory synapses were exclusively located in the dendrites, with a uniform density, and no excitatory synapses were located in the first 40 μm dendritic segments close to soma.

9.1.3 Modelling the consequences of spindles on neocortical pyramidal neurons

9.1.3.1 Model of thalamic postsynaptic potentials in pyramidal neurons

The model of thalamic EPSP/IPSP sequences was based on three sets of constraints: (a) the relative density of excitatory and inhibitory synapses in different regions of the cell, (b) the fraction of these synapses activated by a given stimulus, and (c) the kinetics of the different types of receptors involved. The relative ratios of excitatory and inhibitory terminal densities in different regions of the cell were estimated above; further data are needed to determine the absolute magnitude of the synaptic currents activated by any given stimulus, as shown below.

If thalamic stimulation is subthreshold in the recorded cell, it is likely to be subthreshold in the majority of pyramidal cells in the neighbourhood of the cell. It seems therefore reasonable to assume that the EPSP resulting from thalamic stimulation is essentially due to the direct activation of excitatory synapses by thalamocortical fibres, with polysynaptic contributions making a negligible contribution. Thalamic axons end preferentially in layers I, IV and VI (Herkenham, 1980; Jones, 1985). Based on these anatomical data, we estimated that thalamocortical synapses are distributed 15% in layer I (more than 800 μm above the soma), 60% in layer IV (from 50 μm to 200 μm above the soma), and 25% in layer VI (below 200 μm of the soma); there were no excitatory synapses in any other layers in the model. A schematic of this distribution is shown in Fig. 9.5A.

Since there are no direct inhibitory inputs following thalamic stimulation in neocortical pyramidal cells, IPSPs presumably occur through the feedforward recruitment of interneurons by thalamocortical fibres. Subthreshold thalamic stimulation is likely to fire interneurons due to the higher input resistance of the smaller inhibitory neurons. We therefore assumed that that thalamic stimulation first recruits cortical cells through short bursts of high-frequency action potentials. Each excitatory synapse received a burst of randomized presynaptic pulses at an average rate of 300 Hz (range tested, 100–400 Hz) during 10 ms (range tested, 5–20 ms). Inhibitory synapses received a similar presynaptic pattern, after a 2 ms delay (range tested, 1–4 ms).

These parameters were fit to recordings of EPSP/IPSP sequences following thalamic stimulation in KAc- and KCl-filled pipettes. With the ratios of synapses and the pattern of activation described above, the maximal conductance of excitatory and inhibitory synapses was

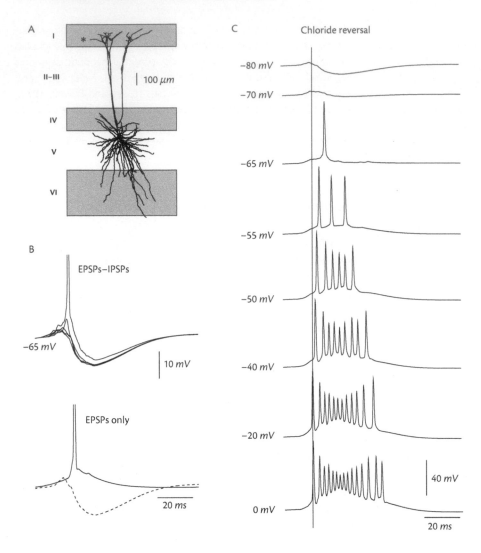

Fig. 9.5: Simulation of postsynaptic potentials and evoked bursts following thalamic stimulation. A: Distribution of direct excitatory inputs in layer V pyramidal cell. Grey shaded areas indicate the limits of cortical layers I, IV and VI, in which excitatory inputs were simulated in the cell. Inhibitory inputs were simulated in all layers, with a higher density in soma and proximal dendrites. B: Simulated EPSP/IPSP sequence following synaptic activation. The top panel indicates three superimposed responses of the cell, which occasionally led to firing (spike truncated) in about 3% of the cases. In bottom panel, the response of the cell to the same EPSPs, in the absence of IPSP, led to firing in 95% of the cases. C: Progressive shift in the reversal of chloride ions changed the response of the cell from EPSP/IPSP sequence (−80 mV), to shunting response (−70 mV), and gradually larger EPSPs and bursts of action potentials with spike inactivation. Reproduced from Contreras et al., 1997c.

constrained by the relative amplitude and time course of EPSP/IPSP sequences and by the typical bursts of action potentials induced by thalamic stimuli in Cl-filled pipettes.

In normal conditions (KAc-filled pipettes), subthreshold thalamic stimuli evoked an EPSP followed by an IPSP (see Fig. 9.1). These conditions could be reproduced by a large range of parameters, provided that IPSPs were delayed by about 2 *ms* following the EPSPs (Fig. 9.5B). However, there is a possibility that the hyperpolarization following the EPSP is due to the activation of dendritic currents without spiking in the soma, such as dendritic $I_{K[Ca]}$ (Sah and Bekkers, 1996; Lang and Paré, 1997).

The responses of pyramidal cells to the same stimulus intensities, recorded by using KCl-filled electrodes, provided more constraints to the model. The same parameters could reproduce the EPSP/IPSP sequence in normal conditions (Fig. 9.5B) and the gradual development of a burst of action potentials with Cl⁻ ions in the cell (Fig. 9.5C). This constraint narrowed down the range of parameters that reproduce these phenomena. It was consistently found that bursts of action potentials, similar to those observed during recordings, are primarily due to the reversal of IPSPs in the somatic region, which produces a powerful inward current near the soma (see also below).

The reversed Cl⁻ currents produced powerful spike bursts, but they were also accompanied by a significant phase advance of action potential firing relative to the peak of the field potential (see Fig. 9.2, right panel). This feature could be reproduced in the model (Fig. 9.5C) assuming that EPSPs and IPSPs do not arrive exactly at the same time at each synapse, leading to a short period of competition between EPSPs and IPSPs in the dendrites. When IPSPs are reversed, instead of competing against each other, EPSPs and IPSPs sum, leading to earlier firing. Assuming that synaptic events occurred with a standard deviation of about 6 *ms* for EPSPs and 4.5 *ms* for IPSPs (plus a 2 *ms* delay), a phase advance of about 10 *ms* was reproduced (Fig. 9.5C), comparable to experimental data. Standard deviations of 4–6 *ms* are compatible with the variability of cellular discharges observed in intracellular recordings of cortical and thalamic neurons *in vivo*, as compared to field potentials (D. Contreras and M. Steriade, personal communication).

9.1.3.2 Mechanisms underlying the 'chloride bursts'

The type and localization of currents underlying Cl⁻-induced bursts were investigated in the model by adding or removing currents in selected regions of the cells. Suppressing IPSPs in dendrites had minimal effects (Fig. 9.6A), but the burst was highly affected by removing proximal IPSPs (Fig. 9.6B). Removing dendritic Ca^{2+} (Fig. 9.6C) or Na^+ (Fig. 9.6D) currents caused minimal changes. Dendritic EPSPs and IPSPs could not account for the burst (Fig. 9.6E–F), but keeping only proximal IPSPs produced bursts similar to those in the control condition (Fig. 9.6G). Finally, injection of an inward triangular ramp current in the soma led to a burst with a morphology similar to that induced by synaptic currents (Fig. 9.6H).

An additional factor is the distribution of Na^+ channels. In the model, Na^+ currents were the major inward current and were located with highest density in the axon and initial segment. In order to determine the influence of this distribution, simulations were performed using uniform Na^+ channel density throughout the neuron. To generate overshooting spikes, the uniform density had to be increased to several times that estimated by Magee and Johnston (1995) in adult pyramidal cells (not shown), a conclusion also reached in other modelling studies (Mainen et al., 1995; Rapp et al., 1996; Paré et al., 1998). Simulations using this higher uniform density had negligible influence on the properties of the bursts described above.

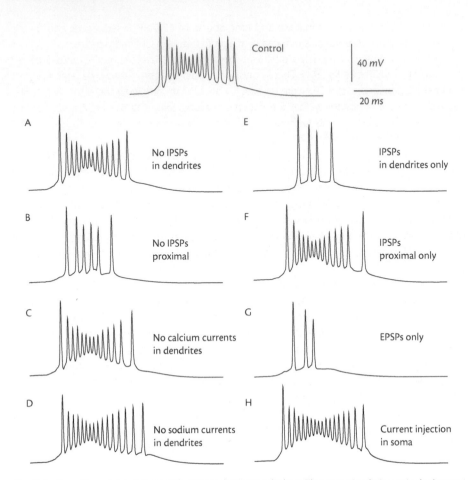

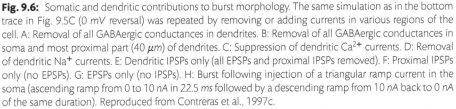

Fig. 9.6: Somatic and dendritic contributions to burst morphology. The same simulation as in the bottom trace in Fig. 9.5C (0 *mV* reversal) was repeated by removing or adding currents in various regions of the cell. A: Removal of all GABAergic conductances in dendrites. B: Removal of all GABAergic conductances in soma and most proximal part (40 μm) of dendrites. C: Suppression of dendritic Ca^{2+} currents. D: Removal of dendritic Na^{+} currents. E: Dendritic IPSPs only (all EPSPs and proximal IPSPs removed). F: Proximal IPSPs only (no EPSPs). G: EPSPs only (no IPSPs). H: Burst following injection of a triangular ramp current in the soma (ascending ramp from 0 to 10 *nA* in 22.5 *ms* followed by a descending ramp from 10 *nA* back to 0 *nA* of the same duration). Reproduced from Contreras et al., 1997c.

These features show that Cl⁻ bursts can be explained by a massive inward current in the somatic region, with little participation of dendritic currents. The perisomatic localization of the current underlying these bursts is also supported by morphological data. In cortical pyramidal cells, the largest density of GABAergic synapses is near the soma; excitatory synapses only occur in dendrites more than 40 μm away from soma (reviewed in DeFelipe and Fariñas, 1992). Therefore, in these cells the reversal of GABAergic IPSPs is likely to cause a strong inward current located mainly in the somatic region.

9.1.3.3 Estimating the synaptic conductances of thalamic inputs

The following procedure was used to estimate the plausible range of glutamatergic and GABAergic conductance densities arising from thalamic inputs. First, the GABAergic conductance density can be estimated based on Cl^--induced bursts of action potentials. With no EPSP present, GABAergic conductance densities that produced bursts of similar morphology as in experiments were in the range of 0.6 to 6 mS/cm^2 in the proximal region. Adding dendritic EPSPs or IPSPs had only a small effect on the shape of the burst (see Fig. 9.6). Therefore, because the Cl^- bursts are dominated by reversed $GABA_A$ conductances in the proximal regions, matching the model to experiments provides a rough estimate of this conductance.

Second, the glutamatergic conductance density can be estimated based on the EPSP-IPSP sequence in normal conditions. We found that dendritic EPSPs significantly affected the EPSP-IPSP sequence in control conditions. EPSPs must be strong enough to compensate for the powerful GABAergic conductance of the IPSP. The density of dendritic AMPA and NMDA conductances that gave rise to EPSP-IPSP sequences of correct amplitude and phase in the soma was in the range of 5–7.5 mS/cm^2, a much narrower range than for IPSPs. These simulations were performed using the layer V morphology shown in Fig. 9.4, but the same range of conductance densities also applied to the layer VI cell.

The shape of the burst and the features of the EPSP-IPSP sequence in the soma varied depending on the values of conductance values given above. For example, GABAergic conductance densities close to 6 mS/cm^2 gave rise to IPSPs of large amplitude in the soma and the Cl^- bursts led to spike inactivation, as sometimes seen during the experiments. The optimal set of conductance densities was around 6.3 mS/cm^2 for dendritic AMPA and NMDA, 1 mS/cm^2 for dendritic GABA, and 3 mS/cm^2 for GABA in soma and proximal dendrites. A ratio of about 1:3 between somatic and dendritic GABAergic conductance corresponds to the ratio of density of inhibitory synapses in soma and dendrites of cortical pyramidal cells (DeFelipe and Fariñas, 1992).

9.1.3.4 Thalamic stimuli evoke calcium entry in pyramidal cell dendrites

Although the details varied, the main qualitative conclusion of the model given above is that the EPSP-IPSP sequences due to thalamic stimulation are characterized by both strong excitatory and strong inhibitory inputs arriving in the cell at approximately the same time. What could be the function of such inputs?

To answer this question, we investigated the spatial profile of membrane potential and Ca^{2+} following thalamic stimulation in the model. Although thalamic stimulation is subthreshold in the soma (Fig. 9.7A), there were significant voltage transients in the dendrites (Fig. 9.7B). In Fig. 9.7, the simulations exhibited Na^+ and Ca^{2+} spikes in the dendrites following simulated thalamic stimulation. However, the proximal inhibition drastically attenuated somatic invasion of these transients. Due to the activation of dendritic Ca^{2+} currents, the voltage transients were accompanied by significant Ca^{2+} entry in the dendrites (Fig. 9.7C). This finding is consistent with the observation that visual inputs can induce Ca^{2+} spikes in visual cortical neurons after intracellular Na+ blockade (Hirsch et al., 1995).

These simulations suggest that during thalamic spindles, cortical pyramidal neurons receive mixed high-conductance excitatory and inhibitory inputs, with excitatory inputs dominating in dendrites and strong inhibition in the soma. This asymmetric distribution of conductances

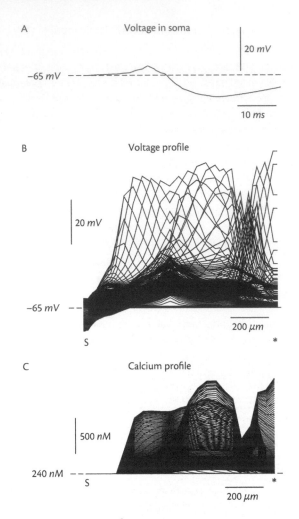

Fig. 9.7: Predicted dendritic voltage and Ca^{2+} transients during synchronized thalamic inputs. EPSP-IPSP sequence following simulated thalamic input is shown in soma and dendrites (same simulation as the top trace of Fig. 9.5C with −80 mV reversal). A: EPSP-IPSP sequence in the soma. B: Profiles of voltage along a path from soma (S) to distal apical dendrite (*; indicated in Fig. 9.5A). Instantaneous profiles taken every 0.2 ms were superimposed. High-amplitude Na^+- and Ca^{2+}-dependent transients were seen in the dendrites but not in the soma. C: Profiles of intracellular Ca^{2+} concentration represented as in B. The intracellular Ca^{2+} concentration was calculated in a thin shell (0.1 μm) beneath the membrane. Reproduced from Contreras et al., 1997c.

provides profound depolarization of the dendrites, while preventing cells from excessive discharges. This input pattern evokes massive Ca^{2+} entry into the dendrites of pyramidal neurons; the physiological consequences of intracellular Ca^{2+} buildup during spindle oscillations may be important (see discussion below).

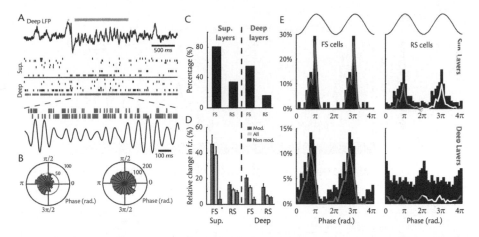

Fig. 9.8: Strong recruitment of interneurons during sleep spindles in rat prefrontal cortex. A–B: Example neuronal recruitment by spindles by layers. A: Deep LFP trace (top) shows a sleep spindle sequence at about 15 Hz (grey bar); below: raster of discriminated units spiking activity in the superficial layers (II/III) and deep layers (V/VI). Firing from two FS cells, one in each layer, is depicted here in colour while RS cell spikes are displayed in black. Note the delta wave preceding the spindles (green asterisk). Bottom panel displays an enlarged picture of the spindle oscillation: filtered deep LFP trace (10–15 Hz) is shown together with the spiking activity of the same FS cells as above. B: Phase histogram of the example FS cells from panel A relative to spindle oscillations (same colour code). C: Percentage of significantly phase modulated cells. D: Average relative change in firing rate during spindle oscillations compared with non spindle SWS for significantly phase-locked cells (*Mod.*, *purple*) or not (*Non mod.*, *grey*) and for all the cells (*All*, *green*) changed to green. Error bars display SEM. E: Distributions of preferred firing phase of cells. Black bars indicate preferred phase for all cells, distributions for significantly phase-modulated cells only are shown in colour lines. Modified from Peyrache et al., 2011.

9.1.3.5 Experimental confirmations of model predictions

Some of the predictions formulated by the models of this section seem to have been confirmed experimentally. A first prediction was that cortical pyramidal neurons should be powerfully inhibited during cortical spindles (Contreras et al., 1997). This prediction has been validated by the finding that inhibitory neurons are highly active during spindles in rats (Fig. 9.8; Peyrache et al., 2011), and during other types of oscillation in human and monkey cortex (Dehghani et al., 2016; Le Van Quyen et al., 2016).

A second prediction of the model was that spindles should strongly induce calcium entry in the dendrites of pyramidal cells, which could activate specific molecular gates such as protein-kinase A, leading to long-term synaptic plasticity (Sejnowski and Destexhe, 2000; Destexhe and Sejnowski, 2001; see Section 9.3.1). This prediction has been verified in experiments where unusually strong calcium signals have been measured in pyramidal cell dendrites during sleep spindles (Fig. 9.9; Seibt et al., 2017). Note that these are the strongest calcium signals seen across all behavioural states (Seibt et al., 2017).

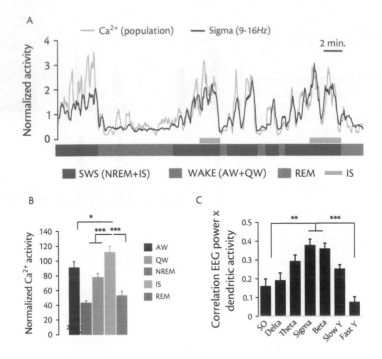

Fig. 9.9: Sleep spindles correlate with increased calcium activity in pyramidal cell dendrites. A. Example of Ca^{2+} activity in population of dendrites of L5 pyramidal neurons and sigma-frequency (9–16 Hz) activity fluctuations for a ~30-min recording segment comprising periods of wakefulness (WAKE=active wake [AW]+quiet wake [QW]), slow wave sleep (SWS=NREM+Intermediate State [IS]) and REM sleep. The periods of transitional state IS is shown by green bars during SWS. Data are represented as trendlines. B. Mean (±s.e.m.) Ca^{2+} activity in populations of dendrites across behavioural states. The largest activity was found for the spindle-rich during sleep. C. Mean (± s.e.m.) Pearson correlation coefficient across animals between dendritic Ca^{2+} activity and EEG power for all frequency bands (SO=0.5–1.5 Hz, delta=1–4 Hz, theta=5–9 Hz, sigma=9–16 Hz, beta=16–30 Hz, slow=30–50 Hz, fast=60–100 Hz) for N =eleven rats. Modified from Seibt et al., 2017.

9.2 Oscillations during natural sleep and wakefulness

In addition to the spindle oscillations investigated above, the electroencephalogram (EEG) exhibits a rich variety of oscillation patterns during wake and sleep. Here we investigate the spatiotemporal distribution of slow oscillations of natural sleep as well as fast oscillations during wake and rapid-eye movement (REM) sleep episodes. We also relate these oscillations to the firing of extracellularly recorded cortical neurons at multiple sites, setting the stage for exploring the possible physiological roles of these oscillations (see details in Destexhe et al., 1999b).

9.2.1 Spatiotemporal coherence of local field potentials during natural wake and sleep states

Multisite local field potentials (LFPs) were recorded using a set of eight equidistant bipolar electrodes in the cerebral cortex (suprasylvian gyrus) of unanesthetized cats. Wake/sleep states were identified using the following criteria: Wake: low-amplitude fast activity in LFPs, high electrooculogram (EOG) and high electromyogram (EMG) activity; Slow-wave sleep: LFPs dominated by high-amplitude slow-waves, low EOG activity and EMG activity present; REM sleep: low-amplitude fast LFP activity, high EOG activity and abolition of EMG activity. During waking and attentive behaviour, LFPs were characterized by low-amplitude fast (15–75 Hz) activity (Fig. 9.10A, Awake). During SWS, LFPs were dominated by high-amplitude slow-wave complexes occurring at a frequency of <1 Hz (Fig. 9.10B, Slow-wave sleep). Slow-wave complexes of higher frequency (1–4 Hz) and spindle waves (7–14 Hz) were also present in SWS. During periods of REM sleep, cortical activity was similar to that observed during awake states (Fig. 9.10C, REM sleep).

The decay of correlations as a function of distance revealed marked differences in large-scale coherence between awake/REM and SWS (Fig. 9.10, right panels). Slow-wave complexes during SWS episodes displayed high spatiotemporal coherence, in contrast with the steeper decline of the correlations with distance during wakefulness and REM sleep. The same patterns of the spatial correlations were observed in different animals and during different wake/sleep episodes in the same animals (Fig. 9.10) (see details in Destexhe et al., 1999b and references therein).

The local coherence of fast oscillations contrasts with the large-scale synchrony of slow-waves. This difference was further investigated by monitoring the evolution of local correlations as a function of time. The maximum of the peak of the crosscorrelation between two neighbouring sites was evaluated in successive 100 ms time windows for the fast oscillations and in 2 s time windows for SWS, in order to have a similar number of oscillation cycles. The representation of the maximal correlation as a function of time is shown in Fig. 9.11. During waking periods (Fig. 9.11A) as well as during REM sleep (Fig. 9.11B), neighbouring electrodes were occasionally synchronized, as shown by correlations close to 1, but only for short periods of time (100–500 ms). Distant electrodes displayed lower correlations values (1–4 in Fig. 9.11A), as did 'shuffled' signals ('Sh'. in Fig. 9.11A).

The coherence was radically different during SWS (Fig. 9.11B): correlations between neighbouring electrodes tended to stay close to 1, and although distant electrodes displayed lower correlations, the coherence was still high (1–4 in Fig. 9.11B). These results indicate that fast oscillations are characterized by brief periods of synchrony between neighbouring electrodes, occurring irregularly and within short time windows, in contrast to SWS, during which slow-wave complexes always appeared coherently over large distances.

9.2.2 Correlations with unit discharges

LFPs and single unit activity were separated by standard procedures (see details in Steriade et al., 1996; Destexhe et al., 1999b). During the waking state, units tended to discharge tonically, as observed previously (Hubel, 1959; Evarts, 1964; Steriade et al., 1974). The relationship between units and LFP was not immediately obvious although there was a tendency to discharge during LFP negativity (see below). During SWS, the pattern of discharge was more phasic and

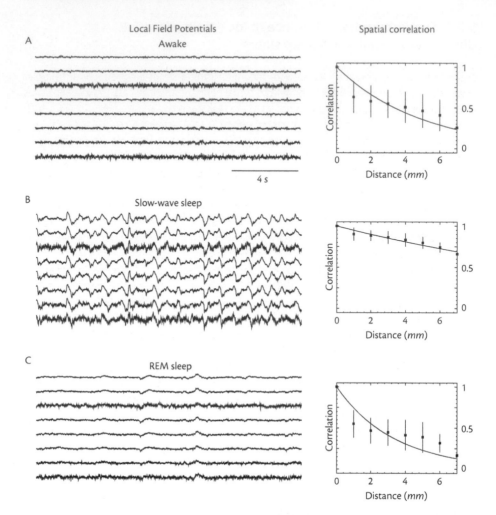

Fig. 9.10: Multisite local field potentials in cat cerebral cortex during natural wake and sleep states. Eight bipolar electrodes (interelectrode distance of 1 *mm*) were inserted into the depth (1 *mm*) of areas 5–7 of cat neocortex (suprasylvian gyrus, areas 5–7); same arrangement of electrodes and recordings as in Fig. 7.6. Local field potentials (LFPs) are shown (left panels) together with a representation of the correlations as a function of distance (Spatial correlations; right panels). A. When the animal was awake, LFPs were characterized by low-amplitude fast activities in the beta/gamma frequency range (15–75 *Hz*). Correlations decayed steeply with distance. B. During SWS, the LFPs were dominated by large-amplitude slow-wave complexes recurring at a slow frequency (<1 *Hz*; up to 4 *Hz*). Correlations stayed high for large distances. C. During episodes of REM sleep, LFPs and correlations had similar characteristics as during wake periods. Modified from Destexhe et al., 1999b.

characterized by periods of silences and of increased firing, as reported previously (Evarts, 1964; Steriade et al., 1974). Positive deflections of slow-wave complexes were almost always associated with a neuronal silence in all units, while negative deflections tended to be correlated with a brief increase of firing (Fig. 9.12A). REM sleep displayed similar activity patterns as in awake animals.

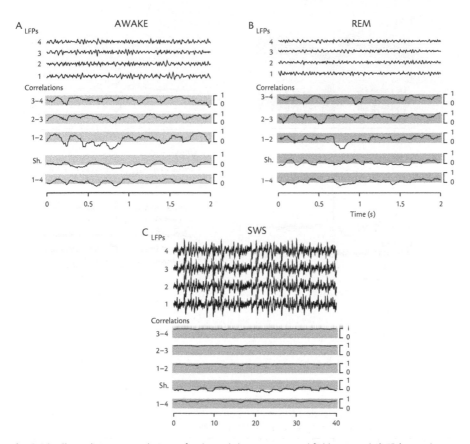

Fig. 9.11: Short-distance correlations of wake and sleep states. Local field potentials (LFPs) recordings in the suprasylvian gyrus are shown together with the maximal cross-correlation (Correlations) calculated between pairs of electrodes (1–2, 2–3, 3–4, 1–4 pairs are shown; 'Sh'. indicates the control correlation obtained between electrode 1 and the same signal taken 20 seconds later). A. Fast oscillations during wake periods. Neighbouring electrodes were occasionally synchronized, as shown by correlations close to 1, but only for short periods of time (100–500 ms). B. Period of REM sleep. Fast oscillations had a similar dynamics as in A, consisting in brief periods of synchrony between neighbouring electrodes, occurring irregularly and within short time windows. C. Period of SWS with similar number of oscillation cycles as in A and B (note difference in time scale). In this case, correlations between neighbouring electrodes stayed close to unity and the synchrony extended to the entire recorded area. Correlations were calculated in successive time windows of 100 ms for A and C and 2 s for B. Modified from Destexhe et al., 1999b.

The temporal modulation of unit firing during SWS was investigated by calculating wave-triggered averages. Superposition of LFPs and unit activity showed that slow-wave complexes almost invariably correlate with a silence in the units (Fig. 9.12A). Wave-triggered averages calculated over 210 complexes revealed a marked modulation of firing rate associated with slow-wave complexes (Fig. 9.12B, Units avg), while the same procedure performed with randomly shuffled spikes did not show any pattern (Fig. 9.12B, control). This analysis demonstrates a drop of firing rate in close correspondence to the depth-positivity of slow-wave complexes, while depth-negativity is associated with an increase of neuronal firing (Fig. 9.12B), in agreement with

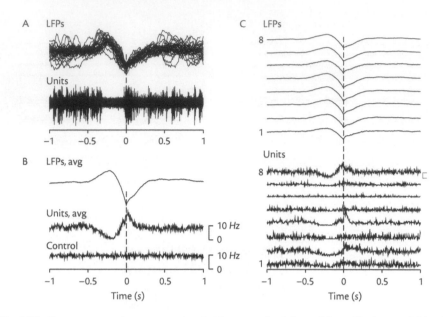

Fig. 9.12: Slow-wave complexes are correlated with a generalized silence followed by increased firing. A. Individual slow-wave complexes were detected numerically during SWS and were aligned with respect to the negative peak of local field potentials (LFPs). The multiunit discharges detected in the same electrode were aligned similarly (Units). B. Wave-triggered averages of field potentials and multiunit discharges. The averaged field potentials (LFPs, avg) were constructed by averaging the LFP over the eight electrodes and over 210 detected slow-wave complexes. The resulting averaged LFP consisted in a slow positivity followed by a sharp negativity. The corresponding multiunit discharges were averaged similarly (Units, avg) and displayed a drop of firing rate correlated with LFP positivity, followed by an increase of firing during the LFP negativity. The same wave-triggered average did not show any modulation of firing rate if performed on randomly shuffled spikes (Control). C. Spatial profile of the relation between units and LFPs. Local field potentials, averaged over 210 slow-wave complexes, are shown for each electrode. The bottom traces show the corresponding wave-triggered averages of multiunit discharge at each electrode. Slow-wave complexes consisted of a widespread drop of firing, correlated with LFP positivity, followed by a synchronized increase of firing, correlated with LFP negativity. These events were synchronous over the entire extent of the cortical area recorded (7 *mm*). Modified from Destexhe et al., 1999b.

previous intracellular observations under anesthesia (Contreras and Steriade, 1995; Steriade et al., 1996) and in awake animals (Steriade et al. 2001; see Fig. 9.15).

The same analysis was also performed as a function of distance (Fig. 9.12C). Wave-triggered averages performed simultaneously between LFP and cells at each electrode showed that the depth-positivity of slow-wave complexes corresponded to a concerted silence in almost all units, while the depth-negativity was correlated with an increased firing. Only one unit was not correlated with LFP (Unit 6 in Fig. 9.12C). In other experiments with fewer electrodes, all units were correlated with the LFP (not shown). This analysis shows that slow-wave complexes are characterized by a generalized decrease of firing occurring over large cortical distances, followed by an increased firing occurring in rebound to each period of silence.

During wake and REM sleep, units fired tonically and their relation with LFPs was not evident at first sight. The same wave-triggered average procedures during fast oscillations of waking periods using the peak negativity of LFP to trigger the averaging procedure computed from a

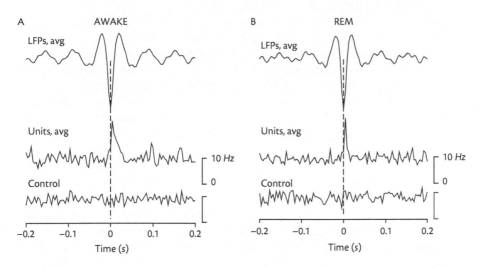

Fig. 9.13: Fast oscillations of wake and REM sleep are weakly correlated with unit firing. A. Relation between local field potentials (LFPs, avg) and multiunit discharges (Units, avg) in periods of wakefulness. Field potentials were filtered between 15–75 *Hz* and the peak negativities were detected. The LFP waveform shown was obtained by averaging over 467 detected events from eight electrodes. The corresponding wave-triggered average of multiunit discharges displayed an increase of firing correlated with the LFP negativity. The same analysis performed on randomly shuffled spikes did not show any pattern (Control). B. Same analysis during periods of REM sleep in the same animal. 1721 detected events were used to compute the averaged LFP. During REM sleep, similar to wake states, cells tended to fire in relation with the negativity of the field potentials during fast oscillations while no increase of firing was seen in the control. Modified from Destexhe et al., 1999b.

total of 467 events showed that units were indeed significantly correlated with LFPs (Fig. 9.13A, Units avg), while the same procedure applied to randomly shuffled spikes did not reveal any pattern (Fig. 9.13A, Control). This analysis therefore demonstrates a marked increase of firing in correspondence with the depth-negative component of fast oscillations, as shown previously in anesthetized animals (Eckhorn et al., 1988; Gray and Singer, 1989; Murthy and Fetz, 1992; Steriade et al., 1996). The same conclusions were also obtained from fast oscillations of REM sleep (Fig. 9.13B).

Thus, during wake and REM sleep, cortical neurons tended to fire in correspondence with the depth-negative component of fast oscillations recorded by the same electrode, but not with distant electrodes (not shown). By contrast, during SWS, neuronal firing was coherent across large distances and consisted of periods of decreased and increased firing, correlated with the depth-positive and depth-negative component of slow-wave complexes, respectively.

9.2.3 Fast oscillations during slow-wave sleep

Although SWS was clearly dominated by high-amplitude, slow-wave activity, on closer scrutiny it appears that brief periods of low-amplitude fast oscillations were also present (Steriade et al., 1996). In LFPs, examining the distribution of dominating frequencies using fine time windows (~0.5 s) revealed that periods of slow-wave complexes, with high power in low (0.1–4 *Hz*) frequency bands, alternated with periods dominated by frequencies in the beta/gamma range (15–75 Hz) (Fig. 9.14). Thus, SWS is composed of slow-wave complexes separated by brief periods of fast oscillations, in agreement with previous observations (Steriade et al., 1996).

Investigating the decay of correlation as a function of distance based on multisite LFPs (see above) revealed that periods of slow-wave complexes displayed high coherence (* in Fig. 9.14), in contrast to the coherence during brief periods of fast oscillations, which is markedly diminished (** in Fig. 9.14). This local analysis of coherence suggests that the brief periods of fast oscillations occurring during SWS have similar characteristics as the 'sustained' fast oscillations of wake or REM sleep episodes.

These brief periods of fast oscillations were further characterized in more detail (Fig. 9.15). The correlations between neighbouring sites during the fast oscillations of SWS fluctuated but stayed high over periods of approximately 100–500 ms (Fig. 9.15B), similar to fast oscillations in wake and REM sleep (see Fig. 9.11A, C). To investigate if this similarity also extended to the relationship between the LFP and unit activity, long periods of SWS were analysed by artificially removing slow-wave complexes and removing the corresponding spikes from the multiunit signals. Performing wave-triggered averages of unit discharges showed that the units tended to fire with higher probability during the depth-negative LFPs (Fig. 9.15C). On the other hand, randomly shuffling the spikes destroyed these relationships (Control in Fig. 9.15C). This analysis therefore shows that during the brief periods of fast oscillations in SWS the statistical relationship between units and LFPs is the same as the 'sustained' fast oscillations of wake and REM sleep periods.

The relation between field potentials and cellular events suggested by the present analysis is summarized in Fig. 9.16. Slow-wave complexes were correlated with periods of silence (or decreased neuronal firing), followed by periods of increased firing, and the same pattern is seen coherently over large cortical distances (≥7 mm). In contrast, fast oscillations were correlated with sustained and tonic firing activity, in which spikes tended to associate with the depth-negative component of LFPs, but only for units adjacent to the LFP-recording electrode, but no relationship was apparent with more distant units. Brief periods of fast oscillations with similar characteristics also appeared during SWS.

The relationship between slow-wave complexes and neuronal silences in the extracellular recordings have recently been confirmed by intracellular recordings in awake and naturally sleeping cats (Fig. 9.17; Steriade et al., 2001). The slow-wave events were always correlated with cellular hyperpolarization, causing a cessation of firing (Fig. 9.17, upper and bottom-left panels). It is also apparent that, during the short periods of low-amplitude fast EEG activity between slow-wave complexes, the intracellular activity was similar to that in the wake state, in agreement with the conclusions of the analysis above. It is thus now clear that slow-wave complexes are associated with cellular hyperpolarization and cessation of firing (Fig. 9.17). However, the mechanisms underlying these slow waves and the significance of their widespread coherence through cerebral cortex (Fig. 9.12) await further exploration. Using computational models, we analyse below possible mechanisms for this type of slow-wave activity.

9.2.4 Origin of cortical slow waves

The above analysis suggests that slow-wave events are generated by a concerted pattern of firing of cortical neurons. Possible mechanisms for generating these slow waves were investigated

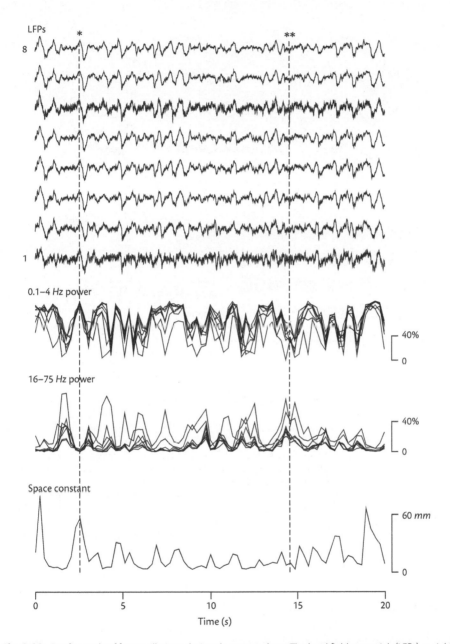

Fig. 9.14: Brief periods of fast oscillations during slow-wave sleep. The local field potentials (LFPs) at eight cortical sites (top curves; same experiment as in Fig 9.10), the relative power of low-frequency (0.1–4 Hz) and fast-frequency (15–75 Hz) components (middle curves), and the space constant of correlation decay with distance (bottom curve) are shown for a 20 s period of SWS. Power spectra and spatial correlations were calculated in successive windows of 0.512 s (128 points). Slow-wave complexes (*) were synchronous over the eight electrodes. Brief periods of fast oscillations (**) were also present and had lower spatial coherence. Modified from Destexhe et al., 1999b.

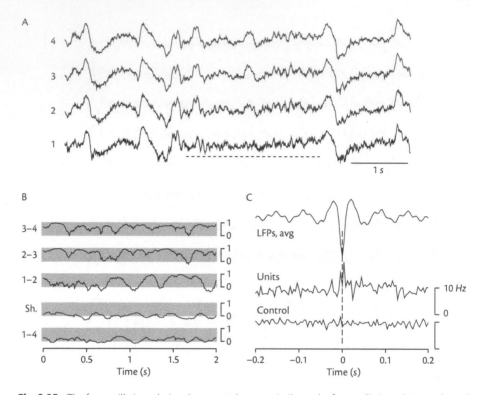

Fig. 9.15: The fast oscillations during slow-wave sleep are similar to the fast oscillations during wake and REM sleep. A. Brief period of fast oscillations (dashed line) during SWS. B. Dynamics of correlations during the period of fast oscillations shown in A, analysed similarly as in Fig. 9.11. Fast oscillations displayed local patterns of synchrony, within short time windows and between neighbouring electrodes, similar to wake and REM periods. C. Relationship between neuronal firing and local field potentials. LFPs and multiunit discharges were then analysed similarly to Fig. 9.13 during periods with fast oscillations selected in an episode of 11 min of SWS (390 events processed). The wave-triggered averaging procedure shows that the negative LFP of fast oscillations of SWS (LFPs, avg) was correlated with an increase of firing (Units, avg). The same analysis based on randomly shuffled spikes did not show any pattern (Control). Spikes were correlated with LFP negativity of fast oscillations (LFP, avg), similar to wake and REM sleep, but did not show any preferred pattern if spikes were randomly shuffled (Control). Modified from Destexhe et al., 1999b.

using a combination of intracellular recordings and computational models (see details in Timofeev et al., 2000).

9.2.4.1 Slow waves in cortical slabs *in vivo*

An *in vivo* preparation was developed to study the mechanisms underlying spontaneous sleep oscillations (Timofeev et al., 2000). Small cortical slabs were isolated by undercutting inputs while preserving the vasculature arising from the pia and upper layers of the cortex. The slabs contained around 10×10^6 neurons (10 $mm \times 6$ mm) from areas 5 and 7 of the suprasylvian gyrus of cats, which were anesthetized with ketamine and xylazine. Under this type of anesthesia, the electrographic pattern in intact cortex consists of a slow oscillation at < 1 Hz (Steriade et al.,

Slow-wave complexes

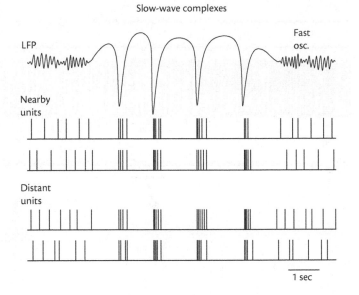

Fast oscillations

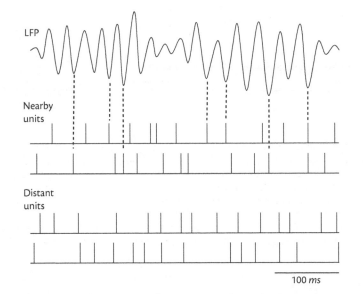

Fig. 9.16: Schematic relationship between oscillations in local field potentials and unit discharges during slow-wave sleep. During SWS (upper panel), slow-wave complexes were correlated with phasic firing activity. Periods of neuronal silence coincided with depth-positivity in LFP, while depth-negative components occurred in coincidence with increased firing in units. The occurrence of periods of decreased and increased firing were synchronous over large cortical distances (>7 mm). During fast oscillations (lower panel), units discharged more tonically, with an increased probability of firing during the depth-negative component of the LFP. The coherence extended over short distances (~1–2 mm) and unit activity was correlated only with the nearby LFP. These results are similar to those for fast oscillations during wake and REM sleep. Modified from Destexhe et al., 1999b.

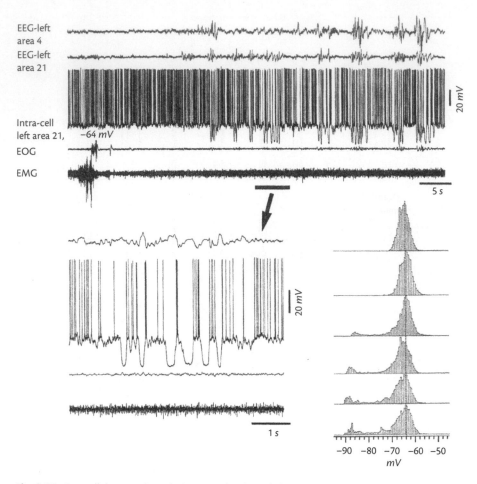

Fig. 9.17: Intracellular recordings during natural wake and sleep states in cats. The five traces in the top panel depict EEG traces recorded from the depth of cortical area 4 (motor) and 21 (visual association), intracellular recording from area 21 neuron (resting membrane potential indicated), electro-oculogram (EOG) and electromyogram (EMG). Part marked by horizontal bar is expanded below (left). Note the relationship between cellular hyperpolarization and slow-wave complexes in the EEG. The bottom right panel shows histograms of membrane potential (10 s epochs) during the period of transition from waking to SWS depicted above. The membrane potential fluctuated around −64 mV during the waking period, and a tail progressively hyperpolarized as the animal entered SWS. From Steriade, 2000, after Steriade et al., 2001.

1993c; Amzica and Steriade, 1995; Contreras and Steriade, 1995), similar to that of the slow oscillation during natural sleep in cats (Steriade et al., 1996; Destexhe et al., 1999b; see above) and humans (Achermann and Borbély, 1997).

Dual and triple simultaneous intracellular recordings as well as local field potential recordings made from neurons in the small cortical slabs showed that the network was in a silent state most of the time. Spontaneously occurring slow-wave oscillations, present in the adjacent intact cortex, were absent. However, the isolated slabs displayed brief active periods separated by long periods of silence, up to 60 s in duration. During these silent periods, 60% of neurons showed nonlinear amplification of low amplitude depolarizing activity.

Examination of intracellular membrane potential fluctuations during the silent state of the slab revealed the presence of small amplitude depolarizing potentials but action potentials were never observed between the bursts (Fig. 9.18). Because all neurons from which stable, long-lasting recordings were obtained never displayed action potentials during the interburst lulls, the depolarizing potentials were most likely spike-independent Spontaneous miniature synaptic potentials (minis), as described at the neuromuscular junction (Fatt and Katz, 1952) and also found in central structures, including the cerebral cortex (Redman, 1990; Paré et al., 1997). After the cessation of the burst, the mini activity was significantly suppressed for 1–2 seconds (Fig. 9.18C) and thereafter progressively recovered.

Multisite field potential and intracellular recordings from small slabs revealed that the onset of spontaneous active periods was systematically delayed at neighbouring locations, suggesting that the active periods were propagating; however, the exact origin of the spontaneous bursts was difficult to assess. It was possible to elicit such bursts by low-intensity electrical stimuli within the slab. Triple simultaneous intracellular recordings showed that the stimulus elicited an initially depth-negative field potential and compound-depolarizing events in all three neurons, with progressively longer latencies in neurons located farther from the stimulating electrode (Timofeev et al., 2000). Based on multisite recordings, the propagation velocity was in the range of 10 to 100 mm/s.

The relatively low frequency of active states recorded in small isolated slabs, compared to the cortically generated slow oscillation (Steriade et al., 1993c; Contreras and Steriade, 1995), could be due to the relatively small number of neurons in the slab. In the isolated suprasylvian gyrus, containing more than 100×10^6 neurons, the frequency of recorded slow activity was similar to the frequency of normal slow oscillations, but spindles were absent because thalamocortical projections were interrupted. Intracellular activity recorded from the isolated gyrus shows that neurons were hyperpolarized during the depth-positive component of the field potential and were depolarized, and fired action potentials, during the depth-negative field potentials, as found in intact cortex (see Fig. 8 in Timofeev et al., 2000).

9.2.4.2 Model of slow waves in the cortex

Because slow waves arise in the context of very large cortical networks, it is necessary to simplify the model of the cortical neuron presented above to its essential elements so that large networks of neurons can be simulated. Each cortical neuron in the network was modelled by two compartments (Mainen and Sejnowski, 1996), with the addition of a persistent sodium current, $I_{Na(p)}$ (Alzheimer et al., 1993; Kay et al., 1998) The persistent sodium current was described by:

$$I_{Na(p)} = \bar{g}_{Na(p)} \, m \, (V - E_{Na}), \tag{9.1}$$

where the maximal conductance is $\bar{g}_{Na(p)} = 0.06-0.07 \ mS/cm^2$ for axo-somatic compartment; $E_{Na} = 50 \ mV$, and the gating variable m is determined by

$$\frac{dm}{dt} = (m_\infty(V) - m)/\tau_m, \tag{9.2}$$

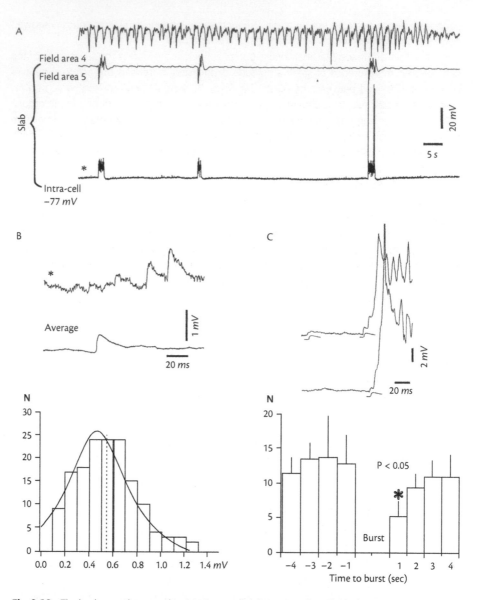

Fig. 9.18: The background neuronal activity in a small slab consists of small depolarizing potentials (SDPs) interrupted by bursts of high-amplitude depolarizing events. A. The three traces represent (from top to bottom) field potentials from area 4 (outside the slab), area 5 (in the slab) and intracellular recording of a neuron within the slab. Segment indicated by a star at the extreme left of the intracellular trace is expanded below (in b) and shows the presence of SDPs. B. Portion indicated by asterisk in (a) and average of 100 SDPs, selected by the maximum of the first derivative of the voltage trace (>4.0 V/s). Below, histogram of amplitude distribution of slow amplitude depolarizing potentials (SDPs). C. Superposition of the onset of two bursts from the cellular trace in the top panel (second and third bursts). The average SDP is shown close to individual SDPs. Below, histogram of temporal distribution of SDPs before and after the high-amplitude bursts. The number of SDPs significantly decreased during 1 s after the high-amplitude bursts. Modified from Timofeev et al., 2000.

$$m_\infty(V) = \frac{1}{1 + exp[-(V + 42)/5)]},$$ (9.3)

$$\tau_m = 0.2 \ ms.$$ (9.4)

Some of the intrinsic parameters (such as the maximal conductances for the persistent sodium current, the high-threshold Ca^{2+} current and the fast voltage-dependent sodium and potassium currents and resting membrane potentials of the neurons) were initialized around the mean with a random variability of about 10% to ensure that the results were robust to parameter differences (for details, see Timofeev et al., 2000).

The network model of the cortex consisted of a one-dimensional two-layer array of PY and IN cells with local connectivity. Each PY cell made AMPA-mediated connections with all other cells within a fixed radius of four cells, which included eight PY cells and nine IN cells, and each IN cell made $GABA_A$-mediated connections with PY cells within the same radius. A simple model was used to describe short-term depression of excitatory and inhibitory synaptic connections (Tsodyks and Markram, 1997; Galarreta and Hestrin, 1998). All AMPA and $GABA_A$ synapses were modelled by first-order activation schemes (see Chapter 5).

Spontaneous miniature EPSPs and miniature IPSPs were simulated using the same equations as for the regular PSPs and their arrival times were modelled by Poisson processes (Stevens, 1993), with time dependent mean rate. When the summation of the miniature EPSPs in one of the PY cells depolarized this cell sufficiently to activate the $I_{Na(p)}$ and to initiate a Na^+ action potential, the activity spread through the network and was maintained by lateral PY-PY excitation and $I_{Na(p)}$. Weak depression of the excitatory interconnections and activation of the Ca^{2+}-dependent K^+ current led to the termination of activity after a few hundred milliseconds. Because any PY neuron could initiate the spontaneous activity and because these events were independent, the total probability of initiation in the network increased with N, the number of cells in the network (see below). As a consequence, the time intervals between patterns of activity in the small networks was highly variable. In larger networks, the larger number of foci where the activity could be initiated resulted in an increase in the frequency of active periods and less variability in their occurrence. Using a network model with Hodgkin-Huxley-like neurons it was not possible to simulate more than a few thousands cells. To overcome this limitation the mean period of spontaneous bursting and its standard deviation were analytically estimated (Timofeev et al., 2000) and compared both with the compartmental simulations and in vivo data (Fig. 9.19).

The number of cells in the isolated slab was close to the minimal number that the analytical model predicted should be needed to generate activity (Fig. 9.19). Further reduction of the slab size led to a dramatic decrease of the average frequency of burst initiation and increase in the variability of the interburst intervals. As the number of neurons and synapses in the network increased, the frequency of active periods increased toward an asymptotic value. For an isolated gyrus the mean period of bursting was predicted to be 4.9 sec with CV = 0.47. For 109 neurons in a 100 cm^2 region of cortex, the predicted frequency of spontaneous bursting was about 0.5 Hz and the CV less than 0.3 (Fig. 9.19b), in good agreement with in vivo recordings. The frequency was nearly independent of size for larger cortical regions because the model included a minimum recovery time following a depolarizing event.

Spontaneous miniature synaptic activity is caused by action-potential independent release of transmitter vesicles and is regulated at the level of single synapses (Salin and Prince, 1996; Paré

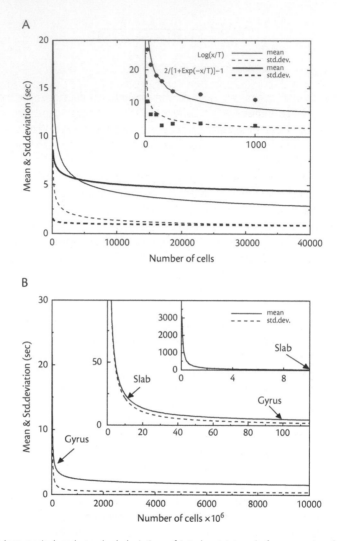

Fig. 9.19: Mean period and standard deviation of interburst intervals for networks of different size. A. Analytical model (curves) and data from simulated Hodgkin-Huxley model with different number of cells (circles: mean; squares: standard deviation). Thinner lines correspond to the logarithmic rate function and thicker lines correspond to the sigmoid rate function (see Appendix). B. Analytical curves based on *in vivo* data. Estimated mean of interburst intervals for a slab (about 10×10^6 neurons) is 24 sec (standard deviation is 21 sec) and for a gyrus (about 100×10^6 neurons) the mean is 4.9 sec (standard deviation 2.3 sec.). Modified from Timofeev et al., 2000.

et al., 1997). The frequency of these minis increases with the probability of evoked release in cortical neurons (Prange and Murphy, 1999). Thus, the synapses with the highest probability of release (Hessler et al., 1993; Rosenmund et al., 1993; Murthy et al., 1997) should make the largest contribution to initiating an action potential in a neuron. Glutamate application at synapses between hippocampal neurons produces long-term potentiation of the frequency of spontaneous miniature synaptic currents (Malgaroli and Tsien, 1992), which suggests that the synapses with the highest rates of spontaneous miniature synaptic currents are the most likely to have been recently potentiated. During sleep the initiation of spikes could therefore occur in neurons having the largest number of recently potentiated synapses.

Since spontaneously occurring minis are amplified by the intrinsic currents in dendrites it may not take a large number of coincident events to initiate a spike despite the hyperpolarized level of the membrane potential. The spiking neuron would further need to recruit additional neurons connected to it; these would also be depolarized as a consequence of minis, so that the recruited network would preferentially include cells that had been recently potentiated. Other factors that would influence recruitment include multiple synaptic boutons (Markram, 1997; Thomson and Deuchars, 1997), spike bursting, which may itself further potentiate recently activated afferent synapses by virtue of the burst in the postsynaptic cell (Paulsen and Sejnowski, 2000), and efferent synapses that were recently potentiated. This interpretation is consistent with the observation that unilateral somatosensory stimulation prior to sleep in humans increases the power of low frequency oscillations, but only on the side of the cortex that received stimulation (Kattler et al., 1994).

The model shows how a single neuron could recruit an avalanche of activity in a selected sub-set of a previously silent recurrent network of cortical neurons and serve as the nucleus for the spread of activity to neighbouring cortical territory. This sequence of events should start from a basal state of inactivity, when the silent cortex is primed for the type of sharp wave activity that has been observed in the hippocampus (Buzsáki, 1986). Thus, the slow oscillations characteristic of the cortex during sleep may be an emergent property of large corticothalamic systems that, surprisingly, may be triggered by activity originating in single neurons. One of the predictions of this model is that the frequency of slow oscillations should be temperature dependent, since the rate of miniature synaptic activity (Barrett et al., 1978) and hence the probability of reaching threshold should increase with temperature. The temperature dependence of the frequency of slow oscillations observed in hibernating hamsters is compatible with this prediction (Deboer, 1998).

The relatively simple cortical network models examined here may not include features of the cerebral cortex needed to fully understand the genesis of slow cortical oscillations. Other explanations may also be compatible with the experimental data; for example, the onset of the silent phase may also be due to intrinsic voltage-dependent currents, as modelled recently (Compte et al., 2000). Nonetheless, the data and models form a consistent picture; future experimental studies are needed to test the predictions and refine the proposed mechanisms.

9.2.4.3 Simplified models of slow waves

A thalamocortical network model consisting of AdEx integrate-and-fire neurons, could generate spindle and slow-wave oscillations with Up and Down states (Destexhe, 2009). This model could also generate a transition between Up/Down states and the sustained irregular activity typical of wakefulness (Fig. 9.20; Destexhe, 2009). The transitions were generated by modulating the adaptation currents in PY cells, mimicking the action of neuromodulators. This model complements previous models of Up/Down states and asynchronous irregular states with integrate and fire neurons (Brunel, 2000).

These minimal models of thalamic and thalamocortical circuits are capable of generating spindle oscillations, slow (Up/Down) oscillations, and sustained irregular activity. These are relatively realistic models because they include the intrinsic properties of thalamic and cortical neurons, conductance-based synapses, and the effect of neuromodulators (Fig. 9.20). In addition, Up/Down states are self-sustained in such networks of AdEx neurons, without the need for additional noise as in other models (see details in Destexhe, 2009). This property allowed us to simulate self-sustained Up/Down states in neuromorphic hardware (Brüderle et al., 2011).

Another type of reduced models was proposed recently. Map models of cortical neurons, which are based on functions that map the state from one discrete time to the next, are fast and efficient, allowing large-scale cortical model of slow-wave sleep and Up/Down states to be efficiently simulated (Komarov et al., 2018).

9.3 A possible function for sleep oscillations

The models and analyses reviewed in this chapter show that the different types of slow-wave oscillations are characterized by distinct spatiotemporal patterns of activity, which suggest that these oscillations may have distinct functions. At this point we depart from the solid shores of what is known and begin to navigate the less certain waters of reorganization and remodelling that might occur in cortical neurons during sleep. This discussion is meant to raise issues and explore new hypotheses.

9.3.1 Sleep oscillations as a trigger for plasticity

The models in combination with the *in vivo* intracellular recordings reviewed in Section 9.1 suggest that during spindling, the strength of thalamic input to neocortical pyramidal cells is stronger than that expected based on the low firing rates of these cells. The first general conclusion is that the thalamic inputs in neocortical pyramidal cells evoke powerful glutamatergic conductances in parallel with strong GABAergic conductances. It is conceivable that this strong GABAergic component is required to prevent the cell from firing by shunting the EPSPs in order to avoid the production of avalanches of excitatory discharges.[1] This conclusion is in agreement with direct measurements during visual inputs in cortical neurons that have revealed large shunting GABAergic conductances (Borg-Graham et al., 1998).

The second conclusion from the model is that the synaptic conductances evoked by thalamic inputs in cortical pyramidal neurons are more than a simple shunt of the EPSPs by IPSPs. Morphological data indicate that excitation and inhibition are not evenly distributed in pyramidal neurons—the dendrites are dominated by excitatory synapses while the soma essentially receives inhibitory synapses (DeFelipe and Fariñas, 1992). This imbalance necessarily implies that the dendrites of pyramidal neurons must experience strong depolarization, but this depolarization is not visible at the somatic level due to the proximal inhibition. Thus the model suggests that during spindling, cortical pyramidal cells receive strong dendritic excitation in parallel with strong inhibition around the soma, preventing the cell from firing.

---

[1] This finding is also consistent with the high sensitivity of the cortical network to alteration of $GABA_A$-mediated inhibition, leading to paroxysmal discharges (see Chapter 8).

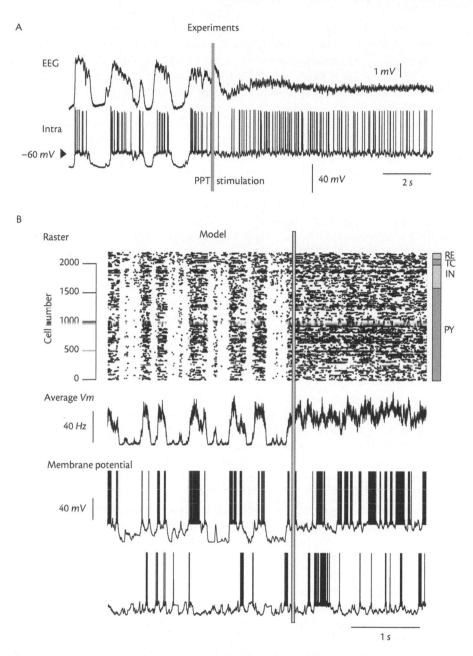

Fig. 9.20: Neuromodulatory modulation of network states and its modelling by thalamocortical networks of integrate and fire neurons. (a) Experiments where the pedonculopontine (PPT) nucleus was stimulated electrically in anesthetized cats (grey bar), leading to the transition from Up/Down states to a self-sustained desynchronized activity (Top: surface EEG; Bottom: intracellular recording of a pyramidal cell in parietal cortex; modified from Rudolph et al., 2005). (b) Same paradigm simulated by a thalamocortical model, where thalamic and cortical neurons were modelled by AdEx mechanisms (Top: raster showing the spiking activity of the different cell types; Middle: average Vm of all cortical neurons; Bottom: example Vm of two pyramidal cells. The action of the PPT was simulated by reducing the adaptation current in pyramidal cells, mimicking the effect of acetylcholine, leading to a similar transition between Up/Down states and asynchronous state (modified from Destexhe, 2009).

It thus seems that, during sleep spindles, the synchronized bursts of high-frequency action potentials in thalamic neurons generate in the cortex ideal conditions for strongly depolarizing the dendrites of cortical neurons while preventing them from firing. What advantage could this type of input have? Reducing the probability for a cortical neuron to fire also reduces the probability of lateral interactions between pyramidal cells in the cortex. Thus, the spindle strongly depolarizes the dendrites of cortical cells receiving direct thalamic inputs without strongly activating other cortical neurons. Because dendritic depolarization is accompanied by Ca^{2+} entry (Yuste and Tank, 1996), synchronized thalamic inputs would provide an ideal signal to trigger a massive Ca^{2+} entry into the dendrites of selected pyramidal neurons without also producing excessive firing in the cortical network. As mentioned in Section 9.1.3.5, experiments have confirmed this prediction by measuring strong calcium signals in the dendrites of pyramidal cells during sleep spindles (Seibt et al., 2017).

What is the consequence of massive Ca^{2+} entry in pyramidal neurons? Calcium-dependent mechanisms are involved in synaptic plasticity (reviewed in Ghosh and Greenberg, 1995). Thus, calcium entry during sleep may serve to prime the synapses for permanent changes. In particular, Calcium calmodulin dependent protein kinase II (CaMKII), an enzyme that is abundant at synapses and is implicated in synaptic plasticity of excitatory synapses in the cortex and elsewhere (Soderling, 1993; Wu and Cline, 1998; Malenka and Nicoll, 1999; Soderling and Derkach, 2000), is not only sensitive to Ca^{2+} but also to the *frequency* of Ca^{2+} spikes (Dekoninck and Shulman, 1998). For Ca^{2+} pulses of 80 *ms*, the optimal frequency is close to 10 *Hz* (Dekoninck and Shulman, 1998), which is precisely within the range of spindle oscillations. Sleep spindles may provide a selective signal to efficiently activate CaMKII in the dendrites of cortical pyramidal neurons.

Another possible consequence of massive Ca^{2+} entry in pyramidal neurons is calcium-dependent gene expression, which is frequency sensitive (Gu and Spitzer 1995, 1997; Li et al., 1998) in the delta range of frequencies (1–4 *Hz*). Calcium that enters dendrites during spindles may accumulate in the endoplasmic reticulum, which forms a continuous compartment within the neuron continuous with the nucleus. Calcium-stimulated calcium release from the endoplasmic reticulum during delta waves may then deliver the calcium signal to the nucleus (Berridge, 1998).

A third possibility is that repeated Ca^{2+} entry may activate a molecular 'gate', opening the door to gene expression. This possibility is based on the observation that repeated high-frequency stimuli, but not isolated stimuli, activate protein kinase A (PKA), an enzyme implicated in long-lasting synaptic changes and long-term memory (Abel et al., 1997). PKA acts like a 'gate' by inhibiting protein phosphatases, which themselves exert a tonic inhibition over biochemical cascades leading to long-term synaptic modifications (Blitzer et al., 1995; Abel et al., 1997; Blitzer et al., 1998; for a model, see Bhalla and Iyengar, 1999; Soderling and Derkach, 2000). The evidence for this hypothetical mechanism, illustrated in Fig. 9.21, is based on observations that activation of PKA alone does not induce synaptic changes, but blocking PKA suppresses long-term synaptic changes in the hippocampus (Frey et al., 1993). All the necessary enzymes for this mechanism are located at or near the synapse (Kennedy et al., 1983; Carr et al., 1992; Dosemeci and Reese, 1993; Mons et al., 1995).

Similar mechanisms could occur during spindles. It may be that the massive Ca^{2+} entry during spindles in the thalamic recipient cortical neurons, repeated at ~10 *Hz* frequency, provides the conditions needed to open a molecular gate, for example mediated by PKA as in Fig. 9.21. Sleep spindles would therefore provide a physiological signal similar to the repeated

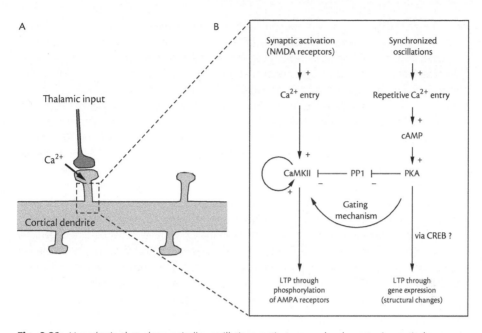

Fig. 9.21: Hypothesis that sleep spindle oscillations activate a molecular gate in cortical neurons. A. Scheme of dendritic branch of a cortical pyramidal neuron receiving direct inputs from the thalamus. The high-frequency burst discharges in synchrony by thalamic neurons provide unusually powerful excitatory inputs to cortical neurons, provoking a massive Ca^{2+} entry in their dendrites. B. Hypothetical molecular gate that could be activated by massive Ca^{2+} entry. The left column represents the pathway linking synaptic activation to long-term changes (LTP), through the CaMKII cascade. However, this pathway is tonically inhibited by protein phosphatases (PP1; right column). Repetitive Ca^{2+} entry, through production of cyclic AMP (cAMP), activates protein kinase A (PKA), which inhibits PP1. cAMP and/or PKA pathways might also lead to permanent changes through gene expression. Activation of PKA therefore releases the inhibition of the CaMKII pathway and 'opens' the route between synaptic activation and long-term changes in the cell. The spindle oscillations could therefore gate the plasticity mechanisms inside neocortical neurons.

tetanus used to induce long-term synaptic changes in slices. It may also provide a 'priming' signal, opening a gate that allows permanent changes to subsequent inputs following the sleep spindles (see below). During subsequent periods of slow-wave activity, the most highly primed thalamic recipient cortical neurons could then recruit additional cortical neurons through the selection mechanism outlined about in Section 9.2.4.

9.3.2 Slow oscillations

Spindles generally appear during the early phase of SWS, which is later dominated by other types of oscillation, such as slow-waves (delta waves at 1–4 Hz or slow oscillations of <1 Hz). These types of oscillations were analysed in Section 9.2 and led to the following conclusions: (a) the slow oscillations of natural sleep are coherent across several millimetres in cortex and are correlated with a concerted decreased/increased firing in units; (b) the fast oscillations of wake and REM sleep are characterized by less global and less stable coherence and their depth-negative EEG components are correlated with an increased probability of unit firing; (c) during slow-wave sleep brief episodes of fast oscillations occur whose spatiotemporal properties are similar to the sustained fast oscillations of activated states.

Brief periods of fast oscillations are present between slow-wave complexes in cat suprasylvian cortex during both ketamine-xylazine anesthesia (Contreras and Steriade, 1995) and natural sleep (Section 9.2; Steriade et al., 1996; Destexhe et al., 1999b). The spatial and temporal coherence of these brief periods of fast oscillations, as well as their relation with unit discharges, are indistinguishable from that during 'sustained' fast oscillations of wake and REM sleep. However, the above analysis only investigated oscillations over distances up to 7 *mm* in cerebral cortex, and further experiments are required to demonstrate that the same conclusions also apply to larger cortical distances.

The observation that highly coherent slow-wave patterns alternate with brief periods of low-coherence fast oscillations can be interpreted in several ways. First, the fast oscillations could constitute ever-present background activity in thalamocortical networks that is regularly interrupted by slow-wave events. Slow-waves would therefore sculpt spatiotemporally coherent events into a background of low-coherent fast oscillations. This possibility is consistent with intracellular recordings in naturally sleeping cats (Fig. 9.17; Steriade et al., 2001).

Alternatively, the fast oscillations could occur as a rebound to slow-wave events. Slow-waves are associated with a widespread hyperpolarization in cortical cells, followed by a rebound depolarization, as occurs in neocortical neurons during ketamine-xylazine anesthesia (Contreras and Steriade, 1995). This rebound has been modelled above as the rapid recruitment of neurons that are brought near threshold by spontaneous miniature depolarizing events (Section 9.2.4; Timofeev et al., 2000). At the cellular level, fast oscillations may occur upon depolarization in neocortical cells due to intrinsic mechanisms (Llinás et al., 1991; Nuñez et al., 1992; Gutfreund et al., 1995; Gray and McCormick, 1996; Steriade et al., 1998). Short periods of fast oscillations may appear during the rebound depolarization as a consequence of intrinsic ionic mechanisms in cortical cells.

A third possible interpretation (Destexhe et al., 1999b) is that slow-wave sleep is part of a cyclic process, where brief periods of processing similar to the awake state (Up states) alternate with periods of synchronized silences (Down states), concomitantly with slow waves. As shown above, during the rebound of the network (depth negativity of the slow wave), cortical cells should receive strong EPSPs followed by IPSPs, which is an ideal signal to trigger a massive calcium entry in the dendrites (see Section 9.1; Contreras et al., 1997c). Slow-waves may therefore be part of a process for establishing permanent changes in the network through calcium-dependent mechanisms (see above). Slow-wave sleep in this scenario could then be part of 'recall' and 'store' sequences, in which the fast oscillations could reflect recalled events experienced previously, which are stored in the network. It was proposed that plasticity thresholds are lower during sleep (El Boustani et al., 2012), which could be a possible mechanism for storing such recalled events in synaptic weights (see Section 9.4.6).

9.3.3 Network reorganization

It is now possible to provide an overall summary of the sequence of events that occurs during the descent into sleep. At sleep onset, the thalamus enters burst mode and generates synchronized spindle oscillations. The synchronized high-frequency bursts of thalamic relay cells during these oscillations provide unusually strong excitatory/inhibitory inputs in pyramidal neurons. The repetition of these inputs at ~10 Hz triggers periodic calcium entry in cortical dendrites and activates intracellular mechanisms, such as CaMKII and/or molecular gates. This process could serve to open the door between synaptic activation and long-term synaptic changes, so that pyramidal neurons are ready to produce permanent changes in response to some specific synaptic events that need to be consolidated.

As sleep deepens, slow waves progressively dominate the electroencephalogram. The above analysis (Section 9.2) has shown that slow-wave complexes represent a highly synchronized network event, and that they alternate with brief episodes of fast oscillations. Our hypothesis is that these brief periods represent 'recall' events, possibly related to hippocampal replay activity (see below). During these events, the activity is not strong enough to produce permanent synaptic changes, but synapses are 'primed' through conventional NMDA-dependent long-term potentiation, similar to the 'synaptic tagging' hypothesis (Frey and Morris, 1998; see also Larson and Lynch, 1986). In a later stage, the network provides the highly synchronized activity needed to mobilize calcium-dependent mechanisms throughout the neuron, trigger protein synthesis and possibly gene expression. The periods of slow-wave events would then activate mechanisms to establish permanent modifications of the primed synapses, and thereby implement a 'store' operation. Alternating slow-wave complexes with brief periods of fast oscillations during slow-wave sleep could thus result in the permanent formation of small sets of strongly interacting neurons. The longer episodes of delta oscillations (1–4 Hz) during slow-wave sleep are accompanied by further entry of calcium in the dendrites and somata of cortical neurons and could provide the long time scales needed to mobilize genetic machinery in the nucleus (Berridge, 1998) and to carry out structural changes. Consistent with this scenario, protein synthesis is significantly increased during slow-wave sleep (Ramm and Smith, 1990).

This hypothetical scheme is also compatible with recent reports showing the involvement of slow-wave sleep in plasticity during development and the consolidation of long-term memory. Traditionally, REM sleep has been considered as a candidate for memory consolidation, primarily because of studies showing that REM sleep deprivation affects long-term memory (Dement, 1960; Fishbein and Gutwein, 1977; Hobson, 1988). However, these studies may have confounded the effect of stress with REM deprivation (Horne and McGrath, 1984; Vertes and Eastman, 2000). It has been shown that performance on visual discrimination tasks is significantly enhanced if the training period is followed by sleep, but that the enhancement correlates most closely with slow-wave sleep (Fowler et al., 1973; Gais et al., 2000; Stickgold et al., 2000). Similar results were reported for ocular dominance plasticity during the critical period in cats (Frank et al., 2001). Animals allowed to sleep for six hours after a period of monocular stimulation developed twice the amount of brain change as those cats with the same stimulation kept awake in the dark for six hours, consistent with earlier observations (Imbert and Buisseret, 1975). Moreover the amount of change was strongly correlated with the amount of slow-wave sleep and not with the amount of REM sleep. This is direct evidence that one function of slow-wave sleep is to help consolidate the effects of waking experience on cortical plasticity, converting memory into more permanent forms, as suggested previously (Moruzzi, 1966; Fowler et al., 1973; Steriade et al., 1993d).

The experimental investigation of memory consolidation during sleep has continued to accelerate (as reviewed in Walker and Stickgold, 2006). It now seems clear that SWS is directly involved in the consolidation of declarative memories and that the density of sleep spindles in human SWS is highly correlated with memory enhancement (Gais et al., 2002). Furthermore, increasing the density of sleep spindles pharmacologically, enhances recall for both neutral (Mednick et al., 2013) and highly arousing emotional stimuli (Kaestner, 2013).

9.4 A computational theory of sleep

The discussion thus far has focused on the biophysical aspects of thalamocortical assemblies that lead to rhythmic activity. We turn now to how the global nature of these rhythms might be used during sleep to help reorganize cortical circuits. Mechanisms for learning and memory have been generally thought to operate during normal behaviours, in the awake state. However, there are computational reasons and experimental evidence, summarized here, for why some of these changes might also occur during sleep states. We first discuss mechanisms for synaptic plasticity and then how memory consolidation during sleep follows as a consequence of constraints on learning in the broader context of computational theories of sleep.

9.4.1 Temporally asymmetric Hebbian plasticity

Long-term memory may be based on mechanisms for synaptic plasticity that are synapse specific and driven by local signals. Donald Hebb in 1949 proposed a Neurophysiological Postulate for how synaptic strengths could be adjusted by an activity-dependent mechanism:

> When an axon of cell A is near enough to excite cell B and repeatedly or persistently takes part in firing it, some growth process or metabolic change takes place in one or both cells such that A's efficiency, as one of the cells firing B, is increased. (Hebb, 1949, p. 62)

The traditional interpretation of this postulate is that synaptic plasticity is based on coincidences between the release of neurotransmitter from a presynaptic terminal and the depolarization of the postsynaptic dendrite. Evidence for a coincidence detection mechanism has been found in the hippocampus, where long-term potentiation (LTP), discovered there in 1973 by Tim Bliss and Terje Lømo, was shown to be Hebbian in the above sense (Kelso et al., 1990): LTP of synapses on hippocampal neurons can be elicited by pairing synaptic input with strong depolarizing current, when neither alone produces a long lasting change, consistent with this interpretation. Furthermore, the induction of LTP at some synapses is controlled by the NMDA receptor, which requires both binding of glutamate and depolarization to allow entry of calcium into the cell. Insofar as the NMDA receptor is a coincidence detector it might even be called a Hebb molecule. The only part that Hebb had apparently not gotten quite right was his statement about the firing of cell B, since LTP could still be induced after fast spiking was abolished by blocking active currents in the postsynaptic neuron, suggesting that cooperativity with other synaptic inputs might be needed to depolarize the dendrite sufficiently to open the NMDA receptor. Another issue is that increases in the strength of a synapse from random coincidences will end inexorably in saturation (Sejnowski, 1977). Hebb suggested that unused synapses might decay, and a form of long-term depression (LTD) induced by low frequency activity might provide such decay from spontaneous activity in the cortex. However, if synaptic strengths are to encode long-term memories it is important to have a mechanism for LTD as specific as that for LTP.

Synapses between pairs of cells can be directly examined with dual intracellular recordings in cortical slices. In an experiment designed to test the importance of relative timing of the presynaptic release of neurotransmitter and the postsynaptic activity to LTP, Markram et al. (1997b) paired stimulation of cell A either 10 *ms* before or after spike initiation in cell B. They found reliable LTP when the presynaptic stimulus preceded the postsynaptic spike, but, remarkably, there was LTD when the presynaptic stimulus immediately followed the postsynaptic spike. Similar results have been found for hippocampal neurons grown in culture (Bi and Poo, 1998; Debanne et al., 1998), between retinal axons and neurons in the optic tectum of frogs (Zhang et al., 1998), and in the electrical line organ of weakly electric fish—this last example is different from the others in that it is of opposite polarity (presynaptic release before the postsynaptic spike causes LTD) (Bell et al., 1997). Thus, temporal asymmetry in synaptic plasticity is widespread in cerebellar as well as cortical structures. In Fig. 9.22, where the time delay between the synaptic stimulus and the postsynaptic spike was varied over a wide range, the window for plasticity is around ±20 *ms* and the transition between LTP and LTD occurs within a time difference of a few *ms*.

This temporally asymmetric form of synaptic plasticity has many nice features. First, it solves the problem of balancing LTD and LTP in a particularly elegant way since chance coincidences should occur about equally with positive and negative relative time delays. Second, when sequences of inputs are repeated in a network of neurons with recurrent excitatory connections, this form of synaptic plasticity will learn the sequence and the pattern of activity in the network will tend to predict future input. This may occur in the visual cortex where simulations of

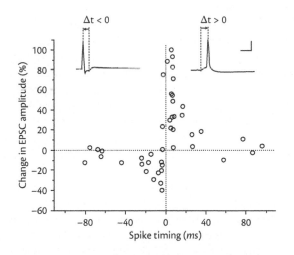

Fig. 9.22: Synaptic modification in cultured hippocampal neurons. The relative timing of the paired postsynaptic spike and excitatory synaptic inputs (inset) determines whether the subsequent change in the excitatory postsynaptic current (EPSC) increases (when the spike follows the synaptic input) or decreases (when the spike precedes the synaptic input). This is a temporally asymmetric form of Hebbian synaptic plasticity. Figure reproduced from Bi and Poo (1998). Calibration: 50 *mV*, 10 *ms*.

cortical neurons can become directionally selective when exposed to moving visual stimuli (Rao and Sejnowski, 2000). Similar models have been proposed for neurons in other brain regions. The temporal window for synaptic plasticity was taken to be a hundred milliseconds in the hippocampus (Blum and Abbott, 1996), where there is evidence that the locations of place cells shift to earlier locations in rats running repetitively through a maze (Mehta et al., 1997), and less than 1 *ms* in a model for learning auditory localization by the relative timing of spikes from two ears (Gerstner et al., 1996).

Is the temporally asymmetric learning algorithm Hebbian? The rapid transition between LTP and LTD at the moment of temporal coincidence does not conform to the traditional view of a Hebbian synapse. Notice that in Hebb's formulation the synapse increases in strength when an axon of cell A is near enough to excite cell B and repeatedly or persistently takes part in firing it. For cell A to take part in firing cell B implies causality, not simple coincidence. Thus, the importance of temporal order is implicit in Hebb's formulation. If cell A produces an excitatory event just before cell B fires a spike then it is likely to have contributed. Hebb did not specify what should happen if cell A fires just after cell B, but weakening is consistent with causality since it is then unlikely for cell A to have caused cell B to fire. The temporally asymmetric learning rule may be more Hebbian than the earlier coincidence version.

For the spike at the soma to influence synapses on distal dendrites, there must be a flow of information from the soma toward the dendrites, which violates the principle of dynamic polarization. This reverse flow of information could not occur without active currents in dendrites, which we now know support exactly the sort of backpropagating action potentials in pyramidal neurons required by the strict form of Hebb's postulate. How the backpropagating spike interacts with the NMDA receptor to produce a knife-edge switch from LTP to LTD is an open research problem (Franks et al., 2000). Recent evidence from the hippocampus of freely moving rats suggests that backpropagation of action potentials occurs *in vivo* and decreases with experience in a new context (Quirk et al., 2001).

The temporally asymmetric Hebbian learning rule is equivalent to the temporal difference learning algorithm in reinforcement learning (Rao and Sejnowski, 2001), which can be used to make predictions and implement classical conditioning (Montague and Sejnowski, 1994). The unconditioned stimulus in a classical conditioning experiment must occur before the reward for the stimulus-reward association to occur. This is reflected in the temporal difference learning algorithm by a postsynaptic term that depends on the time derivative of the postsynaptic activity level. The goal is for the synaptic input to predict future reward: if the reward is greater than predicted the postsynaptic neuron is depolarized and the synapse strengthens, but if the reward is less than predicted, the postsynaptic neuron is hyperpolarized and the synapse decreases in strength. There is evidence in primates that the transient output from dopamine neurons in the ventral tegmental area carry information about the reward predicted from a sensory stimulus (Schultz et al., 1997) and in bees, an octopaminergic neuron has a similar role (Hammer and Menzel, 1995). The temporal window for classical conditioning is several seconds, much longer than the window for LTP/LTD observed at cortical and hippocampal synapses. A circuit of neurons in the basal ganglia and frontal cortex may be needed to extend the computation of temporal differences to such long time intervals (Berns and Sejnowski, 1998). The implementation of same temporal-difference learning algorithm in different types of learning systems in different parts of the brain suggests that the temporal order of input stimuli is a useful source of information about causal dependence in many different learning contexts and over a range of time scales.

9.4.2 Thalamocortical assemblies

If relative timing of spikes has a major influence on the strengths of synapses in the cortex, then spike timing is likely to be internally regulated. In particular, local inhibitory interneurons, such as basket cells that can induce rebound spiking in many cortical pyramidal cells, may have an important function in regulating the timing of spikes within a column of neurons (Ritz and Sejnowski, 1997). The relative order of spikes in a population of neurons could also be used to encode information about objects in the world (Hopfield, 1995). In particular, the first neuron in a population to spike in response to a sensory stimulus will have an advantage since its synapses will be the first to activated and more likely to be strengthened compared to synapses from other neurons that spike later (Van Rullen et al., 1998; Abbott and Song, 1999). Such cell assemblies in primary visual cortex could account for our ability to distinguish vernier offset of visual stimuli with arc second resolution, an order of magnitude smaller than the diameter of a photoreceptor (Miller and Zucker, 1999).

We can now be more specific about what a cell assembly might be and how it might provide insight into the role of sleep oscillations in the consolidation of cortical memory traces. Consider first the recruitment of neurons in visual cortex in response to a flashed visual stimulus in the wake state. Stimulus specificity of single neurons is already present in the first few spikes of a response (Thorpe and Gautrais, 1997). If the most selective neurons come to threshold first, they could recruit neurons through local connectivity within a column. Inhibitory neurons would also be recruited, which would limit the total size of the assembly that represents the stimulus in the column. At the same time, temporally asymmetric Hebbian plasticity would strengthen the sequence of neurons responding to the stimulus. Noise in the system would produce a different order of firing each presentation, which would result in reciprocal connection among the neurons most selective for the repeated stimulus. This defines a cortical assembly by construction.

Sensory experience may not produce a sufficiently strong tetanization of cortical synapses to produce LTP, particularly in a primary sensory areas where the threshold for plasticity is set high, but it may be sufficient to prime synapses in an assembly that later become consolidated during slow-wave sleep. It would be dangerous to change the strengths of synapses in a feedforward system during the processing itself; a safer policy is to statistically sample the inputs over an extended time period and wait until the cortex is in no longer processing sensory information before making irreversible changes to the network. The reason why we need a sleep phase for consolidating memories may be to reduce the relay of sensory information through the thalamus.

LTP through temporally asymmetric Hebbian plasticity in reduced preparations depends on the relative timing of the presynaptic and postsynaptic spikes, the frequency of pairing, and the total number of paired spikes. Thalamocortical sleep oscillations provide conditions that allow the relative phases of neurons in an assembly to be consistently maintained over many pairing. This is a consequence of the spatio-temporal coherence observed during slow-wave sleep. This mechanism could be used to strengthen connections between neurons in an assembly. If too many neurons are activated then assemblies that share neurons will interfere with each other and lose their identity. This type of learning would correspond to a form of priming, or sensory learning, in which the statistical structure of sensory input would alter the response properties of cortical neurons. However, there are other types of memory that can be distinguished behaviourally following lesions in specific brain structures.

9.4.3 Sleep and memory consolidation

The theory of memory consolidation was proposed more than a century ago by Muller and Pilzecker (1900). This theory postulated that temporary memories are initially formed and consolidated to a permanent form during a later stage. Experiments conducted during the last decades support this theory (reviewed in McGaugh 2000). The evidence is particularly clear for declarative memories, which need a period of time to consolidate, or reach a more permanent status. The fact that sleep might be the stage during which a form of memory consolidation occurs has been widely explored (Jenkins, 1924; Dement, 1960; Dewan, 1968; Fowler et al., 1973; Hobson, 1988; Buzsáki, 1989, 1996, 1998; Sejnowski, 1998; Sutherland and McNaughton, 2000; Stickgold et al., 2000). In particular, slow-wave sleep has been shown to correlate with the consolidation phase of long-term memory (Fowler et al., 1973; Gais et al., 2000; Stickgold et al., 2000; see above).

These experiments should be considered together with behavioural evidence from retrograde amnesia. Lesions limited to the hippocampus and surrounding regions produce memory impairment in monkeys and humans, but only recent memories are impaired while remote memories are intact (Zola-Morgan and Squire, 1990; Alvarez et al., 1995). These experiments suggest that the hippocampal formation is required for memory storage for only a limited time period after learning. As time passes, its role in memory diminishes, and a more permanent memory gradually develops, probably in neocortex, that is independent of the hippocampal formation (Zola-Morgan and Squire, 1990; Squire and Zola-Morgan, 1991; McClelland et al., 1995; Bontempi et al., 1999; Kapur and Brooks, 1999; Gluck and Myers, 2000). The hippocampus may not be the site of storage for the information before it is consolidated, but may instead facilitate the associative links between information stored in different parts of the cerebral cortex and other parts of the brain.

Until recently, the activity that may occur in the cortex during consolidation could only be inferred indirectly from lesion experiments, but recordings from freely moving rats during wake and sleep states corroborate the idea that the hippocampus and the neocortex interact during sleep (Buzsaki, 1989, 1996; Wilson and McNaughton, 1994; Siapas and Wilson, 1998; Nadasdy et al., 1999; Sutherland and McNaughton, 2000). Neurons in the rat hippocampus respond to places in the environment (O'Keefe and Nadel, 1978). Changes occur in the correlations between hippocampal place cells in freely moving rats as a consequence of learning a new environment (Wilson and McNaughton, 1994). In these experiments, neurons that had neighbouring place fields and fired together during exploration of a new environment became more highly correlated during subsequent sleep episodes in comparison with activity during previous sleep episodes. The correlated firing of neurons in the hippocampus during sleep may be a 'played back' version of newly acquired experiences to the neocortex through feedback projections (Marr, 1971; Buzsáki, 1989; Chrobak and Buzsáki, 1994; McClelland et al., 1995; Siapas and Wilson, 1998). Thus, the neocortex during the wake state provides the hippocampal formation with a detailed description of the days events; during sleep, the hippocampus may recapitulate some version of these events to the neocortex, where permanent memory representations are gradually formed over repeated episodes of synchronized slow-wave activity.

Why would the brain go to so much trouble to recreate memories in the cortex? Cortical representations of objects and events are widely distributed in cerebral cortex; for example, the representation of the shape of a violin might be stored in the visual cortex, the sounds made by a violin in the auditory cortex, how it is grasped in the somatosensory cortex, and how it is played in the motor cortex (Damasio and Tranel, 1993). Problems arise when new experiences and objects must be integrated with existing information that is widely distributed. Learning algorithms designed for artificial neural networks that use such distributed representations can suffer from 'catastrophic interference' when new information is stored in the same neural circuits as old information (McClelland et al., 1995). Therefore, the brain must solve two problems during learning: where to make the changes needed to create new links between existing memories, and how to make changes that are compatible with previously stored memories.

Before consolidation, according to this explanation, the lack of direct connections between distant brain areas prevent parts of the memory trace to reactivate other parts that represent a unique object or episode. During sleep, indirect connections are formed within the neocortex that allow the memory traces in different parts of the brain to become re-excited without the need of the hippocampus (Alvarez and Squire, 1994; McClelland et al., 1995). After learning, activity in the hippocampus is no longer needed to reactive a memory, and in the process, the elements of the specific memory have been integrated into the general knowledge store by virtue of repeated reactivations. This type of theory might be called the completion model for memory consolidation since the purpose of sleep in this model is to improve associative pattern completion in a sparsely connected system of networks in the neocortex.

All of these different mechanisms may contribute to memory consolidation. During conscious experience, latent memories are formed throughout the cortex together with special links in the hippocampal formation that allow top-down retrieval to occur. During sleep spindles these patterns are activated and during slow-wave sleep changes occur in the neocortex that strengthen internal links. According to this view, spindle oscillations would mobilize the machinery needed for memory consolidation. In the deeper phases of SWS, the activity alternates between brief periods of fast oscillations, and periods of slow-wave activity (Fig. 9.12). We propose that during brief periods of fast oscillations, the hippocampal formation activate memories stored in the neocortex ('replay'); during the slow waves that follow, a highly synchronized network activity strengthens the synapses that were just activated, perhaps even creating new synaptic connections. This hypothesis predicts that special correlations between hippocampal and cortical activities should occur during the brief periods of fast oscillations in slow-wave sleep.

9.4.4 Reciprocal interactions between the hippocampus and the neocortex

During the transition from wakefulness to sleep, the cerebral cortex becomes less responsive to external sensory inputs and less concerned with actively gathering information. The thalamus, which during the wake state relays sensory information from the periphery to the cortex, becomes less of a relay and more of a mirror, as feedback connection from the cortex to the thalamus become capable of entraining thalamic neurons through synchronous bursting. In a sense, during sleep, the cortex no longer listens to the outside world, but rather to itself. Feedback connections from the cortex to the thalamus become as important as feedforward ones from the thalamus to the cortex and information flows freely in both directions.

Periodically during sleep, periods of slow-wave activity in the cortex alternate with extended periods during with there are high-frequency 30–80 Hz oscillations in the cortex and rapid eye movements, called paradoxical or REM sleep, which is indistinguishable from the activity that occurs in the cortex during attentive waking states. An even higher frequency synchrony of action potentials in the 100–200 Hz range, called ripples, occurs in the hippocampus during slow-wave sleep. The spatial extent of synchrony between the firing of neurons in these high-frequency states is more spatially localized than during the low-frequency oscillations. Inhibitory neurons in the thalamus and the cortex are of particular importance in producing synchrony and in controlling the spatial extent of the coherent populations during high-frequency oscillations.

Synchrony and other network properties could be exploited for controlling the flow of information between brain areas and for deciding where to store important information. We have already seen that *ms* time delays between synaptic release of glutamate and the backpropagating action potential in the postsynaptic cell can control the sign of change in the synaptic strength. Synchronization can also enhance the signal-to-noise ratio of a message encoded as a train of action potentials (Hô and Destexhe, 2000; Hô et al., 2000; Salinas and Sejnowski, 2000) but it can also reduce the amount of information that can be encoded (Ritz and Sejnowski, 1997) since perfectly synchronous firing in a pool of neurons signals a single event. However, the amount of mutual information that a pyramidal neurons can convey about its synaptic inputs in small deviations from synchrony in its spike train are high when the neuron is entrained (Hopfield, 1995; Tiesinga et al., 2001).

What is the function of global coherence and temporal synchrony during slow wave sleep? In the hippocampus, theta oscillations occur when the hippocampus is in a state that is primed for receiving and retaining information from the neocortex: the inputs to the hippocampus report on the detailed state of the cortex while the animal is actively exploring the world and the synapses in the hippocampus are particularly susceptible to changes in efficacy (Huerta and Lisman, 1996). In contrast, during SWS, the hippocampus bombards the cortex with activity rather than the sensory world. Recently stored information in the hippocampus appears to be 'played back' to the neocortex during SWS, but at a different rate and in combinations that may not have occurred simultaneously during the day (Quin et al., 1987; Chrobak and Buzsaki, 1996; Buzsaki, 1998; Nadasdy et al., 1999; Louie and Wilson, 2000; Nadasdy, 2000). This information is a distillation of recent sensory impressions and cortical states that are activated during REM sleep. The information stored in the hippocampus is probably not a literal copy of information stored in the neocortex, which has a much larger capacity, but rather is an abstraction of that information, a pointer that is capable of reactivating that information in the neocortex through feedback connection from the hippocampus.

Feedback from the hippocampus to the neocortex also takes the form of sharp waves, brief bursts of activity that occur at intervals of 0.3–3 Hz (Buzsáki, 1986). Not much is known about the mechanisms that initiate sharp waves in the hippocampus or elsewhere (see Paré et al., 1995). It is thought that spontaneous activity in CA3 hippocampal neurons, perhaps primed by minis (Timofeev et al., 2000), ignites an assembly of reciprocally connected neurons to discharge in less than 100 *ms*. The temporary associations between neurons formed during the day, perhaps at different times, may be recapitulated during the brief bursts of activity that then can imprint traces of these associations on the neocortex, which is in a receptive state of slow-wave activity.

When the neocortex switches from slow-wave sleep to rapid eye-movement sleep, charac-terized by high-frequency activity, the hippocampus switches from sharp wave activity to a theta rhythm. During REM, the cortex may activate recently formed associations between neurons, as indicated by functional imaging in humans (Maquet et al., 2000). These associations may lead to changes in the connection strengths of neuron in the hippocampus, while it is in a theta state.[2] This reciprocal activation and reactivation occurs repeatedly during sleep. In rats that are awake, a similar alternation between theta activity and sharp waves in the hippocampus also occurs on a faster time scale, during exploration and quietly attentive activities such as grooming and feeding, respectively. This suggests that some aspects of memory consolidation may take place during periods of alertness. In humans, the inward-directed periods may correspond to daydreaming, or contemplative periods when thoughts visit the mind unbidden.

Although speculative, this framework is consistent with what is currently known about sleep oscillations, as we have shown throughout this monograph. It is also consistent with models of learning that require a 'sleep' phase for long-term learning of generative representations (Ackley et al., 1985; Gardner-Medwin 1989; Hinton et al., 1995). The key insight is that slow-wave sleep is a specific state in which information is consolidated by activating Ca^{2+}-mediated intracel-lular cascades in pyramidal neurons. Implementing such a massive Ca^{2+} entry and network reorganization must necessarily be performed during a state in which normal processing—such as sensory processing should not occur. This, ultimately, may be the reason why we sleep.

9.4.5 Recent developments on hippocampal-neocortical relations

In rodents, spindles are often associated with brief bursts of activity and 200 *Hz* oscillations, called sharp-waves ripples (SPWRs), which originate in the hippocampus (Buzsaki, 1992) and which may also trigger cortical Down-to-Up state transitions, as well as K-complexes and low-voltage spindles (Johnson et al., 2010). It was also shown that SPWRs are a replay of activity patterns related to past experiences (Wilson and McNaughton, 1994) and are involved in forming long-term of memories (Girardeau et al., 2009). During SWS, monosynaptic projection between hippocampus and medial prefrontal cortex (mPFC) replay activity patterns formed during previous learning experiences (Jay and Witter, 1991; Euston et al., 2007; Peyrache et al., 2009). How this transfer occurs has been unclear.

There is now substantial evidence that the dialogue between the hippocampus and the neocortex during sleep involves oscillations (Maingret et al., 2016). These authors electrically stimulated the cortex immediately after hippocampal SWRs, which led to a bout of cortical spindle and delta oscillations. They found that this augmented hippocampal-cortical coordi-nation resulted in better performance in a memory task the next day. These results provide direct evidence that coordination between hippocampus and cortex is sufficient for memory consolidation during sleep and that this coordination most likely operates through spindles and delta oscillations during slow wave sleep in neocortex. *In vivo* recordings from the cortex in rodents during the Down state revealed sparse activity, called 'delta spikes', immediately following SPWR events (Todorova and Zugaro, 2019). Following learning, these neurons formed

---

[2] Other theories have been proposed that also involve 'reprogramming' the brain during REM sleep (see Jouvet, 1998).

synchronously firing assemblies. These studies suggest that the purpose of reducing activity during the Down state may be to eliminate irrelevant neurons while new assemblies are formed, thereby reducing interference with other assemblies.

Alternatively, it was proposed that the Up and Down state dynamics may directly allow synaptic plasticity to occur in neocortex (El Boustani et al., 2012), as developed in the next section.

9.4.6 Up/Down states and synaptic plasticity

We have seen above that replay, possibly arising from hippocampal-neocortical interactions, may underlie memory consolidation, but how replayed patterns are stored in neocortex remains a mystery. Up/Down state transitions during SWS may be involved. This was examined in a thalamocortical of replay that implemented neuronal mechanism for how an interaction between hippocampal input mediated by sharp wave-ripple events, cortical slow oscillations and synaptic plasticity could lead to consolidation of memories through preferential replay of cortical cell spike sequences during slow-wave sleep (Wei et al., 2016). An external input, mimicking hippocampal ripples, delivered to the cortical network resulted in input-specific changes of synaptic weights, which persisted after stimulation was removed. These synaptic changes promoted replay of specific firing sequences of the cortical neurons. They found that the spatiotemporal pattern of Up-state propagation determined the changes of synaptic strengths between neurons.

Another model of replay proposed a link between synaptic plasticity and Up/Down state dynamics, which may be relevant to the memory consolidation during sleep (El Boustani et al., 2012). This model focused on spike timing dependent plasticity (STDP), which has been intensely investigated, both experimentally and theoretically, because it provides a biologically grounded model for learning with spiking neurons (Feldman, 2012). However, STDP suffers from the fact that it is sensitive to spontaneous activity, and any pattern of synaptic weights acquired during learning, will be invariably lost if the system is subject to spontaneous activity (Morrison et al., 2007), a situation called the 'catastrophic forgetting' effect.

To avoid such a catastrophic forgetting, a metaplastic STDP (mSTDP) rule was proposed with local floating plasticity thresholds, which was resistant to sustained and irregular spontaneous activity because the threshold for plasticity automatically sets to high values (El Boustani et al., 2012). However, if the activity is not sustained, but consists of periods of spiking interleaved with periods of silence (as in Up/Down states), then plasticity can occur. This suggests that the purpose of the 'Down' states may be to relax the floating thresholds, thereby enabling the activity during the 'Up' states to efficiently trigger long-term plasticity at neocortical synapses. In the context of replay during SWS, replay separated by periods of silence is an efficient way to store these patterns (El Boustani et al., 2012). This constitutes a biophysically plausible mechanism allowing replayed patterns to be stored in neocortex during sleep, while remaining stable during wakefulness. It also provides a possible explanation for why we need a separate, disconnected brain state to store memories.

9.4.7 Computational models of sleep

Since there are several types of learning and memory, there may well be several types of memory consolidation that occur during different phases of sleep. Several neural network models of associative memory have included a 'sleep phase' to make the storage of memories acquired by Hebbian mechanisms more efficient (Crick and Mitchison, 1983; Hopfield et al., 1983; Ackley et al., 1985; Gardner-Medwin, 1989). In particular, these models identify REM sleep as a phase when 'reverse' or anti-learning may be taking place. First, the cortex is in a state that is indistinguishable from the awake state, based on EEG recordings, and is as metabolically active as in the awake state; second, visual images and dream narratives can be recalled when a sleeper is wakened during REM sleep; third, there is strong cholinergic activation of thalamocortical circuits, which activates second-messenger systems inside neurons; finally, upon awakening dreams fade quickly from conscious memory and little of the dreams can be recalled later.

In a related theory of unsupervised learning, the cortex stores probability densities for sensory states in a hierarchy of layers; that is, the higher areas of the cortex encode higher-order statistical regularities in the sensory inputs and in the absence of sensory input can generate activity in earlier stages of processing with the same statistics using feedback connections. By generating ideal input patterns, the feedforward connections responsible for recognition can be accurately trained to improve processing at earlier stages (Hinton et al., 1995). Conversely, when the brain is awake, the sensory inputs drive the feedforward system, during which the weights on the feedback connections can be altered, thereby improving the generative model. This two-phase process produces an internal, hierarchical representation of experiences that are economical. The feedback connections in this model are used to generate prototypical input patterns. The learning mechanisms needed are biologically possible since, unlike previous learning algorithms for hierarchical networks that required a detailed teacher and error backpropagation, this wake-sleep model only depends on locally available signals and there is no teacher other than the sensory experience.

The wake-sleep learning algorithm attempts to capture the statistics of sensory inputs with an internal code that is capable of representing component features that are common to many objects. Because these statistical components are not apparent without comparing many sensory experiences, the training process is gradual, in the sense that only small changes are made during any one wake-sleep cycle. Although the feedback connections are not used during the awake or feedforward phase of the algorithm, it is possible to view them as representing a prior probability distribution on complex brain states. Thus, if sensory input is locally ambiguous, it may be possible for the feedback connections from higher levels in the hierarchy help disambiguate them (Hinton and Ghahramani, 1997; Lewicki and Sejnowski, 1997).

The wake-sleep model is limited to a passive, unsupervised form of learning that is entirely driven by the statistics of sensory states (Hinton and Sejnowski, 1999). Not all sensory inputs are equally important, and some tasks might require special representations. Attentional mechanisms may modulate the learning rate with significance of the stimulus. There may also be biases in cortical representations at birth that are specified during development, which could incorporate a prior probability distribution for the world. There is also a goal-directed reinforcement learning system in the brain that involves subcortical as well as cortical structures (Montague et al., 1996). Unsupervised wake-sleep learning and other forms of learning could work together, biasing, shifting and adapting cortical representations to insure survival in complex and uncertain environments.

The computational explanations offered above for memory consolidation during sleep are neither mutually exclusive nor exhaustive. They nonetheless have the virtue of making specific predictions for some of the puzzling physiological phenomena summarized above. For the completion hypothesis, feedback projections from higher cortical areas to lower ones are used during sleep for the purpose of reactivating assemblies of neurons that had previously been used to represent specific episodic memories. In this case, plasticity should strengthen connections that encourage the same patterns to reoccur in the future. In the case of the generative hypothesis, ideal patterns of activity are instantiated in lower cortical areas for the purpose of altering the feedforward connections. Experimental tests could distinguish between these two predictions.

9.4.8 Conclusion: a scenario for memory consolidation during sleep

Summing up, we would like to conclude by proposing a scenario that accounts for the most recent experimental results on memory consolidation during sleep, as reviewed above. The most recent experiments tend to corroborate our hypothesis on the mechanisms underlying memory consolidation. These experiments and models lead to the following scenario: the hippocampus receives inputs from the top of cortical hierarchies that induce synaptic plasticity for specific events and objects experienced during attentive wakefulness. During the early stages of SWS, spindles occur in the thalamocortical system due to the decreased neuromodulatory drive. Sleep spindles mobilize the molecular machinery needed for long-term plasticity through a massive calcium entry in dendrites of pyramidal cells. In the deeper phases of SWS, the patterns of activity stored in the hippocampus are replayed in sharp-wave ripple events. Many of these events trigger a transition from Down states to Up states in neocortex, where the replayed information would be stored (or consolidated) through spike-time dependent plasticity. Thus, latent memories are activated in the neocortex by replay during sleep, and induce longer-term changes in intrinsic and synaptic conductances during Up states. The mechanism of the permanent change could be related to a lowering of the threshold for plasticity due to the presence of Down states (El Boustani et al., 2012), which would explain why Up and Down states are invariably seen in all mammals during sleep.

This hypothetical scenario for memory consolidation during sleep is consistent with all of the known electrophysiological properties of sleep oscillations. One can see a mechanistic explanation of the replay by hippocampus, and the consolidation in neocortex, in the same line as originally proposed by Buzsaki (1989), but with substantial differences, for example the role of spindles and the Up/Down states. This scenario not only gives for the first time an explanation for the presence of Up/Down states in mammalian neocortex during SWS, but it also predicts that suppressing Down states during SWS should prevent memory consolidation. It also provides a possible mechanistic explanation for why we need to sleep.

9.5 Summary

In this final chapter, we have broadened the focus of our investigation to include sleep oscillations with the lowest and the highest frequencies, and to explore possible physiological roles for these oscillations. In the first part of the chapter, the impact of thalamic spindles on cortical neurons was examined using cortical models. These models predict that high-frequency synchronized burst discharges in the thalamus produce unusually powerful excitatory inputs in the dendrites of cortical pyramidal cells in conjunction with strong inhibitory inputs around the soma. The powerful dendritic excitation evoked massive Ca^{2+} entry in the dendrites, which can activate Ca^{2+}-dependent molecular cascades leading to long-term synaptic changes and gene expression. Spindle oscillations could therefore open the door to permanent changes in cortical networks.

In the second part, the spatiotemporal dynamics of cortical oscillations was examined during wake and sleep states. Slow-wave sleep consists in slowly recurring waves characterized by large-scale spatiotemporal synchrony in cortical neurons. These slow-wave complexes alternate with brief periods of fast oscillations, similar to the sustained fast oscillations that occur during the wake state. We propose that the alternating fast and slow waves consolidate information acquired previously during wakefulness. Slow-wave sleep would thus begin by spindle oscillations that open molecular gates to plasticity, then proceed by iteratively 'recalling' and 'storing' information primed in neural assemblies. This scenario provides biophysically plausible mechanisms consistent with the growing behavioural evidence that sleep serves to consolidate memories, as discussed in the last part of this final chapter.

Ionic bases of the membrane potential

A living neuron has a *membrane potential*, which depends on the gradient of ions across the membrane and the selective ionic permeabilities of the membrane. We review here the physico-chemical properties of membranes, then show how ionic mechanisms establish an electrical potential difference across the membrane. Neuronal signalling takes advantage of the membrane potential to transform information from chemical to electrical signals and to transmit information by all-or-none action potentials.

A.1 Water and phospholipid membranes

When the electrons that form the chemical bonds in a molecule are unevenly shared amongst the atoms, the molecule becomes polarized. Oxygen has a high affinity for electrons, a tendency called electronegativity, so oxygen-containing molecules typically have a nonuniform distribution of electrons. For example, the electrons forming the bond between the O and H atoms in a water molecule (H_2O) are more densely distributed around the oxygen atom. Consequently, the oxygen atom bears a small net negative charge while hydrogen atoms are positively charged. Bipolar water molecules form a *polar* medium.

Bipolar molecules like water tend to associate with each other to stabilize these net charges. Oxygen atoms tend to stay close to the hydrogen atoms of neighbouring molecules, creating *hydrogen bonds*. Water molecules also tend to stabilize other types of bipolar molecules or charged ions. In aqueous solution, an ion is surrounded by a *hydration shell*, stabilized by water molecules oriented around the ionic charge (Fig. A.1, top). These pictures are somewhat misleading because the water molecules undergo Brownian motion and water molecules in the shell are in constant reorganization; however, the hydration shell for each ion has a well-defined average radius.

Nonpolar molecules, lacking electronic charge redistribution, do not have a tendency to create hydrogen bonds. As a consequence they strongly avoid water; this *hydrophobic* property is as important as the hydrogen bonds in organizing molecules. For example, long carbohydrate chains—$(CH_2)_n$—such as fatty acids are hydrophobic molecules and, in addition to avoiding water, have a tendency to associate with each other. As a consequence, hydrophobic molecules in aqueous solution tend to avoid water molecules and associate with each other to form a distinct phase, much as of oil separates in water. The entropy of the system provides the organizing principle.

Phospholipids, which make membranes, have both polar and nonpolar parts and are have properties that are between these polar properties and nonpolar extremes. The phosphate group on a phospholipid molecule is highly polar and in aqueous solution will tend to associate with water molecules; in contrast, the carbohydrate chains are highly nonpolar and tend to associate with the hydrophobic regions on other phospholipid molecules. This leads to geometric arrangements, like micelles, in which lipids aggregate in the centre of a ball with the phosphate groups on the outside facing the aqueous solution, and the phospholipid bilayer, a similar arrangement in two dimensions (Fig. A.1). The phospholipid bilayer is the primary structure that forms cellular membranes.

A.2 Establishing a membrane potential

A lipid bilayer membrane has a highly negatively charged surface that attracts a positive layer of ions. It is energetically unfavourable for an ion to cross a membrane, since the charged ion would need to enter a hydrophobic medium. Because only a small proportion of ions have the required energy, pure phospholipid membranes have an extremely low permeability. This extremely important property makes it possible for the cell to regulate ion fluxes across the membrane and maintain a membrane potential, as discussed below.

Consider a membrane with NaCl one side and KCl on the other side (Fig. A.2A). Assume that there exists a channel in the membrane that is selectively permeable to K^+ ions. With respect to K^+ ions, there are two compartments: the first one has K^+ ions while the second one does not. Therefore, diffusion will drive K^+ ions from the first compartment into the second to equilibrate the K^+ concentrations (Fig. A.2A). However, only K^+ ions cross the membrane, so for each ion crossing there is a positive charge added to the second compartment and a positive charge subtracted from compartment one. Additional K^+ ions face a positively charged solution. Thus, there are two opposing forces: diffusion pushes K^+ ions into the second compartment, while a repulsive electrostatic force impedes K^+ ions from entering (Fig. A.2B). It does not take many ions, or much time, to reach an equilibrium between these two tendencies. At equilibrium, there is a stable electrical potential difference between the two compartments due to the net difference of charge, called the equilibrium potential for K^+.

This example illustrates how the selective permeability of the membrane for an ion establish a difference ionic concentration and a membrane potential. The formal relationship between the ionic concentrations and the membrane potential was established by Nernst:

$$E_K = \frac{RT}{ZF} \ln \frac{[K^+]_o}{[K^+]_i} \tag{A.1}$$

where R is the gas constant, T is temperature in degrees Kelvin, Z is the valence of the ion ($Z = 1$ for K^+ ions, $Z = -1$ for Cl^- ions, etc), F is the Faraday constant, $[K^+]_o$ and $[K^+]_i$ are the concentration of K^+ ions outside and inside of the membrane, respectively.

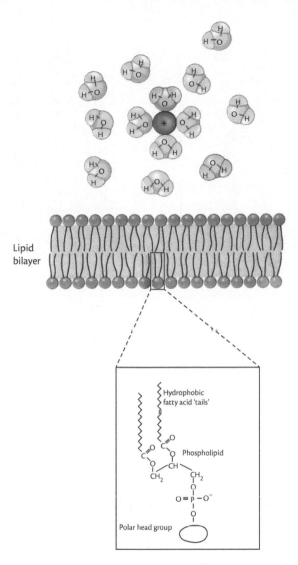

Fig. A.1: Bilayer structure of phospholipids in aqueous solution. Phospholipids (inset) have a hydrophobic tail and a polar head. The hydrophobic tails join to form a two-dimensional bilayer in which only the polar groups are in contact with water molecules. From Kandel et al., 1995.

E_K is also called the *reversal potential* for K^+ ions. A similar Nernst potential can be defined for Na^+, Cl^- and Ca^{2+}. The standard extracellular and intracellular concentrations of these ions, as well as their reversal potential, are listed in Table 1.1.

If the membrane is selectively permeable to several ionic species, the membrane potential depends on a complex equilibrium between all of the ionic fluxes. The resting membrane potential is given by the Goldman-Hodgkin-Katz equation:

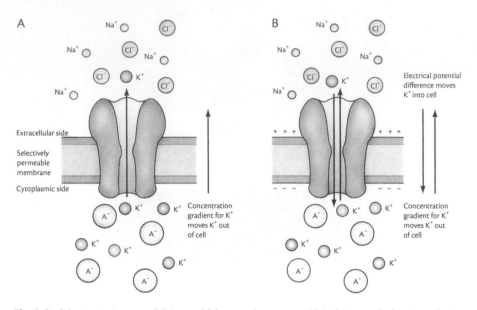

Fig. A.2: Selective ionic permeabilities establish a membrane potential. In this example there is a selective permeability to K^+ ions. Initially, two solutions of different ionic composition are on each side of the membrane. Initially, there is an equal number of positive and negative ions in both compartments (NaCl extracellular and KCl intracellular), with no net charge and no potential difference. A. Because of the difference in K^+ concentration, K^+ ions diffuse across the membrane towards the extracellular compartment. B. The K^+ that cross the membrane produce a net charge difference; this creates a repulsive electrostatic force that retards further entry of K^+ ions. The equilibrium between these two tendencies results in the establishment of an electrical potential difference across the membrane. From Kandel et al., 1995.

Table 1.1 Concentrations and reversal potentials of main ionic species. Intracellular and extracellular concentrations are taken from various measurements in hart muscle, squid axon and mammalian central neurons (see Johnston and Wu, 1995; Kandel et al., 1995).

Ion	extracellular concentration (mM)	intracellular concentration (mM)	reversal potential 36°C)
K^+	5–20	140–400	−105 to −80mV
Na^+	145–460	5–50	+50 to +90 mV
Cl^-	110–540	4–100	−90 to −40mV
Ca^{2+}	1–2	1–2e-4	+110 to +130 mV

$$V_r = \frac{RT}{ZF} \ln\left[\frac{P_K[K^+]_o + P_{Na}[Na^+]_o + P_{Cl}[Cl^-]_i}{P_K[K^+]_i + P_{Na}[Na^+]_i + P_{Cl}[Cl^-]_o}\right] \qquad (A.2)$$

where P_K, P_{Na} and P_{Cl} are the permeabilities of K^+, Na^+ and Cl^- ions, respectively.

For typical permeabilities found in neurons, the resting membrane potential V_r is in the range $-80\ mV$ to $-60\ mV$ because the permeability of K^+ ions dominates the contributions from other ions.

A.3 Passive properties of neuronal membranes

Selective membrane permeability is regulated by *ion channels*; these are transmembrane proteins that establish a pore specific to one or several ionic species. Some ion channels can be gated. For example, specialized ion channels can be opened by mechanical actions, such as stretching, or by binding to a ligand, and others have conductances that depend on the membrane potential. Specific examples are examined in detail in Chapters 2 and 5. *Leak channels* are considered below.

A.3.1 Leak channels

Conceptually, the simplest type of ion channel is a pore specific to one ion that is always open. Such channels cause a permanent leak of ions across the membrane and are called *leak channels*. The presence of leak channels for a given ion type confers to the membrane a permeability for this ion, and therefore, as described above, contributes to the membrane potential. Leak channels for K^+, Na^+ and Cl^- contribute to the resting membrane potential according to Eq. A.2.

The equivalent electrical circuit for a membrane represents a leak channel as a battery (the reversal potential) in series with a resistance (inverse of the conductance). The membrane potential will tend to the equilibrium potential of each ion. Membranes also have dielectric properties, which are taken into account with a membrane capacitance (C_m) in the equivalent circuit. This linear model of the membrane is only an approximation to the nonlinear Goldman-Hodgkin-Katz equation (Eq. A.2).

The membrane potential V of circuit A in Fig. A.3 obeys the equation:

$$C_m \frac{dV}{dt} = -g_{Na}(V - E_{Na}) - g_K(V - E_K) - g_{Cl}(V - E_{Cl}) \qquad (A.3)$$

where g_{Na}, g_K and g_{Cl} are the conductance densities (in mS/cm^2), of Na^+, K^+ and Cl^-, respectively, and C_m is the specific membrane capacitance (in $\mu F/cm^2$) . E_{Na}, E_K and E_{Cl} are their respective reversal potentials. According to this equation, the equilibrium potential from the leak channels is given by:

$$E_L = \frac{g_{Na}E_{Na} + g_K E_K + g_{Cl}E_{Cl}}{g_{Na} + g_K + g_{Cl}} \qquad (A.4)$$

which is also the resting membrane potential (V_r) in this simple case.

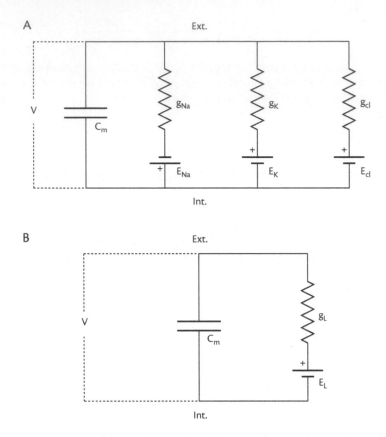

Fig. A.3: Equivalent circuit of the membrane. A. Equivalent circuit for a membrane with specific permeability for three types of ions: Na^+, K^+ and Cl^-. B. Equivalent circuit where the three ionic conductances were merged into a single leak current. See text for description of symbols.

As the leak Na^+, K^+ and Cl^- channels are characterized by constant conductances, they can be merged into a single current, called the *leak current*. This current is represented by a simpler equivalent circuit (Fig. A.3B) and is described by the equation:

$$C_m \frac{dV}{dt} = -g_L (V - E_L) \qquad \text{(A.5)}$$

where the leak current is $I_L = g_L (V - E_L)$ and the leak conductance g_L is the sum of all leak conductances for all ion species ($g_L = g_{Na} + g_K + g_{Cl}$ here).

If an additional current is injected through an intracellular electrode, then I_{inj} (in $\mu A/cm^2$) is added to the leak current equation:

$$C_m \frac{dV}{dt} = -g_L (V - E_L) + I_{inj}. \qquad \text{(A.6)}$$

In this case, the resting potential is given by.

$$V_r = \frac{g_L\, E_L + I_{inj}}{g_L} \tag{A.7}$$

According to this convention, positive current injection will lead to more positive values of V_r and therefore will *depolarize* the membrane, whereas negative injected currents will *hyperpolarize* the membrane.

The electrical circuit representation of the membrane is a useful model for modelling the dynamical changes that occur to the membrane when current is injected in the cell. The leak currents and membrane capacitance endow the membrane with the passive properties of an RC (Resistance-Capacitance) circuit (Fig. A.4). Injection of positive or negative current pulses results in a 'passive' exponential relaxation towards a new equilibrium value for the potential.

The equivalent circuit of Fig. A.3B accounts for exponential relaxation of the membrane potential towards its equilibrium value. Solving Eq. A.6 from an initial condition V_0 gives:

$$V(t) = V_0 + (V_r - V_0)\, \exp[-(t - t_0)/\tau_m] \tag{A.8}$$

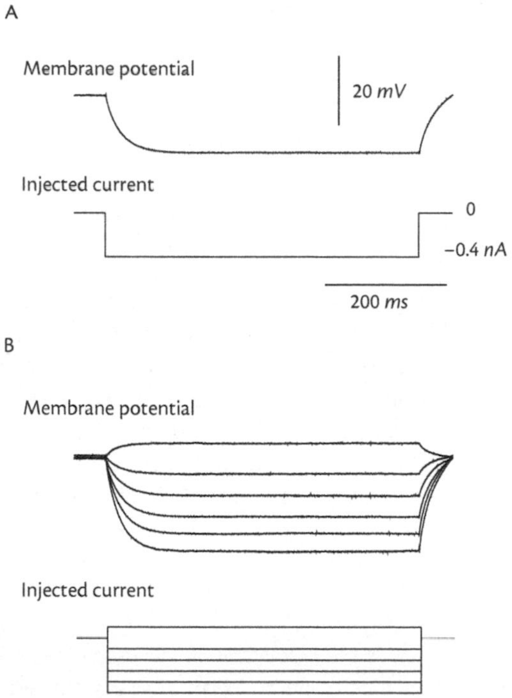

A

Membrane potential

20 mV

Injected current

0

−0.4 nA

200 ms

B

Membrane potential

Injected current

Fig. A.4: Passive properties of neuronal membranes. A. Injection of a current pulse in an intracellularly recorded neuron results in a change of the membrane potential and an exponential relaxation to a steady-state value. B. Injection of families of current pulses in the same neuron (pulse amplitudes of −0.5, −0.4, −0.3, −0.1 and 0.1 nA). Data from intracellularly recorded neocortical pyramidal neuron *in vivo* (courtesy of D. Paré).

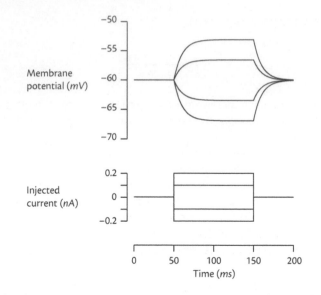

Fig. A.5: Model of passive properties. The membrane potential is shown (top traces) for four different pulses of injected current (indicated in bottom traces). The circuit of Fig. A.3B was simulated with $g_L = 0.1mS/cm^2$, $E_L = -60mV$, $C_m = 1\mu F/cm^2$ and a membrane area of about 29000 μm^2. Pulse amplitudes for injected current were of $-0.2, -0.1, 0.1$ and 0.2 nA.

where $V_0 = V(t_0)$, V_r is the equilibrium value (given by Eq. A.7), and $\tau_m = C_m/g_L$ is the *membrane time constant*. The time course of the membrane potential is shown in Fig. A.5.

A.3.2 Time constant

The membrane time constant, τ_m, is an important parameter and can be determined from experimental recordings (Fig. A.4). τ_m depends on the ratio between the specific membrane capacitance and the leak conductance density. Note that the value of τ_m is independent of the surface of the membrane since it is the ratio between two densities. Note also that the capacitance C_m is a universal property of neuronal membranes and is around 1 $\mu F/cm^2$ (see Johnston and Wu, 1995; Kandel et al., 1995); τ_m, however, depends on the leak conductance g_L. The larger g_L, the faster the membrane relaxes towards equilibrium (Fig. A.6A).

A.3.3 Input resistance

Another useful passive property is the *membrane input resistance*, R_{in}. Rather than describe the conductance density of the membrane in mS/cm^2, it is easier to estimate the membrane resistance, usually expressed in $M\Omega$. Experimentally, R_{in} can be determined by measuring the voltage deflection ΔV (measured in mV) produced by current injection ΔI (measured in nA), leading to $R_{in} = \Delta V / \Delta I$. More generally, R_{in} is most precisely determined from the slope of the linear region in the I-V relationship.

The time constant is independent of the size of the cell and is therefore an intrinsic property of the membrane depending only on conductance and capacitance expressed as densities. The membrane input resistance, however, depends on the size of the cell. In the case of a passive membrane:

$$R_{in} = \frac{1}{S\,g_L} \tag{A.9}$$

where S is the total membrane surface in cm^2. This is illustrated in Fig. A.6B, where R_{in} is shown to depend on the cell surface while the time constant remains unchanged. More generally, the input resistance depends on all conductances present in the membrane, according to:

$$R_{in} = \frac{1}{S\,\sum_i g_i} \tag{A.10}$$

where g_i is the conductance density of current i.

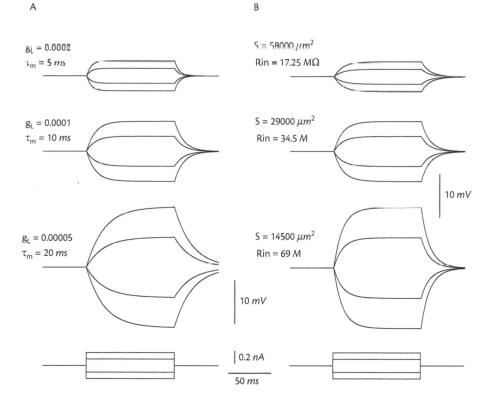

A

$g_L = 0.0002$
$\tau_m = 5\ ms$

$g_L = 0.0001$
$\tau_m = 10\ ms$

$g_L = 0.00005$
$\tau_m = 20\ ms$

10 mV

B

$S = 58000\ \mu m^2$
Rin = 17.25 MΩ

$S = 29000\ \mu m^2$
Rin = 34.5 M

$S = 14500\ \mu m^2$
Rin = 69 M

10 mV

0.2 nA

50 ms

Fig. A.6: Time constant and input resistance. A. The time constant is a measure of the membrane conductance. Same circuit as in Fig. A.5 with three different values of the membrane conductance. The bottom traces indicate the injected current. Leak conductances used are indicated (g_L) as well as the corresponding values of the membrane time constant (τ_m). B. The membrane input resistance depends on the size of the cell. Same simulation as in Fig. A.5 with three different values of the total membrane surface (S).

The concept of membrane potential and the passive properties of neuronal membranes will be used throughout the this book. In particular, the dependence of some ionic permeabilities to the membrane potential is a core mechanism underlying cellular excitability, as presented in Chapter 2.

Optimized algorithms for simulating synaptic currents

In this Appendix, we outline how kinetic models of postsynaptic receptors, combined with brief release of transmitter, can lead to analytic expressions for the synaptic current (Destexhe et al., 1994c, 1994d; Lytton, 1996). These expressions lead to efficient algorithms for simulating synaptic interaction and a considerable improvement in computation time, which may be critical in large-scale network simulations.

B.1 Single synapse

Consider a synaptic current described by the following first-order equation:

$$\frac{dr}{dt} = \alpha\,[T]\,(1-r) - \beta\,r$$

$$I_{syn} = \bar{g}_{syn}\,r\,(V - E_{syn}) \qquad \text{(B.1)}$$

where $[T]$ represents the transmitter concentration in the synaptic cleft; binding of transmitter to the receptor leads to a channel opening with rate constants α and β; r is the fraction of receptors in the open state, $\bar{g}_{syn}$ is the maximal conductance and E_{syn} is the reversal potential of the synaptic current I_{syn}.

If the transmitter time course can be approximated by a pulse, Eq. B.1 can be solved analytically during each phase of the pulse during which $[T]$ is constant (Destexhe et al., 1994c). Define the following two variables:

$$r_\infty = \frac{\alpha\,T_{max}}{\alpha\,T_{max} + \beta} \quad \text{and} \quad \tau_r = \frac{1}{\alpha\,T_{max} + \beta}$$

where T_{max} is the maximal concentration of the transmitter during the pulse (usually, $T_{max} = 1\ mM$).

The analytical expression for the fraction of open receptors r for each phase of the pulse can be calculated as follows:

1. When the pulse is on ($t_0 < t < t_1$), $[T] = T_{max}$ and r is given by:

$$r(t - t_0) = r_\infty + (r(t_0) - r_\infty) \exp[-(t - t_0)/\tau_r] \qquad (B.2)$$

2. When the pulse is off ($t > t_1$), $[T] = 0$, and r is given by:

$$r(t - t_1) = r(t_1) \exp[-\beta (t - t_1)] \qquad (B.3)$$

At each integration step Δt, the update rule for the state variable is:

$$\begin{aligned} r &= r_\infty + (r - r_\infty) \exp[-\Delta t/\tau_r] & \text{if } [T] > 0 \\ r &= r \exp[-\beta \, \Delta t] & \text{if } [T] = 0 \end{aligned} \qquad (B.4)$$

The computational advantages of two-state kinetic models of synaptic currents are, first, that no differential equation needs to be solved, and, second, at each time step, only one exponential term is evaluated, independently of the number of spikes received by the synapse. This exponential term can be precalculated, leading to further increase in computational efficiency.

B.2 Multiple synapses

The algorithm can be further optimized in the case of multiple synapses (Lytton, 1996). Suppose that the same postsynaptic compartment receives N identical synaptic contacts from N different sources. The synaptic current at each individual contact is:

$$I_i = \bar{g}_{syn} \, r_i \, (V - E_{syn}) \qquad (B.5)$$

where r_i is the fraction of open receptors at synapse i. Thus, the total synaptic current is:

$$I = \sum_i I_i = \bar{g}_{syn} \sum_i r_i \, (V - E_{syn}) . \qquad (B.6)$$

The calculation of the currents at these synapses can be greatly simplified after merging them. The N synapses are first separated into those that are releasing, N_{on}, and all the others (the vast majority), that are decaying, $N_{off} = N - N_{on}$. The state variables for these two populations of synapses are merged together, leading to:

1. For all releasing synapses,

$$R_{on} = \sum_{i \text{ releasing}} r_i$$

and is updated according to:

$$R_{on} = N_{on}\, r_{\infty} + (R_{on} - N_{on}\, r_{\infty})\, \exp[-\Delta t/\tau_r].\qquad\text{(B.7)}$$

2. For all decaying synapses,

$$R_{off} = \sum_{i\ \text{decaying}} r_i$$

and is updated according to:

$$R_{off} = R_{off}\, \exp[-\beta\, \Delta t].\qquad\text{(B.8)}$$

Whenever a pulse of transmitter begins or ends, the variables R_{on} and R_{off} must be changed accordingly. This is easy to do since the value of r_i at any time can be calculated from its value at the time it last changed. When a spike occurs at a synapse i, these values are calculated as follows:

$$\begin{aligned}
r_i(t) &= r_i(t_0)\, \exp[-\beta\, (t - t_0)] \\
R_{on} &= R_{on} + r_i(t) \\
R_{off} &= R_{off} - r_i(t)
\end{aligned}\qquad\text{(B.9)}$$

where t_0 is the time of the preceding event that occurred at that synapse.

When the pulse of transmitter ends, the values are again updated:

$$\begin{aligned}
r_i(t) &= r_{\infty} + (r_i(t_1) - r_{\infty})\, \exp[-(t - t_1)/\tau_r] \\
R_{on} &= R_{on} - r_i(t) \\
R_{off} &= R_{off} + r_i(t)
\end{aligned}\qquad\text{(B.10)}$$

where t_1 is the time at which the pulse of transmitter began at synapse i.

This algorithm (Lytton, 1996) greatly minimizes the number of exponentials calculated at each integration step. Because most synapses spend most of their time in the decaying phase, they can be merged into a single equation (Eq. B.8), leading to a marked reduction of the computation time. Benchmarks indicate that this algorithm is much faster than all other existing methods (see Lytton, 1996). Calculation of alpha functions, even when optimized (Srinivasan and Chiel, 1993), would require at least one exponential to be calculated at each time step for *each synapse*. The speedup by this algorithm is significant for systems with many synapses, especially with large network simulations.

Note that this algorithm can also be extended to multiple synapses with different conductances by including a multiplicative factor for each r_i in Eqs. B.7, B.8 according to its conductance value (Lytton, 1996).

Data available on the Internet

C.1 NEURON simulation programs

Computer programs and their associated input files are available to simulate the results shown in some of the figures of the present manuscript. Most of these simulations use the NEURON program, a publicly available simulation environment for multicompartment models and networks of neurons (Hines, 1993; Hines and Carnevale, 1997; see http://www.neuron.yale.edu). The mod files needed to replicate some of the models in this book can be found at the Internet addresses:

http://www.cnl.salk.edu/~alain/

and

http://cns.iaf.cnrs-gif.fr.

Programs are available for the following models:

1. *Simulations of the bursting behavior of thalamic relay and reticular neurons.* Both detailed compartmental models of thalamic neurons and simplified models are included, as described in Chapters 3 and 4.
2. *Simulations of synaptic transmission at AMPA, NMDA, GABA$_A$ and GABA$_B$ receptors.* Both simplified kinetic models and detailed Markov models are included for each of these receptor types (see Chapter 5).
3. *Network simulations.* These files implement networks of reticular neurons and networks of thalamic relay and reticular cells, as described in Chapter 6).

Using these files, together with NEURON, the reader should be able to reproduce many of the results reported in the figures of the book, as well as figures in published article. Some knowledge of how to use NEURON is needed before these models can be simulated, and tutorials can be found on the NEURON web site. The reader is encouraged to perform these simulations since they provide a much more dynamic and vivid introduction to the principles and concepts in this book. In addition, the reader can modify these files to build on these results and explore new hypothesis. Electronic copies of the original articles upon which this book is based are also available at the above Internet addresses.

C.2 Computer-generated animations

These Internet sites also provide access to a series of computer-generated animations of the models, as described in this book and in the original articles. These movies are valuable because they visualize important features of the dynamical systems that cannot be shown in static figures.

The following mpeg movies are available:

1. *Dendritically generated bursts in thalamic reticular neurons.* The animations show the dynamics of somato-dendritic interactions in RE cells during the genesis of bursts of action potentials (see Chapter 4).

2. *Spatiotemporal dynamics of activity in the thalamic reticular network.* These animations illustrate the spatiotemporal spread of activity during simulated oscillations in network of RE cells, showing spiral-like and waxing-and-waning patterns (see Chapter 6).

3. *Spatiotemporal dynamics of thalamic relay-reticular networks.* Propagating patterns of spindle and bicuculline-induced oscillations in networks of TC and RE neurons based on ferret thalamic slice experiments (see Chapter 6).

4. *Dynamics of extracellular field potentials in the thalamus.* Spatiotemporal dynamics of oscillatory activity recorded by multiple electrodes in the thalamus *in vivo* (see Chapter 7).

Each of the movies is keyed to figures in this book or to figures in the original research articles; electronic copies of the original articles are available on the Internet at the above addresses.

Summary

During sleep, the mammalian brain actively generates an orderly progression of low frequency, oscillatory states: spindle oscillations appear shortly after sleep onset and even lower frequency delta oscillations arise as the brain falls into a state of deep sleep. These oscillations are prominent in the electroencephalogram, recorded directly from the surface of the scalp; however, the underlying neural mechanisms and the purpose of the oscillations have remained illusive. This is changing rapidly as we learn more about the intrinsic properties of neurons in the thalamus and the cerebral cortex and how their interactions give rise to sleep oscillations. These complex interactions can be explored with computational models that simulate sleep oscillations.

In this book on 'thalamocortical assemblies', the molecular components and ionic mechanisms underlying sleep oscillations are reviewed, including the properties of ion channels and synaptic receptors present in thalamic and cortical neurons as well as their patterns of interconnectivity. These properties are examined and assembled in successive chapters that focus on building accurate computational models of these processes. The scope of the book is broad, from how ion channels generate membrane excitability, how different types of ion channels interact within a single neuron to generate particular patterns of responsiveness, how different types of neurons interact at the network level to generate the complexity of patterns seen in the living brain, and finally how oscillations could play a role in processes such as memory consolidation.

The high quality and precision of experimental data upon which the models are based has allowed a coherent framework to be built that spans three orders of spatial scale from microns to millimetres and six orders of temporal scale from microseconds to seconds. This foundation is a milestone in the study of dynamical activity in central brain systems and allows the possible function of thalamocortical oscillations in memory consolidation to be explored in the final chapter of the book.

SHORT SUMMARY:

The mammalian brain generates a wide range of oscillations during sleep, from spindle oscillations that appear shortly after sleep onset to delta oscillations during deeper stages of sleep. This book reviews the mechanisms underlying these oscillations and their physiological consequences in the context of a computational framework. Many spatial and temporal scales are spanned, from how ion channels generate membrane excitability, how different types of ion channels interact within a single neuron to generate patterns of responsiveness, how different types of neurons interact at the network level to generate the complexity of patterns seen in the living brain, and finally how oscillations may be involved in memory consolidation. This coherent framework for the genesis of sleep oscillations is a milestone in the study of dynamic activity in the thalamus and cortex.

References

Abbott, L. F. and Sejnowski, T. J. (eds.) (1999) *Neural Codes and Distributed Representations: Foundations of Neural Computation.* MIT Press, Cambridge, MA.

Abbott, L. F. and Song, S. (1999) Temporally asymmetric Hebbian learning, spike timing and neuronal response variability. *Adv. Neural Information Processing Systems* **11**: 69–75.

Abel, T., Nguyen, P. V., Barad, M., Deuel, T. A. S., Kandel, E. R. and Bourtchouladze, R. (1997) Genetic demonstration of a role for PKA in the late phase of LTP and in hippocampus-based long-term memory. *Cell* **88**: 615–626.

Achermann, P. and Borbély, A. (1997) Low-frequency (<1 *Hz*) oscillations in the human sleep EEG. *Neuroscience* **81**: 213–222.

Ackley, D. H., Hinton, G. E. and Sejnowski, T. J. (1985) A learning algorithm for Boltzmann Machines. *Cogn. Sci.* **9**: 147–169.

Adams, D. J., Smith, S. J. and Thompson, S. H. (1980) Ionic currents in molluscan soma. *Annual Rev. Neurosci.* **3**: 141–167.

Adrian, E. D. (1941) Afferent discharges to the cerebral cortex from peripheral sense organs. *J. Physiol.* **100**: 159–191.

Ahlsen, G., Grant, K. and Lindström, S. (1982) Monosynaptic excitation of principal cells in the lateral geniculate nucleus by corticofugal fibers. *Brain Res.* **234**: 454–458.

Akasu, T. and Tokimasa, T. (1992) Cellular metabolism regulating H and M currents in bullfrog sympathetic ganglia. *Canadian J. Physiol. Pharmacol.* **70** Suppl: S51–S55.

Albowitz, B. and Kuhnt, U. (1993) Spread of epileptiform potentials in the neocortical slice: recordings with voltage-sensitive dyes. *Brain Research* **631**: 329–333.

Aldrich, R. W. (1981) Inactivation of voltage-gated delayed potassium currents in molluscan neurons. *Biophys. J.* **36**: 519–532.

Aldrich, R. W. and Stevens, C. F. (1987) Voltage-dependent gating of single sodium channels from mammalian neuroblastoma cells. *J. Neurosci.* **7**: 418–431.

Aldrich R. W., Corey, D. P. and Stevens, C. F. (1983) A reinterpretation of mammalian sodium channel gating based on single channel recording. *Nature* **306**: 436–441.

Allman, J. M. (1999) *Evolving Brains.* Freeman & Co., New York.

Alonso, A. and Llinás, R. R. (1989) Subthreshold Na$^+$-dependent theta-like rhythmicity in stellate cells of entorhinal cortex layer II. *Nature* **342**: 175–177.

Alvarez, P. and Squire, L. R. (1994) Memory consolidation and the medial temporal lobe: a simple network model. *Proc. Natl. Acad. Sci. USA* **91**: 7041–7045.

Alvarez, P., Zola-Morgan, S. and Squire, L. R. (1995) Damage limited to the hippocampal region produces long-lasting memory impairment in monkeys. *J. Neurosci.* **15**: 3796–3807.

Alzheimer, C., Schwindt, P. C. and Crill, W. E. (1993) Modal gating of Na+ channels as a mechanism of persistent Na+ current in pyramidal neurons from rat and cat sensorimotor cortex. *J. Neurosci.* **13**: 660–673.

Amzica, F. and Steriade, M. (1995) Short- and long-range neuronal synchronization of the slow (<1 *Hz*) cortical oscillation. *J. Neurophysiol.* **73**: 20–38.

Andersen, O. and Koeppe, R. E. (1992) Molecular determinants of channel function. *Physiol. Rev.* **72**: S89–S158.

Andersen, P. and Andersson, S. A. (1968) *Physiological Basis of the Alpha Rhythm.* Appelton Century Crofts, New York.

Andersen, P. and Eccles, J. C. (1962) Inhibitory phasing of neuronal discharge. *Nature* **196**: 645–647.

Andersen, P. and Rutjord, T. (1964) Simulation of a neuronal network operating rhythmically through recurrent inhibition. *Nature* **204**: 289–190.

Andersen, P. and Sears, T. A. (1964) The role of inhibition in phasing of spontaneous thalamo-cortical discharge. *J. Physiol.* **173**: 459–480.

Andersen, P., Andersson, S. A. and Lømo, T. (1967) Nature of thalamo-cortical relations during spontaneous barbiturate spindle activity. *J. Physiol.* **192**: 283–307.

Anderson, M. P., Mochizuki, T., Xie, J., Fischler, W., Manger, J. P., Talley, E. M., Scammell, T. E. and Tonegawa, S. (2005) Thalamic Cav3.1 T-type Ca2+ channel plays a crucial role in stabilizing sleep. *Proc. Natl. Acad. Sci. USA* **102**: 1743–1748.

Andrade, R., Malenka, R. C. and Nicoll, R. A (1986) A G protein couples serotonin and GABA$_B$ receptors to the same channels in hippocampus. *Science (Wash.)* **234**: 1261–1265.

Andrillon, T., Nir, Y., Staba, R. J., Ferrarelli, F., Cirelli, C., Tononi, G. and Fried, I. (2011) Sleep spindles in humans: insights from intracranial EEG and unit recordings. *J. Neurosci.* **31**: 17821–17834.

Arakaki, T., Mahon, S., Charpier, S., Leblois, A. and Hansel, D. (2016) The role of striatal feedforward inhibition in the maintenance of absence seizures. *J. Neurosci.* **36**: 9618–9632.

Armstrong, C. M. (1969) Inactivation of the potassium conductance and related phenomena caused by quaternary ammonium ion injection in squid axons. *J. Gen. Physiol.* **54**: 553–575.

Armstrong, C. M. (1981) Sodium channels and gating currents. *Physiol. Rev.* **62**: 644–683.

Armstrong, C. M. (1992) Voltage-dependent ion channels and their gating. *Physiol. Rev.* **72**: S5–S13.

Armstrong, C. M. and Hille, B. (1998) Voltage-gated ion channels and electrical excitability. *Neuron* **20**: 371–380.

Astori, S., Wimmer, R. D., Prosser, H. M., Corti, C., Corsi, M., Liaudet, N., Volterra, A., Franken, P., Adelman, J. P. and Luthi, A. (2011) The Ca(V)3.3 calcium channel is the major sleep spindle pacemaker in thalamus. *Proc. Natl. Acad. Sci. USA* **108**: 13823–13828.

Atkins, P. W. (1986) *Physical Chemistry.* Freeman, New York, third edition.

Avanzini, G., M. de Curtis, F. Panzica and R. Spreafico. (1989) Intrinsic properties of nucleus reticularis thalami neurones of the rat studied in vitro. *J. Physiol.* **416**: 111–122.

Avendaño, C., Rausell, E. and Reinoso-Suarez, F. (1985) Thalamic projections to areas 5a and 5b of the parietal cortex in the cat: a retrograde horseradish peroxidase study. *J. Neurosci.* **5**: 1446–1470.

Avendaño, C., Rausell, E., Perez-Aguilar, D., Isorna, S. (1988) Organization of the association cortical afferent connections of area 5: a retrograde tracer study in the cat. *J. Comp. Neurol.* **278**: 1–33.

Avoli, M. and Gloor, P. (1981) The effect of transient functional depression of the thalamus on spindles and bilateral synchronous epileptic discharges of feline generalized penicillin epilepsy. *Epilepsia* **22**: 443–452.

Avoli, M. and Gloor, P. (1982) Role of the thalamus in generalized penicillin epilepsy: observations on decorticated cats. *Exp. Neurol.* **77**: 386–402.

Avoli, M., Gloor, P., Kostopoulos, G. and Gotman, J. (1983) An analysis of penicillin-induced generalized spike and wave discharges using simultaneous recordings of cortical and thalamic single neurons. *J. Neurophysiol.* **50**: 819–837.

Avoli, M., Gloor, P., Kostopoulos, G. and Naquet, R. (eds.) (1990) *Generalized Epilepsy.* Birkhäuser, Boston, MA.

Babb, T. L., Prectorius, J. K., Kupfer, W. R. and Brown, W. J. (1988) Distribution of glutamate-decarboxylase immunoreactive neurons and synapses in the rat and monkey hippocampus: light and electron microscopy. *J. Comp. Neurol.* **278**: 121–138.

Babloyantz, A. and Destexhe, A. (1986) Low dimensional chaos in an instance of epileptic seizure. *Proc. Natl. Acad. Sc. USA* **83**: 3513–3517.

Bal, T. and McCormick, D. A. (1993) Mechanisms of oscillatory activity in guinea-pig nucleus reticularis thalami *in vitro*: a mammalian pacemaker. *J. Physiol.* **468**: 669–691.

Bal, T. and McCormick, D. A. (1996) What stops synchronized thalamocortical oscillations? *Neuron* **17**: 297–308.

Bal, T., Debay, D. and Destexhe, A. (2000) Cortical feedback controls the frequency and synchrony of oscillations in the visual thalamus. *J. Neurosci.* **20**: 7478–7488.

Bal, T., von Krosigk, M. and McCormick, D. A. (1995a) Synaptic and membrane mechanisms underlying synchronized oscillations in the ferret LGNd *in vitro*. *J. Physiol.* **483**: 641–663.

Bal, T., von Krosigk, M. and McCormick, D. A. (1995b) Role of the ferret perigeniculate nucleus in the generation of synchronized oscillations *in vitro*. *J. Physiol.* **483**: 665–685.

Bal, T., Foust, A. J., Casale, A. E. and McCormick, D. A. (2012) Dendritic processing in the GABAergic neurons of the thalamic reticular nucleus. *Soc. Neurosci. Abstracts.*

Baranyi, A., Szente, M. B. and Woody, C. D. (1993) Electrophysiological characterization of different types of neurons recorded *in vivo* in the motor cortex of the cat. II. Membrane parameters, action potentials, current-induced voltage responses and electrotonic structures. *J. Neurophysiol.* **69**: 1865–1879, 1993.

Barrett, J. N. Motoneuron dendrites: role in synaptic integration. *Fed. Proc.* **34**: 1398–1407, 1975.

Barrett, J. N. and Crill, W. E. (1974) Influence of dendritic location and membrane properties on the effectiveness of synapses on cat motoneurones. *J. Physiol.* **239**: 325–345.

Barrett, E. F., Barrett, J. N., Botz, D., Chang, D. B. and Mahaffey, D. (1978) Temperature-sensitive aspects of evoked and spontaneous transmitter release at the frog neuromuscular junction. *J. Physiol.* **279**: 253–273.

Bartol, T. M., Land, B. R., Salpeter, E. E. and Salpeter, M. M. (1991) Monte Carlo simulation of miniature endplate current generation in the vertebrate neuromuscular junction. *Biophys. J.* **59**: 1290–1307.

Bartol, T. M. and Sejnowski, T. J. (1993) Model of the quantal activation of NMDA receptors at a hippocampal synaptic spine. *Soc. Neurosci. Abstracts* **19**: 1515.

Bartol, T. M., Keller, D. X., Kinney, J. P., Bajaj, C., Harris, K. M., Sejnowski, T. J. and Kennedy, M. B. (2015) Computational reconstitution of spine calcium transients from individual proteins. *Front. Synaptic Neurosci.* **7**: 17.

Bazhenov, M., Timofeev, I., Steriade, M. and Sejnowski, T. J. (1998a) Cellular and network models for intrathalamic augmenting responses during 10-*Hz* stimulation. *J. Neurophysiol.* **79**: 2730–2748.

Bazhenov, M., Timofeev, I., Steriade, M. and Sejnowski, T. J. (1998b) Computational models of thalamocortical augmenting responses. *J. Neurosci.* **18**: 6444–6465.

Bazhenov, M., Timofeev, I., Steriade, M. and Sejnowski, T. J. (1999) Self-sustained rhythmic activity in the thalamic reticular nucleus mediated by depolarizing GABA$_A$ receptor potentials. *Nature Neurosci.* **2**: 168–174.

Bazhenov, M., Timofeev, I., Steriade, M. and Sejnowski, T. J. (2004) Potassium model for slow (2–3 *Hz*) *in vivo* neocortical paroxysmal oscillations. *J. Neurophysiol.* **92**: 1116–1132.

Bell, C. C., Han, V. Z., Sugawara, Y. and Grant, K. (1997) Synaptic plasticity in a cerebellum-like structure depends on temporal order. *Nature* **387**: 278–281.

Benardo, L. S. (1994) Separate activation of fast and slow inhibitory postsynaptic potentials in rat neocortex in vitro. *J. Physiol.* **476**: 203–215.

Berger, H. (1929) Über den zeitlichen verlauf der negativen schwankung des nervenstroms. *Arch. Ges. Physiol.* **1**: 173.

Bernander, O., Douglas, R. J., Martin, K. A., Koch, C. (1991) Synaptic background activity influences spatiotemporal integration in single pyramidal cells. *Proc. Natl. Acad. Sci. USA* **88**: 11569–11573.

Berns, G. S. and Sejnowski, T. J. (1998) A computational model of how the basal ganglia produce sequences. *J. Cogn. Neurosci.* **10**: 108–121.

Berridge, M. J. (1998) Neuronal calcium signaling. *Neuron* **21**: 13–26.

Berridge, M. J. and Irvine, R. F. (1989) Inositol phosphates and cell signalling. *Nature* **341**: 197–205.

Bezanilla, F. (1985) Gating of sodium and potassium channels. *J. Membr. Biol.* **88**: 97–111.

Bhalla, U. S. and Iyengar, R. (1999) Emergent properties of networks of biological signaling pathways. *Science* **283**: 381–387.

Bi, G. and Poo, M. (1998) Activity-induced synaptic modifications in hippocampal culture: dependence on spike timing, synaptic strength and cell type. *J. Neurosci.* **18**: 10464–10472.

Bishop, G. H. (1936) The interpretation of cortical potentials. *Cold Sping Harbor Symp. Quant. Biol.* **4**: 305–319.

Black, J. A., Kocsis, J. D. and Waxman, S. G. (1990) Ion channel organization of the myelinated fiber. *Trends Neurosci.* **13**: 48–54.

Blaustein, M. P. (1988) Calcium transport and buffering in neurons. *Trends Neurosci.* **11**: 438–443.

Bliss, T. and Collingridge, G. L. (1993) A synaptic model of memory: long-term potentiation in the hippocampus. *Nature* **361**: 31–39.

Blitzer, R. D., Wong, T., Nouranifar, R., Iyengar, R. and Landau, E. M. (1995) Postsynaptic cAMP pathway gates early LTP in hippocampal CA1 region. *Neuron* **15**: 1403–1414.

Blitzer, R. D., Connor, J. H., Brown, G. P., Wong, T., Shenolikar, S., Iyengar, R. and Landau, E. M. (1998) Gating of CaMKII by cAMP-regulated protein phosphatase activity during LTP. *Science* **280**: 1940–1942.

Bloomfield, S. A., Hamos, J. E. and Sherman, S. M. (1987) Passive cable properties and morphological correlates of neurones in the lateral geniculate nucleus of the cat. *J. Physiol.* **383**: 653–692.

Blum, K. I. and Abbott, L. F. (1996) A model of spatial map formation in the hippocampus of the rat. *Neural Computation* **8**: 85–93.

Blumenfeld, H. and McCormick, D. A. (2000) Corticothalamic inputs control the pattern of activity generated in thalamocortical networks. *J. Neurosci.* **20**: 5153–5162.

Boland, L. M. and Bean, B. P. (1993) Modulation of N-type calcium channels in bullfrog sympathetic neurons by luteinizing hormone-releasing hormone: kinetics and voltage dependence. *J. Neurosci.* **13**: 516–533.

Bonjean, M., Baker, T., Lemieux, M., Timofeev, I., Sejnowski, T. J. and Bazhenov, M. (2011) Corticothalmamic feedback controls sleep spindle duration *in vivo*. *J. Neurosci.* **31**: 9124–9134.

Bonjean, M., Baker, T., Bazhenov, M., Cash, S., Halgren, E. and Sejnowski, T. J. (2012) Interactions between core and matrix thalamocortical projections in human sleep spindle synchronization. *J. Neurosci.* **32**: 5250–5263.

Bonnet, V. and Bremer, F. (1938) Étude des potentiels électriques de la moelle épinière faisant suite chez la grenouille spinale à une ou deux vollées d'influx centripètes. *C. R. Soc. Biol. Paris* **127**: 806–812.

Bontempi, B., Laurent-Demir, C., Destrade, C. and Jaffard, R. (1999) Time-dependent reorganization of brain circuitry underlying long-term memory storage. *Nature* **400**: 671–675.

Borg-Graham, L. J. (1991) Modeling the nonlinear conductances of excitable membranes. In: *Cellular and Molecular Neurobiology: A Practical Approach*. ed. Wheal, H. and Chad, J. Oxford University Press, New York, pp. 247–275.

Borg-Graham, L. J., Monier, C. and Frégnac, Y. (1998) Visual input evokes transient and strong shunting inhibition in visual cortical neurons. *Nature* **393**: 369–373.

Bourassa, J., Deschênes, M. (1995) Corticothalamic projections from the primary visual cortex in rats: a single fiber study using biocytin as an anterograde tracer. *Neuroscience* **66**: 253–263.

Brazier, M. A. B. (1961) *A History of the Electrical Activity of the Brain*. Pitman, London.

Bremer, F. (1938a) Effets de la déafférentation complète d'une région de l'écorce cérébrale sur son activité électrique. *C.R. Soc. Biol. Paris* **127**: 355–359.

Bremer, F. (1938b) L'activité électrique de l'écorce cérébrale. *Actualités Scientifiques et Industrielles* **658**: 3–46.

Bremer, F. (1949) Considérations sur l'origine et la nature des 'ondes' cérébrales. *Electroencephalogr. Clin. Neurophysiol.* **1**: 177–193.

Bremer. F., Brihaye, J. and André-Balisaux, G. (1956) Physiologie et pathologie du corps calleux. *Schweiz. Arch. Neurol. Psychiat.* **78**: 31–87.

Bremer, F. (1958) Cerebral and cerebellar potentials. *Physiol. Reviews* **38**: 357–388.

Brette, R. and Gerstner, W. (2005) Adaptive exponential integrate-and-fire model as an effective description of neuronal activity. *J. Neurophysiol.* **94**: 3637–3642.

Briska, A. M., Uhlrich, D. J. and Lytton, W. W. (2003) Computer model of passive signal integration based on whole-cell in vitro studies of rat lateral geniculate nucleus. *Eur. J. Neurosci.* **17**: 1531–1541.

Brock, L. G., Coombs, J. S. and Eccles, J. C. (1952) The recording of potential from motoneurones with an intracellular electrode. *J. Physiol.* **117**: 431–460.

Brown, D. A. (1990) G-proteins and potassium currents in neurons. *Annu. Rev. Physiol.* **52**: 215–242.

Brown, A. M. and Birnbaumer, L. (1990) Ionic channels and their regulation by G protein subunits. *Annu. Rev. Physiol.* **52**: 197–213.

Brüderle, D., Petrovici, M. A., Vogginger, B., Matthias Ehrlich, M., Pfeil, T., Millner, S., Grübl, A., Wendt, K., Müller, E., Schwartz, M.-O., Husmann de Oliveira, D., Jeltsch, S., Fieres, J., Schilling, M., Müller, P., Breitwieser, O., Petkov, V., Muller, L. E., Davison, A. P., Krishnamurthy, P., Kremkow, J., Lundqvist, M., Muller, E., Partzsch, J., Scholze, S., Zühl, L., Mayr, C., Destexhe, A., Diesmann, M., Potjans, T. C., Lansner, A., Schüffny, R., Schemmel, J. and Meier, K. (2011) A comprehensive workflow for general-purpose neural modeling with highly configurable neuromorphic hardware systems. *Biol. Cybernetics* **104**: 263–296.

Brunel, N. (2000) Dynamics of sparsely connected networks of excitatory and inhibitory spiking neurons. *J. Computational. Neurosci.* **8**: 183–208.

Budde, T., Mayer, R. and Pape, H. C. (1992) Different types of potassium outward current in relay neurons acutely isolated from the rat lateral geniculate nucleus. *Eur. J. Neurosci.* **4**: 708–722.

Budde, T., Biella, G., Munsch, T. and Pape, H. C. (1997) Lack of regulation by intracellular Ca^{2+} of the hyperpolarization-activated cation current in rat thalamic neurones. *J. Physiol.* **503**: 79–85.

Bullock, T. H. and McClune, M. C. (1989) Lateral coherence of the electrocorticogram: a new measure of brain synchrony. *Electroencephalogr. Clin. Neurophysiol.* **73**: 479–498.

Burgard, E. C. and Hablitz, J. J. (1993) NMDA receptor-mediated components of miniature excitatory synaptic currents in developing rat neocortex. *J. Neurophysiol.* **70**: 1841–1852.

Burke, W. and Sefton, A. J. (1966) Inhibitory mechanisms in lateral geniculate nucleus of rat. *J. Physiol.* **187**: 231–246.

Busch, C. and Sakmann, B. (1990) Synaptic transmission in hippocampal neurons: numerical reconstruction of quantal IPSCs. *Cold Spring Harbor Symp. Quant. Biol.* **55**: 69–80.

Bush, P. and Sejnowski, T. J. (1991) Simulations of a reconstructed cerebellar Purkinje cell based on simplified channel kinetics. *Neural Computation* **3**: 321–332.

Bush, P. and Sejnowski, T. J. (1993) Reduced compartmental models of neocortical pyramidal cells. *J. Neurosci. Methods* **46**: 159–166.

Buzsáki, G. (1986) Hippocampal sharp waves: their origin and significance. *Brain Res.* **398**: 242–252.

Buzsáki, G. (1989) Two-stage model of memory trace formation: a role for 'noisy' brain states. *Neuroscience* **31**: 551–570.

Buzsáki, G. (1996) The hippocampo-neocortical dialogue. *Cerebral Cortex* **6**: 81–92.

Buzsáki, G. (1998) Memory consolidation during sleep: a neurophysiological perspective. *J. Sleep Res.* **7** (Suppl. 1): 17–23.

Buzsáki, G., Bickford, R. G., Ponomareff, G., Thal, L. J., Mandel, R., Gage, F. H. (1988) Nucleus basalis and thalamic control of neocortical activity in the freely moving rat. *J. Neurosci.* **8**: 4007–4026.

Buzsáki, G., Smith, A., Berger, S., Fisher, L. J. and Gage, F. H. (1990) Petit mal epilepsy and parkinsonian tremor: hypothesis of a common pacemaker. *Neuroscience* **36**: 1–14.

Buzsáki, G., Horvath, Z., Urioste, R., Hetke, J. and Wise, K. (1992) High-frequency network oscillation in the hippocampus. *Science* **256**: 1025–1027.

Carbone, E. and Lux, H. D. (1984a) A low voltage-activated, fully inactivating Ca channel in vertebrate sensory neurones. *Nature* **310**: 501–502.

Carbone, E. and Lux, H. D. (1984b) A low voltage-activated calcium conductance in embryonic chick sensory neurons. *Biophys. J.* **46**: 413–418.

Carr, D. W., Stofko-Hahn, R. E., Fraser, I. D., Cone, R. D. and Scott, J. D. (1992) Localization of the cAMP-dependent protein kinase to the postsynaptic densities by A-kinase anchoring proteins. Characterization of AKAP 79. *J. Biol. Chem.* **267**: 16816–16823.

Castro-Alamancos, M. A. (1999) Neocortical synchronized oscillations induced by thalamic disinhibition *in vivo*. *J. Neurosci. Online* **19**: RC27.

Castro-Alamancos, M. A. and Connors, B. W. (1996a) Spatiotemporal properties of short-term plasticity in sensorimotor thalamocortical pathways of the rat. *J. Neurosci.* **16**: 2767–2779.

Castro-Alamancos, M. A. and Connors, B. W. (1996b) Cellular mechanisms of the augmenting response: short-term plasticity in a thalamocortical pathway. *J. Neurosci.* **16**: 7742–7756.

Catterall, W. A. (1992) Cellular and molecular biology of voltage-gated sodium channels. *Physiol. Rev.* **72**: S15–48.

Catterall, W. A. (2000) From ionic currents to molecular mechanisms: the structure and function of voltage-gated sodium channels. *Neuron* **26**: 13–25.

Cauller, L. J. and Connors, B. W. (1992) Functions of very distal dendrites: experimental and computational studies of Layer I synapses on neocortical pyramidal cells. In: *Single Neuron Computation*, ed. McKenna, T., Davis, J. and Zornetzer, S. F. Academic Press, Boston, pp. 199–229.

Celentano, J. J. and Wong, R. K. (1994) Multiphasic desensitization of the $GABA_A$ receptor in outside-out patches. *Biophys. J.* **66**: 1039–1050.

Chabala, L. D. (1984) The kinetics of recovery and development of potassium channel inactivation in perfused squid giant axons. *J. Physiol.* **356**: 193–220.

Chagnac-Amitai, Y. and Connors, B. W. (1989) Horizontal spread of synchronized activity in neocortex and its control by GABA-mediated inhibition. *J. Neurophysiol.* **61**: 747–758.

Chanson, M., Chandross, K. J., Rook, M. B., Kessler, J. A., Spray, D. C. (1993) Gating characteristics of a steeply voltage-dependent gap junctional channel in rat Schwann cells. *J. Gen. Physiol.* **102**: 925–946.

Charpier, S., Deniau, J. M., Mahon, S., Leresche, N., Hughes, S. W. and Crunelli, V. (1998) *In vivo* intracellular recordings in cortical neurons during spontaneous spike and wave discharges in a genetic model of absence epilepsy. *Soc. Neurosci. Abstracts* **24**: 1210.

Chausson, P., Leresche, N. and Lambert, R. C. (2013) Dynamics of intrinsic dendritic calcium signaling during tonic firing of thalamic reticular neurons. *PLoS One* **8**: e72275.

Chen, C. and Hess, P. (1990) Mechanisms of gating of T-type calcium channels. *J. Gen. Physiol.* **96**: 603–630.

Chrobak, J. J. and Buzsáki, G. (1994) Selective activation of deep layer (V–VI) retrohippocampal cortical neurons during hippocampal sharp waves in the behaving rat. *J. Neurosci.* **14**: 6160–6170.

Chrobak, J. J. and Buzsaki, G. (1996) High-frequency oscillations in the output networks of the hippocampal-entorhinal axis of the freely behaving rat. *J. Neurosci.* **16**: 3056–3066.

Clark, J. A. and Amara, S. G. (1994) Stable expression of a neuronal gamma-aminobutyric acid transporter, GAT-3, in mammalian cells demonstrates unique pharmacological properties and ion dependence. *Molec. Pharmacol.* **46**: 550–557.

Clay, J. R. (1989) Slow inactivation and reactivation of the potassium channel in squid axons. *Biophys. J.* **55**: 407–414.

Clay, J. R. and DeFelice, L. J. (1983) Relationship between membrane excitability and single channel open-close kinetics. *Biophys. J.* **42**: 151–157.

Clements, J. D. (1996) Transmitter time course in the synaptic cleft: its role into central synaptic function. *Trends Neurosci.* **19**: 163–171.

Clements, J. D. and Westbrook, G. L. (1991) Activation kinetics reveal the number of glutamate and glycine binding sites on the NMDA receptor. *Neuron* **7**: 605–613.

Clements, J. D., Lester, R. A. J., Tong, J., Jahr, C. and Westbrook, G. L. (1992) The time course of glutamate in the synaptic cleft. *Science* **258**: 1498–1501.

Cole, K. S. (1949) Dynamic electrical characteristics of the squid axon membrane. *Arch. Sci. Physiol.* **3**: 253–258.

Cole, K. S. and Curtis, H. J. (1939) Electrical impedance of the squid giant axon during activity. *J. Gen. Physiol.* **22**: 649–670.

Colquhoun, D. and Hawkes, A. G. (1977) Relaxation and fluctuations of membrane currents that flow through drug-operated channels. *Proc. Roy. Soc. Lond. Ser. B* **199**: 231–262.

Colquhoun, D. and Hawkes, A. G. (1981) On the stochastic properties of single ion channels. *Proc. Roy. Soc. Lond. Ser. B* **211**: 205–235.

Colquhoun, D., Jonas, P. and Sakmann, B. (1992) Action of brief pulses of glutamate on AMPA/KAINATE receptors in patches from different neurons of rat hippocampal slices. *J. Physiol.* **458**: 261–287.

Compte, A., Sanchez-Vives, M. V., McCormick, D. A. and Wang, X. J. (2000) A network model of generation and propagation of slow (<1 *Hz*) oscillations in the cerebral cortex. *Soc. Neurosci. Abstracts* **26**: 1967.

Connor, J. A. and Stevens, C. F. (1971a) Inward and delayed outward membrane currents in isolated neural somata under voltage clamp. *J. Physiol.* **213**: 1–19.

Connor, J. A. and Stevens, C. F. (1971b) Voltage clamp studies of a transient outward membrane current in gastropod neural somata. *J. Physiol.* **213**: 21–30.

Connor, J. A. and Stevens, C. F. (1971c) Prediction of repetitive firing behaviour from voltage clamp data on an isolated neurone soma. *J. Physiol.* **213**: 31–53.

Connors, B. W. and Gutnick, M. J. (1990) Intrinsic firing patterns of diverse neocortical neurons. *Trends Neurosci.* **13**: 99–104.

Connors, B. W., Gutnick, M. J. and Prince, D. A. (1982) Electrophysiological properties of neocortical neurons *in vitro*. *J. Neurophysiol.* **48**: 1302–1320.

Constantine-Paton, M., Cline, H. T. and Debski, E. (1990) Patterned activity, synaptic convergence, and the NMDA receptor in developing visual pathways. *Ann. Rev. Neurosci.* **13**: 129–154.

Contreras, D. (1996) Oscillatory properties of cortical and thalamic neurons and the generation of synchronized rhythmicity in the corticothalamic network. PhD Thesis, Laval University, Québec, Canada.

Contreras, D. and Steriade, M. (1995) Cellular basis of EEG slow rhythms: a study of dynamic corticothalamic relationships. *J. Neurosci.* **15**: 604–622.

Contreras, D. and Steriade, M. (1996) Spindle oscillation in cats: the role of corticothalamic feedback in a thalamically-generated rhythm. *J. Physiol.* **490**: 159–179.

Contreras, D., R. Curro Dossi and M. Steriade (1993) Electrophysiological properties of cat reticular thalamic neurones *in vivo*. *J. Physiol.* **470**: 273–294.

Contreras, D., Destexhe, A., Sejnowski, T. J. and Steriade, M. (1996) Control of spatiotemporal coherence of a thalamic oscillation by corticothalamic feedback. *Science* **274**: 771–774.

Contreras, D., Destexhe, A. and Steriade, M. (1997a) Spindle oscillations during cortical spreading depression in naturally sleeping cats. *Neuroscience* **77**: 933–936.

Contreras, D., Destexhe, A., Sejnowski, T. J. and Steriade, M. (1997b) Spatiotemporal patterns of spindle oscillations in cortex and thalamus. *J. Neuroscience* **17**: 1179–1196.

Contreras, D., Destexhe, A. and Steriade, M. (1997c) Intracellular and computational characterization of the intracortical inhibitory control of synchronized thalamic inputs *in vivo. J. Neurophysiology* **78**: 335–350.

Contreras, D., Sugimori, M. and Llinás, R. (1997d) Afferent stimulation frequency determines spatial distribution of excitation in neocortex. A voltage-sensitive dye study. *Soc. Neurosci. Abstracts* **23**: 1005.

Coulter, D. A., Huguenard, J. R. and Prince, D. A. (1989) Calcium currents in rat thalamocortical relay neurones: kinetic properties of the transient, low-threshold current. *J. Physiol.* **414**: 587–604.

Creutzfeldt, O., Watanabe, S. and Lux, H. D. (1966a) Relation between EEG phenomena and potentials of single cortical cells. I. Evoked responses after thalamic and epicortical stimulation. *EEG Clin. Neurophysiol.* **20**: 1–18.

Creutzfeldt, O., Watanabe, S. and Lux, H. D. (1966b) Relation between EEG phenomena and potentials of single cortical cells. II. Spontaneous and convulsoid activity. *EEG Clin. Neurophysiol.* **20**: 19–37.

Crick, F. and Mitchison, G. (1983) The function of dream sleep. *Nature* **304**: 111–114.

Crunelli, V., Leresche, N., Paravelas, J. (1987) Membrane properties of morphologically identified X and Y cells in the lateral geniculate nucleus of the cat in vitro. *J. Physiol.* **390**: 243–256.

Crunelli, V., Lightowler, S. and Pollard, C. E. (1989) A T-type Ca^{2+} current underlies low-threshold Ca^{2+} potentials in cells of the cat and rat lateral geniculate nucleus. *J. Physiol.* **413**: 543–561.

Cucchiaro, J. B., Uhlrich, D. J. and Sherman, S. M. (1991) Electron-microscopic analysis of synaptic input from the perigeniculate nucleus to the A-laminae of the lateral geniculate nucleus in cats. *J. Comp. Neurol.* **310**: 316–336.

Curro Dossi, R., A. Nunez and M. Steriade. (1992) Electrophysiology of a slow (0.5–4 *Hz*) intrinsic oscillation of cat thalamocortical neurones *in vivo. J. Physiol.* **447**: 215–234.

Damasio, A. R. and Tranel, D. (1993) Nouns and verbs are retrieved with differently distributed neural systems. *Proc. Natl. Acad. Sci. USA* **90**: 4957–4956.

Dang-Vu, T. T., Bonjean, M., Schabus, M., Boly, M., Darsaud, A., Desseilles, M., Degueldre, C., Balteau, E., Phillips, C., Luxen, A., Sejnowski, T. J. and Maquet, P. (2011) Interplay between spontaneous and induced brain activity during human non-rapid eye movement sleep. *Proc. Natl. Acad. Sci. USA* **108**: 15438–15443.

Danober, L., Deransart, C., Depaulis, A., Vergnes, M. and Marescaux, C. (1998) Pathophysiological mechanisms of genetic absence epilepsy in the rat. *Progr. Neurobiol.* **55**: 27–57.

David, F., Schmiedt, J. T., Taylor, H. L., Orban, G., Di Giovanni, G., Uebele, V. N., Renger, J. J., Lambert, R. C., Leresche, N. and Crunelli, V. (2013) Essential thalamic contribution to slow waves of natural sleep. *J. Neurosci.* **33**: 19599–19610.

Davies, C. H., Davies, S. N. and Collingridge, G. L. (1990) Paired-pulse depression of monosynaptic GABA-mediated inhibitory postsynaptic responses in rat hippocampus. *J. Physiol.* **424**: 513–531.

Debanne, D., Gahwiler, B. H. and Thompson, S. M. (1998) Long-term synaptic plasticity between pairs of individual CA3 pyramidal cells in rat hippocampal slice cultures. *J. Physiol.* **507**: 237–247.

Deboer, T. (1998) Brain temperature dependent changes in the elecgtroencephalogram power spectrum of humans and animals. *J. Sleep Res.* **7**: 254–262.

DeFelipe, J. and Fariñas, I. (1992) The pyramidal neuron of the cerebral cortex: morphological and chemical characteristics of the synaptic inputs. *Progress Neurobiol.* **39**: 563–607.

Dehghani, N., Peyrache A., Telenczuk, B., Le Van Quyen, M., Halgren, E., Cash, S. S., Hatsopoulos, N. G. and Destexhe, A. (2016) Dynamic balance of excitation and inhibition in human and monkey neocortex. *Sci. Reports* **6**: 23176.

De Koninck, Y. and Mody, I. (1994) Noise analysis of miniature IPSCs in adult rat brain slices: properties and modulation of synaptic $GABA_A$ receptor channels. *J. Neurophysiol.* **71**: 1318–1335.

De Koninck, P. and Schulman, H. (1998) Sensitivity of CaM kinase II to the frequency of Ca^{2+} oscillations. *Science* **279**: 227–230.

de la Pena, E. and Geijo-Barrientos, E. (1996) Laminar organization, morphology and physiological properties of pyramidal neurons that have the low-threshold calcium current in the guinea-pig frontal cortex. *J. Neurosci.* **16**: 5301–5311.

Dement, W. (1960) The effect of dream deprivation. *Science* **131**: 1705–1707.

Depannemaecker, D., Carlu, M., Boutéé, J. and Destexhe, A. (2022) A Model for the propagation of seizure activity in normal brain tissue. *eNeuro* **9**: 0234–21.

Derbyshire, A. J., Rempel, B., Forbes, A. and Lambert, E. F. (1936) The effects of anethetics on action potentials in the cerebral cortex of the cat. *Am. J. Physiol.* **116**: 577–596.

Deschênes, M., Paradis, M., Roy, J. P., Steriade, M. (1984) Electrophysiology of neurons of lateral thalamic nuclei in cat: resting properties and burst discharges. *J. Neurophysiol.* **55**: 1196–1219.

Deschênes, M., Madariaga-Domich, A. and Steriade, M. (1985) Dendrodendritic synapses in the cat reticularis thalami nucleus: a structural basis for thalamic spindle synchronization. *Brain Res.* **334**: 165–168.

Deschênes, M. and Hu, B. (1990) Electrophysiology and pharmacology of the corticothalamic input to lateral thalamic nuclei: an intracellular study in the cat. *Eur. J. Neurosci.* **2**: 140–152, 1990.

De Schutter, E. and Bower, J. M. (1994) An active membrane model of the cerebrellar purkinje cell. II. Simulation of synaptic responses. *J. Neurophysiol.* **71**: 401–419.

Destexhe, A. (1992) Nonlinear dynamics of the rhythmical activity of the brain (in French), Doctoral Dissertation, Université Libre de Bruxelles, Brussels.

Destexhe, A. (2009) Self-sustained asynchronous irregular states and Up/Down states in thalamic, cortical and thalamocortical networks of nonlinear integrate-and-fire neurons. *J. Computational Neurosci.* **27**: 493–506.

Destexhe, A. and Babloyantz, A. (1991) Pacemaker-induced coherence in cortical networks. *Neural Computation* **3**: 145–154.

Destexhe, A. and Babloyantz, A. (1992) Cortical coherent activity induced by thalamic oscillations. In: *Neural Network Dynamics*, ed. Taylor, J. G., Caianello, E. R., Cotterill, R. M. J. and Clark, J. W., Springer-Verlag, Berlin, pp. 234–249.

Destexhe, A. and Babloyantz, A. (1993) A model of the inward current I_h and its possible role in thalamocortical oscillations. *NeuroReport* **4**: 223–226.

Destexhe, A., Babloyantz, A. and Sejnowski, T. J. (1993a) Ionic mechanisms for intrinsic slow oscillations in thalamic relay neurons. *Biophysical Journal* **65**: 1538–1552.

Destexhe, A., McCormick, D. A. and Sejnowski, T. J. (1993b) A model for 8–10 *Hz* spindling in interconnected thalamic relay and reticularis neurons. *Biophysical Journal* **65**: 2474–2478.

Destexhe, A., Contreras, D., Sejnowski, T. J. and Steriade, M. (1993c) A model of spindle rhythmicity in the isolated thalamic reticular nucleus. In: *Institute for Neural Computation Technical Report Series* no. INC-9308.

Destexhe, A., Contreras, D., Sejnowski, T. J. and Steriade, M. (1994a) A model of spindle rhythmicity in the isolated thalamic reticular nucleus. *J. Neurophysiology* **72**: 803–818.

Destexhe, A., Contreras, D., Sejnowski, T. J. and Steriade, M. (1994b) Modeling the control of reticular thalamic oscillations by neuromodulators. *NeuroReport* **5**: 2217–2220.

Destexhe, A., Mainen, Z. F. and Sejnowski, T. J. (1994c) An efficient method for computing synaptic conductances based on a kinetic model of receptor binding. *Neural Computation* **6**: 14–18.

Destexhe, A., Mainen, Z. F. and Sejnowski, T. J. (1994d) Synthesis of models for excitable membranes, synaptic transmission and neuromodulation using a common kinetic formalism. *J. Computational Neuroscience* **1**: 195–230.

Destexhe, A. and Sejnowski, T. J. (1995) G-protein activation kinetics and spill-over of GABA may account for differences between inhibitory responses in the hippocampus and thalamus. *Proc. Natl. Acad. Sci. USA* **92**: 9515–9519.

Destexhe, A., Bal, T., McCormick, D. A. and Sejnowski, T. J. (1996a) Ionic mechanisms underlying synchronized oscillations and propagating waves in a model of ferret thalamic slices. *J. Neurophysiology* **76**: 2049–2070.

Destexhe, A., Contreras, D., Steriade, M., Sejnowski, T. J. and Huguenard, J. R. (1996b) *In vivo*, in vitro and computational analysis of dendritic calcium currents in thalamic reticular neurons. *J. Neuroscience* **16**: 169–185.

Destexhe, A. (1997) Conductance-based integrate and fire models. *Neural Computation* **9**: 503–514.

Destexhe, A. and Sejnowski, T. J. (1997) Synchronized oscillations in thalamic networks: insights from modeling studies. In: *Thalamus*, ed. Steriade, M., Jones, E. G. and McCormick, D. A. Elsevier, Amsterdam, pp. 331–371.

Destexhe, A. (1998) Spike-and-wave oscillations based on the properties of GABA$_B$ receptors. *J. Neurosci.* **18**: 9099–9111.

Destexhe, A., Contreras, D. and Steriade, M. (1998a) Mechanisms underlying the synchronizing action of corticothalamic feedback through inhibition of thalamic relay cells. *J. Neurophysiol.* **79**: 999–1016.

Destexhe, A., Mainen, M. and Sejnowski, T. J. (1998b) Kinetic models of synaptic transmission. In: *Methods in Neuronal Modeling* (2nd edition), ed. Koch, C. and Segev, I. MIT Press, Cambridge, MA, pp. 1–26.

Destexhe, A., Neubig, M., Ulrich, D. and Huguenard, J. R. (1998c) Dendritic low-threshold calcium currents in thalamic relay cells. *J. Neurosci.* **18**: 3574–3588.

Destexhe, A. (1999) Can GABA$_A$ conductances explain the fast oscillation frequency of absence seizures in rodents? *Eur. J. Neurosci.* **11**: 2175–2181.

Destexhe, A. (2000a) Modeling corticothalamic feedback and the gating of the thalamus by the cerebral cortex. *J. Physiol. (Paris)* **94**: 391–410.

Destexhe, A. (2000b) Kinetic models of membrane excitability and synaptic interactions. In: *Computational Models of Molecular and Cellular Interactions*, ed. Bower, J. and Bolouri, H. MIT Press, Cambridge, MA, pp. 225–262.

Destexhe, A. and Sejnowski, T. J. (2001) *Thalamocortical Assemblies*. Oxford University Press (*Monographs of the Physiological Society*), Oxford.

Destexhe, A. and Paré, D. (1999) Impact of network activity on the integrative properties of neocortical pyramidal neurons *in vivo. J. Neurophysiol.* **81**: 1531–1547.

Destexhe, A., Contreras, D. and Steriade, M. (1999a) Cortically-induced coherence of a thalamic-generated oscillation. *Neuroscience* **92**: 427–443.

Destexhe, A., Contreras, D. and Steriade, M. (1999b) Spatiotemporal analysis of local field potentials and unit discharges in cat cerebral cortex during natural wake and sleep states. *J. Neurosci.* **19**: 4595–4608.

Destexhe, A. and Huguenard, J. R. (2000a) Nonlinear thermodynamic models of voltage-dependent currents. *J. Computational Neuroscience* **9**: 259–270.

Destexhe, A. and Huguenard, J. R. (2000b) Which formalism to use for modeling voltage-dependent conductances? In: *Computational Neuroscience: Realistic Modeling for Experimentalists*, ed. DeSchutter, E. CRC Press, pp. 129–157.

Destexhe, A. Contreras, D. and Steriade, M. (2001) LTS cells in cerebral cortex and their role in generating spike-and-wave oscillations. *Neurocomputing* **38**: 555–563.

Dewan, E. (1968) Tests of the programming (P) hypothesis for REM. *Psychophysiology* **4**: 365–366.

DiFrancesco, D. (1985) The cardiac hyperpolarization-activated current I_f, origins and developments. *Prog. Biophys. Molec. Biol.* **46**: 163–183.

DiFrancesco, D. (1999) Dual allosteric modulation of pacemaker (f) channels by cAMP and voltage in rabbit SA node. *J. Physiol.* **515**: 367–376.

DiFrancesco, D. and Mangoni, M. (1994) Modulation of single hyperpolarization-activated channels (i(f)) by cAMP in the rabbit sino-atrial node. *J. Physiol.* **474**: 473–482.

DiFrancesco, D. and Totora, P. (1991) Direct activation of cardiac pacemaker channels by intracellular cyclic AMP. *Nature* **351**: 145–147.

Dodge, F. A., Jr and Rahamimoff, R. (1967) Co-operative action a calcium ions in transmitter release at the neuromuscular junction. *J. Physiol.* **193**: 419–432.

Domich, L., Oakson, G. and Steriade, M. (1986) Thalamic burst patterns in the naturally sleeping cat: a comparison between cortically projecting and reticularis neurones. *J. Physiol. Lond.* **379**: 429–449.

Dosemeci, A. and Reese, T. S. (1993) Inhibition of endogeneous phosphatase in a postsynaptic density fraction allows extensive phosphorylation of the major postsynaptic density protein. *J. Neurochem.* **61**: 550–555.

Doyle, D. A., Cabral, J. M., Pfuetzner, R. A., Kuo, A. L., Gulbis J. M., Cohen, S. L., Chait, B. T. and MacKinnon, R. (1998) The structure of the potassium channel: molecular basis of K+ conduction and selectivity. *Science* **280**: 69–77.

Drover, J. D., Schiff, N. D. and Victor, J. D. (2010) Dynamics of coupled thalamocortical modules. *J. Comput. Neurosci.* **28**: 605–616.

Dutar, P. and Nicoll, R. A. (1988) A physiological role for GABA$_B$ receptors in the central nervous system. *Nature* **332**: 156–158.

Eccles, J. C. (1951) Interpretation of action potentials evoked in the cerebral cortex. *J. Neurophysiol.* **3**: 449–464.

Eckhorn, R., Bauer, R., Jordan, W., Brosch, M., Kruse, W., Munk, M., Reitboek, H. J. (1988) Coherent oscillations: a mechanism of feature linking in the visual cortex? Multiple electrode and correlation analyses in the cat. *Biol. Cybernetics* **60**: 121–130.

Edmonds, B. and Colquhoun, D. (1993) Rapid decay of averaged single-channel NMDA receptor activations recorded at low agonist concentration. *Proc. Roy. Soc. Lond. Ser. B* **250**: 279–286.

Edwards, F. A., Konnerth, A. and Sakmann, B. (1990) Quantal analysis of inhibitory synaptic transmission in the dentate gyrus of rat hippocampal slices: a patch-clamp study. *J. Physiol.* **430**: 213–249.

Egelman, D. M. and Montague, P. R. (1999) Calcium dynamics in the extracellular space of mammalian neural tissue. *Biophys. J.* **76**: 1856–1867.

El Boustani, S., Yger, P., Fregnac, Y. and Destexhe, A. (2012) Stable learning in stochastic network states. *J. Neurosci.* **32**: 194–214.

Erickson, K. R., O. K. Ronnekleiv and M. J. Kelly (1993) Electrophysiology of guinea-pig supraoptic neurones: role of a hyperpolarization-activated cation current in phasic firing. *J. Physiol.* **460**: 407–425.

Erisir, A., VanHorn, S. C., Sherman, S. M. (1997a) Relative numbers of cortical and brainstem inputs to the lateral geniculate nucleus. *Proc. Natl. Acad. Sci. USA* **94**: 1517–1520.

Erisir, A., VanHorn, S. C., Bickford, M. E., Sherman, S. M. (1997b) Immunocytochemistry and distribution of parabrachial terminals in the lateral geniculate nucleus of the cat: a comparison with corticogeniculate terminals. *J. Comp. Neurol.* **377**: 535–549.

Euston, D. R., Tatsuno, M. and McNaughton, B. L. (2007) Fast-forward playback of recent memory sequences in prefrontal cortex during sleep. *Science* **318**: 1147–1150.

Evarts, E. V. (1964) Temporal patterns of discharge of pyramidal tract neurons during sleep and waking in the monkey. *J. Neurophysiol.* **27**: 152–171.

Eyring, H. (1935) The activated complex in chemical reactions. *J. Chem. Phys.* **3**: 107–115.

Eyring, H., Lumry, R., Woodbury, J. W. (1949) Some applications of modern rate theory to physiological systems. *Record Chem. Prog.* **10**: 100–114.

Fatt, P. and Katz, B. (1952) Spontaneous subthreshold activity at motor nerve endings. *J. Physiol.* **117**: 109–128.

Fariñas, I. and DeFelipe, J. (1991a) Patterns of synaptic input on corticocortical and corticothalamic cells in the visual cortex. I. The cell body. *J. Comp. Neurol.* **304**: 53–69.

Fariñas, I. and DeFelipe, J. (1991b) Patterns of synaptic input on corticocortical and corticothalamic cells in the visual cortex. II. The axon initial segment. *J. Comp. Neurol.* **304**. 70–77.

Feldman, D. E. (2012) The spike-timing dependence of plasticity. *Neuron* **75**: 556–571.

Fishbein, W. and Gutwein, B. M. (1977) Paradoxical sleep and memory storage processes. *Behavior. Biol.* **19**: 425–464.

FitzGibbon, T., Tevah, L. V., Jervie-Sefton, A. (1995) Connections between the reticular nucleus of the thalamus and pulvinar-lateralis posterior complex: a WGA-HRP study. *J. Comp. Neurol.* **363**: 489–504.

Fitzhugh, R. (1965) A kinetic model of the conductance changes in nerve membrane. *J. Cell. Comp. Physiol.* **66**: 111–118.

Fowler, M. J., Sullivan, M. J. and Ekstrand, B. R. (1973) Sleep and memory. *Science* **179**: 302–304.

Frank, M. G., Issa, N. P. and Stryker, M. P. (2001) Sleep enhances plasticity in the developing visual cortex. *Neuron* **30**: 275–287.

Franke, C., Hatt, H. and Dudel, J. (1987) Liquid filament switch for ultrafast exchanges of solutions at excised patches of synaptic membrane of crayfish muscle. *Neurosci. Lett.* **77**: 199–204.

Frankenhaeuser, B. and Hodgkin, A. L. (1957) The action of calcium on the electrical properties of squid axons. *J. Physiol.* **137**: 218–244.

Franks, K. M., Bartol, T. M., Poo, M. and Sejnowski, T. J. (2000) High spatial and temporal resolution estimates of calcium dynamics in dendritic spines using MCELL simulations. *Soc. Neurosci. Abstracts* **26**: 1122.

Freund, T. F., Martin, K. A., Soltesz, I., Somogyi, P., Whitteridge, D. (1989) Arborisation pattern and postsynaptic targets of physiologically identified thalamocortical afferents in striate cortex of the macaque monkey. *J. Comp. Neurol.* **289**: 315–336.

Frey, U. and Morris, R. G. (1998) Synaptic tagging: implications for late maintenance of hippocampal long-term potentiation. *Trends Neurosci.* **21**: 181–188.

Frey, U., Huang, Y. Y. and Kandel, E. R. (1993) Effects of cAMP simulate a late phase of LTP in hippocampal CA1 neurons. *Science* **260**: 1661–1664.

Frohlich, F., Bazhenov, M., Timofeev, I. and Sejnowski, T. J. (2005) Maintenance and termination of neocortical oscillations by dynamic modulation of intrinsic and synaptic excitability. *Thalamus Related Syst.* **3**: 147–156.

Frohlich, F., Bazhenov, M., Timofeev. I., Steriade, M. and Sejnowski, T. J. (2006) Slow state transitions of sustained neural oscillations by activity-dependent modulation of intrinsic excitability. *J. Neurosci.* **26**: 6153–6162.

Fuentealba, P., Crochet, S., Timofeev, I., Bazhenov, M. and Sejnowski, T. J. (2004) Experimental evidence and modeling studies support a synchronizing role for electrical coupling in the cat thalamic reticular neurons *in vivo*. *Eur. J. Neurosci.* **20**: 111–119.

Fuentealba, P., Timofeev, I., Bazhenov, M., Sejnowski, T. J. and Steriade, M. (2005) Membrane bistability in thalamic reticular neurons during spindle oscillations. *J. Neurophysiol.* **93**: 294–304.

Gais, S., Plihal, W., Wagner, U. and Born, J. (2000) Early sleep triggers memory for early visual discrimination skills. *Nat. Neurosci.* **3**: 1335–1339.

Gais, S., Molle, M., Helms, K. and Born, J. (2002) Learning-dependent increases in sleep spindle density. *J. Neurosci.* **22**: 6830–6834.

Galarreta, M. and Hestrin, S. (1998) Frequency-dependent synaptic depression and the balance of excitation and inhibition in the neocortex. *Nat. Neurosci.* **1**: 587–594.

Galligan, J. J., H. Tatsumi, K. Z. Shen, A. Surprenant and R. A. North. (1990) Cation current activated by hyperpolarization (I_h) in guinea pig enteric neurons. *Am. J. Physiol.* **259**: G966–972.

Garcia, J. W. (2017) An improved event-driven model of presynaptic dynamics for large-scale simulations of biophysically realistic and diverse synapses. *PhD Thesis*, University of California San Diego.

Gardner-Medwin, A. R. (1989) Doubly modifiable synapses: a model of short and long term auto-associative memory. *Proc. Roy. Soc. London Ser. B* **238**: 137–154.

Geiger, J. R., Melcher, T., Koh, D. S., Sakmann, B., Seeburg, P. H., Jonas, P. and Monyer, H. (1995) Relative abundance of subunit mRNAs determines gating and Ca2+ permeability of AMPA receptors in principal neurons and interneurons in rat CNS. *Neuron* **15**: 193–204.

Gerstner, W., Kempter, R., van Hemmen, J. L. and Wagner, H. (1996) A neural learning rule for sub-millisecond temporal coding. *Nature* **383**: 76–78.

Getting, P. A. (1989) Emerging principles governing the operation of neuronal networks. *Annual Rev. Neurosci.* **12**: 185–204.

Ghosh, A., Greenberg, M. E. (1995) Calcium signaling in neurons: molecular mechanisms and cellular consequences. *Science* **268**: 239–247.

Gibbs, J. W., Berkow-Schroeder, G. and Coulter, D. A. (1996) GABA$_A$ receptor function in developing rat thalamic reticular neurons: whole cell recordings of GABA-mediated currents and modulation by clonazepam. *J. Neurophysiol.* **76**: 2568–2579.

Girardeau, G., Benchenane, K., Wiener, S. I., Buzsaki, G. and Zugaro, M. B. (2009) Selective suppression of hippocampal ripples impairs spatial memory. *Nat. Neurosci.* **12**: 1222–1223.

Gloor, P. and Fariello, R. G. (1988) Generalized epilepsy: some of its cellular mechanisms differ from those of focal epilepsy. *Trends Neurosci.* **11**: 63–68.

Gloor, P., Pellegrini, A. and Kostopoulos, G. K. (1979) Effects of changes in cortical excitability upon the epileptic bursts in generalized penicillin epilepsy of the cat. *Electroencephalogr. Clin. Neurophysiol.* **46**: 274–289.

Gloor, P., Quesney, L. F. and Zumstein, H. (1977) Pathophysiology of generalized penicillin epilepsy in the cat: the role of cortical and subcortical structures. II. Topical application of penicillin to the cerebral cortex and subcortical structures. *EEG Clin. Neurophysiol.* **43**: 79–94.

Gluck, M. A. and Myers, C. E. (2000) *Gateway to Memory: An Introduction to Neural Network Modeling of the Hippocampus and Learning*. MIT Press, Cambridge, MA.

Golard, A. and Siegelbaum, S. A. (1993) Kinetic basis for the voltage-dependent inhibition of N-type calcium current by somatostatin and norepinephrine in chick sympathetic neurons. *J. Neurosci.* **13**: 3884–3894.

Goldman, D. E. (1943) Potential, impedance and rectification in membranes. *J. Gen. Physiol.* **27**: 37–60.

Golomb, D. and Rinzel, J. (1993) Dynamics of globally coupled inhibitory neurons with heterogeneity. *Phys. Rev. E* **48**: 4810–4814.

Golomb, D. and Rinzel, J. (1994) Clustering in globally coupled inhibitory neurons. *Physica D* **72**: 259–282.

Golomb, D., Wang, X. J. and Rinzel, J. (1994) Synchronization properties of spindle oscillations in a thalamic reticular nucleus model. *J. Neurophysiol.* **72**: 1109–1126.

Golomb, D., Wang, X. J. and Rinzel, J. (1996) Propagation of spindle waves in a thalamic slice model. *J. Neurophysiol.* **75**: 750–769.

Golshani, P., Liu, X. B. and Jones, E. G. (2001) Differences in quantal amplitude reflect GluR4-subunit number at corticothalamic synapses on two populations of thalamic neurons. *Proc. Natl. Acad. Sci. USA* **98**: 4172–4177.

Gonzalo-Ruiz, A. and Lieberman, A. R. (1995) Topographic organization of projections from the thalamic reticular nucleus to the anterior thalamic nuclei in the rat. *Brain Res. Bull.* **37**: 17–35.

Gray, C. M. and McCormick, D. A. (1996) Chattering cells: superficial pyramidal neurons contributing to the generation of synchronous oscillations in the visual cortex. *Science* **274**: 109–113.

Gray, C. M. and Singer, W. (1989) Stimulus-specific neuronal oscillations in orientation columns of cat visual cortex. *Proc. Natl. Acad. Sc. USA* **86**: 1698–1702.

Grenier, F., Timofeev, I. and Steriade, M. (1998) Leading role of thalamic over cortical neurons during postinhibitory rebound excitation. *Proc. Natl. Acad. Sci. USA* **95**: 13929–13934.

Gu, X. and Spitzer, N. C. (1995) Distinct aspects of neuronal differentiation encoded by frequency of spontaneous Ca^{2+} transients. *Nature* **375**: 784–787.

Gu, X. and Spitzer, N. C. (1997) Breaking the code: regulation of neuronal differentiation by spontaneous calcium transients. *Devel. Neurosci.* **19**: 33–41.

Gulbis, J. M., Zhou, M., Mann, S. and MacKinnon, R. (2000) Structure of the cytoplasmic beta subunit–T1 assembly of voltage-dependent K+ channels. *Science* **289**: 123–127.

Gutfreund, Y., Yarom, Y. and Segev, I. (1995) Subthreshold oscillations and resonant frequency in guinea-pig cortical neurons–physiology and modeling. *J. Physiol.* **483**: 621–640.

Gutkind, J. S. (1998) The pathways connecting G protein-coupled receptors to the nucleus through divergent mitogen-activated protein kinase cascades. *J. Biol. Chem.* **273**: 1839–1842.

Hagiwara, N. and Irisawa, H. (1989) Modulation by intracellular Ca^{2+} of the hyperpolarization-activated inward current in rabbit single sino-atrial node cells. *J. Physiol.* **409**: 121–141.

Hagler, D. J. Jr, Ulbert, I., Wittner, L., Erss, L., Madsen, J. R., Devinsky, O., Doyle, W., Fabo, D., Cash, S. S. and Halgren, E. (2018) Heterogeneous origins of human sleep spindles in different cortical layers. *J Neurosci.* **38**: 3013–3025.

Hammer, M. and Menzel, R. (1995) Learning and memory in the honeybee. *J. Neurosci.* **15**: 1617–1630.

Harris-Warrick, R. M. and Marder, E. (1991) Modulation of neural networks for behavior. *Annual Rev. Neurosci.* **14**: 39–57.

Harris, K. M. and Landis, D. M. (1986) Membrane structure at synaptic junctions in area CA1 of the rat hippocampus. *Neuroscience* **19**: 857–872.

Harris, A. L., Spray, D. C. and Bennett, M. V. L. (1981) Kinetic properties of a voltage-dependent junctional conductance. *J. Gen. Physiol.* **77**: 95–117.

Hebb, D. O. (1949) *Organization of Behavior: A Neuropsychological Theory.* John Wiley and Sons, New York.

Herkenham, M. (1980) Laminar organization of thalamic projections to the rat neocortex. *Science* **207**: 532–535.

Hernandez-Cruz, A. and Pape, H. C. (1989) Identification of two calcium currents in acutely dissociated neurons from the rat lateral geniculate nucleus. *J. Neurophysiol.* **61**: 1270–1283.

Hersch, S. M. and White, E. L. (1981) Thalamocortical synapses on corticothalamic projections neurons in mouse Sml cortex: electron microscopic demonstration of a monosynaptic feedback loop. *Neurosci. Lett.* **24**: 207–210.

Hertz, L. Functional interactions between neurons and astrocytes I. (1979) Turnover and metabolism of putative amino acid transmitters. *Progress Neurobiol.* **13**: 277–323.

Hessler, N. A., Shirke, A. M. and Malinow, R. (1993) The probability of transmitter release at a mammalian central synapse. *Nature* **366**: 569–572.

Hestrin, S. (1992) Activation and desensitization of glutamate-activated channels mediating fast excitatory synaptic currents in the visual cortex. *Neuron* **9**: 991–999.

Hestrin, S. (1993) Different glutamate receptor channels mediate fast excitatory synaptic currents in inhibitory and excitatory cortical neurons. *Neuron* **11**: 1083–1091.

Hestrin, S. and Sah, P. and Nicoll, R. A. (1990) Mechanisms generating the time course of dual component excitatory synaptic currents recorded in hippocampal slices. *Neuron* **5**: 247–253.

Hill, T. L. and Chen, Y. D. (1972) On the theory of ion transport across nerve membranes. VI. Free energy and activation free energies of conformational change. *Proc. Natl. Acad. Sci. USA* **69**: 1723–1726.

Hill, S. and Tononi, G. (2005) Modeling sleep and wakefulness in the thalamocortical system. *J. Neurophysiol.* **93**: 1671–1698.

Hille, B. (1992) *Ionic Channels of Excitable Membranes*. Sinauer Associates INC, Sunderland, MA.

Hindmarsh, J. L. and Rose, R. M. (1994a) A model for rebound bursting in mammalian neurons. *Phil. Trans. Roy. Soc. Lond. B* **346**: 129–150.

Hindmarsh, J. L. and Rose, R. M. (1994b) A model of intrinsic and driven spindling in thalamocortical neurons. *Phil. Trans. Roy. Soc. Lond. B* **346**: 165–183.

Hines, M. (1993) NEURON—A program for simulation of nerve equations. In: *Neural Systems: Analysis and Modeling*, ed. Eeckman, F. Kluwer Academic Publishers, Norwell, MA, pp. 127–136.

Hines, M. L. and Carnevale, N. T. (1997) The NEURON simulation environment. *Neural Computation* 9: 1179–1209.

Hinton, G. E. and Ghahramani, Z. (1997) Generative models for discovering sparse distributed representations. *Phil. Trans. Roy. Soc. Lond. Ser. B* **352**: 1177–1190.

Hinton, G. E. and Sejnowski, T. J. (1999) *Unsupervised Learning: Foundations of Neural Computation*, MIT Press, Cambridge, MA.

Hinton, G. E., Dayan, P., Frey, B. J. and Neal, R. M. (1995) The 'wake-sleep' algorithm for unsupervised neural networks. *Science* **268**: 1158–1161.

Hirsch, J. A., Alonso, J. M. and Reid, R. C. (1995) Visually evoked calcium action potentials in cat striate cortex. *Nature* **378**: 612–616.

Hô, N. and Destexhe, A. (2000) Synaptic background activity enhances the responsiveness of neocortical pyramidal neurons. *J. Neurophysiol.* **84**: 1488–1496.

Hô, N., Kröger, H. and Destexhe, A. (2000) Consequences of correlated synaptic bombardment on the responsiveness of neocortical pyramidal neurons. *Neurocomputing* **32**: 155–160.

Hobson, J. A. (1988) *The Dreaming Brain*. Basic Books, New York.

Hodgkin, A. L. and Huxley, A. F. (1952) A quantitative description of membrane current and its application to conduction and excitation in nerve. *J. Physiol.* **117**: 500–544.

Hodgkin, A. L. and Katz, B. (1949) The effect of sodium ions on the electrical activity of the giant axon of the squid. *J. Physiol.* **108**: 37–77.

Holmes, W. R. and Woody, C. D. (1989) Effects of uniform and non-uniform synaptic 'activation-distributions' on the cable properties of modeled cortical pyramidal neurons. *Brain Res.* **505**: 12–22.

Hopfield, J. J. (1995) Pattern recognition computation using action potential timing for stimulus representation. *Nature* **376**: 33–36.

Horn, R. and Lange, K. (1983) Estimating kinetic constants from single channel data. *Biophys. J.* **43**: 207–223.

Horn, R. and Vandenberg, C. A. (1984) Statistical properties of single sodium channels. *J. Gen. Physiol.* **84**: 505–535.

Horne, J. A. and McGrath, M. J. (1984) The consolidation hypothesis for REM sleep function: stress and other confounding factors—a review. *Biol. Psychol.* **18**: 165–184.

Hosford, D. A., Clark, S., Cao, Z., Wilson, W. A., Jr, Lin, F. H., Morrisett, R. A., Huin, A. (1992) The role of $GABA_B$ receptor activation in absence seizures of lethargic (lh/lh) mice. *Science* **257**: 398–401.

Hosford, D. A., Wang, Y. and Cao, Z. (1997) Differential effects mediated by $GABA_A$ receptors in thalamic nuclei of lh/lh model of absence seizures. *Epilepsy Res.* **27**: 55–65.

Hoshi, T. and Aldrich, R. W. (1988) Gating kinetics of four classes of voltage-dependent K^+ channels in pheochromocytoma cells. *J. Gen. Physiol.* **91**: 107–131.

Hosoya, Y., Yamada, M., Ito, H. and Kurachi, Y. (1996) A functional model for G-protein activation of the muscarinic K^+ channel in guinea pig atrial myocites. *J. Gen. Physiol.* **108**: 485–495.

Hu, B., Steriade, M., Deschênes, M. (1989) The effects of brainstem peribrachial stimulation on perigeniculate neurons: the blockage of spindle waves. *Neuroscience* **31**: 1–12.

Hubel, D. (1959) Single-unit activity in striate cortex of unrestrained cats. *J. Physiol.* **147**: 226–238.

Huerta, P. T. and Lisman, J. E. (1996) Synaptic plasticity during the cholinergic theta-frequency oscillation in vitro. *Hippocampus* **6**: 58–61.

Huertas, M. A. and Smith, G. D. (2006) A multivariate population density model of the dLGN/PGN relay. *J. Comput. Neurosci.* **21**: 171–189.

Huguenard, J. R. and McCormick, D. A. (1992) Simulation of the currents involved in rhythmic oscillations in thalamic relay neurons. *J. Neurophysiol.* **68**: 1373–1383.

Huguenard, J. R. and Prince, D. A. (1991) Slow inactivation of a TEA-sensitive K current in acutely isolated rat thalamic relay neurons. *J. Neurophysiol.* **66**: 1316–1328.

Huguenard, J. R. and Prince, D. A. (1992) A novel T-type current underlies prolonged calcium-dependent burst firing in GABAergic neurons of rat thalamic reticular nucleus. *J. Neurosci.* **12**: 3804–3817.

Huguenard, J. R. and Prince, D. A. (1994a) Clonazepam suppresses $GABA_B$-mediated inhibition in thalamic relay neurons through effects in nucleus reticularis. *J. Neurophysiol.* **71**: 2576–2581.

Huguenard, J. R. and Prince, D. A. (1994b) Intrathalamic rhythmicity studied *in vitro*: nominal T-current modulation causes robust anti-oscillatory effects. *J. Neurosci.* **14**: 5485–5502.

Huguenard, J. R., Coulter, D. A. and Prince, D. A. (1991) A fast transient potassium current in thalamic relay neurons: kinetics of activation and inactivation. *J. Neurophysiol.* **66**: 1305–1315.

Huguenard, J. R., Hamill, O. P., Prince, D. A. (1988) Developmental changes in Na+ conductances in rat neocortical neurons: appearance of a slowly inactivating component. *J. Neurophysiol.* **59**: 778–795.

Huguenard, J. R., Hamill, O. P., Prince, D. A. (1989) Sodium channels in dendrites of rat cortical pyramidal neurons. *Proc. Natl. Acad. Sci. USA* **86**: 2473–2477.

Huntley, G. W., Vickers, J. C. and Morrison, J. H. (1994) Cellular and synaptic localization of NMDA and non-NMDA receptor subunits in neocortex: organizational features related to cortical circuitry, function and disease. *Trends Neurosci.* **17**: 536–543.

Imbert, M. and Buisseret, P. (1975) Receptive field characteristics and plastic properties of visual cortical cells in kittens reared with or without visual experience. *Exp. Brain Res.* **22**: 25–36.

Inoue, M., Duysens, J., Vossen, J. M. H., Coenen, A. M. L. (1993) Thalamic multiple-unit activity underlying spike-wave discharges in anesthetized rats. *Brain Res.* **612**: 35–40.

Isaacson, J. S., Solis, J. M. and Nicoll, R. A. (1993) Local and diffuse synaptic actions of GABA in the hippocampus. *Neuron* **10**: 165–175.

Ito, H., Tung, R. T., Sugimoto, T., Kobayashi, I., Takahashi, K., Katada, T., Ui, M. and Kurachi, Y. (1992) On the mechanism of G-protein beta gamma subunit activation of the muscarinic K^+ channel in guinea pig atrial cell membrane. Comparison with the ATP-sensitive K+ channel. *J. Gen. Physiol.* **99**: 961–983.

Izhikevich, E. M. and Edelman, G. M. (2008) Large-scale model of mammalian thalamocortical systems. *Proc. Natl. Acad. Sci. USA* **105**: 3593–3598.

Jahn, R. and Sudhof, T. C. (1999) Membrane fusion and exocytosis. *Annual Review Biochem.* **68**: 863–911.

Jahnsen, H. and Llinás, R. R. (1984a) Electrophysiological properties of guinea-pig thalamic neurons: an *in vitro* study. *J. Physiol.* **349**: 205–226.

Jahnsen, H. and Llinás, R. R. (1984b) Ionic basis for the electroresponsiveness and oscillatory properties of guinea-pig thalamic neurons *in vitro. J. Physiol.* **349**: 227–247.

Jahr, C. E. (1992) High probability of opening of NMDA receptor channels by L-glutamate. *Science* **255**: 470–472.

Jahr, C. E. and Stevens, C. F. (1990a) A quantitative description of NMDA receptor-channel kinetic behavior. *J. Neurosci.* **10**: 1830–1837.

Jahr, C. E. and Stevens, C. F. (1990b) Voltage dependence of NMDA-activated macroscopic conductances predicted by single-channel kinetics. *J. Neurosci.* **10**: 3178–3182.

Jan, L. Y. and Jan, Y. N. (1992) Structural elements involved in specific K^+ channel functions. *Annu. Rev. Physiol.* **54**: 537–555.

Jasper, H. and Kershman, J. (1941) Electroencephalographic classification of the epilepsies. *Arch. Neurol. Physchiat.* **45**: 903–943.

Jay, T. M. and Witter, M. P. (1991) Distribution of hippocampal ca1 and subicular efferents in the prefrontal cortex of the rat studied by means of anterograde transport of phaseolus vulgaris-leucoagglutinin. *J. Compar. Neurol.* **313**: 574–586.

Jenkins, J. C. and Dallenbach, K. M. (1924) Obliviscence during sleep and waking. *Amor. J. Psychol.* **35**: 605–612.

Johnson, F. H., Eyring, H. and Stover, B. J. (1974) *The Theory of Rate Processes in Biology and Medicine.* John Wiley and Sons, New York.

Johnson, L. A., Euston, D. R., Tatsuno, M. and McNaughton, B. L. (2010) Stored-trace reactivation in rat prefrontal cortex is correlated with down-to-up State fluctuation density. *J. Neurosci.* **30**: 2650–2661.

Johnston, D. and Wu, S. M. (1995) *Foundations of Cellular Neurophysiology.* Bradford Book/MIT Press, Cambridge, MA.

Jonas, P., Major, G. and Sakmann, B. (1993) Quantal components of unitary EPSCs at the mossy fibre synapse on CA3 pyramidal cells of rat hippocampus. *J. Physiol.* **472**: 615–663.

Jonas, P., Racca, C., Sakmann, B., Seeburg, P. H. and Monyer, H. (1994) Differences in Ca2+ permeability of AMPA-type glutamate receptor channels in neocortical neurons caused by differential GluR-B subunit expression. *Neuron* **12**: 1281–1289.

Jones, E. G. (1985) *The Thalamus.* Plenum Press, New York.

Jones, E. G. and Powell, T. P. S. (1969) Morphological variations in the dendritic spines of the neocortex. *J. Cell. Sci.* **5**: 509–529.

Jones, E. G. and Powell, T. P. S. (1970) Electron microscopy of the somatic sensory cortex of the cat. I. Cell types and synaptic organization. *Philos. Trans. R. Soc. Lond.* **B 257**: 1–11.

Jouvet, M. (1998) Paradoxical sleep as a programming system. *J. Sleep Res.* **7** (Suppl. 1): 1–5.

Kaestner, E. J., Wixted, J. T. and Mednick, S. C. (2013) Pharmacologically increasing sleep spindles enhances recognition for negative and high-arousal memories. *J. Cogn. Neurosci.* **25**: 1597–1610.

Kamondi, A. and Reiner, P. B. (1991) Hyperpolarization-activated inward current in histaminergic tuberomammillary neurons of the rat hypothalamus. *J. Neurophysiol.* **66**: 1902–1911.

Kandel, E. R. (1976) *Cellular Basis of Behavior.* Freeman, San Francisco.

Kandel, E. R., Schwartz, J. H. and Jessel, T. M. (1995) *Essentials of Neural Science and Behavior.* Appleton and Lange, Norwalk, CT.

Kao, C. Q. and Coulter, D. A. (1997) Physiology and pharmacology of corticothalamic stimulation-evoked responses in rat somatosensory thalamic neurons in vitro. *J. Neurophysiol.* **77**: 2661–2676.

Kapur, N. and Brooks, D. J. (1999) Temporally-specific retrograde amnesia in two cases of discrete bilateral hippocampal pathology. *Hippocampus* **9**: 247–254.

Karst, H., Joels, M. and Wadman, W. J. (1993) Low-threshold calcium current in dendrites of the adult rat hippocampus. *Neurosci. Lett.* **164**: 154–158.

Kattler, H., Dijk, D. J. and Borbély, A. A. (1994) Effect of unilateral somatosensory stimulation prior to sleep on the sleep EEG in humans. *J. Sleep Res.* **3**: 159–164.

Katz, B. (1966) *Nerve, Muscle and Synapse.* McGraw Hill Book Co., New York.

Kay, A. R., Sugimori, M. and Llinás, R. R. (1998) Kinetic and stochastic properties of a persistent sodium current in mature guinea pig cerebellar Purkinje cells. *J. Neurophysiol.* **80**: 1167–1179.

Keller, B. U., Hartshorne, R. P., Talvenheimo, J. A., Catterall, W. A. and Montal, M. (1986) Sodium channels in planar lipid bilayers. Channel gating kinetics of purified sodium channels modified by batrachotoxin. *J. Gen. Physiol.* **88**: 1–23.

Kelso, S. R., Ganong, A. H. and Brown, T. H. (1986) Hebbian synapses in hippocampus. *Proc. Nat. Acad. Sci. USA* **83**: 5326–5330.

Kennedy, M. B., Bennett, M. K. and Erondu, N. E. (1983) Biochemical and immunochemical evidence that the 'major postsynaptic density protein' is a subunit of a calmodulin-dependent protein kinase. *Proc. Natl. Acad. Sci. USA* **80**: 7357–7361.

Kerr, R., Bartol, T. M., Kaminsky, B., Dittrich, M., Chang, J. C. J., Baden, S., Sejnowski, T. J. and Stiles, J. R. (2008) Fast Monte-Carlo simulation methods for biological reaction-diffusion systems in solution and on surfaces. *SIAM J. Sci. Comput.* **30**: 3126–3149.

Kienker, P. (1989) Equivalence of aggregated Markov models of ion-channel gating. *Proc. Roy. Soc. Lond. Ser. B* **236**: 269–309.

Kim, U., Bal, T. and McCormick, D. A. (1995) Spindle waves are propagating synchronized oscillations in the ferret LGNd in vitro. *J. Neurophysiol.* **74**: 1301–1323.

Kim, U., Sanchez-Vives, M. V. and McCormick, D. A. (1997) Functional dynamics of GABAergic inhibition in the thalamus. *Science* **278**: 130–134.

Kim, D., Song, U., Keum, S., Lee, T., Jeong, M.-J., Kim, S.-S., McEnery, M. W. and Shin, H. S. (2001) Lack of the burst firing of thalamocortical relay neurons and resistance to absence seizures in mice lacking the alpha-1G T-type Ca2+ channels. *Neuron* **31**: 35–45.

Kinney, J. P., Spacek, J., Bartol, T. M., Bajaj, C. L., Harris, K. M. and Sejnowski, T. J. (2013) Extracellular sheets and tunnels modulate glutamate diffusion in hippocampal neuropil. *J. Comp. Neurol.* **521**: 448–464.

Klee, M. R., Offenloch, K. and Tigges, J. (1965) Cross-correlation analysis of electroencephalographic potentials and slow membrane transients. *Science* **147**: 519–521.

Koch, C. (1999) *Biophysics of Computation.* Oxford University Press, Oxford.

Koch, C. and Segev, I. (eds.) (1998) *Methods in Neuronal Modeling* (2nd edition). MIT Press, Cambridge, MA.

Koh, D. S., Geiger, J. R., Jonas, P. and Sakmann, B. (1995) Ca(2+)-permeable AMPA and NMDA receptor channels in basket cells of rat hippocampal dentate gyrus. *J. Physiol.* **485**: 383–402.

Komarov, M., Krishnan, G., Chauvette, S., Rulkov, N., Timofeev, I. and Bazhenov, M. (2018) New class of reduced computationally efficient neuronal models for large-scale simulations of brain dynamics. *J. Comput. Neurosci.* **44**: 1–24.

Kopell, N. and LeMasson, G. (1994) Rhythmogenesis, amplitude modulation, and multiplexing in a cortical architecture. *Proc. Natl. Acad. Sci. USA* **91**: 10586–10590.

Kostopoulos, G., Avoli, M. and Gloor, P. (1983) Participation of cortical recurrent inhibition in the genesis of spike and wave discharges in feline generalized epilepsy. *Brain Res.* **267**: 101–112.

Kostopoulos, G., Gloor, P., Pellegrini, A., Siatitsas, I. (1981a) A study of the transition from spindles to spike and wave discharge in feline generalized penicillin epilepsy: EEG features. *Exp. Neurol.* **73**: 43–54.

Kostopoulos, G., Gloor, P., Pellegrini, A. and Gotman, J. (1981b) A study of the transition from spindles to spike and wave discharge in feline generalized penicillin epilepsy: microphysiological features. *Exp. Neurol.* **73**: 55–77.

Krishnan, G. P., Chauvette, S., Shamie, I., Soltani, S., Timofeev, I., Cash, S. S., Halgren, E. and Bazhenov, M. (2016) Cellular and neurochemical basis of sleep stages in the thalamocortical network. *eLife*: e18607.

Krishnan, G. P., Rosen, B. Q., Chen, J.-Y., Muller, L., Sejnowski, T. J., Cash, S. S., Halgren, E. and Bazhenov, M. (2018) Thalamocortical and intracortical laminar connectivity determines sleep spindle properties. *PLoS Comput. Biol.* **14**: e1006171.

Krnjevic, K., Pumain, R. and Renaud, L. (1971) The mechanism of excitation by acetylcholine in the cerebral cortex. *J. Physiol.* **215**: 247–268.

Labarca, P., Rice, J. A., Fredkin, D. R. and Montal, M. (1985) Kinetic analysis of channel gating. Application to the cholinergic receptor channel and the chloride channel from *Torpedo Californica. Biophys. J.* **47**: 469–478.

Landisman, C. E., Long, M. A., Beierlein, M., Deans, M. R., Paul, D. L. and Connors, B. W. (2002) Electrical synapses in the thalamic reticular nucleus. *J. Neurosci.* **22**: 1002–1009.

Landry, P. and Deschênes, M. (1981) Intracortical arborizations and receptive fields of identified ventrobasal thalamocortical afferents to the primary somatic sensory cortex in the cat. *J. Comp. Neurol.* **199**: 345–371.

Lang, E. J. and Paré, D. (1997) Synaptic and synaptically activated intrinsic conductances underlie inhibitory potentials in cat lateral amygdaloid projection neurons *in vivo. J. Neurophysiol.* **77**: 353–363.

Larkman, A. U. (1991) Dendritic morphology of pyramidal neurons of the visual cortex of the rat. III. Spine distributions. *J. Comp. Neurol.* **306**: 332–343.

Larson, J. and Lynch, G. (1986) Induction of synaptic potentiation in hippocampus by patterned stimulation involves two events. *Science* **232**: 985–988.

Latorre, R., Oberhauser, A., Labarca, P. and Alvarez, O. (1989) Varieties of calcium-activated potassium channels. *Annu. Rev. Physiol.* **51**: 385–399.

Lee, J. H., Daud, A. N., Cribbs, L. L., Lacerda, A. E., Pereverzev, A., Klockner, U., Schneider, T. and Perez-Reyes, E. (1999) Cloning and expression of a novel member of the low voltage-activated T-type calcium channel family. *J. Neurosci.* **19**: 1912–1921.

Lee, J., Kim, D. and Shin, H. S. (2004) Lack of delta waves and sleep disturbances during non-rapid eye movement sleep in mice lacking alpha1G-subunit of T-type calcium channels. *Proc. Natl. Acad. Sci. USA* **101**: 18195–18199.

Lee, J., Song, K., Lee, K., Hong, J., Lee, H., Chae, S., Cheong, E. and Shin, H. S. (2013) Sleep spindles are generated in the absence of T-type calcium channel-mediated low-threshold burst firing of thalamocortical neurons. *Proc. Natl. Acad. Sci. USA* **110**: 20266–20271.

Legendre, P., Rosenmund, C. and Westbrook, G. L. (1993) Inactivation of NMDA channels in cultured hippocampal neurons by intracellular calcium. *J. Neurosci.* **13**: 674–684.

LeMasson, G., Renaud-LeMasson, S., Sharp, A., Abbott, L. F. and Marder, E. (1992) Real time interaction between a model neuron and the crustacean somatogastric nervous system. *Soc. Neurosci. Abs.* **18**: 1055.

Leresche, N., Jassik-Gerschenfeld, D., Haby, M., Soltesz, I. and Crunelli, V. (1990) Pacemaker-like and other types of spontaneous membrane potential oscillations in thalamocortical cells. *Neurosci. Lett.* **113**: 72–77.

Leresche, N., Lightowler, S., Soltesz, I., Jassik-Gerschenfeld, D. and Crunelli, V. (1991) Low-frequency oscillatory activities intrinsic to rat and cat thalamocortical cells. *J. Physiol.* **441**: 155–174.

Lester, R. A. and Jahr, C. E. (1992) NMDA channel behavior depends on agonist affinity. *J. Neurosci.* **12**: 635–643.

Lester, R. A., Clements, J. D., Westbrook, G. L. and Jahr, C. E. (1990) Channel kinetics determine the time course of NMDA receptor-mediated synaptic currents. *Nature* **346**: 565–567.

Le Van Quyen, M., Muller, L., Telenczuk, B., Cash, S. S., Halgren, E., Hatsopoulos, N. G., Dehghani, N. and Destexhe, A. (2016) High-frequency oscillations in human and monkey neocortex during the wake-sleep cycle. *Proc. Natl. Acad. Sci. USA* **113**: 9363–9368.

Levitt, D. G. (1989) Continuum model of voltage-dependent gating. *Biophys. J.* **55**: 489–498.

Lewicki, M. S. and Sejnowski, T. J. (1997) Bayesian unsupervised learning of higher order structure. *Adv. Neural Information Processing Systems* **9**: 529–535.

Li, W., Llopis, J., Whitney, M., Zlokarnik, G. and Tsien, R. Y. (1998) Cell-permanent caged InsP3 ester shows that Ca^{2+} spike frequency can optimize gene expression. *Nature* **392**: 936–941.

Liebovitch, L. S. and Sullivan, J. M. (1987) Fractal analysis of a voltage-dependent potassium channel from cultured mouse hippocampal neurons. *Biophys. J.* **52**: 979–988.

Liebovitch, L. S. and Toth, T. I. (1991) A model of ion channel kinetics using deterministic chaotic rather than stochastic processes. *J. Theor. Biol.* **148**: 243–267.

Lindström, S. (1982) Synaptic organization of inhibitory pathways to principal cells in the lateral geniculate nucleus of the cat. *Brain Res.* **234**: 447–453.

Liu, Z., Vergnes, M., Depaulis, A., Marescaux, C. (1991) Evidence for a critical role of GABAergic transmission within the thalamus in the genesis and control of absence seizures in the rat. *Brain Res.* **545**: 1–7.

Liu, Z., Vergnes, M., Depaulis, A. and Marescaux, C. (1992) Involvement of intrathalamic *GABA$_B$* neurotransmission in the control of absence seizures in the rat. *Neuroscience* **48**: 87–93.

Liu, X. B., Honda, C. N., Jones, E. G. (1995) Distribution of four types of synapse on physiologically identified relay neurons in the ventral posterior thalamic nucleus of the cat. *J. Comp. Neurol.* **352**: 69–91.

Liu, X. B., Jones, E. G. (1999) Predominance of corticothalamic synaptic inputs to thalamic reticular nucleus neurons in the rat. *J. Comp. Neurol.* **414**: 67–79.

Llinás, R. R. and Yarom, Y. (1981) Electrophysiology of mammalian inferior olivary neurones in vitro. Different types of voltage-dependent ionic conductances. *J. Physiol.* **315**: 549–567.

Llinás, R. R. (1988) The intrinsic electrophysiological properties of mammalian neurons: a new insight into CNS function. *Science* **242**: 1654–1664.

Llinás, R. R. (1999) *The squid giant synapse: a model for chemical transmission.* Oxford University Press, Oxford.

Llinás, R. R. and Geijo-Barrientos, E. (1988) In vitro studies of mammalian thalamic and reticularis thalami neurons. In: *Cellular Thalamic Mechanisms*, ed. Bentivoglio, M. and Spreafico, R. Elsevier, Amsterdam, pp. 23–33.

Llinás, R. R. and Jahnsen, H. (1982) Electrophysiology of thalamic neurones in vitro. *Nature* **297**: 406–408.

Llinás, R. R. and Sugimori, M. (1980a) Electrophysiological properties of *in vitro* Purkinje cell somata in mammalian cerebellar slices. *J. Physiol.* **305**: 171–195.

Llinás, R. R. and Sugimori, M. (1980b) Electrophysiological properties of *in vitro* Purkinje cell dendrites in mammalian cerebellar slices. *J. Physiol.* **305**: 197–213.

Llinás, R. R. and Yarom, Y. (1981a) Electrophysiology of mammalian inferior olivary neurones *in vitro*. Different types of voltage-dependent ionic conductances. *J. Physiol.* **315**: 549–567.

Llinás, R. R. and Yarom, Y. (1981b) Properties and distribution of ionic conductances generating electroresponsiveness of mammalian inferior olivary neurones *in vitro*. *J. Physiol.* **315**: 569–584.

Llinás, R. R., Grace, A. R. and Yarom, Y. (1991) *In vitro* neurons in mammalian cortical layer 4 exhibit intrinsic oscillatory activity in the 10 to 50 Hz frequency range. *Proc. Natl. Acad. Sci. USA* **88**: 897–901.

Llinás, R., Ribary, U., Joliot, M. and Wang, X. J. (1994) Content and context in temporal thalamocortical binding. In: *Temporal Coding in the Brain*, ed. Buzsáki, G., Llinás, R., Singer, W., Berthoz, A. and Christen, Y. Springer, Berlin, pp. 251–272.

Louie, K. and Wilson, M. A. (2001) Temporally structured replay of awake hippocampal ensemble activity during rapid eye movement sleep. *Neuron* **29**: 145–156.

Lubenov, E. V. and Siapas, A. G. (2009) Hippocampal theta oscillations are travelling waves. *Nature* **459**: 534–539.

Lüthi, A. and McCormick, D. A. (1998) Periodicity of thalamic synchronized oscillations: the role of Ca^{2+}-mediated upregulation of I_h. *Neuron* **20**: 553–563.

Lüthi, A. and McCormick, D. A. (1999) Modulation of a pacemaker current through Ca^{2+}-induced stimulation of cAMP production. *Nature Neurosci.* **2**: 634–641.

Lüthi, A., Bal, T. and McCormick, D. A. (1998) Periodicity of thalamic spindle waves is abolished by ZD7288, a blocker of I_h. *J. Neurophysiol.* **79**: 3284–3289.

Lytton, W. W. (1996) Optimizing synaptic conductance calculation for network simulations. *Neural Computation* **8**: 501–509.

Lytton, W. W. and Sejnowski, T. J. (1991) Simulations of cortical pyramidal neurons synchronized by inhibitory interneurons. *J. Neurophysiol.* **66**: 1059–1079.

Lytton, W. W. and Sejnowski, T. J. (1992) Computer model of ethosuximide's effect on a thalamic cell. *Annals Neurol.* **32**: 131–139.

Lytton, W. W., Destexhe, A. and Sejnowski, T. J. (1996) Control of slow oscillations in the thalamocortical neuron: a computer model. *Neuroscience* **70**: 673–684.

Lytton, W. W., Contreras, D., Destexhe, A. and Steriade, M. (1997) Dynamic interactions determine partial thalamic quiescence in a computer network model of spike-and-wave seizures. *J. Neurophysiol.* **77**: 1679–1696.

MacNeil, S., Lakey, T. and Tomlinson, S. (1985) Calmodulin regulation of adenylate cyclase activity. *Cell Calcium* **6**: 213–216.

Magee, J. C. and Johnston, D. (1995a) Characterization of single voltage-gated Na+ and Ca2+ channels in apical dendrites of rat CA1 pyramidal neurons. *J. Physiol.* **487**: 67–90.

Magee, J. C. and Johnston, D. (1995b) Synaptic activation of voltage-gated channels in the dendrites of hippocampal pyramidal neurons. *Science* **268**: 301–304.

Magleby, K. L. and Stevens, C. F. (1972) A quantitative description of end-plate currents. *J. Physiol.* **223**: 173–197.

Maingret, N., Girardeau, G., Todorova, R., Goutierre, M. and Zugaro, M. (2016) Hippocampo-cortical coupling mediates memory consolidation during sleep. *Nat. Neurosci.* **19**: 959–964.

Mainen, Z. F. and Sejnowski, T. J. (1995) Reliability of spike timing in neocortical neurons. *Science* **268**: 1503–1506.

Mainen, Z. F. and Sejnowski, T. J. (1996) Influence of dendritic structure on firing pattern in model neocortical neurons *Nature* **382**: 363–366.

Mainen, Z. F., Joerges, J., Huguenard, J. R. and Sejnowski, T. J. (1995) A model of spike initiation in neocortical pyramidal neurons. *Neuron* **15**: 1427–1439.

Mainen, Z. F., Malinow, R. and Svoboda, K. (1999) Synaptic calcium transients in single spines indicate that NMDA receptors are not saturated. *Nature* **399**: 151–155.

Major, G., Larkmann, A. U., Jonas, P., Sakmann, B. and Jack, J. J. B. (1994) Detailed passive cable models of whole-cell recorded CA3 pyramidal neurons in rat hippocampal slices. *J. Neurosci.* **14**: 4613–4638.

Mak-McCully, R. A., Deiss, S. R., Rosen, B. Q., Jung, K.-Y., Sejnowski, T. J., Bastuji, H., Rey, M., Cash, S. S., Bazhenov, M. and Halgren, E. (2014) Synchronization of isolated downstates (K-complexes) may be caused by cortically-induced disruption of thalamic spindling. *PLOS Comp. Biol.* **10**: e1003855.

Malenka, R. C. and Nicoll, R. A. (1999) Long-term potentiation—a decade of progress? *Science* **285**: 1870–1874.

Malerba, P. and Bazhenov, M. Circuit mechanisms of hippocampal reactivation during sleep. *Neurobiol. Learn. Memory* **S1074-7427**: 30103–30105.

Malgaroli, A. and Tsien, R. W. (1992) Glutamate-induced long-term potentiation of the frequency of miniature synaptic currents in cultured hippocampal neurons. *Nature* **357**: 134–139.

Maquet, P., Laureys, S., Peigneux, P., Fuchs, S., Petiau, C., Phillips, C., Aerts, J., Del Fiore, G., Degueldre, C., Meulemans, T., et al. (2000) Experience-dependent changes in cerebral activation during human REM sleep. *Nat. Neurosci.* **3**: 831–836.

Marban, E., Yamagishi, T. and Tomaselli, G. F. (1998) Structure and function of voltage-gated ion channels. *J. Physiol.* **508**: 647–657.

Marcus, E. M. and Watson, C. W. (1966) Bilateral synchronous spike wave electrographic patterns in the cat: interaction of bilateral cortical foci in the intact, the bilateral cortical-callosal and adiencephalic preparations. *Arch. Neurol.* **14**: 601–610.

Markram, H. (1997) A network of tufted layer 5 pyramidal neurons. *Cereb. Cortex* **7**: 523–533.

Markram, H. and Tsodyks, M. (1996) Redistribution of synaptic efficacy between neocortical pyramidal neurons. *Nature* **382**: 807–810.

Markram, H., Lubke, J., Frotscher, M. and Sakmann, B. (1997) Regulation of synaptic efficacy by coincidence of postsynaptic APs and EPSPs. *Science* **275**: 213–215.

Marom, S. and Abbott, L. F. (1994) Modeling state-dependent inactivation of membrane currents. *Biophys. J.* **67**: 515–520.

Marr, D. (1971) Simple memory: a theory for the archicortex. *Phil. Trans. Roy. Soc. (Lond.)* **262**: 23–81.

Matsumura, M., Cope, T. and Fetz, E. E. (1988) Sustained excitatory synaptic input to motor cortex neurons in awake animals revealed by intracellular recording of membrane potentials. *Exp. Brain Res.* **70**: 463–469.

McLachlan, R. S., Avoli, M. and Gloor, P. (1984) Transition from spindles to generalized spike and wave discharges in the cat: simultaneous single-cell recordings in the cortex and thalamus. *Exp. Neurol.* **85**: 413–425.

McBain, C. and Dingledine, R. (1992) Dual-component miniature excitatory synaptic currents in rat hippocampal CA3 pyramidal neurons. *J. Neurophysiol.* **68**: 16–27.

McCormick, D. A. (1991) Functional properties of a slowly inactivating potassium current in guinea-pig dorsal lateral geniculate relay neurons. *J. Neurophysiol.* **66**: 1176–1189.

McCormick, D. A. (1992) Neurotransmitter actions in the thalamus and cerebral cortex and their role in neuromodulation of thalamocortical activity. *Prog. Neurobiol.* **39**: 337–388.

McCormick, D. A. and Hashemiyoon, R. (1998) Thalamocortical neurons actively participate in the generation of spike-and-wave seizures in rodents. *Soc. Neurosci. Abstracts* **24**: 129.

McCormick, D. A. and Huguenard, J. R. (1992) A model of the electrophysiological properties of thalamocortical relay neurons. *J. Neurophysiol.* **68**: 1384–1400.

McCormick, D. A. and Pape, H. C. (1990a) Properties of a hyperpolarization-activated cation current and its role in rhythmic oscillations in thalamic relay neurones. *J. Physiol.* **431**: 291–318.

McCormick, D. A. and Pape, H.C. (1990b) Noradrenergic modulation of a hyperpolarization-activated cation current in thalamic relay neurones. *J. Physiol.* **431**: 319–342.

McCormick, D. A. and Prince, D. A. (1986) Mechanisms of action of acetylcholine in the guinea-pig cerebral cortex *in vitro*. *J. Physiol.* **375**: 169–194.

McCormick, D. A. and Wang, Z. (1991) Serotonin and noradrenaline excite GABAergic neurones of the guinea-pig and cat nucleus reticularis thalami. *J. Physiol.* **442**: 235–255.

McCormick, D. A. and Williamson, A. (1989) Convergence and divergence of neurotransmitter action in human cerebral cortex. *Proc. Natl. Acad. Sci. USA* **86**: 8098–8102.

McCormick, D. A., Wang, Z. and Huguenard, J. (1993) Neurotransmitter control of neocortical neuronal activity and excitability. *Cerebral Cortex* **3**: 387–398.

McClelland, J. L., McNaughton, B. L. and O'Reilly, R. C. (1995) Why there are complementary learning systems in the hippocampus and neocortex: insights from the successes and failures of connectionist models of learning and memory. *Psychol. Review* **102**: 419–457.

McGaugh, J. L. (2000) Memory—a century of consolidation. *Science* **287**: 248–251.

McKernan, R.M. and Whiting, P. J. (1996) Which GABA$_A$-receptor subtypes really occur in the brain? *Trends Neurosci.* **19**: 139–143.

McManus, O. B. (1991) Calcium-activated potassium channels: regulation by calcium. *J. Bioenergetics and Biomembranes* **23**: 537–560.

McManus, O. B., Weiss, D. S., Spivak, C. E., Blatz, A. L. and Magleby, K. L. (1988) Fractal models are inadequate for the kinetics of four different ion channels *Biophys. J.* **54**: 859–870.

McMullen, T. A. and N. Ly. (1988) Model of oscillatory activity in thalamic neurons: role of voltage and calcium-dependent ionic conductances. *Biol. Cybernetics.* **58**: 243–259.

Mednick, S. C., McDevitt, E. A., Walsh, J. K., Wamsley, E., Paulus, M., Kanady, J. C. and Drummond, S. P. (2013) The critical role of sleep spindles in hippocampal-dependent memory: a pharmacology study. *J. Neurosci.* **33**: 4494–4504.

Meeren, H. K., Pijn, J. P., Van Luijtelaar, E. L., Coenen, A. M. and Lopes da Silva, F. H. (2002) Cortical focus drives widespread corticothalamic networks during spontaneous absence seizures in rats. *J. Neurosci.* **22**: 1480–1495.

Mehta, M. R., Barnes, C. A. and McNaughton, B. L. (1997) Experience-dependent, asymmetric expansion of hippocampal place fields. *Proc. Natl. Acad. Sci. USA* **94**: 8918–8921.

Miller, D. A. and Zucker, S. W. (1999) Computing with self-excitatory cliques: a model and an application to hyperacuity-scale computation in visual cortex. *Neural Comput.* **11**: 21–66.

Miles, R. and Wong, R. K. S. (1984) Unitary inhibitory synaptic potentials in the guinea-pig hippocampus in vitro. *J. Physiol.* **356**: 97–113.

Millhauser, G. L., Saltpeter, E. E. and Oswald, R. E. (1988) Diffusion models of ion-channel gating and the origin of power-law distributions from single-channel recording. *Proc. Natl. Acad. Sci. USA* **85**: 1503–1507.

Minderhoud, J. M. (1971) An anatomical study of the efferent connections of the thalamic reticular nucleus. *Exp. Brain Res.* **112**: 435–446.

Mody, I., Dekoninck, Y., Otis, T. S. and Soltesz, I. (1994) Bridging the cleft at GABA synapses in the brain. *Trends Neurosci.* **17**: 517–525.

Molinoff, P. B., Williams, K., Pritchett, D. B. and Zhong, J. (1994) Molecular pharmacology of NMDA receptors: modulatory role of NR2 subunits. *Prog. Brain Res.* **100**: 39–45.

Mons, N., Harry, A., Dubourg, P., Premont, R. T., Iyengar, R. and Cooper D. M. (1995) Immunohistochemical localization of adenylyl cyclase in rat brain indicates a highly selective concentration at synapses. *Proc. Natl. Acad. Sci. USA* **92**: 8473–8477.

Montague, P. R. and Sejnowski, T. J. (1994) The predictive brain: temporal coincidence and temporal order in synaptic learning mechanisms. *Learning and Memory* **1**: 1–33.

Montague, P. R., Dayan, P. and Sejnowski, T. J. (1996). A framework for mesencephalic dopamine systems based on predictive Hebbian learning. *J. Neurosci.* **16**: 1936–1947.

Monyer, H., Burnashev, N., Laurie, D. J., Sakmann, B. and Seeburg, P. H. (1994) Developmental and regional expression in the rat brain and functional properties of four NMDA receptors. *Neuron* **12**: 529–540.

Morin, D. and Steriade, M. (1981) Development from primary to augmenting responses in primary somatosensory cortex. *Brain Res.* **205**: 49–66.

Morison, R. S. and Dempsey, E. W. (1943) Mechanisms of thalamocortical augmentation and repetition. *American J. Physiol.* **138**: 297–308.

Morison, R. S. and Bassett, D. L. (1945) Electrical activity of the thalamus and basal ganglia in decorticate cats. *J. Neurophysiol.* **8**: 309–314.

Morrison, A., Aertsen, A., and Diesmann, M. (2007) Spike-timing-dependent plasticity in balanced random networks. *Neural Comput.* **19**: 1437–1467.

Moruzzi, G. (1966) The functional significance of sleep with particular regard to the brain mechanisms underlining consciousness. In: *Brain and Conscious Experience*, ed. Eccles, J. C. Springer, New York, pp. 345–379.

Mosbacher, J., Schoepfer, R., Monyer, H., Burnashev, N., Seeburg, P. H. and Ruppersberg, J. P. (1994) A molecular determinant for submillisecond desensitization in glutamate receptors. *Science* **266**: 1059–1062.

Muhlethaler, M. and Serafin, M. (1990) Thalamic spindles in an isolated and perfused preparation in vitro. *Brain Res.* **524**: 17–21.

Mulle, C., Madariaga, A. and Deschênes, M. (1986) Morphology and electrophysiological properties of reticularis thalami neurons in cat: *in vivo* study of a thalamic pacemaker. *J. Neurosci.* **6**: 2134–2145.

Müller, W. and Lux, H. D. (1993) Analysis of voltage-dependent membrane currents in spatially extended neurons from point-clamp data. *J. Neurophysiol.* **69**: 241–247.

Muller, G. E. and Pilzecker, A. (1900) Experimentalle Beiträge zur Lehre vom Gedächtnis. *Zeitschrift Psychol.* **1**: 1–300.

Muller, L. and Destexhe, A. (2012) Propagating waves in thalamus, cortex and the thalamocortical system: experiments and models. *J. Physiol. Paris* **106**: 222–238.

Muller, L., Reynaud, A., Chavane, F. and Destexhe, A. (2014) The stimulus-evoked population response in visual cortex of awake monkey is a propagating wave. *Nature Commun.* **5**: 3675.

Muller, L., Piantoni, S., Koller, D., Cash S. S., Halgren, E. and Sejnowski, T. J. (2016) Rotating waves during human sleep spindles organize global patterns of activity during the night *eLife* e17267.

Muller, L., Chavane, F., Reynolds, J. and Sejnowski, T. J. (2010) Cortical travelling waves: mechanisms and computational principles. *Nature Reviews Neurosci.* **19**: 255–268.

Mungai, J. M. (1967) Dendritic patterns in the somatic sensory cortex of the cat. *J. Anat.* **101**: 403–418.

Munsch, T., Budde, T. and Pape, H. C. (1997) Voltage-activated intracellular calcium transients in thalamic relay cells and interneurons. *NeuroReport* **11**: 2411–2418.

Murthy, V. N. and Fetz, E. E. (1992) Coherent 25- to 35-*Hz* oscillations in the sensorimotor cortex of awake behaving monkeys. *Proc. Natl. Acad. Sci. USA* **89**: 5670–5674.

Murthy, V. N., Sejnowski, T. J. and Stevens, C. F. (1997) Heterogeneous release properties of visualized individual hippocampal synapses. *Neuron* **18**: 599–612.

Nadasdy, Z. (2000) Spike sequences and their consequences. *J. Physiol. (Paris)* **94**: 505–524.

Nadasdy, Z., Hirase, H., Czurko, A., Csicsvari, J. and Buzsaki, G. (1999) Replay and time compression of recurring spike sequences in the hippocampus. *J. Neurosci.* **19**: 9497–9507.

Nadkarni, S., Bartol, T. M., Stevens, C. F., Sejnowski, T. J. and Levine, H. (2012) Short-term synaptic plasticity constrains spatial organization of a hippocampal presynaptic terminal. *Proc. Natl. Acad. Sci. USA* **109**: 14657–14662.

Neher, E. (1992) Ion channels for communication between and within cells. *Science* **256**: 498–502.

Newberry, N. R. and Nicoll, R. A. (1985) Comparison of the action of baclofen with gamma-aminobutyric acid on rat hippocampal pyramidal cells in vitro. *J. Physiol.* **360**: 161–185.

Niedermeyer, E. and Lopes da Silva, F. (eds.) (1998) *Electroencephalography* (4th edition). Williams and Wilkins, Baltimore, MD.

Nowak, L., Bregestovski, P., Ascher, P., Herbet, A. and Prochiantz, A. (1984) Magnesium gates glutamate-activated channels in mouse central neurones. *Nature* **307**: 462–465.

Nunez, P. L. (1981) *Electric Fields of the Brain. The Neurophysics of EEG*. Oxford University Press, Oxford.

Nuñez, A., Amzica, F. and Steriade, M. (1992) Voltage-dependent fast (20–40 *Hz*) oscillations in long-axoned neocortical neurons. *Neuroscience* **51**: 7–10.

Nuñez, A., Curro-Dossi, R., Contreras, D. and Steriade, M. (1992) Intracellular evidence for incompatibility between spindle and delta oscillations in thalamocortical neurons of cat. *Neuroscience* **48**: 75–85.

Ohara, P. T. and Lieberman, A. R. (1985) The thalamic reticular nucleus of the adult rat: experimental anatomical studies. *J. Neurocytol.* **14**: 365–411.

Ohishi, H., Shigemoto, R., Nakanishi, S. and Mizuno, N. (1993) Distribution of the mRNA for a metabotropic glutamate receptor (mGluR3) in the rat brain: an in situ hybridization study. *J. Comp. Neurol.* **335**: 252–266.

Otis, T. S. and Mody, I. (1992a) Modulation of decay kinetics and frequency of $GABA_A$ receptor-mediated spontaneous inhibitory postsynaptic currents in hippocampal neurons. *Neuroscience* **49**: 13–32.

Otis, T. S. and Mody, I. (1992b) Differential activation of $GABA_A$ and $GABA_B$ receptors by spontaneously released transmitter. *J. Neurophysiol.* **67**: 227–235.

Otis, T. S., De Koninck, Y. and Mody, I. (1992) Whole-cell recordings of evoked and spontaneous GABAB responses in hippocampal slices. *Pharmacol. Commun.* **2**: 75–83.

Otis, T. S., Y. De Koninck and I. Mody. (1993) Characterization of synaptically elicited $GABA_B$ responses using patch-clamp recordings in rat hippocampal slices. *J. Physiol.* **463**: 391–407.

Pape, H. C. (1996) Queer current and pacemaker: the hyperpolarization-activated cation current in neurons. *Annual Rev. Physiol.* **58**: 299–327.

Paré, D., Dong, J. and Gaudreau, H. (1995) Amygdalo-entorhinal relations and their reflection in the hippocampal formation: generation of sharp sleep potentials. *J. Neurosci.* **15**: 2482–2503.

Paré, D., Lang, E. J. and Destexhe, A. (1998) Inhibitory control of somatic and dendritic sodium spikes in neocortical pyramidal neurons *in vivo*: an intracellular and computational study. *Neuroscience* **84**: 377–402.

Paré, D., Lebel, E. and Lang, E. J. (1997) Differential impact of miniature synaptic potentials on the somata and dendrites of pyramidal neurons *in vivo*. *J. Neurophysiol.* **78**: 1735–1739.

Partridge, L. D. and Swandulla, D. (1988) Calcium-activated non-specific cation channels. *Trends Neurosci.* **11**: 69–72.

Patel, J., Fujisawa, S., Beranyi, A., Royer, S. and Buzsaki, G. (2012) Traveling theta waves along the entire septotemporal axis of the hippocampus. *Neuron* **75**: 410–417.

Patneau, D. K. and Mayer, M. L. (1991) Kinetic analysis of interactions between kainate and AMPA: evidence for activation of a single receptor in mouse hippocampal neurons. *Neuron* **6**: 785–798.

Paulsen, O. and Sejnowski, T. J. (2000) Natural patterns of activity and long-term synaptic plasticity. *Curr. Opin. Neurobiol.* **10**: 172–179.

Paz, J. T., Chavez, M., Saillet S, Deniau, J. M. and Charpier, S. (2007) Activity of ventral medial thalamic neurons during absence seizures and modulation of cortical paroxysms by the nigrothalamic pathway. *J. Neurosci.* **27**: 929–941.

Pedroarena, C. and Llinás, R. (1997) Dendritic calcium conductances generate high-frequency oscillation in thalamocortical neurons. *Proc. Natl. Acad. Sci. USA* **94**: 724–728.

Pellegrini, A., Musgrave, J. and, Gloor, P. (1979) Role of afferent input of subcortical origin in the genesis of bilaterally synchronous epileptic discharges of feline generalized epilepsy. *Exp. Neurol.* **64**: 155–173.

Perez-Pinzon, M. A., Tao, L. and Nicholson, C. (1995) Extracellular potassium, volume fraction, and tortuosity in rat hippocampal CA1, CA3, and cortical slices during ischemia. *J. Neurophysiol.* **74**: 565–573.

Perkel, D. H. and Mulloney, B. (1974) Motor pattern production in reciprocally inhibitory neurons exhibiting postinhibitory rebound. *Science* **185**: 181–183.

Perozo, E. and Bezanilla, F. (1990) Phosphorylation affects voltage gating of the delayed rectifier K+ channel by electrostatic interactions. *Neuron* **5**: 685–690.

Peters, A. and Kaiserman-Abramof, I. R. (1970) The small pyramidal neuron of the rat cerebral cortex. The perikaryon, dendrites and spines. *Am. J. Anat.* **127**: 321–356.

Peters, A., Palay, S. L. and Webster, H. F. (1981) *The Fine Structure of the Nervous System.* Oxford University Press, New York.

Peyrache, A., Khamassi, M., Benchenane, K., Wiener, S. I. and Battaglia, F. P. (2009) Replay of rule-learning related neural patterns in the prefrontal cortex during sleep. *Nat. Neurosci.* **12**: 919–926.

Peyrache, A., Battaglia, F. and Destexhe, A. (2011) Inhibition recruitment in prefrontal cortex during sleep spindles and gating of hippocampal inputs. *Proc. Natl. Acad. Sci. USA* **108**: 17207–17212.

Peyrache, A., Dehghani, N., Eskandar, E. N., Madsen, J. R., Anderson, W. S., Donoghue, J. S., Hochberg, L. R., Halgren, E., Cash, S. S. and Destexhe, A. (2012) Spatiotemporal dynamics of neocortical excitation and inhibition during human sleep. *Proc. Natl. Acad. Sci. USA* **109**: 1731–1736.

Pinault, D., Bourassa, J. and Deschênes, M. (1995) The axonal arborization of single thalamic reticular neurons in the somatosensory thalamus of the rat. *Eur. J. Neurosci.* **7**: 31–40.

Pinault, D., Smith, Y. and Deschênes, M. (1997) Dendrodendritic and axosomatic synapses in the thalamic reticular nucleus of the adult rat. *J. Neurosci.* **17**: 3215–3233.

Pinault, D., Leresche, N., Charpier, S., Deniau, J. M., Marescaux, C., Vergnes, M. and Crunelli, V. (1998) Intracellular recordings in thalamic neurones during spontaneous spike and wave discharges in rats with absence epilepsy. *J. Physiol.* **509**: 449–456.

Polack, P. O., Guillemain, I., Hu, E., Deransart, C., Depaulis, A. and Charpier, S. (2007) Deep layer somatosensory cortical neurons initiate spike-and-wave discharges in a genetic model of absence seizures. *J. Neurosci.* **27**: 6590–6599.

Pollard, C. E. and V. Crunelli. (1988) Intrinsic membrane currents in projection cells of the cat and rat lateral geniculate nucleus. *Neurosci. Lett.* **32**: S39.

Pollen, D. A. (1964) Intracellular studies of cortical neurons during thalamic induced wave and spike. *Electroencephalogr. Clin. Neurophysiol.* **17**: 398–404.

Prange, O. and Murphy, T. H. (1999) Correlation of miniature synaptic activity and evoked release probability in cultures of cortical neurons. *J. Neurosci.* **19**: 6427–6438.

Press, W. H., Flannery, B. P., Teukolsky, S. A. and Vetterling, W. T. (1986) *Numerical Recipes. The Art of Scientific Computing.* Cambridge University Press, Cambridge, MA.

Prevett, M. C., Duncan J. S., Jones, T., Fish, D. R., Brooks, D. J. (1995) Demonstration of thalamic activation during typical absence seizures during $H_2^{15}O$ and PET. *Neurology* **45**: 1396–1402.

Prince, D. A. and Farrell, D. (1969) 'Centrencephalic' spike-wave discharges following parenteral penicillin injection in the cat. *Neurology* **19**: 309–310.

Puigcerver, A., Van Luijtenaar, E. J. L. M., Drinkenburg, W. H. I. M. and Coenen, A. L. M. (1996) Effects of the $GABA_B$ antagonist CGP-35348 on sleep-wake states, behaviour and spike-wave discharges in old rats. *Brain Res. Bull.* **40**: 157–162.

Qin, Y. L., McNaughton, B. L., Skaggs, W. E. and Barnes, C. A. (1997) Memory reprocessing in corticocortical and hippocampocortical neuronal ensembles. *Phil. Trans. Roy. Soc. Lond. Ser. B* **352**: 1525–1533.

Quirk, M. C., Blum, K. I. and Wilson, M. A. (2001) Experience-dependent changes in extracellular spike amplitude may reflect regulation of dendritic action potential back-propagation in rat hippocampal pyramidal cells. *J. Neurosci.* **21**: 240–248.

Rall, W. (1967) Distinguishing theoretical synaptic potentials computed for different soma-dendritic distributions of synaptic inputs. *J. Neurophysiol.* **30**: 1138–1168.

Rall, W. (1969) Time constants and electrotonic length of membrane cylinders and neurons. *Biophys. J.* **9**: 1483–1508.

Rall, W. (1995) *The Theoretical Foundation of Dendritic Function*, ed. Segev, I., Rinzel, J. and Shephert, G. M. MIT Press, Cambridge, MA.

Rall, W., Burke, R. E., Holmes, W. R., Jack, J. J., Redman, S. J. and Segev, I. (1992) Matching dendritic neuron models to experimental data. *Physiol. Reviews* **72**: S159–S186.

Ralston, B. and Ajmone-Marsan, C. (1956) Thalamic control of certain normal and abnormal cortical rhythms. *EEG Clin. Neurophysiol.* **8**: 559–582.

Raman, I. M., Zhang, S. and Trussell, L. O. (1994) Pathway-specific variants of AMPA receptors and their contribution to neuronal signaling. *J. Neurosci.* **14**: 4998–5010.

Ramón y Cajal, S. (1909) *Histologie du Système Nerveux de l'Homme et des Vertébrés* (translated by Azoulay, L.). Maloine, Paris.

Ramm, P. and Smith, C. T. (1990) Rates of cerebral protein synthesis are linked to slow wave sleep in the rat. *Physiology and Behavior* **48**: 749–753.

Rao, R. P. N. and Sejnowski, T. J. (2000) Predictive sequence learning in recurrent neocortical circuits. *Adv. Neural Information Processing Systems* **12**: 164–170.

Rao, R. P. N. and Sejnowski, T. J. (2001) Spike-timing dependent Hebbian plasticity as temporal difference learning. *Neural Computation* **13**, in press, pp. 2221–2237.

Rapp, M., Segev, I. and Yarom, Y. (1994) Physiology, morphology and detailed passive models of guinea-pig cerebellar Purkinje cells. *J. Physiol.* **474**: 101–118.

Rapp, M., Yarom, Y. and Segev, I. (1992) The impact of parallel fiber background activity on the cable properties of cerebellar Purkinje cells. *Neural Computation* **4**: 518–533.

Rapp, M., Yarom, Y. and Segev, I. (1996) Modeling back propagating action potential in weakly excitable dendrites of neocortical pyramidal cells. *Proc. Natl. Acad. Sci. USA* **93**: 11985–11990.

Rausell, E. and Jones, E. G. (1995) Extent of intracortical arborization of thalamocortical axons as a determinant of representational plasticity in monkey somatic sensory cortex. *J. Neurosci.* **15**: 4270–4288.

Redman, S. (1990) Quantal analysis of synaptic potentials in neurons of the central nervous system. *Physiol. Rev.* **70**: 165–198.

Renaud-Le Masson, S., Le Masson, G., Marder, E. and Abbott, L. F. (1993) Hybrid circuits of interacting computer model and biological neurons. In: *Advanced in Neural Information Processing Systems*, Vol. 5, ed. Hanson, S. J., Cowan, J. D. and Giles, G. L. Morgan Kaufmann Publishers, San Mateo, CA, pp. 813–819.

Renshaw, B., Forbes, A. and Morison, B. R. (1940) Activity of isocortex and hippocampus: electrical studies with microelectrodes. *J. Neurophysiol.* **3**: 74–105.

Reuveni, I., Friedman, A., Amitai, Y. and Gutnick, M. J. (1993) Stepwise repolarization from Ca^{2+} plateaus in neocortical pyramidal cells: evidence for nonhomogeneous distribution of HVA Ca^{2+} channels in dendrites. *J. Neurosci.* **13**: 4609–4621.

Rhodes, P. A. and Llinás, R. R. (2005) A model of thalamocortical relay cells. *J Physiol*. **565**: 765–781.

Ribary, U., Ioannides, A. A., Singh, K. D., Hasson, R., Bolton, J. P., Lado, F., Mogilner, A. and Llinás, R. (1991) Magnetic field tomography of coherent thalamocortical 40-*Hz* oscillations in humans. *Proc. Natl. Acad. Sci. USA* **88**: 11037–11041.

Rinzel, J. A. formal classification of bursting mechanisms in excitable systems. (1987) In: *Mathematical Topics in Population Biology, Morphogenesis and Neurosciences*, ed. Teramoto, E. and Yamaguti, M. Springer-Verlag, Berlin, pp. 267–281.

Ritz, R. and Sejnowski, T. J. (1997) Synchronous oscillatory activity in sensory systems: new vistas on mechanisms. *Curr. Opin. Neurobiol*. **7**: 536–546.

Robertson, R. T. and Cunningham, T. J. (1981) Organization of corticothalamic projections from parietal cortex in cat. *J. Comp. Neurol*. **199**: 569–585.

Ropert, N., Miles, R. and Korn, H. (1990) Characteristics of miniature inhibitory postsynaptic currents in CA1 pyramidal neurones of rat hippocampus. *J. Physiol*. **428**: 707–722.

Rose, R. M. and Hindmarsh, J. L. (1985) A model of a thalamic neuron. *Proc. R. Soc. Lond. B* **225**: 161–193.

Rose, R. M. and Hindmarsh, J. L. (1989) The assembly of ionic currents in a thalamic neuron. I. The three-dimensional model. *Proc. R. Soc. Lond. B Biol. Sci*. **237**: 267–288.

Rosenmund, C., Clements, J. D. and Westbrook, G. L. (1993) Nonuniform probability of release at a hippocampal synapse. *Science* **262**: 754 757.

Rothberger, F. (1931) cited in: Bremer, F. (1938b) L'activité électrique de l'écorce cérébrale. *Actualités Scientifiques et Industrielles* **658**: 46.

Rovó, Z., Mátyás, F., Barthó, P., Slézia, A., Lecci, S., Pellegrini, C., Astori, S., Dávid, C., Hangya, B., Lüthi, A. and Acsády, L. (2014) Phasic, nonsynaptic GABA-A receptor-mediated inhibition entrains thalamocortical oscillations. *J. Neurosci*. **34**: 7137–7147.

Roy, J. P., Clercq, M., Steriade, M. and Deschênes, M. (1984) Electrophysiology of neurons in lateral thalamic nuclei in cat: mechanisms of long-lasting hyperpolarizations. *J. Neurophysiol*. **51**: 1220–1235.

Rudolph, M., Pelletier, J.-G., Paré, D. and Destexhe, A. (2005) Characterization of synaptic conductances and integrative properties during electrically-induced EEG-activated states in neocortical neurons *in vivo*. *J. Neurophysiol*. **94**: 2805–2821, 2005.

Rumelhart, D. E., McClelland, J. E. and the PDP Research Group (1986) *Parallel Distributed Processing: Explorations in the Microstructure of Cognition*. MIT Press, Cambridge, MA.

Sah, P. and Bekkers, J. M. (1996) Apical dendritic location of slow afterhyperpolarization current in hippocampal pyramidal neurons: implications for the integration of long-term potentiation. *J. Neurosci*. **16**: 4537–4542.

Sakmann, B. (1992) Elementary steps in synaptic transmission revealed by currents through single ion channels. *Science* **256**: 503–512.

Sakmann, B. and Neher, E. (eds.) (1983) *Single-Channel Recording*. Plenum Press, New York.

Sakmann, B. and Neher, E. (eds.) (1995) *Single-Channel Recording* (2nd edition). Plenum Press, New York.

Salin, P. A. and Prince, D. A. (1996) Spontaneous GABA$_A$ receptor-mediated inhibitory currents in adult rat somatosensory cortex. *J. Neurophysiol*. **75**: 1573–1588.

Salinas, E. and Sejnowski, T. J. (2000) Impact of correlated synaptic input on output firing rate and variability in simple neuronal models. *J. Neurosci*. **20**: 6193–6209.

Sanchez, J. A., Dani, J. A., Siemen, D. and Hille, B. (1986) Slow permeation of organic cations in acetylcholine receptor channels. *J. Gen. Physiol*. **85**: 985–1001.

Sanchez-Vives, M. V. and McCormick, D. A. (1997a) Functional properties of perigeniculate inhibition of dorsal lateral geniculate nucleus thalamocortical neurons in vitro. *J. Neurosci.* **17**: 8880–8893.

Sanchez-Vives, M. V., Bal, T. and McCormick, D. A. (1997b) Inhibitory interactions between perigeniculate GABAergic neurons. *J. Neurosci.* **17**: 8894–8908.

Sanchez-Vives, M. V. and McCormick, D. A. (2000) Cellular and network mechanisms of rhythmic recurrent activity in neocortex. *Nature Neurosci.* **3**: 1027–1034.

Sanderson, K. J. (1971) The projection of the visual field to the lateral geniculate and medial interlaminar nuclei in the cat. *J. Comp. Neurol.* **143**: 101–108.

Sansom, M. S. P., Ball, F. G., Kerry, C. J., Ramsey, R. L. and Usherwood, P. N. R. (1989) Markov, fractal, diffusion, and related models of ion channel gating. A comparison with experimental data from two ion channels. *Biophys. J.* **56**: 1229–1243.

Scheibel, M. E. and Scheibel, A. B. (1966a) The organization of the nucleus reticularis thalami: a Golgi study. *Brain Res.* **1**: 43–62.

Scheibel, M. E. and Scheibel, A. B. (1966b) Patterns of organization in specific and nonspecific thalamic fields. In: *The Thalamus*, ed. Purpura, D. P. and Yahr, M. Columbia University Press, New York, pp. 13–46.

Scheibel, M. E. and Scheibel, A. B. (1967) Structural organization of nonspecific thalamic nuclei and their projection toward cortex. *Brain Res.* **6**: 60–94.

Scheibel, M. E. and Scheibel, A. B. (1972) Specialized organization patterns within the nucleus reticularis thalami of the cat. *Exp. Neurol.* **34**: 316–322.

Schlag, J. and Waszak, M. (1971) Electrophysiological properties of units of the thalamic reticular complex. *Exp. Neurol.* **32**: 79–97.

Schultz, W., Dayan, P. and Montague, P. R. (1997) A neural substrate of prediction and reward. *Science* **275**: 1593–1599.

Seibt, J., Richard, C. J., Sigl-Glockner, J., Takahashi, N., Kaplan, D. I., Doron, G., de Limoges, D., Bocklisch, C. and Larkum, M. E. (2017) Cortical dendritic activity correlates with spindle-rich oscillations during sleep in rodents. *Nature Commun.* **25**: 684.

Seidenbecher, T., Staak, R. and Pape, H. C. (1998) Relations between cortical and thalamic cellular activities during absence seizures in rats. *Eur. J. Neurosci.* **10**: 1103–1112.

Sejnowski, T. J. (1977) Storing covariance with nonlinearly interacting neurons. *J. Math. Biol.* **4**: 203–211.

Sejnowski, T. J. (1998) The computational neuroethology of sleep. In: *New Neuroethology on the Move*, ed. Elsner, N. and Wehner, R. Georg Theime Verlag, Stuttgart, New York (Vol. 1), pp. 127–144.

Sejnowski, T. J. and Destexhe, A. (2000) Why do we sleep? *Brain Research* **886**: 208–223.

Sejnowski, T. J., Bazhenov, M., Timofeev, I. and Frohlich, F. (2008) Cellular and network mechanisms of electrographic seizures. *Drug Discovery Today* **5**: 45–57.

Selbie, L. A. and Hill, S. J. (1998) G protein-coupled-receptor cross-talk: the fine-tuning of multiple receptor-signalling pathways. *Trends Pharmacol. Sci.* **19**: 87–93.

Selverston, A. I. (ed.) (1985) *Model Neural Networks and Behavior*. Plenum Press, New York.

Sheeba, J. H., Stefanovska, A. and McClintock, P. V. (2008) Neuronal synchrony during anesthesia: a thalamocortical model. *Biophys. J.* **95**: 2722–2727.

Sherman, S. M. and Guillery, R. W. (2001) *Exploring the Thalamus*. Academic Press, New York.

Siapas, A. G. and Wilson, M. A. (1998) Coordinated interactions between hippocampal ripples and cortical spindles during slow-wave sleep. *Neuron* **21**: 1123–1128.

Silva, L. R., Amitai, Y. and Connors, B. W. (1991) Intrinsic oscillations of neocortex generated by layer 5 pyramidal neurons. *Science* **251**: 432–435.

Silver, R. A., Traynelis, S. F. and Cull-Candy, S. G. (1992) Rapid time-course miniature and evoked excitatory currents at cerebellar synapses *in situ*. *Nature* **355**: 163–166.

Smith, K. A. and Fisher, R. S. (1996) The selective GABA$_B$ antagonist CGP-35348 blocks spike-wave bursts in the cholesterol synthesis rat absence epilepsy model. *Brain Res.* **729**: 147–150.

Smith, G. D. and Sherman, S. M. (2002) Detectability of excitatory versus inhibitory drive in an integrate-and-fire-or-burst thalamocortical relay neuron model. *J. Neurosci.* **22**: 10242–10250.

Snead, O. C. (1992) Evidence for GABA$_B$-mediated mechanisms in experimental generalized absence seizures. *Eur. J. Pharmacol.* **213**: 343–349.

Soderling, T. R. (1993) Calcium/calmodulin-dependent protein kinase II: role in learning and memory. *Mol. Cell. Biochem.* **127/128**: 93–101.

Soderling, T. R. and Derkach, V. A. (2000) Postsynaptic protein phosphorylation and LTP. *Trends Neurosci.* **23**: 75–80.

Soltesz, I. and Crunelli, V. (1992) GABA$_A$ and pre- and post-synaptic GABA$_B$ receptor-mediated responses in the lateral geniculate nucleus. *Progress Brain Res.* **90**: 151–169.

Soltesz, I. S., Lightowler, N., Leresche, D., Jassik Gerschenfeld, C. E., Pollard and V. Crunelli. (1991) Two inward currents and the transformation of low frequency oscillations of rat and cat thalamocortical cells. *J. Physiol.* **441**: 175–197.

Song, I., Kim, D., Choi, S., Sun, M., Kim, Y. and Shin, H. S. (2004) Role of the alpha1G T-type calcium channel in spontaneous absence seizures in mutant mice. *J. Neurosci.* **24**: 5249–5257.

Sperk, G., Furtinger, S., Schwarzer, C. and Pirker, S. (2004) GABA and its receptors in epilepsy. *Adv. Exp. Med. Biol.* **548**: 92–103.

Spreafico, R., de Curtis, M., Frassoni, C. and Avanzini, G. (1988) Electrophysiological characteristics of morphologically identified reticular thalamic neurons from rat slices. *Neuroscience* **27**: 629–638.

Spruston, N., Schiller, Y., Stuart, G. and Sakmann, B. (1995) Activity-dependent action potential invasion and calcium influx into hippocampal CA1 dendrites. *Science* **268**: 297–300.

Squire, L. R. and Zola-Morgan, S. (1991) The medial temporal lobe memory system. *Science* **253**: 1380–1386.

Srinivasan, R. and Chiel, H. J. (1993) Fast calculation of synaptic conductances. *Neural Computation* **5**: 200–204.

Staak, R. and Pape, H. C. (2001) Contribution of GABA(A) and GABA(B) receptors to thalamic neuronal activity during spontaneous absence seizures in rats. *J. Neurosci.* **21**: 1378–1384.

Standley, C., Ramsey, R. L. and Usherwood, P. N. R. (1993) Gating kinetics of the quisqualate-sensitive glutamate receptor of locust muscle studied using agonist concentration jumps and computer simulations. *Biophys. J.* **65**: 1379–1386.

Staubli, U., Ambros-Ingerson, J. and Lynch, G. (1992) Receptor changes and LTP: an analysis using aniracetam, a drug that reversibly modifies glutamate (AMPA) receptors. *Hippocampus* **2**: 49–58.

Steriade, M. (1974) Interneuronal epileptic discharges related to spike-and-wave cortical seizures in behaving monkeys. *Electroencephalogr. Clin. Neurophysiol.* **37**: 247–263.

Steriade, M. (1978) Cortical long-axoned cells and putative interneurons during the sleep-waking cycle. *Behav. Brain Sci.* **3**: 465–514.

Steriade, M. (2000) Corticothalamic resonance, states of vigilance, and mentation. *Neuroscience* **101**: 243–276.

Steriade, M. and Amzica, F. (1996) Intracortical and corticothalamic coherency of fast spontaneous oscillations. *Proc. Natl. Acad. Sci. USA* **93**: 2533–2538.

Steriade, M. and Contreras, D. (1995) Relations between cortical and thalamic cellular events during transition from sleep patterns to paroxysmal activity. *J. Neurosci.* **15**: 623–642.

Steriade, M. and Contreras, D. (1998) Spike-wave complexes and fast components of cortically generated seizures. I. Role of neocortex and thalamus. *J. Neurophysiol.* **80**: 1439–1455.

Steriade, M. and Deschênes, M. (1984) The thalamus as a neuronal oscillator. *Brain Res. Rev.* **8**: 1–63.

Steriade, M. and Deschênes, M. (1988) Intrathalamic and brainstem-thalamic networks involved in resting and alert states. In: *Cellular Thalamic Mechanisms*, ed. Bentivoglio, M. and Spreafico, R. Elsevier, Amsterdam, pp. 51–76.

Steriade, M. and Llinás, R. R. (1988) The functional states of the thalamus and the associated neuronal interplay. *Physiol. Reviews* **68**: 649–742.

Steriade, M. and McCarley, R. W. (1990) *Brainstem Control of Wakefulness and Sleep*. Plenum Press, New York.

Steriade, M. and Timofeev, I. (1997) Short-term plasticity during intrathalamic augmenting responses in decorticated cats. *J. Neurosci.* **17**: 3778–3795.

Steriade, M., Amzica, F. and Contreras, D. (1996) Synchronization of fast (30–40 *Hz*) spontaneous cortical rhythms during brain arousal. *J. Neurosci.* **16**: 392–417.

Steriade, M., Curro Dossi, R. and Contreras, D. (1993a) Electrophysiological properties of intralaminar thalamocortical cells discharging rhythmic (approximately 40 *Hz*) spike-bursts at approximately 1000 *Hz* during waking and rapid eye movement sleep. *Neuroscience* **56**: 1–9.

Steriade, M., Curró Dossi, R. and Nuñez, A. (1991) Network modulation of a slow intrinsic oscillation of cat thalamocortical neurons implicated in sleep delta waves: cortical potentiation and brainstem cholinergic suppression. *J. Neurosci.* **11**: 3200–3217.

Steriade, M., Deschênes, M., Domich, L. and Mulle, C. (1985) Abolition of spindle oscillations in thalamic neurons disconnected from nucleus reticularis thalami. *J. Neurophysiol.* **54**: 1473–1497.

Steriade, M., Deschênes, M. and Oakson, G. (1974) Inhibitory processes and interneuronal apparatus in motor cortex during sleep and waking. I. Background firing and responsiveness of pyramidal tract neurons and interneurons. *J. Neurophysiol.* **37**: 1065–1092.

Steriade, M., Domich, L. and Oakson, G. (1986) Reticularis thalami neurons revisited: activity changes during shifts in states of vigilance. *J. Neurosci.* **6**: 68–81.

Steriade, M., Domich, L., Oakson, G. and Deschênes, M. (1987) The deafferented reticular thalamic nucleus generates spindle rhythmicity. *J. Neurophysiol.* **57**: 260–273.

Steriade, M., Jones, E. G. and Llinás, R. R. (1990) *Thalamic Oscillations and Signalling*. John Wiley & Sons, New York.

Steriade, M., Jones, E. G. and McCormick, D. A. (eds.) (1997) *Thalamus*. Elsevier, Amsterdam.

Steriade, M., McCormick, D. A. and Sejnowski, T. J. (1993b) Thalamocortical oscillations in the sleeping and aroused brain. *Science* **262**: 679–685.

Steriade, M., Nunez, A. and Amzica, F. (1993c) A novel slow (< 1 *Hz*) oscillation of neocortical neurons *in vivo*: depolarizing and hyperpolarizing components. *J. Neurosci.* **13**: 3252–3265.

Steriade, M., Contreras, D., Curró Dossi, R. and Nunez, A. (1993d) The slow (< 1 *Hz*) oscillation in reticular thalamus and thalamocortical neurons. Scenario of sleep rhythms generation in interacting thalamic and neocortical networks. *J. Neurosci.* **13**: 3284–3299.

Steriade, M., Timofeev, I., Durmüller, N. and Grenier, F. (1998) Dynamic properties of corticothalamic neurons and local cortical interneurons generating fast rhythmic (30–40 *Hz*) spike bursts. *J. Neurophysiol.* **79**: 483–490.

Steriade, M., Timofeev, I. and Grenier, F. (2001) Natural waking and sleep states: a view from inside neocortical neurons. *J. Neurophysiol.* **85**: 1969–1985.

Steriade, M., Wyzinski, P. and Apostol, V. (1972) Corticofugal projections governing rhythmic thalamic activity. In: *Corticothalamic Projections and Sensorimotor Activities*, ed. Frigyesi, T. L., Rinvik, E. and Yahr, M. D. Raven Press, New York, pp. 221–272.

Stevens, C. F. (1978) Interactions between intrinsic membrane protein and electric field. *Biophys. J.* **22**: 295–306.

Stevens, C. F. (1993) Quantal release of neurotransmitter and long-term potentiation. *Cell* **72**: 55–63.

Stevens, C. F. and Wang, Y. (1995) Facilitation and depression at single central synapses. *Neuron* **14**: 795–802.

Stickgold, R., Whidbee, D., Schirmer, B., Patel, V. and Hobson, J. A. (2000) Visual discrimination task improvement: a multi-step process occurring during sleep. *J. Cogn. Neurosci.* **12**: 246–254.

Stiles, J. R., Bartol, T. M., Salpeter, M. M., Salpeter, E. E. and Sejnowski, T. J. (2000) Synaptic variability: new insights from reconstructions and Monte Carlo simulations with MCell. In: *Synapses*, ed. Cowan, W. M., Sudhof, T. C. and Stevens, C. F. Johns Hopkins University Press, Baltimore, pp. 681–731.

Stiles, J. R., Van Helden, D., Bartol, T. M., Salpeter, E. E. and Salpeter, M. M. (1996) Miniature endplate current rise times less than 100 microseconds from improved dual recordings can be modeled with passive acetylcholine diffusion from a synaptic vesicle. *Proc. Natl. Acad. Sci. USA* **93**: 5747–5752.

Strassberg, A. F. and DeFelice, L. J. (1993) Limitations of the Hodgkin-Huxley formalism: effects of single channel kinetics on transmembrane voltage dynamics. *Neural Computation* **5**: 843–855.

Stratford, K., Mason, A., Larkman, A., Major, G. and Jack, J. (1989) The modeling of pyramidal neurones in the visual cortex. In: *The Computing Neuron*, ed. Durbin, A. Miall, C. and Mitchison, G. Addison-Wesley, Workingham, UK, pp. 296–321.

Stricker, C., Field, A. C. and Redman, S. J. (1996) Statistical analysis of amplitude fluctuations in EPSCs evoked in rat CA1 pyramidal neurones in vitro. *J. Physiol.* **490**: 419–441.

Stuart, G. and Spruston, N. (1998) Determinants of voltage attenuation in neocortical pyramidal neuron dendrites. *J. Neurosci.* **18**: 3501–3510.

Stuart, G. J. and Sakmann, B. (1994) Active propagation of somatic action potentials into neocortical pyramidal cell dendrites. *Nature* **367**: 69–72.

Sutherland, G. R. and McNaughton, B. (2000) Memory trace reactivation in hippocampal and neocortical neuronal ensembles. *Curr. Opin. Neurobiol.* **10**: 180–186.

Suzuki, S. and Rogawski, M. A. (1989) T-type calcium channels mediate the transition between tonic and phasic firing in thalamic neurons. *Proc. Natl. Acad. Sci. USA* **86**: 7228–7232.

Tang, M. J. and Gilman, A. G. (1991) Type-specific regulation of adenylyl cyclase by G protein beta gamma subunits. *Science* **254**: 1500–1503.

Thompson, S. M. (1994) Modulation of inhibitory synaptic transmission in the hippocampus. *Progress Neurobiol.* **42**: 575–609.

Thompson, S. M. and Gähwiler, B. H. (1992) Effects of the GABA uptake inhibitor tiagabine on inhibitory synaptic potentials in rat hippocampal slice cultures. *J. Neurophysiol.* **67**: 1698–1701.

Thomson, A. M. and Destexhe, A. (1999) Dual intracellular recordings and computational models of slow IPSPs in rat neocortical and hippocampal slices. *Neuroscience* **92**: 1193–1215.

Thomson, A. M. and Deuchars, J. (1994) Temporal and spatial properties of local circuits in neocortex. *Trends Neurosci.* **17**: 119–126.

Thomson, A. M. and Deuchars, J. (1997) Synaptic interactions in neocortical local circuits: dual in cellular recordings in vitro. *Cerebral Cortex* **6**: 510–522.

Thomson, A. M. and West, D. C. (1991) Local-circuit excitatory and inhibitory connections in slices of the rat thalamus. (Abstract) *J. Physiol.* **438**: 113P.

Thorpe, S. J. and Gautrais, J. (1997) Rapid visual processing using spike asynchrony. *Adv. Neural Information Processing Systems* **9**: 901–907.

Tiesinga, P. H. E., Fellous, J.-M., Jose, J. V. and Sejnowski, T. J. (2001) Optimal information transfer in synchronized neocortical neurons. *Neurocomputing* **38**: 397–402.

Timofeev, I. and Steriade, M. (1996) The low-frequency rhythms in the thalamus of intact-cortex and decorticated cats. *J. Neurophysiol.* **76**: 4152–4168.

Timofeev, I., Contreras, D. and Steriade, M. (1996) Synaptic responsiveness of cortical and thalamic neurones during various phases of slow sleep oscillation in cat. *J. Physiol.* **494**: 265–278.

Timofeev, I., Grenier, F., Bazhenov, M., Sejnowski, T. J. and Steriade, M. (2000) Origin of slow cortical oscillations in deafferented cortical slabs. *Cerebral Cortex* **10**: 1185–1199.

Timofeev, I., Sejnowski, T. J., Bazhenov, M., Chauvette, S. and Grand, L. (2013) Age dependency of trauma-induced neocortical epileptogenesis. *Front. Cell. Neurosci.* **7**: 154.

Todorova, R. and Zugaro, M. (2019) Isolated cortical computations during delta waves support memory consolidation. *Science* **366**: 377–381.

Toth, T. and Crunelli, V. (1992) Computer simulations of the pacemaker oscillations of thalamocortical cells. *NeuroReport* **3**: 65–68.

Traub, R. D. and Llinás, R. R. (1979) Hippocampal pyramidal cells: significance of dendritic ionic conductances for neuronal function and epileptogenesis. *J. Neurophysiol.* **42**: 476–496.

Traub, R. D. and Miles, R. (1991) *Neuronal Networks of the Hippocampus.* Cambridge University Press, Cambridge.

Traynelis, S. F., Silver, R. A. and Cull-Candy, S. G. (1993) Estimated conductance of glutamate receptor channels activated during EPSCs at the cerebellar mossy fiber-granule cell synapse. *Neuron* **11**: 279–289.

Tsakiridou, E., Bertollini, L., de Curtis, M., Avanzini, G. and Pape, H. C. (1995) Selective increase in T-type calcium conductance of reticular thalamic neurons in a rat model of absence epilepsy. *J. Neurosci.* **15**: 3110–3117.

Tsien, R. W. and Noble, D. (1969) A transition state theory approach to the kinetics of conductances in excitable membranes. *J. Membr. Biol.* **1**: 248–273.

Tsodyks, M. and Markram, H. (1997) The neural code between neocortical pyramidal neurons depends on neurotransmitter release probability. *Proc. Natl. Acad. Sci. USA* **94**: 719–723.

Tsodyks, M., Pawelzik, K. and Markram, H. (1998) Neural networks with dynamic synapses. *Neural Computation* **10**: 821–835.

Uchimura, N., Cherubini, E. and North, R. A. (1990) Cation current activated by hyperpolarization in a subset of rat nucleus accumbens neurons. *J. Neurophysiol.* **64**: 1847–1850.

Uebele, V. N., Nuss, C. E., Fox, S. V., Garson, S. L., Cristescu, R., Doran, S. M., Kraus, R. L., Santarelli, V. P., Li, Y., Barrow, J. C., Yang, Z. Q., Schlegel, K. A., Rittle, K. E., Reger, T. S., Bednar, R. A., Lemaire, W., Mullen, F. A., Ballard, J. E., Tang, C., Dai, G., et al. (2009) Positive allosteric interaction of structurally diverse T-type calcium channel antagonists. *Cell Biochem. Biophys.* **55**: 81–93.

Uhlrich, D. J., Cucchiaro, J. B., Humphrey, A. L. and Sherman, S. M. (1991) Morphology and axonal projection patterns of individual neurons in the cat perigeniculate nucleus. *J. Neurophysiol.* **65**: 1528–1541.

Ulrich, D. and Huguenard, J. R. (1996) GABA$_B$ receptor-mediated responses in GABAergic projection neurones of rat nucleus reticularis thalami in vitro. *J. Physiol.* **493**: 845–854.

Ulrich, D. and Huguenard, J. R. (1997a) Nucleus-specific chloride homeostasis in the thalamus. *J. Neurosci.* **17**: 2348–2354.

Ulrich, D. and Huguenard, J. R. (1997b) GABA$_A$-receptor-mediated rebound burst firing and burst shunting in thalamus. *J. Neurophysiol.* **78**: 1748–1751.

Unwin, N. (1989) The structure of ion channels in membranes of excitable cells. *Neuron* **3**: 665–676.

Updyke, B. V. (1981) Projections from visual areas of the middle suprasylvian sulcus onto the lateral posterior complex and adjacent thalamic nuclei in cat. *J. Comp. Neurol.* **201**: 477–506.

Vandenberg, C. A. and Bezanilla, F. (1991) A model of sodium channel gating based on single channel, macroscopic ionic, and gating currents in the squid giant axon. *Biophys. J.* **60**: 1511–1533.

VanDongen, A. M. J., Codina, J., Olate, J., Mattera, R., Joho, R., Birnbaumer, L. and Brown, A. M. (1988) Newly identified brain potassium channels gated by the guanine nucleotide binding protein G$_o$. *Science* **242**: 1433–1437.

van Ginneken, A. C. G. and Giles, W. (1991) Voltage-clamp measurements of the hyperpolarization-activated inward current I_f in single cells from rabbit sino-atrial node. *J. Physiol.* **434**: 57–83.

Van Rullen, R., Gautrais J., Delorme, A. and Thorpe, S. (1998) Face processing using one spike per neurone. *Biosystems* **48**: 229–239.

Varela, J. A., Sen, K., Gibson, J., Fost, J., Abbott, L. F. and Nelson, S. B. (1997) A quantitative description of short-term plasticity at excitatory synapses in layer 2/3 of rat primary visual cortex. *J. Neurosci.* **17**: 7926–7940.

Vergnes, M., Marescaux, C., Micheletti, G., Depaulis, A., Rumbach, L. and Warter, J. M. (1984) Enhancement of spike and wave discharges by GABAmimetic drugs in rats with spontaneous petit-mal-like epilepsy. *Neurosci. Lett.* **44**: 91–94.

Vergnes, M. and Marescaux, C. (1992) Cortical and thalamic lesions in rats with genetic absence epilepsy. *J. Neural Transmission* **35** (Suppl.): 71–83.

Verzeano, M. (1972) Pacemakers, synchronization, and epilepsy. In: *Synchronization of EEG Activity in Epilepsies*, ed. Petsche, H. and Brazier, M. A. B. Springer, Berlin, pp. 154–158.

Verzeano, M. and Negishi, K. (1960) Neuronal activity in cortical and thalamic networks. A study with multiple microelectrodes. *J. Gen. Physiol.* **43**: 177–195.

Verzeano, M., Laufer, M., Spear, P. and McDonald, S. (1965) L'activité des réseaux neuroniques dans le thalamus du singe. *Actualités Neurophysiologiques* **6**: 223–251.

Vijayan, S. and Kopell, N. J. (2012) Thalamic model of awake alpha oscillations and implications for stimulus processing. *Proc. Natl. Acad. Sci. USA* **109**: 18553–18558.

Volgushev, M., Chen, J. Y., Ilin, V., Goz, R., Chistiakova, M. and Bazhenov, M. (2016) Partial breakdown of input specificity of STDP at individual synapses promotes new learning. *J. Neurosci.* **36**: 8842–8855.

von Krosigk, M. and McCormick, D. A. (1992) Mechanisms of frequency dependent facilitation of corticothalamic EPSPs. *Soc. Neurosci. Abstracts* **18**: 140.

von Krosigk, M., Bal, T. and McCormick, D. A. (1993) Cellular mechanisms of a synchronized oscillation in the thalamus. *Science* **261**: 361–364.

Waldmeier, P. C. and Baumann, P. A. (1990) Presynaptic GABA receptors. *Ann. N. Y. Acad. Sci.* **604**: 136–151.

Walker, M. P. and Stickgold, R. (2006) Sleep, memory, and plasticity. *Annu. Rev. Psychol.* **57**: 139–166.

Wallenstein, G. V. (1994a) A model of the electrophysiological properties of nucleus reticularis thalami neurons. *Biophys. J.* **66**: 978–988.

Wallenstein, G. V. (1994b) The role of thalamic I_GABAB in generating spike-wave discharges during petit mal seizures. *Neuroreport* **5**: 1409–1412.

Wallenstein, G. V. (1996) Adenosinic modulation of 7–14 *Hz* spindle rhythms in interconnected thalamic relay and nucleus reticularis neurons. *Neuroscience* **73**: 93–98.

Wang, X. J. (1994) Multiple dynamical modes of thalamic relay neurons: rhythmic bursting and intermittent phase-locking. *Neuroscience* **59**: 21–31.

Wang, X. J. and Rinzel, J. (1992) Alternating and synchronous rhythms in reciprocally inhibitory model neurons. *Neural Computation* **4**: 84–97.

Wang, X. J. and J. Rinzel. (1993) Spindle rhythmicity in the reticularis thalami nucleus—synchronization among inhibitory neurons. *Neurosci.* **53**: 899–904.

Wang, X. J., Golomb, D. and Rinzel, J. (1995) Emergent spindle oscillations and intermittent burst firing in a thalamic model: specific neuronal mechanisms. *Proc. Natl. Acad. Sci. USA* 92: 5577–5581.

Wang, Z. and McCormick, D. A. (1993) Control of firing mode of corticotectal and corticopontine layer V burst-generating neurons by norepinephrine, acetylcholine, and 1S,3R-ACPD. *J. Neurosci.* **13**: 2199–2216.

Warren, R. A., Agmon, A. and Jones, E. G. (1994) Oscillatory synaptic interactions between ventroposterior and reticular neurons in mouse thalamus *in vitro*. *J. Neurophysiol.* **72**: 1993–2003.

Wathey, J. C., Nass, M. M. and Lester, H. A. (1979) Numerical reconstruction of the quantal event at nicotinic synapses. *Biophys. J.* **27**: 145–164.

Wei, Y., Krishnan, G. P. and Bazhenov, M. (2016) Synaptic mechanisms of memory consolidation during sleep slow oscillations. *J. Neurosci.* **36**: 4231–4247.

White, E. L. (1986) Termination of thalamic afferents in the cerebral cortex. In: *Cerebral Cortex*, Vol. 5, ed. Jones, E. G. and Peters, A. Plenum Press, New York, pp. 271–289.

White, E. L. (1989) *Cortical Circuits*. Birkhauser, Boston, MA.

White, E. L. and Hersch, S. M. (1982) A quantitative study of thalamocortical and other synapses involving the apical dendrites of corticothalamic cells in mouse SmI cortex. *J. Neurocytol.* **11**: 137–157.

Widen, K. and Ajmone Marsan, C. (1960) Effects of corticopetal and corticofugal impulses upon single elements of the dorsolateral geniculate nucleus. *Exp. Neurol.* **2**: 468–502.

Williams, D. (1953) A study of thalamic and cortical rhythms in Petit Mal. *Brain* **76**: 50–69.

Williams, S. R. and Stuart, G. J. (1999) Dendritic branch points compromise action potential backpropagation in thalamocortical neurons. *Soc. Neurosci. Abstracts* **25**: 1741.

Williams, S. R. and Stuart, G. J. (2000) Action potential backpropagation and somato-dendritic distributions of ion channels in thalamocortical neurons. *J. Neurosci.* **20**: 1307–1317.

Willis, A. M., Slater, B. J., Gribkova, E. D. and Llano, D. A. (2015) Open-loop organization of thalamic reticular nucleus and dorsal thalamus: a computational model. *J. Neurophysiol.* **114**: 2353–2367.

Wilson, M. A. and McNaughton, B. L. (1994) Reactivation of hippocampal ensemble memories during sleep. *Science* **265**: 676–679.

Wolfart, J., Debay, D., Le Masson, G., Destexhe, A. and Bal, T. (2005) Synaptic background activity controls spike transfer from thalamus to cortex. *Nature Neurosci.* **8**: 1760–1767.

Wu, G. Y. and Cline, H. T. (1998) Stabilization of dendritic arbor structure *in vivo* by CaMKII. *Science* **279**: 222–226.

Xiang, Z., Greenwood, A. C. and Brown, T. (1992) Measurement and analysis of hippocampal mossy-fiber synapses. *Soc. Neurosci. Abstracts* **18**: 1350.

Yamada, W. M. and Zucker, R. S. (1992) Time course of transmitter release calculated from simulations of a calcium diffusion model. *Biophys. J.* **61**: 671–682.

Yamada, W. N., Koch, C. and Adams, P. R. (1989) Multiple channels and calcium dynamics. In: *Methods in Neuronal Modeling*, ed. Koch, C. and Segev, I. MIT Press, Cambridge, MA, pp. 97–134.

Yamada, M., Jahangir, A., Hosoya, Y., Inanobe, A., Katada, T. and Kurachi, Y. (1993) GK* and brain G beta gamma activate muscarinic K^+ channel through the same mechanism. *J. Biol. Chem.* **268**: 24551–24554.

Yang, N., George, A. L. and Horn, R. (1996) Molecular basis of charge movement in voltage-gated sodium channels. *Neuron* **16**: 113–122.

Yarom, Y. (1991) Rhythmogenesis in a hybrid system—interconnecting an olivary neuron to an analog network of coupled oscillators. *Neuroscience* **44**: 263–275.

Yen, C. T., Conley, M., Hendry, S. H. and Jones, E. G. (1985) The morphology of physiologically identified GABAergic neurons in the somatic sensory part of the thalamic reticular nucleus in the cat. *J. Neurosci.* **5**: 2254–2268.

Yingling, C. D. and Skinner, J. E. (1977) Gating of thalamic input to the cerebral cortex by nucleus reticularis thalami. *Prog. Clin. Neurophysiol.* **1**: 70–96.

Yousif, N. and Denham, M. (2007) The role of cortical feedback in the generation of the temporal receptive field responses of lateral geniculate nucleus neurons: a computational modelling study. *Biol Cybern.* **97**: 269–277.

Yuste, R. and Tank, D. W. (1996) Dendritic integration in mammalian neurons, a century after Cajal. *Neuron* **16**: 701–716.

Zaza, A., Maccaferri, G., Mangoni, M. and DiFrancesco, D. (1991) Intracellular calcium does not directly modulate cardiac pacemaker (if) channels. *Pflug. Archiv. Eur. J. Physiol.* **419**: 662–664.

Zhang, L. I., Tao, H. W., Holt, C. E., Harris, W. A. and Poo, M. (1998) A critical window for cooperation and competition among developing retinotectal synapses. *Nature* **395**: 37–44.

Zhang, S. and Trussell, L. O. (1994) Voltage clamp analysis of excitatory synaptic transmission in the avian nucleus magnocellularis. *J. Physiol.* **480**: 123–136.

Zhang, S. J., Huguenard, J. R. and Prince, D. A. (1997) $GABA_A$-receptor mediated Cl^- currents in rat thalamic reticular and relay neurons. *J. Neurophysiol.* **78**: 2280–2286.

Zhou, Q., Godwin, D. W., O'Malley, D. M. and Adams, P. R. (1997) Visualisation of calcium influx through channels that shape the burst and tonic firing modes of thalamic relay cells. *J. Neurophysiol.* **77**: 2816–2825.

Zola-Morgan, S. M. and Squire, L. R. (1990) The primate hippocampal formation: evidence for a time-limited role in memory storage. *Science* **250**: 288–290.

Zomorrodi, R., Kroger, H. and Timofeev, I. (2008) Modeling thalamocortical cell: impact of ca channel distribution and cell geometry on firing pattern. *Front. Comput. Neurosci.* **2**: 5.

Zukin, R. S. and Bennett, M. V. (1995) Alternatively spliced isoforms of the NMDARI receptor subunit. *Trends Neurosci.* **18**: 306–313.

Index